INTRODUCTION TO
MINERAL EXPLORATION

Introduction to Mineral Exploration

EDITED BY
ANTHONY M. EVANS

WITH CONTRIBUTIONS FROM
WILLIAM L. BARRETT
TIMOTHY BELL
ANTHONY M. EVANS
JOHN MILSOM
CHARLES J. MOON
BARRY C. SCOTT
MICHAEL K.G. WHATELEY

Blackwell Science

© 1995 by
Blackwell Science Ltd
Editorial Offices:
Osney Mead, Oxford OX2 0EL
25 John Street, London WC1N 2BL
23 Ainslie Place, Edinburgh EH3 6AJ
238 Main Street, Cambridge
   Massachusetts 02142, USA
54 University Street, Carlton
   Victoria 3053, Australia

Other Editorial Offices:
Arnette Blackwell SA
   1, rue de Lille, 75007 Paris
   France

Blackwell Wissenschafts-Verlag GmbH
   Kurfürstendamm 57
   10707 Berlin, Germany

   Feldgasse 13, A-1238 Wien
   Austria

First published 1995

Printed and bound in Great Britain
at the Alden Press Limited,
Oxford and Northampton

DISTRIBUTORS

Marston Book Services Ltd
PO Box 87
Oxford OX2 0DT
(*Orders*: Tel: 01865 791155
       Fax: 01865 791927
       Telex: 837515)

North America
Blackwell Science, Inc.
238 Main Street
Cambridge, MA 02142
(*Orders*: Tel: 800 215-1000
       617 876-7000
      Fax: 617 492-5263)

Canada
Oxford University Press
70 Wynford Drive
Don Mills
Ontario M3C 1J9
(*Orders*: Tel: 416 441 2941)

Australia
Blackwell Science Pty Ltd
54 University Street
Carlton, Victoria 3053
(*Orders*: Tel: 03 9347-0300
      Fax: 03 9349-3016)

A catalogue record for this title
is available from the British Library

ISBN 0-632-02427-5

Library of Congress
Cataloging-in-Publication Data

Evans, Anthony M.
  Introduction to mineral exploration
  William L. Barrett . . . [et al.].
      p.   cm.
  Includes bibliographical references
  and index.
  ISBN 0-632-02427-5
  1. Prospecting.
  I. Barrett, William L. II. Title.
  TN270.E93 1995
  622'. 1–dc20
                          95-16538
                              CIP

Contents

Contributors

WILLIAM L. BARRETT *Tarmac Quarry Products Ltd, Millfields Road, Ettingshall, Wolverhampton WV4 6JP.* At present he is Geological Manager, Estates and Environment Department of Tarmac but he has enjoyed a varied career in Applied Geology. His first degree was taken at the University of Wales and he obtained his doctorate from Leeds University. Much of his early career was spent in base metal exploration and underground mining in East Africa and this culminated with a spell as Chief Geologist at Kilembe Mines. Latterly he has been concerned with all aspects of Tarmac's work in the field of industrial minerals.

TIMOTHY BELL *Surpac Software International, G.P.T. Business Park, Technology Drive, Beeston, Nottingham NG9 2ND.* Graduated from West London College with a first class degree (cum laude) in Geology and Geography before obtaining an MSc degree with distinction in Mineral Exploration and Mining Geology at Leicester University in 1989. He joined CP Holdings (opencast coal contractors) as geologist in charge of computer applications in the mining division before moving to SURPAC Software International where he is Technical Manager in their Nottingham office.

ANTHONY M. EVANS *19 Hall Drive, Burton on the Wolds, Loughborough LE12 5AD.* Formerly Senior Lecturer in charge of the MSc Course in Mining Geology and Mineral Exploration and the BSc Course in Applied Geology at Leicester University. A graduate of Liverpool University in Physics and Geology he obtained a PhD at Queen's University, Ontario and worked in Canada for the Ontario Department of Mines spending much time investigating uranium and other mineralization. In Europe he has carried out research on many aspects of mineralization and industrial mineralogy and is the editor and author of a number of books and papers in these fields. For a number of years he was Vice-President of the International Society for Geology Applied to Mineral Deposits.

JOHN MILSOM *Department of Geological Sciences, University College, Gower St, London WC1E 6BT.* Senior Lecturer in Exploration Geophysics at University College. He graduated in Physics from Oxford University in 1961 and gained the Diploma of Imperial College in Applied Geophysics the following year. He joined the Australian Bureau of Mineral Resources as a geophysicist and worked in the airborne, engineering and regional gravity sections. He obtained a PhD from Imperial College for his research on the geophysics of eastern Papua and joined the staff there in 1971. Then came a spell as Chief Geophysicist for Barringer Research in Australia followed by one as an independent consultant when his clients included the UNDP, the Dutch Overseas Aid Ministry, and BP Minerals. In 1981 he joined the staff at Queen Mary College to establish the Exploration Geophysics BSc Course which later transferred to University College. He is the author of a textbook on Field Geophysics.

CHARLES J. MOON *Department of Geology, University of Leicester, Leicester LE1 7RH.* A graduate in Mining Geology from Imperial College, London in 1972, he returned some years later to take his PhD in Applied Geochemistry. He has had wide experience in the fields of mineral deposit assessment and exploration all over the globe. He has worked for Consolidated Gold Fields searching for tin deposits in Cornwall, the South African Geological Survey, Esso Minerals Africa Inc., with whom he was involved in the discovery of three uranium deposits. Besides his general expertise in mineral exploration he has developed the use of raster and vector based GIS systems and courses in these techniques, particularly in ARC/INFO for exploration integration. His present position is that of a lecturer in Mineral Exploration at Leicester University.

BARRY C. SCOTT *Knoyle House, Vicarage Way, Gerrards Cross, Bucks SL9 8AS.* His most recent position was Lecturer in Mining Geology and Mineral Exploration at Leicester University from which he retired in 1993. Previously he had occupied senior positions with a number of mining companies and academic institutions. These included Borax Consolidated Ltd, Noranda Mines Ltd, G. Wimpey PLC, the Royal School of Mines and Imperial College. Dr Scott obtained his BSc and PhD from Imperial College and his MSc from Queen's University, Ontario. He has been a Member of Council of the Institution of Mining and Metallurgy for many years, its President and is now its Treasurer.

MICHAEL K.G. WHATELEY *Department of Geology, University of Leicester, Leicester LE1 7RH*. Lecturer in Mining Geology and Mineral Exploration at Leicester University. He graduated in Geology and Geography from London University in 1968, gained an MSc from the University of the Witwatersrand in 1980 and a PhD from Leicester University in 1994 for his work on the computer modelling of the Leicestershire Coalfield. He worked in diamond exploration in Botswana and on gold mines near Johannesburg before joining Hunting Geology and Geophysics in South Africa. He then joined Utah International Inc. in charge of their Fuel Division in southern Africa. In 1982 he returned to the UK to work for Golder Associates before taking up his present post.

Preface

It is a matter of regret how little time is devoted to teaching mineral exploration in first degree university and college courses in the United Kingdom and in many other countries. As a result the few available textbooks tend to be expensive because of the limited market. By supplying our publisher with camera ready copy we have attempted to produce a cheaper book which will help to redress this neglect of one of the most important aspects of applied geology. We hope that lecturers will find here a structure on which to base a course syllabus and a book to recommend to their students.

In writing such a book it has to be assumed that the reader has a reasonable knowledge of the nature of mineral deposits. Nevertheless some basic facts are discussed in the early chapters.

The book is in two parts. In the first we discuss the principles of mineral exploration and in the second, case histories of selected deposit types. In both of these we carry the discussions right through to the production stage. This is because if mining projects are to progress successfully from the exploration and evaluation phases to full scale production, then exploration geologists must appreciate the criteria used by mining and mineral processing engineers in deciding upon appropriate mining and processing methods; and by the financial frater-nity in assessing the economic viability of a proposed mining operation. We have tried to cover these and many other related subjects to give the reader an overall view of mineral exploration. As this is only an introductory textbook we have given some guidance with regard to further reading rather than leaving the student to choose at random from references given in the text.

This textbook arises largely from the courses in mineral exploration given in the Geology Department of Leicester University during the last three decades and has benefitted from the ideas of our past and present colleagues and to them, as well as to our students and our friends in industry, we express our thanks. They are too many to name but we would like in particular to thank Dave Rothery for his help with Chapter 6; Martin Hale for discussion of Chapter 8; Nick Laffoley, Don May and Brendon Monahan for suggestions which have improved Chapter 9; and John Rickus for a helping hand with Chapter 12. Steve Baele of US Borax gave us good advice and Sue Button and Clive Cartwright performed many miracles of draughting in double quick time. In addition we are grateful to the Director-General of MTA for permission to reproduce parts of the maps used to prepare Figs 12.2 & 12.3, and to the Director-General of TKI for permission to publish the data on Soma.

Units and Abbreviations

Note on units

With few exceptions the units used are all SI (Système International), which has been in common use by engineers and scientists since 1965. The principal exceptions are: (a) for commodity prices still quoted in old units, such as troy ounces for precious metals and the short ton (= 2000 lb); (b) when there is uncertainty about the exact unit used, e.g. tons in certain circumstances might be short or long (2240 lb) and (c) Chapter 15 where all the original information was collected in American units, but many metric equivalents are given.

SI prefixes and suffixes commonly used in this text are k = kilo-, 10^3; M = mega-, 10^6 (million); G = giga-, 10^9 (billion is never used as it has different meanings on either side of the Atlantic).

Some abbeviations used in the text

AAS	Atomic absorption spectrometry
ASTM	American Society for Testing Materials
CCT	Computer compatible tapes
CIF	Carriage, insurance and freight
CIPEC	Conseil Inter-governmental des Pays Exportateurs de Cuivre (Intergovernmental Council of Copper Exporting Countries)
CIS	Commonwealth of Independent States (includes many of the countries formerly in the USSR)
d.c.	direct current
DCF ROR	Discounted cash flow rate of return
DTH	Down-the-hole (logs)
DTM	Digital terrain model
EEC	see EU
EIS	Environmental impact statement
e.m.	electromagnetic
ERTS	Earth Resources Technology Satellite
EU	European Union sometimes still referred to as EEC
FOB	Freight on board
FOV	Field of view
GA	Golder Associates
GIS	Geographical Information Systems
GRD	Ground resolution distance
HRV	High resolution visible
ICP-ES	Inductively coupled plasma emission spectroscopy
IFOV	Instantaneous field of view
IGRF	International Geomagnetic Reference Field
INPUT	Induced pulse transient system
IP	Induced polarization
IR	Infrared
ITC	International Tin Council
MS	Multispectral
MSS	Multispectral scanners
MTA	Maden Tetkik ve Arama
NAF	North Anatolian Fault
NPV	Net present value
OECD	Organization for Economic Co-operation and Development
OPEC	Organization of Petroleum Exporting Countries
PFE	Per cent frequency effect
PGE	Platinum group elements
PGM	Platinum group metals
ppb	Parts per billion
ppm	Parts per million
REE	Rare earth elements
RMR	Rock mass rating
ROM	Run-of-mine
RQD	Rock quality designation
SEM	Scanning electron microscopy
SI units	Système International Units
SLAR	Side-looking airborne radar
TDRS	Tracking and Data Relay System
TEM	Transient electromagnetics
TKI	Turkiye Komur Isletmekeri Kurumu
TM	Thematic mapper
t p.a.	tonnes per annum
t p.d.	tonnes per diem
USGS	United States Geological Survey
USSR	The geographical area that once made up the now disbanded Soviet Republics
VLF	Very low frequency
VMS	Volcanic-associated massive sulphide (deposits)
XRF	X-ray fluorescence

Part I—Principles

1 Ore, Mineral Economics and Mineral Exploration

ANTHONY M. EVANS

1.1 INTRODUCTION

A large economic mineral deposit, e.g. 50 Mt underlying an area of 2 km^2 is minute in comparison with the earth's crust and in most countries the easily found deposits cropping out at the surface have nearly all been found. The deposits we now search for are largely concealed by weathered and leached outcrops, drift, soil or some other cover and require sophisticated exploration methods to find them. We usually refer to the target material in these deposits as ore, unless we use a more specific term such as coal, gas, oil or water. So what is ore? And what sort of ore should we seek? To answer these questions we must start with some knowledge of mineral economics.

1.2 MINERAL ECONOMICS

1.2.1 Ore

The word ore was for many decades restricted to material from which a metal or metals could be won at a profit. An economically mineable ore deposit is normally called an orebody. The words ore and orebody have however been undergoing slow and confusing transitions in their meanings which are still not complete, and the tyro must read the context carefully to discern the sense in which a particular writer is using these words. They are still used by some geologists in the old above-mentioned sense, in which case the term ore minerals would refer only to those from which metals are extracted. Nowadays the term ore is generally used to include industrial minerals and the term orebody extended to include economic deposits of industrial minerals and bulk materials.

'Industrial minerals have been defined as any rock, mineral or other naturally occurring substance of economic value, exclusive of metallic ores, mineral fuels and gemstones' (Noetstaller 1988). They are therefore minerals where either the mineral itself, e.g. asbestos, baryte, or the oxide or some other compound derived from the mineral has an industrial application (end use). They include rocks such as granite, sand, gravel, limestone that are used for constructional purposes (these are often referred to as aggregates or bulk materials), as well as more valuable minerals with specific chemical or physical properties like fluorite, phosphate, kaolinite and perlite. Industrial minerals are also frequently and confusingly called non-metallics (e.g. Harben & Bates 1984) although they can contain and be the source of metals, e.g. sodium derived from the industrial mineral halite. On the other hand many 'metallic ores' such as bauxite, ilmenite, chromite and manganese minerals are also important raw materials for industrial mineral end uses. In view of all this, how is ore now defined? Two very useful discussions of this subject are to be found in Lane (1988) and Taylor (1989). Taylor's discussion is an easier introduction for the beginner; Lane should be read by all industrial and mining geologists and advanced students. Taylor favours a wide and inclusive definition that will survive being 'blown about by every puff of economic wind' such as changes in market demand, commodity prices, mining costs, taxes, environmental legislation and other factors: 'ore is rock that may be, is hoped to be, will be, is or has been mined; and from which something of value may be (or has been) extracted'. This is very similar to the official UK Institution of Mining and Metallurgy definition: 'Ore is a solid naturally-occurring mineral aggregate of economic interest from which one or more valuable constituents may be recovered by treatment'. Both these definitions cover ore minerals *and* industrial minerals and imply extension of the term orebody to include economic deposits of industrial minerals and rocks. This is the sense in which these terms will normally be employed in this book, except that they will be extended to include the instances where the whole rock, e.g. granite, limestone, salt, is utilized and not just a part of it. Lane prefers the use of the term mineralized ground for such comprehensive usage of the word ore as that in the definitions by Taylor and the I.M.M., and he would restrict ore to describing material in the ground that can be extracted to the overall economic benefit of a particular mining operation, governed by the financial determinants at the time of examination.

A definition about which there is little argument is that of gangue. This is simply the unwanted material, minerals or rock, with which ore minerals are usually intergrown. Mines commonly possess processing plants in which the raw ore is milled before the separation of the ore minerals from the gangue minerals by various processes, which provide ore concentrates, and tailings which are made up of the gangue.

1.2.2 The relative importance of ore and industrial minerals

Metals always seem to be the focus of attention for various reasons such as their use in warfare, rapid and cyclical changes in price, occasional occurrence in very rich deposits (gold bonanzas for example), with the result that the great importance of industrial minerals to our civilization is overlooked, and yet, in the form of flints and stone axes, bricks, pottery, etc., these were the first earth resources to be exploited by human beings. Today industrial minerals permeate every segment of our society (McVey 1989). They occur as components in durable and non-durable consumer goods. In many industrial activities and products, from the construction of buildings to the manufacture of ceramic tables or sanitary ware, the use of industrial minerals is obvious but often unappreciated. With numerous other goods, ranging from books to pharmaceuticals, the consumer is frequently unaware that industrial minerals play an essential role.

In developed countries such as the UK and USA industrial mineral production is far more important than metal production from both the tonnage and financial viewpoints, as in fact it is on a worldwide basis (see Table 1.1). This importance has become even greater in recent years because of the marked difference in growth rates between these two commodity groups (see Table 1.2). The world production of some individual mineral commodities ranked in order of tonnage produced is given in Table 1.3, and that of some other metals in Table 1.4.

Graphs of world production of the traditionally important metals (see Figs 1.1–1.3) show interesting trends. The world's appetite for the major metals appeared to be almost insatiable after World War Two and post-war production increased with great rapidity; however in the mid seventies an abrupt slackening in demand occurred triggered by the coeval oil crisis but clearly continuing up to the present day. These curves suggest that consumption of major metals is following a wave pattern in which the various crests may not be far off in time. Lead, indeed, may be over the crest. Various factors are probably at work here: recycling; more economical use of metals; and substitution by ceramics and plastics—

Table 1.1 Tonnage and value of mineral products in 1983 (from Noetstaller 1988)

Category	World production (1000 t)	%	Value of output (million $US)	%
Industrial minerals	11 798 630	72	129 147	40
Solid fuel minerals	4 004 287	24	122 285	38
Metals and ores*	543 581	4	39 007	13
Precious minerals	14	1	30 341	9
Totals	16 346 512	100	320 781	100

*Iron is included in this figure as iron ore.

industrial minerals are much used as a filler in plastics. Production of plastics rose by a staggering 1529% between 1960 and 1985 and a significant fraction of the demand behind this is attributable to metal substitution. In Table 1.5 the increases in production of selected metals and industrial minerals provide a striking contrast and one that explains why for some years now many large metal mining companies have been moving into industrial mineral production. An example is the RTZ Corporation, probably now the world's largest mining company, which in 1993 derived 27.9% of its net profit from industrial mineral operations compared with 72.1% from metal mining.

Are we soon to pass onwards from the Iron Age into a ceramic–plastic age? Readers are urged to monitor this

Table 1.2 Average growth rates in world production. (Source: *The Economist World Business Cycles* 1982)

Product	1966–1973	1973–1980
Crude oil	70%	700%
Industrial minerals	29%	16%
Metals	54%	7%

Table 1.3 World production of some mineral commodities in 1987. Metals are in italics. (Compiled from various sources and with considerable help from Mr D.E. Highley of the British Geological Survey)

Rank	Commodity	Tonn-age (Mt)	Rank	Commodity	Tonn-age (Mt)
1	Aggregates	10 250	23	Fluorite	4.8
2	Coal	4656	24	Feldspar	4.6
3	Crude oil	2838	25	Baryte	4.2
4	Portland cement	1033	26	*Titania*	4.2
5	*Pig iron*	508	27	Asbestos	4.1
6	Clay	400	28	Fuller's earth	3.6
7	Silica	200	29	*Lead*	3.4
8	Salt	177	30	Nepheline-syenite	3.2
9	Phosphate	144	31	Borates	2.7
10	Gypsum	84	32	Perlite	2.4
11	Sulphur	54	33	Diatomite	1.9
12	Potash	31	34	*Zirconium minerals*	0.85
13	Sodium carbonate (trona)	30	35	*Nickel*	0.80
14	*Manganese ore*	22	36	Graphite	0.63
15	Kaolin	21	37	Vermiculite	0.54
16	*Aluminium*	16.2	38	Sillimanite minerals	0.50
17	Magnesite	12.3	39	*Magnesium*	0.33
18	*Chromium ores and concentrates*	10.8	40	Mica	0.27
19	*Copper*	8.7	41	*Tin*	0.19
20	Talc	7.4	42	Strontium minerals	0.18
21	*Zinc*	7.2	43	Wollastonite	0.12
22	Bentonite	6.4			

Table 1.4 World production of selected metals in 1987

Molybdenum	0.089 Mt	Lithium	0.007 Mt
Antimony	0.059 Mt	Mercury	0.006 Mt
Tungsten	0.040 Mt	Silver	14 133 t
Uranium	0.038 Mt	Gold	1 610 t
Vandium	0.032 Mt	PGM	264 t
Cadmium	0.019 Mt		

Table 1.5 Increases in world production of some metals and industrial minerals, 1973–1988; metals in italics. (Recycled metal production is not included.)

Aluminium	28%	*Cobalt*	35%	*Copper*	16%
Diatomite	29.1	Feldspar	81.5	*Gold*	8.8
Gypsum	37.6	*Iron ore*	12	*Lead*	−5.5
Mica	18.9	*Molyb-denum*	8.3	*Nickel*	16.8
Phosphate	42.5	*PGM*	80.8	Potash	39.1
Silver	13.9	Sulphur	19	Talc	44
Tantalum	143	*Tin*	−9.8	Trona	44
Zinc	26.7				

tentative prophecy by keeping these graphs up-to-date using data from the same or a similar source, which includes production from the former eastern bloc countries as well as that from other countries; be warned, some compilations ignore this former production but still pose as world production figures. A factor of small but growing importance is the demand for non-ferrous metals in the non-OECD countries; this has grown by over 6% per annum during the present decade, compared with less than 1% in the OECD countries and it may increase sufficiently in the coming decade to influence present trends in demand for these metals. This demand too should be monitored. Finally, although the increase in demand for the major, high tonnage production metals is decreasing at the present time, the future is bright for

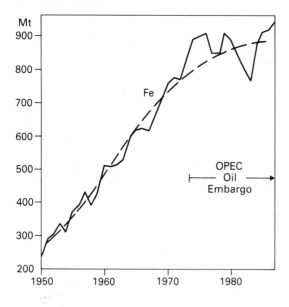

Fig. 1.1 World production of iron ore from 1950–1987. (Source: Lofty *et al.* (1989). General trend superimposed.)

certain minor, low tonnage metals such as cobalt, PGM, REE, tantalum and titanium.

1.2.3 Commodity prices—the market mechanism

Most mineral trading takes place within the market economies of the non-communist world and the prices of minerals or mineral products are governed by the factors of supply and demand. If consumers want more of a mineral product than is being supplied at the current price, this is indicated by their 'bidding up' the price, thus increasing the profits of companies supplying that product. As a result, resources in the form of capital investment are attracted into the industry and supply expands. On the other hand if consumers do not want a particular product its price falls, producers make a loss and resources leave the industry.

WORLD MARKETS

Modern transport leads to many commodities having a world market, a price change in one part of the world affects the price in the rest of the world. Such commodities include wheat, cotton, rubber, gold, silver and base metals. These commodities have a wide demand, are capable of being transported and the costs of transport

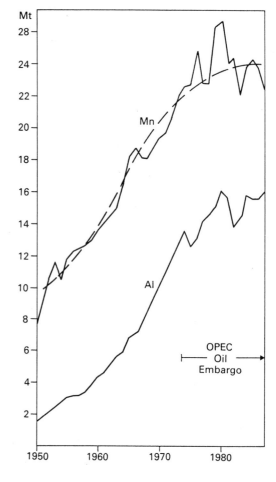

Fig. 1.2 World production of manganese and aluminium from 1950–1987. (Source: Lofty *et al.* (1989). General trend for manganese superimposed.)

are small compared with the value of the commodity. The market for diamonds is worldwide but that for bricks is local.

Over the last few centuries formal organized markets have developed. In these markets buying and selling takes place in a recognized building, business is governed by agreed rules and conventions and usually only members are allowed to engage in transactions. Base metals are traded on the London Metal Exchange, gold and silver on the London Bullion Market. Similar markets exist in many other countries e.g. the New York Commodity Exchange—Comex. Because these markets are composed of specialist buyers and sellers and are in

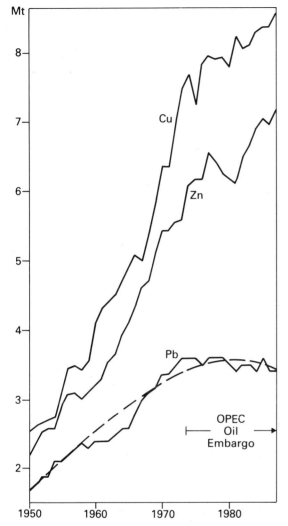

Fig. 1.3 World production of copper, zinc and lead from 1950–1987. (Source: Lofty *et al.* (1989). General trend for lead superimposed.)

concerned may agree on a contract price in advance of production, with clauses to allow for price changes because of such factors as inflation or currency exchange rate fluctuations. Contracts of this nature are still very common in the cases of iron and uranium and industrial minerals production. Whatever the form of sale is to be, the mineral economists of a mining company must try to forecast demand for, and hence the price of, a possible mine product, well in advance of mine development and such considerations will usually play a decisive role in formulating a company's mineral exploration strategy. A useful recent discussion of mineral markets can be found in Gocht *et al.* (1988).

FORCES DETERMINING PRICES

(a) *Demand and supply*. Demand may change over a short period of time for a number of reasons. Where one commodity substitutes to a significant extent for another and the price of this latter falls then the substituting commodity becomes relatively expensive and less of it is bought. Copper and aluminium are affected to a degree in this way. A change in technology may increase the demand for a metal e.g. the use of titanium in jet engines, or decrease it e.g. tin—development of thinner layers of tin on tinplate and substitution (see Table 1.5). The expectation of future price changes or shortages will induce buyers to increase their orders to have more of a commodity in stock.

Supply refers to how much of a commodity will be offered for sale at a given price over a set period of time. This quantity depends on the price of the commodity and the conditions of supply. High prices stimulate supply and investment by suppliers to increase their output. A fall in prices has the opposite effect and some mines may be closed completely or put on a care-and-maintenance basis in the hope of better times in the future. Conditions of supply may change fairly quickly through: (1) changes due to abnormal circumstances such as natural disasters, war, other political events, fire, strikes at the mines of big suppliers; (2) improved techniques in exploitation; (3) discovery and exploitation of large new orebodies.

(b) *Government action*. Governments can act to stabilize or change prices. Stabilization may be attempted by building up a stockpile, although the mere building up of a substantial stockpile increases demand and may push up the price! With a substantial stockpile in being, sales from the stockpile can be used to prevent prices rising significantly and purchases for the stockpile may be used to prevent or moderate price falls. As commod-

constant communication with each other, prices are sensitive to any change in worldwide supply and demand.

The prices of some metals on Comex and the London Metal Exchange are quoted daily by many newspapers whilst more comprehensive guides to current metal and mineral prices can be found in the *Engineering and Mining Journal*, *Industrial Minerals*, the *Mining Journal* and other technical journals. Short- and long-term contracts between buyer and seller may be based on these fluctuating prices. On the other hand, the parties

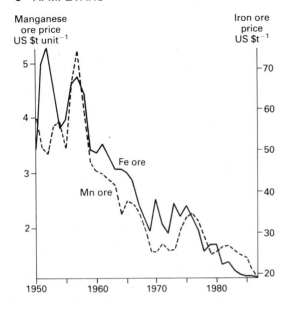

Fig. 1.4 Yearly average price of iron and manganese ores for 1950–1988. The iron ore price is for 61.5% (of iron) Brazilian ore CIF German ports, expressed in constant 1980 US dollars. (Prior to 1965, Liberian ore.) The manganese ore price is for 46–48% Indian ore CIF US ports, expressed in 1980 US dollars per metric unit (10 kg manganese content in the ore). (Source: Lofty *et al.* 1989.)

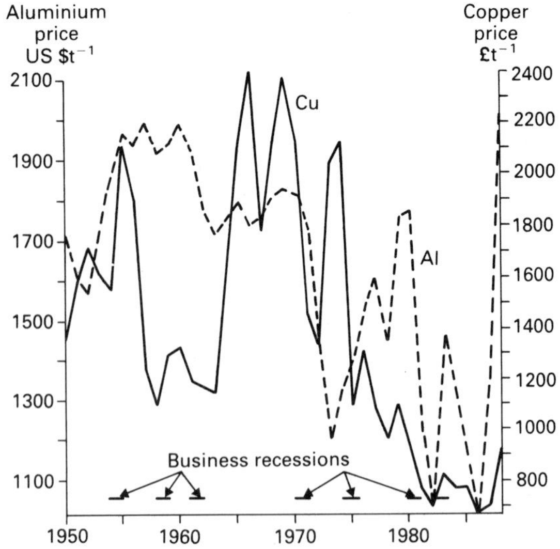

Fig. 1.5 Yearly average producer price of unalloyed aluminium ingot on the New York Market expressed in constant 1980 US dollars and the yearly average price of electrolytic wire bars of copper on the London Metal Exchange expressed in constant 1980 pounds sterling. Both graphs cover the period 1950–1988. (Source: Lofty *et al.* 1989.)

ity markets are worldwide it is in most cases impossible for one country acting on its own to control prices. Groups of countries have attempted to exercise control over tin (ITC) and copper (CIPEC) in this way but with little success and, at times, signal failure.

Stockpiles may also be built up by governments for strategic reasons and this, as mentioned above, can push up prices markedly. Stockpiling policies of some leading industrialized nations are discussed by Morgan (1989). An action that will increase consumption of platinum and rhodium is the adoption of new regulations on car exhausts by the EU countries. The worldwide effort to diminish harmful exhaust emissions resulted in a record industrial purchase of 1.7 million ounces of platinum in 1993. Comparable actions by governments stimulated by environmental lobbies will no doubt occur in the coming years.

(c) *Recycling* is already having a significant effect on some product prices. Economic and particularly environmental considerations will lead to increased recycling of materials in the immediate future. Recycling will prolong resource life and reduce mining wastes and smelter effluents. Partial immunity from price rises, shortages of primary materials or actions by cartels will follow. A direct economic and environmental bonus is that energy requirements for recycled materials are usually much lower than for treating ores—in the case of aluminium 80% less electricity is needed. In the USA the use of ferrous scrap as a percentage of total iron consumption rose from 35 to 42% over the period 1977–87 and aluminium from 26 to 37%; but copper has remained mainly in the range 40–45% and zinc 24–29% (Kaplan 1989). Of course the potential for recycling some materials is much greater than for others. Contrary to metals the potential for recycling industrial minerals is much lower and is confined to a few commodities such as bromine, fluor-compounds, industrial diamonds, iodine and feldspar and silica in the form of glass; so prices will be less affected by this factor (Noetstaller 1988).

(d) *Substitution and new technology* may both lead to a diminution in demand. We have already seen great changes such as the development of longer lasting car batteries that use less lead, substitution of copper and plastic for lead water pipes and a change to lead-free

US $t^{-1}

Fig. 1.6 Yearly average domestic prices of pig lead and prime western grade zinc for 1950–1988 on the New York Market expressed in constant 1980 US dollars. (Source: Lofty *et al.* 1989.)

petrol; all factors that have contributed to a downturn in the demand for lead (Fig. 1.3). But a factor affecting all metals was the OPEC shock in 1973 (Figs 1.1, 1.2, 1.3) which led to huge increases in the prices of oil and other fuels, pushed demand towards materials having a low sensitivity to high energy costs and favoured the use of lighter and less expensive substitutes for metals (Cook 1987).

In the past, base metal producers have spent vast sums of money on exploration, mine development and production but have paid too little attention to the defence and development of markets for their products (Davies 1987, Anthony 1988). Producers of aluminium, plastics and ceramics on the other hand have promoted research for new uses including substitution for metals. Space will allow the citing of only a few examples. Tank armour is now frequently made of multilayer composites—metal, ceramic and fibres, ceramic based engine components are already widely used in automobiles and it has been forecast that by 2030 A.D. 90% of engines used in cars, aeroplanes and power stations will be made from novel ceramics. A useful article on developments in ceramic technology is by Wheat (1987).

METAL AND MINERAL PRICES

(a) *Metals*. Metal prices are erratic and hard to predict (Figs 1.5 & 1.6). In the short run prices fluctuate in response to unforeseen news affecting supply and demand, e.g. strikes at large mines or smelters, unexpected increases in warehouse stocks. This makes it difficult to determine regular behaviour patterns for some metals. Over the intermediate term (several decades) the prices clearly respond to rises and falls in world business activity which is some help in attempts at forecasting price trends (Figs 1.5 & 1.6). The OPEC shock of 1973, which has been mentioned above, besides setting off a severe recession led to less developed countries building up huge debts to pay for the increased costs of energy. This led to their reducing their living standards and purchasing fewer durable goods. At the same time many metal-producing, developing countries such as Chile, Peru, Zambia and Zaïre increased production irrespective of metal prices to earn hard currencies for debt repayment. A further aggravation from the supply and price point of view has been the large number of significant mineral discoveries since the advent of modern exploration methods in the fifties (Fig. 1.7). Metal explorationists have to a considerable extent become victims of their own success. It should be noted that the fall off in non-gold discoveries from 1976 onwards is largely due to the difficulty explorationists now have in finding a viable deposit in an increasingly unfavourable economic climate.

Despite an upturn in price for many metals during the last few years the general outlook is not promising for most of the traditional metals, in particular iron, manganese (Fig. 1.4), lead (Fig. 1.6), tin and tungsten. Some of the reasons for this prognostication have been discussed above. It is the minor metals such as titanium, tantalum and others that are likely to have a brighter future. For a more bullish view on the major metals see Green (1989) and for price trends over a longer period (1880–1980) see Slade (1989). Nevertheless we may still decide to make one of the traditional metals our exploration target in which case we must endeavour to discover readily accessible, high grade, big tonnage orebodies, preferably in a politically stable, developed country; high quality deposits of good address as Morrissey (1986) has put it.

This is a tall order but well exemplified recently by the discovery at Neves-Corvo in southern Portugal of a base metal deposit with 27.5 Mt grading 8.66% Cu, 7.2 Mt grading 4.5% Cu + 2.48% Zn and 32.9 Mt grading 5.7% Zn + 1.14% Pb. In addition tin values in part of the

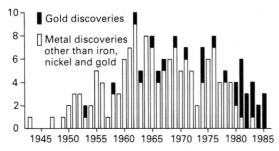

Fig. 1.7 Significant metallic orebody discoveries in non-communist countries. Significant discoveries are relatively low cost producers that have a potential to generate over US$1 000M in gross revenue. (After Cook 1987.)

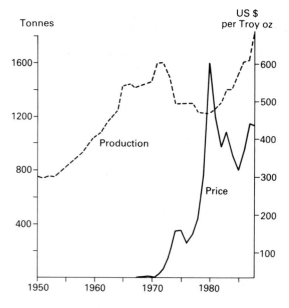

Fig. 1.8 World production of gold from 1950-1988 and the actual yearly average price in US dollars per troy ounce (i.e. no correction for inflation). (Source: Lofty *et al.* (1989) and the *Mining Annual Review* of 1989.)

deposit are sufficient to give a production of 5 000 t p.a. tin metal in concentrates. Annual copper output in 1991 was planned to be 135 000 t from the treatment of just 1.6 Mt of ore. This can be compared with Palabora, RSA, one of the world's great copper mines, which produced 111 692 t copper in 1988 from 29.23 Mt of ore. Clearly some plums are still there for the finding!

Gold has had a different history since World War Two. From 1934 to 1972 the price of gold remained at $US 35 per troy ounce. In 1971 President Nixon removed the fixed link between the dollar and gold and left market demand to determine the daily price. The following decade saw gold soar to a record price of $US 850 an ounce, a figure inconceivable at the beginning of the seventies; it then fell back to a price little higher in real terms than that of the 1930s (Fig. 1.8). (Using the US Consumer Price Index the equivalent of $35 in 1935 would have been $289 in 1987, so that at $400 oz^{-1} the gold price would only be up by 38% in real terms on the 1935 price.) Citizens of many countries were again permitted to hold gold either as bars or coinage and many have invested in the metal. Unfortunately for those attempting to predict future price changes, demand for this metal is not determined so much by industrial demand but by fashion and sentiment — two notoriously variable and unpredictable factors! The main destinations of gold at the present day are carat jewellery and bars for investment purposes. Bar hoarding, led by Japan and Taiwan, jumped by 77% in 1988 compared with the previous year and reached a record 474 t! Carat jewellery in that year absorbed 1 484 t and industrial users only took up 199 t; dentistry accounted for another 59 t (Jacks 1989).

The rise in the price of gold since 1971 has led to a great increase in prospecting and the discovery of many large deposits (Fig. 1.7). This trend is continuing at an increasing rate and gold production since reaching a low point in 1979 has been increasing rapidly (Fig. 1.8)—will fashion and sentiment absorb such an annually sustained increase in supply? Many mining companies have adopted a cautious approach and are not opening new deposits without being sure that they could survive on a price of around $US 250 per ounce, whilst others are putting more emphasis in their exploration budgets on base metals.

(b) *Industrial minerals*. Most industrial minerals can be traded internationally; exceptions are the low value commodities such as sand, gravel and crushed stone which have a low unit value and are mainly produced for local markets. However, minor deviations from this statement are beginning to appear such as crushed granite being shipped from Scotland to the USA, sand from Western Australia to Japan and filtration sand and water from the UK to Saudi Arabia! Lower middle unit value minerals from cement to salt can be moved over intermediate to long distances provided they are shipped in bulk by low cost transport. Nearly all industrial minerals of higher unit value are internationally tradeable, even when shipped in small lots.

Minerals with a low unit value will increase greatly in cost to the consumer with increasing distance to the

Table 1.6 Long term price trends of some major industrial minerals (based on Noetstaller 1988). Prices are ex-mine or processing plant.

Commodity	Average annual price in constant US$ t⁻¹		
	1965	1975	1983
Asbestos	313	273	276
Baryte	81	76	78
Diatomite	160	151	204
Feldspar	29	33	35
Fluorspar	144	148	181
Graphite flake	336	495	606
Gypsum	12	9	9
Kaolin	75	77	90
Mica (ground)	153	122	130
Perlite	25	24	33
Phosphate rock	21	43	24
Potash	132	171	146
Sand and gravel	3.16	2.94	3.25
Stone (crushed)	4.57	4.23	4.27
Stone (dimension)	122	133	139
Sulphur	64	53	91
Talc	69	33	94
Vermiculite	52	72	96

place of use. Consequently low unit value commodities are normally of little or no value unless available close to a market. Exceptions to this rule may arise in special circumstances e.g. the south-eastern sector of England (including London) where demand for aggregates cannot now be met from local resources. Considerable additional supplies now have to be brought in by rail and road over distances in excess of 150 km. For high unit value minerals like industrial diamonds, sheet mica and graphite, location is largely irrelevant.

Like metals, industrial minerals respond to changes in the intensity of business activities, but as a group to nothing like the extent shown by metals and their prices are generally much more stable (see Table 1.6). A more comprehensive table is given by McCarl (1989). One reason for the greater stability of many industrial mineral prices is their use or partial use in consumer non-durables for which consumption remains comparatively stable during recessions, e.g. potash, phosphates, sulphur for fertilizer production; diatomite, fluorspar, iodine, kaolin, limestone, salt, sulphur, talc, etc. used in chemicals, paint, paper, rubber and so on. The value of an industrial mineral depends largely on its end use and the amount of processing it has undergone; with more precise specifications of chemical purity, crystalline perfection, physical form, hardness, etc. the price goes up; for this reason many minerals have quite a price range, e.g. kaolin to be used as coating clay on paper is four times the price of kaolin for pottery manufacture.

Individual commodities show significant price variations related to supply and demand, e.g. potash over the last forty years. When supplies were plentiful, such as after the completion of several large Canadian mines, prices were depressed, whereas when demand has outstripped supply prices have shot up.

According to Noetstaller (1988) already discovered world reserves of most industrial minerals are adequate to meet the expected demand up to at least 2000 A.D. and so no significant increases in real long-term prices are expected. Exceptions to this are likely to be sulphur, baryte, talc and pyrophyllite. Growth rates are expected to rise steadily, rates exceeding 4% p.a. are forecast for nine industrial minerals and 2–4% for twenty-nine others. These figures may well prove to be conservative estimates. Contrary to metals, the recycling potential of industrial minerals is, with some exceptions, low and competing substitutional materials are frequently less efficient (e.g. calcite for kaolinite as a cheaper paper filler) or more expensive.

Before choosing an industrial mineral target for our exploration programme, particularly one with a low or middle unit value, we must make certain that there is an available or potential market for our product, which is not so distant from our exploration area that transport costs will prevent the enterprise being financially viable. In a nutshell look for the market first and then find a nearby deposit, preferably alongside a railway line or good road!

1.3 IMPORTANT FACTORS IN THE ECONOMIC RECOVERY OF MINERALS

1.3.1 Principal steps in the exploration for, and exploition of, an orebody

These may be briefly summarized as follows:

(a) *mineral exploration*: to discover an orebody;

(b) *feasibility study*: to prove its commercial viability;

(c) *mine development*: establishment of the entire infrastructure;

(d) *mining*: extraction of ore from the ground;

(e) *mineral processing (ore dressing)*: milling of the ore, separation of ore minerals from gangue, separation of the ore minerals into concentrates, e.g. copper concentrate; separation and refinement of industrial mineral products;

(f) *smelting*: recovering metals from the mineral concentrates;

(g) *refining*: purifying the metal;

(h) *marketing*: shipping the product (or metal concentrate if not smelted and refined at the mine) to the buyer, e.g. custom smelter, manufacturer.

The exploration step can be subdivided as follows:

(i) *choice of potential target*: study of demand, supply, commodity price trends, available markets, exploration cost, draw up budget;

(ii) *reconnaissance stage*: will start with a literature search and progress to a review of available remote sensing and photogeological data leading to selection of favourable areas, initial field reconnaissance and land acquisition, probably followed by airborne surveys, geological mapping and prospecting, geochemical and geophysical surveys;

(iii) *initial follow-up stage*: detailed geological mapping and detailed geochemical and geophysical surveys, trenching and pitting, limited drilling, some assaying;

(iv) *detailed follow-up stage*: drilling, assaying, mineralogical examination of the ore to ascertain ease or otherwise of mineral processing and likely recoverability.

It is at the end of this stage, or that of the feasibility study, that the explorationists hand over to the mining geologists, mineral processors and geotechical and mining engineers to implement steps (c) to (h). None of this will be achieved in the twinkling of an eye and it will be costly! Typical time spans and costs might be: stage (i) one to two years, 0.25M US$; (ii) two years, 0.5–1.5M US$; (iii) two to three years, 2.5–50M US$; (iv) two years, 2.5–50M US$; (v) feasibility study and raising of finance, one to two years, 0.5–25M US$ (excluding actual capital cost for mine construction). Some of these stages will overlap, but this is unlikely to reduce the time involved and it can be expected that around 12 years will elapse between the start of the exploration programme and commencement of mine production. In a number of cases the lead-in time has been less, but this has usually been the result of lady luck playing a hand, the involvement of other favourable factors or a deliberate search for deposits (particularly of gold) which would have short lead-in times. The above time periods and costs are adapted from Crowson in Tilton *et al.* (1988) which has a wealth of data on mineral exploration.

1.4 SOME FACTORS GOVERNING THE CHOICE OF EXPLORATION AREAS

1.4.1 Location

Geographical factors may determine whether or not an orebody is economically viable. In a remote location there may be no electric power supply, roads, railways, houses, schools, hospitals, etc. All or some of these infrastructural elements will have to be built, the cost of transporting the mine product to its markets may be very high and wages will have to be high to attract skilled workers.

1.4.2 Environmental considerations

New mines bring prosperity to the areas in which they are established but they are bound to have an environmental impact. The new mine at Neves-Corvo in southern Portugal has a total labour force of about 900. When it is remembered that one mine job creates about three indirect jobs in the community in service and construction industries, the impact is clearly considerable. Impacts of this and even much smaller size have led to conflicts over land use and opposition to the exploitation of mineral deposits by environmentalists, particularly in the more populous of the developed countries. The resolution of such conflicts may involve the payment of compensation and the eventual cost of rehabilitating mined out areas, or the abandonment of projects; ' . . . whilst political risk has been cited as a barrier to investment in some countries, environmental risk is as much of a barrier, if not a greater in others.' (Sel. Cttee 1982). Opposition by environmentalists to exploration and mining was partially responsible for the abandonment of a major copper mining project in the UK in 1973. As Woodall (1992) has remarked; in the 1990s explorationists must not only prove their projects to be economically viable but they must also make them socially and therefore politically acceptable.

These new requirements have now led to a time gap, often of several years in developed countries, between the moment when a newly found deposit is proved to be economically mineable, and the time when the complex regulatory environment has been dealt with and governmental approval for development obtained. During this time gap and until production occurs no return is being obtained on the substantial capital invested during the exploration phase.

1.4.3 Taxation

Greedy governments may demand so much tax that mining companies cannot make a reasonable profit. On the other hand, some governments have encouraged mineral development with taxation incentives such as a waiver on tax during the early years of a mining operation. This proved to be a great attraction to mining companies in the Irish Republic in the 1960s and brought considerable economic gains to that country.

When a company only operates one mine, then it is particularly true that dividends to shareholders should represent in part a return of capital, for once an orebody is under exploitation it has become a *wasting asset* and one day there will be no ore, no mine and no further cash flow. The company will be wound up and its shares will have no value. In other words, all mines have a limited life and for this reason should not be taxed in the same manner as other commercial undertakings. When this fact is allowed for in the taxation structure of a country, it can be seen to be an important incentive to investment in mineral exploration and mining in that country.

1.4.4 Political factors

Many large mining houses will not now invest in politically unstable countries. Fear of nationalization with perhaps very inadequate or even no compensation is perhaps the main factor. Nations with a history of nationalization generally have poorly developed mining industries. Possible political turmoil, civil strife and currency controls may all combine to increase greatly the financial risks of investing in certain countries. Periodical reviews of political risks in various countries are prepared and published by specialized companies and references to these can be found in Noetstaller (1988). Useful articles on the subject are Anon. (1985a) and Anon. (1985b).

1.5 RATIONALE OF MINERAL EXPLORATION

1.5.1 Introduction

Most people in the West are environmentalists at heart whether engaged in the mineral extraction industry or some other employment, unfortunately many are of the 'nimby' (not in my backyard) variety! These and many other people fail to realize, or will not face up to, the fact that it is Society that creates the demand for minerals, not the mining and quarrying companies, who are now too often pilloried (section 1.4.2), and financially and legally handicapped when they are simply responding to *our* desire and demand for houses, washing machines, cars with roads to drive them on, guns and atom bombs with which to kill ourselves and so on.

Two stark facts that the majority of ordinary folk and too few politicians understand are first that orebodies are wasting assets (section 1.4.3) and second that they are not evenly distributed throughout the earth's crust. It is the depleting nature of their orebodies that plays a large part in leading mining companies into the field of exploration although it must be pointed out that exploration *per se* is not the only way to extend the life of a mining company. New orebodies, or a share in them, can be acquired by financial arrangements with those who own them, or by making successful takeover bids. Many mining companies welcome the chance to spread the risk by allowing others to buy into their finds (farming out is the commercial term), particularly if these are in politically sensitive or unstable countries.

The chances of success in exploration are tiny. Only generalizations can be made but the available statistics suggest that a success rating of less than a tenth of a per cent is the norm and only in favourable circumstances will this rise above one per cent. With such a high element of risk it might be wondered why any risk capital is forthcoming for mineral exploration! The answer is that successful mining can provide a much higher profitability than can be obtained from most other industrial ventures. Destroy this inducement and investment in mineral exploration will decline and a country's future mineral production will suffer. This is why a growing trend today to penalize mining companies through heavy taxation for what are seen as their 'excessive profits' is to be deplored. A number of examples of such governmental policies over the last two decades could be cited, which show that the net result is that mining companies transfer their activities to countries pursuing more enlightened taxation policies.

1.5.2 Exploration productivity

Tilton *et al.* (1988) drew attention to this important economic measure of mineral exploration success and rightly pointed out that it is even more difficult to assess this factor than it is to determine trends in exploration expenditures. It is a measure that should be assessed on the global, national and company scale because, as Tilton *et al.* emphasize, a growing scarcity of a natural commodity is implied if a larger real expenditure is, on average, required over time to find additional ore reserves of a given quality.

For a company success requires a reasonable financial return on its exploration investment and exploration productivity can be determined by dividing the expected financial return by the exploration costs, *after* these have been adjusted to take account of inflation. For Society, on the global or national scale, the calculation is much more involved but has been attempted by a number of

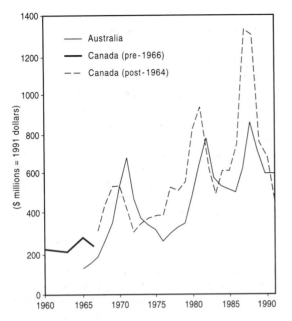

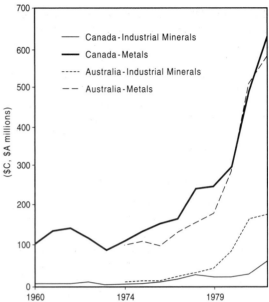

Fig. 1.9 Recent exploration expenditure in Australia (thinner solid line) and Canada (thicker solid line and pecked line). Expenditure for both countries has been converted to the equivalent values in 1991 US dollars. The break in the Canadian data is due to a change in reporting methods. (Sources: Crowson (1988), Canadian Minerals Yearbook and Australian Bureau of Statistics.)

Fig. 1.10 Comparative total exploration expenditure for metals *and* industrial minerals (including coal) in Australia and Canada. The ordinate represents *current* Australian and Canadian dollars, i.e. no corrections have been made for inflationary effects. Also shown are the graphs for industrial mineral exploration in these countries. Canadian data are shown with solid lines and the Australian with pecked lines. Australian expenditure is for companies only, Canadian includes government agencies. (Source: Crowson 1988.)

workers. The study by Mackenzie and Woodall (1988) of Australian and Canadian productivity is extremely penetrating and worthy of much more discussion than there is space for here. It should be emphasized that this analysis and others in the literature are concerned almost exclusively with metallic deposits and comparable studies in the industrial mineral sector still wait to be made.

Mackenzie and Woodall studied *base metal* exploration only and compared the period 1955–78 for Australia with 1946–77 for Canada. They drew some striking conclusions. Australian exploration was found to have been uneconomic but Canadian financially very favourable. Although exploration expenditures in Canada were double those in Australia the resulting number of economic discoveries were eight times greater. Finding an economic deposit in Australia cost four times as much and took four times longer to discover and assess as one in Canada. By contrast the deposits found in Australia were generally three times larger that those found in Canada. As comparable exploration expertise was used

in the two countries then either Australia is endowed with fewer and larger deposits, or the different surface blankets (glacial versus weathered lateritic) render exploration, particularly geophysical, more difficult in Australia especially when it comes to searching for small and medium sized deposits. The latter is the more probable reason for the difference in productivity which suggests that improved exploration methods in the future may lead to the discovery of many more small and medium sized deposits in Australia.

Mackenzie & Dogget (in Woodall 1992) have shown that the *average* cost of finding and proving up *economic* metallic deposits in Australia over the period 1955–86 was about $A 51 million or $A 34 million when discounted at the start of exploration. The average reward discounted in the same manner is $A 35 million — little better than a breakeven situation. Dissecting these data shows that the average exploration expenditure per deposit is gold $A 17M, nickel $A 19M and base metals $A 219M. By contrast the expected deposit value at the start of exploration is gold $A 24M and nickel $A 42 M, which provided the reasonable rates of return of 21% and 19% on the capital invested, whereas on average $53M more has been invested in finding and proving up an economic base metal deposit than has been realized from its subsequent exploitation. Clearly gold and nickel have on average been wealth creating whilst base metal exploration has not!

1.5.3 Exploration expenditure

Some indication has been given in section 1.3.1 of the cost of individual mineral exploration programmes; here we will consider some statistics on mineral exploration expenditure in Australia and Canada (Fig. 1.9). These reflect three troughs and three peaks in expenditure during the last three decades. The troughs correlate closely with the business recessions of the early seventies, eighties and nineties and the peaks with the intervening boom years.

In Fig. 1.10 the comparative total expenditures on exploration for metallic and industrial minerals (including coal) in Australia and Canada are shown and the sums spent on industrial minerals alone. Both countries are important producers of industrial minerals despite their considerably smaller expenditure on exploration for these commodities. The greater expenditure on exploration for metals seems to arise from a number of factors: firstly it is in general cheaper to prospect for industrial minerals and the success rate is higher; secondly metal exploration, particularly for gold, attracts more high risk investment.

These two figures (1.9 & 10) show a general increase in exploration expenditure over recent decades and Cook (1987) records that expenditure in the non-communist world rose from about $400M in 1960 to over $900M in 1980 (constant 1982 US$). This increase is partially due to the use of more costly and sophisticated exploration methods, for example it has been estimated that of deposits found in Canada before 1950, 85% were found by conventional prospecting. The percentage then dropped as follows: 46% in 1951–55, 26% in 1956–65, 10% in 1966–75 and only 4% during 1971–75.

1.5.4 Economic influences

The optimum target (section 1.2.3, METAL AND MINERAL PRICES) in the metals sector must nowadays be a high quality deposit of good address—and if the address is not aristocratic then an acceptable exploration target will have to be of exceptional quality. This implies that for base metal exploration in a remote inland part of Australia, Africa or South America a veritable mammoth of an orebody must be sought with a hundred or so million tonnes of high grade ore or an even greater tonnage of near surface, lower grade material. Only high unit value industrial minerals could conceivably be explored for in such an environment. Anything less valuable would probably not be mineable under existing transportation, infrastructure and production costs.

In existing mining districts smaller targets can be selected, especially close to working mines belonging to the same company, or a company that might purchase any new finds. These finds can be particularly important to a company operating a mine that has only a few years of ore reserves left and, in such a case with an expensive milling plant and perhaps a smelter to supply, an expensive saturation search may be launched to safeguard the future of the operation. Expenditure may then be much higher for a relatively small target than could be justified for exploration in virgin territory.

1.6 FURTHER READING

An excellent summary of the definitions and usages of the terms ore, ore reserves and ore resources is to be found in the paper by Taylor (1989). Gocht *et al.* (1988) *International Mineral Economics* and Strauss' *Trouble in the Third Kingdom* provide recent discussions of mineral markets and Noetstaller's *Industrial Minerals: A Technical Review* is a prolific source of information on the economics of exploring for and exploiting industrial mineral deposits. Many aspects of the philosophical background are to be found in the various papers in *World Mineral Exploration—Trends and Economic Issues* edited by Tilton *et al.* (1988) and a chatty and down to earth discussion of mineral resources in a book of that title by J.A. Wolfe (1984) published by Chapman & Hall.

2

The Mineralogy of Economic Deposits

ANTHONY M. EVANS

2.1 INTRODUCTION

Economic mineral deposits consist of every gradation from bulk materials or aggregates, in which all the rock or mineral is of commercial value, to deposits of precious metals (gold, silver, PGM) from which only a few ppm (or ppb in the case of diamond deposits) are separated and sold. Particular minerals may constitute ore in some deposits, e.g. quartz in silica sands, but be thrown away as gangue in the working of others, such as auriferous quartz veins. Thus the presentation of lists of ore and gangue minerals without any provisos, as given in some textbooks, can be very misleading to the beginner, and may lead him or her to adopt a wasteful mental approach to the examination of mineral deposits, i.e. what do we recover and what do we throw away? Rather than, how can we process everything we are going to mine and market the products at a profit? There are few mineral operations where all the material mined is gainfully disposed of and at the present day fortunes are being made out of the waste left by previous mining and smelting operations, but not usually by the company that dumped it! A good example of a 'waste not, want not' mine is the King's Mountain Operation in the Tin–Spodumene Belt of North Carolina which is the world's most important lithium producing area (Kunasz 1982). The spodumene occurs in micaceous granite–pegmatites and in the mill the ore is processed to produce chemical spodumene and ceramic spodumene concentrates, mica and feldspar concentrates and a quartz–feldspar mix marketed as sandspar. The amphibolite host rock is crushed, sized and sold as road aggregate. Of course such comprehensive exploitation of all the material mined is not possible in isolated locations, but too often the potential of waste material is overlooked. I hope therefore that for this and other reasons, which will be touched upon later in this chapter, the reader will realize that omission of a *comprehensive* mineralogical examination of a potential orebody and its wall rocks may mean that additional valuable materials in the ore are overlooked and the presence of deleterious substances not detected, which may prove calamitous if the project is ever brought to the production stage.

Ore minerals may be native metals (elements) of which gold and silver are examples or compounds of metals with sulphur, arsenic, tellurium, etc., such as lead sulphide—the mineral galena—or they may be carbonates, silicates, borates, phosphates and so on. There are few common minerals that do not have an economic value in some mineralogical context or other. Some of the more important ore minerals are listed in Table 2.1 and those which are often classified as gangue minerals in Table 2.2. Ore minerals may be classed as primary or hypogene if they were deposited during the original period of rock formation or mineralization, and as secondary or supergene if they were formed during a later period of mineralization that was usually associated with weathering and other near surface processes, leading to precipitation of the secondary minerals from descending solutions. When secondary mineralization is superposed on primary mineralization the grade increases and this is termed supergene enrichment.

2.2 MINERALOGICAL INVESTIGATIONS

Before looking at the many methods that may be used, and in the small space available little more can be done than provide an amplified list, it is worth while emphasizing the economic importance of these investigations by discussing briefly the importance of mineralogical form and undesirable constituents.

(a) *Mineralogical form*. The properties of a mineral govern the ease with which existing technology can extract and refine certain metals and this may affect the cut-off grade. Thus nickel is far more readily recovered from sulphide than from silicate ores and sulphide ores can be worked down to about 0.5%, whereas silicate ores must assay about 1.5% to be economic.

Tin may occur in a variety of silicate minerals such as andradite and axinite, from which it is not recoverable, as well as in its main ore mineral form, cassiterite. Aluminium is of course abundant in many silicate rocks, but it must be in the form of hydrated aluminium oxides, the rock called bauxite, for economic recovery. The mineralogical nature of the ore will also place limits on the maximum possible grade of the concentrate. For

Table 2.1 Some of the more important ores and ore minerals

Commodity and principal ore minerals	Mineral formulae	% Metal	Primary	Supergene	Remarks	Major uses
Aggregates					Many different rocks and minerals used	Civil engineering, building
Antimony					Often a by-product of base metal mining	Alloys
stibnite	Sb_2S_3	72	X			
tetrahedrite	$(Cu,Fe)_{12}Sb_4S_{13}$	36	X			
jamesonite	$Pb_4FeSb_6S_{14}$	33	X			
Arsenic					Normally a by-product of metal mining	Herbicide, alloys, wood preservative
arsenopyrite	FeAsS	46	X			
enargite	Cu_3AsS_4	19	X			
löllingite	$FeAs_2$	73	X			
tennantite	$(Cu,Fe)_{12}As_4S_{13}$	13	X			
Asbestos					Dangerous materials, usage declining	Insulation
chrysotile	$(Mg,Fe)_3Si_2O_5(OH)_4$		X			
riebeckite	$Na_3(Fe,Mg)_4FeSi_8O_{22}(OH)_2$		X			
Baryte	$BaSO_4$		X			Drilling muds, filler
Bauxite		≈ 39			Industrial mineral uses increasing	Aluminium production
boehmite	AlO(OH)			X		
diaspore	AlO(OH)			X		
gibbsite	$Al(OH)_3$			X		
Bentonite and fuller's earth	Montmorillonite group minerals		X X	X X		Bleaching clays, drilling muds, binders
Beryllium					Bertrandite is now more important than beryl	Alloys, electronics
bertrandite	$Be_4Si_2O_7(OH)_2$		X			
beryl	$Be_3Al_2Si_6O_{18}$		X			
Bismuth					A by-product of various metal mining operations	Pharmaceutical industry, alloys
bismuthinite	Bi_2S_3	81	X			
Borates			X		Various unusual minerals	Glass wool, borosilicate glasses, detergents agriculture
Chromium					Depends on Cr v. Al content. Steel industry, refractories, chemicals	
chromite	$(Fe,Mg)(Cr,Al)_2O_4$	68(Cr)	X			
Coal	C,O&H		X			Energy source

continued on p. 18

Table 2.1 *(Continued)*

Commodity and principal ore minerals	Mineral formulae	% Metal	Primary	Supergene	Remarks	Major uses
Cobalt					By-product of some copper mines	High temperature alloys, magnets, tool steels
carrollite	$CuCo_2S_4$	56	X			
cobaltiferous pyrite	$(Fe,Co)S_2$	Variable	X			
Common clay	See shale					
Copper					Many other ores of copper	Copper production
bornite	Cu_5FeS_4	63	X			
chalcocite	Cu_2S	80	X	X		
chalcopyrite	$CuFeS_2$	34	X			
covellite	CuS	66	X	X		
Diamond	C		X		Considerable production of synthetic and industrial diamonds	Jewellery, cutting and grinding tools
Diatomite	$SiO_2.nH_2O$		X			Filters, filler, abrasive
Feldspar	$NaAlSi_3O_8$		X			Ceramics, glass making
	$KAlSi_3O_8$		X			
Fluorspar	CaF_2		X		Fluorspar is the raw ore, fluorite the pure mineral	Flux in steel making, fluorochemicals
Gold					Many other tellurides may occur in gold ores	Jewellery, hoarding, dentistry
native gold	Au	90–100	X			
calaverite	$AuTe_2$	39	X			
sylvanite	$(Au,Ag)Te_2$	variable	X			
Graphite	C		X			Steel making, refractories, foundries
Gypsum	$CaSO_4.2H_2O$		X			Plasterboard, insulation
Iron						Iron production
hematite	Fe_2O_3	72	X	X		
magnetite	Fe_3O_4	70	X			
siderite	$FeCO_3$	48	X	X		

Table 2.1 *(Continued)*

Commodity and principal ore minerals	Mineral formulae	% Metal	Primary	Supergene	Remarks	Major uses
Kaolin					A constituent of many clays: ball clay, refractory clay, etc.	Paper manufacture, coating clay, filler, extender
kaolinite	$Al_4Si_4O_{10}(OH)_8$		X	X		
Limestone	$CaCO_3$		X		One of the world's most widely used materials	Constructional, agricultural, chemical & metallurgical industries
Lithium					Li brines are now important producers	Ceramics, glass, enamels, Li salts, Li chemicals, batteries
amblygonite	Li					
lepidolite	$K(LiAl)_3(Si,Al)_4O_{10}(F,OH)_2$		X			
petalite	$LiAlSi_4O_{10}$					
spodumene	$LiAlSi_2O_6$					
Lead					Galena may be argentiferous	Lead production
galena	PbS	86	X			
Magnesite	$MgCO_3$		X		Marketed mainly as magnesia	Refractories, animal feedstuffs, special cements
Manganese					Most important ferro-alloy metal	Steel making. Over 1 Mt p.a. is used for industrial mineral purposes
pyrolusite	MnO_2	63	X	X		
psilomenane	$(BaMn)Mn_4O_8(OH)_2$	50	X	X		
braunite	Mn_7SiO_{12}	64	X	X		
manganite	MnO.OH	62	X	X		
Mercury						Mercury production
cinnabar	HgS	86	X			
Mica					Marketed as sheet or ground mica	Electrical insulator, furnace windows, wallpaper, paints, plasterboard
muscovite	$KAl_2(AlSi_3O_{10})(OH)_2$		X			
phlogopite	$KMg_3(AlSi_3O_{10})(F,OH)_2$		X			
Molybdenum					Mined as the principal metal or as a by-product	Molybdenum production
molybdenite	MoS_2	60	X			
wulfenite	$PbMoO_4$	26	X			
Nepheline–syenite			X		Composed of nepheline, albite and microcline	Container and sheet glass, whitewares, glazes

continued on p. 20

Table 2.1 *(Continued)*

Commodity and principal ore minerals	Mineral formulae	% Metal	Primary	Supergene	Remarks	Major uses
Nickel						Nickel production
pentlandite	$(Fe,Ni)_9S_8$	28(max)	X			
garnierite	$(Ni,Mg)_3Si_2O_5(OH)_4$	<20		X		
Perlite	Rhyolitic composition		X		A volcanic glass that expands on heating	Insulation board, plaster, concrete
PGM					Several other PGM minerals may be present in ores	Catalysts, electrical industry, jewellery
nat. platinum	Pt	≈100	X			
sperrylite	$PtAs_2$	57	X			
braggite	$(Pt,Pd,Ni)S$	variable	X			
laurite	$(Ru,Ir,Os)S_2$	variable	X			
Phosphate rock						Fertilizer (90%), detergents, animal feedstuffs
apatite	$Ca_5(PO_4)_3(F,OH)$	P=18	X	X		
Pyrophyllite	$Al_2Si_4O_{11}.H_2O$		X			Ladle linings in steel mills, ceramics, insecticides
Potash						About 95% goes into fertilizer manufacture, rest for soaps, glass, ceramics, etc.
sylvite	KCl		X			
carnallite	$KCl.MgCl_2.6H_2O$		X			
kainite	$4KCl.4MgSO_4.11H_2O$		X			
langbeinite	$K_2SO_4.2MgSO_4$		X			
REE					Many more REE in these minerals than shown in formulae	Catalysts, glass, ceramics, television tubes, permanent magnets, etc.
bastnäsite	$(Ce,La)(CO_3)F$		X			
parisite	$(Ce,La)_2Ca(CO_3)_3F_2$		X			
monazite	$(Ce,La,Nd,Th)PO_4$		X			
Salt						Innumerable! Over half of all production used in the chemical industry, deicing agent, food preservative
halite	NaCl		X			
Shale and common clay	Clay and mica minerals, quartz		X		60–80% of all 'clay' mined falls into this category	Bricks, tiles, sewer pipes, cement, lightweight aggregate

Table 2.1 *(Continued)*

Commodity and principal ore minerals	Mineral formulae	% Metal	Primary	Supergene	Remarks	Major uses
Silica sand	SiO_2				Sand and gravel working taken worldwide represents one of the most important mining industries	Building, civil engineering, glass manufacture
Sillimanite minerals						Refractories account for 90% of production
andalusite	Al_2SiO_5		X			
kyanite	Al_2SiO_5		X			
sillimanite	Al_2SiO_5		X			
Silver					By-product of many base metal mines particularly Pb-Zn	Photography, electrical and electronic industries, sterling ware, jewellery, etc.
nat. silver	Ag	100	X	X		
acanthite	Ag_2S	87	X			
argentiferous tetrahedrite	$(Cu,Fe,Ag)_{12}Sb_4S_{13}$		X			
cerargyrite and many other minerals	AgCl	75	X			
Sodium carbonate					Bulk of soda ash is (trona) produced synthetically in Solvay plants	Production of soda ash, Na_2CO_3, for glass manufacture, sodium chemicals, paper production, etc.
trona	$Na_2CO_3.NaHCO_3.2H_2O$					
Strontium minerals					Celestite is the only economically important Sr mineral. Used in industry as $SrCO_3$	Pyrotechnics, colour TV tubes, permanent magnets, greases, soaps
celestite	$SrSO_4$		X			
Sulphur					Main source is crude oil. Native sulphur 30%, pyrite 16%	Sulphuric acid production, fertilizers, etc.
nat. sulphur	S	100	X	X		
pyrite	FeS_2					
Talc					Has very many important uses	Filler in paints, plastics, paper, rubber, porcelain, tiles, cosmetics
talc	$Mg_3Si_4O_{10}(OH)_2$		X			
Tin					Cassiterite is the main ore mineral	Tin-plate, solder, alloys, etc.
cassiterite	SnO_2	78	X			
stannite	Cu_2FeSnS_4	27	X			

continued on p. 22

Table 2.1 *(Continued)*

Commodity and principal ore minerals	Mineral formulae	% Metal	Primary	Supergene	Remarks	Major uses
Titania					Both an important metal and indust-rial material in form TiO_2	White pigment in paint, plastics, rubber, paper, etc.
ilmenite	$FeTiO_3$	TiO_2=45–65	X			
rutile	TiO_2	TiO_2=90–98	X			
Tungsten						Cutting materials, steels, electric light bulbs
scheelite	$CaWO_4$	64	X			
wolframite	$(Fe,Mn)WO_4$	61	X			
Uranium					Many other ore minerals	Nuclear-powered generators, spec-ial steels, milit-ary purposes
uraninite-pitchblende	UO_2	88	X			
uranophane	$Ca(UO_2)_2Si_2O_7.6H_2O$	56		X		
carnotite	$K_2(UO_2)_2(VO_4)_2.3H_2O$	55	X	X		
torbernite	$Cu(UO_2)_2(PO_4)_2.8H_2O$	51	X	X		
Vanadium					The main producer is now the RSA from Ti–V–magnet-ites in the Bushveld Complex	Vanadium steels, super alloys
carnotite	$K_2(UO_2)_2(VO_4)_2.3H_2O$	13	X	X		
tyuyamunite	$Ca(UO_2)(VO_4)_2.5-8H_2O$	15	X	X		
montroseite	$VO(OH)$	61	X			
roscoelite	vanadium mica		X			
magnetite	$(Fe,Ti,V)_2O_4$	V=1.5–2.1	X			
Vermiculite	$(Mg,Fe,Al)_3(Al,Si)_4O_{10}(OH)_2.4H_2O$		X	X	Expands on heating	Acoustical and thermal insulator, carrier for fertilizer and agricultural chemicals
Wollastonite	$CaSiO_3$		X			Ceramics, filler, extender
Zeolites			X		Uses multiplying rapidly, much of the demand supplied by synthetic zeo-lites	Cements, paper filler, ion exchange resins, dietary supplement for ani-mals, molecular 'sieve', catalyst
Zinc						Alloy castings, gal-vanizing iron and steel, copper-based alloys, e.g. brass
hemimorphite	$Zn_4(OH)_2Si_2O_7.H_2O$	54		X		
smithsonite	$ZnCO_3$	52		X		
sphalerite	$(Zn,Fe)S$	up to 67	X			
Zirconium					Most zircon is con-verted to ZrO_2 for industrial use	Refractories, foundry sands, nuclear reactors
baddeleyite	ZrO_2	74	X			
zircon	$ZrSiO_4$	50	X			

Table 2.2 List of common gangue minerals

Name	Composition	Primary	Supergene
Quartz	SiO_2	X	
Chert	SiO_2	X	
Limonite	$Fe_2O_3.nH_2O$		X
Calcite	$CaCO_3$	X	X
Dolomite	$CaMg(CO_3)_2$	X	X
Ankerite	$Ca(Mg,Fe)(CO_3)_2$	X	X
Baryte	$BaSO_4$	X	
Gypsum	$CaSO_4.2H_2O$	X	X
Feldspar	All types	X	
Fluorite	CaF_2	X	
Garnet	Andradite most common	X	
Chlorite	Several varieties	X	
Clay minerals	Various	X	X
Pyrite	FeS_2	X	
Marcasite	FeS_2	X	
Pyrrhotite	$Fe_{1-x}S$	X	
Arsenopyrite	FeAsS	X	

2.2.1 Sampling

Mineralogical investigations will lose much of their value if they are not based on systematic and adequate sampling of all the material that might go through the processing plant, i.e. mineralized ground *and* wall rock. The basics of sound sampling procedures are discussed in Chapter 9. The material which the mineralogist will have to work on can vary from solid, coherent rock through rock fragments and chips with accompanying fines to loose sand. Where there is considerable variation in the size of particles in the sample it is advantageous to screen (sieve) the sample to obtain particles of roughly the same size, as these screened fractions are much easier to sample than the unsized material.

The mineralogist will normally subsample the primary samples obtained by geologists from the prospect to produce a secondary sample and this in turn may be further reduced in bulk to provide the working sample using techniques discussed in Jones (1987).

2.2.2 Mineral identification

Initial investigations should be made using the naked eye, the hand lens and a stereobinocular microscope to: (a) determine the ore types present and (b) select representative specimens for thin and polished section preparation. At this stage uncommon minerals may be identified in the hand specimen by using the determinative charts in mineralogical textbooks such as Berry *et al.* (1983) or the more comprehensive method in Jones (1987).

The techniques of identifying minerals in thin section are taught to all geologists and in polished sections to most, and will not be described here. For polished section work the reader is referred to Craig and Vaughan (1981) and Ineson (1989). What is neglected too often these days is the powerful tool of examining crushed samples of rocks and ores in immersion oils and identifying transparent minerals by determining their refractive indices. Loose materials such as beach sands are best studied this way. Using a few thin sections as controls the variation in mineralogical assemblages through a large volume of rock can be rapidly studied using crushed samples and refractive index determinations are much quicker and vastly cheaper than X-ray methods. Nevertheless the last named have their place and to obtain samples for X-ray analysis from thin sections it is recommended that cover slips are normally left off, and for microscopic examination immersion oil of n=1.54 is applied to the section's surface.

example, in an ore containing native copper it is theoretically possible to produce a concentrate containing 100% Cu but, if the ore mineral was chalcopyrite ($CuFeS_2$), the principal source of copper, then the best concentrate would only contain 34.5% Cu.

(b) *Undesirable substances.* Deleterious substances may be present in both ore and gangue minerals. For example, tennantite ($Cu_{12}As_4S_{13}$) in copper ores can introduce unwanted arsenic and sometimes mercury into copper concentrates. These, like phosphorus in iron concentrates and arsenic in nickel concentrates, will lead to custom smelters imposing financial penalties. The ways in which gangue minerals may lower the value of an ore are very varied. For example, an acid leach is normally employed to extract uranium from the crushed ore, but if calcite is present, there will be excessive acid consumption and the less effective alkali leach method may have to be used. Some primary tin deposits contain appreciable amounts of topaz which, because of its hardness, increases the abrasion of crushing and grinding equipment, thus raising the operating costs.

To summarize, the information we will be looking for in a sample includes some, or all, of the following: (i) the bulk chemical composition; (ii) the minerals present; (iii) the proportions of each of these and their chemical compositions; (iv) their grain size; (v) their textures and mineral locking patterns; (vi) any changes in these features from one part of an orebody to another.

Using these simple and, if necessary, more sophisticated methods, all the minerals in the samples must be identified.

X-RAY DIFFRACTION

This is the absolute method of mineral identification, the diffraction pattern recorded on photographic film is unique to any particular mineral, an analogy being a human fingerprint. For identification purposes the powder method is used. A sample must be ground into particles <50 μm in size, mixed with a drop of collodion solution (clear nail varnish) and rolled into a spindle about 0.2 mm diameter (or a small sphere) and mounted in a powder camera. When only a few mineral grains are available in a polished or thin section, one or two fragments can be prised out using a steel needle or diamond point and picked up with a tiny drop of collodion solution on the tip of a glass fibre. A spotty but generally usable powder photograph can be obtained. The theory and methods employed for the powder method, single minerals, rocks, special cases such as clay minerals and other applications including quantitative analysis can be found in Azaroff & Buerger (1958), Berry *et al.* (1983), Hutchison (1974), Jones (1987), Tucker (1988) and Zussman (1977).

ELECTRON PROBE MICROANALYZER

With this equipment a beam of high energy electrons is focused on to about 1–2 μm^2 of the surface of a polished section or a polished thin section. Some of the electrons are reflected and provide a photographic image of the surface whilst other electrons penetrate to depths of 1–2 μm and excite the atoms of the mineral causing them to give off characteristic X-radiation which can be used to identify the elements present and measure their amounts. With this chemical information and knowledge of some optical properties reasonable inferences can be made concerning the identity of minute grains and inclusions. In addition important element ratios such as Fe:Ni in pentlandite and Sb:As in tetrahedrite–tennantite, and small amounts of possible by-products and their mineralogical location can be determined. Useful references are Goldstein *et al.* (1981), Reed (1993) and Zussman (1977).

SCANNING ELECTRON MICROSCOPY (SEM)

SEM is of great value in the three-dimensional examination of surfaces at magnifications from ×20–100 000.

Textures and porosity can be studied and, with an analytical facility, individual grains can be analysed and identified *in situ*. Excellent microphotographs can be taken and, with a tilting specimen stage, stereographic pairs can be produced. This equipment is particularly valuable in the study of limestones, sandstones, shales, clays and placer materials. The method is discussed in Goldstein *et al.* (1981) and Tucker (1988).

DIFFERENTIAL THERMAL ANALYSIS

This method is principally used for clay and clay-like minerals which undergo dehydration and other changes on heating. Measurement of the differences in temperature between an unknown specimen and reference material during heating allow of the determination of the position and intensity of exothermic and endothermic reactions. Comparison with the behaviour of known materials aids in the identification of extremely fine-grained particles that are difficult to identify by other methods (Hutchison 1974).

AUTORADIOGRAPHY

Radioactive minerals emit alpha and beta particles which can be recorded on photographic film or emulsions in contact with the minerals, thus revealing their location in a rock or ore. In ores this technique often shows up the presence of ultra-fine-grained radioactive material whose presence might otherwise go unrecorded. The technique is simple and cheap and suitable for use on hand specimens and thin and polished sections (Robinson 1952, Zussman 1977).

CATHODOLUMINESCENCE

Luminescence (fluorescence and phosphorescence) is common in the mineral kingdom. In cathodoluminescence the exciting radiation is a beam of electrons and a helpful supplement to this technique is ultra-violet fluorescence microscopy. Both techniques are used on the microscopic scale to study transparent minerals. Minerals with closely similar optical properties or which are very fine-grained can be readily differentiated by their different luminescent colours, e.g. calcite *v.* dolomite, feldspar *v.* quartz, halite *v.* sylvite. Features not seen in thin sections using white light may appear, thin veins, fractures, authigenic overgrowths, growth zones in grains, etc. A good description of the apparatus required and the method itself is given in Tucker (1988).

2.2.3 **Quantitative analysis**

GRAIN SIZE AND SHAPE

The *recovery* is the percentage of the *total* metal or industrial mineral contained in the ore that is recovered in the concentrate; a recovery of 90% means that 90% of the metal in the ore passes into the concentrate and 10% is lost in the tailings. It might be thought that if one were to grind ores to a sufficiently fine grain size then complete separation of mineral phases might occur to make 100% recovery possible. In the present state of technology this is not the case, as most mineral processing techniques fail in the ultra-fine size range. Small mineral grains and grains finely intergrown with other minerals are difficult or impossible to recover in the processing plant, and recovery may be poor. Recoveries from primary (bedrock) tin deposits are traditionally poor, ranging over 40–80% with an average around 65%, whereas recoveries from copper ores usually lie in the range 80–90%. Sometimes fine grain size and/or complex intergrowths may preclude a mining operation. The McArthur River deposit in the Northern Territory of Australia contains 200 Mt grading 10% zinc, 4% lead, 0.2% copper and 45 ppm silver with high grade sections running up to 24% zinc and 12% lead. This enormous deposit of base metals has remained unworked since its discovery in 1956 because of the ultra-fine grain size and despite years of mineral processing research on the 'ore'.

Grain size measurement methods for loose materials, e.g. gravels and sands, placer deposits or clays vary, according to grain size, from calipers on the coarsest fragments, through sieving and techniques using settling velocities, to those dependent upon changes in electrical resistance as particles are passed through small electrolyte-filled orifices. These methods are described in Tucker (1988) and other books on sedimentary petrography.

Less direct methods have to be employed with solid specimens because a polished or thin section will only show random profiles through the grains (Fig. 2.1). Neither grain size, nor shape, nor sorting can be measured directly from a polished or thin section and when the actual grains are of different sizes microscopic measurements using micrometer oculars (Hutchison 1974) or other techniques invariably *overestimate* the proportion of small grains present. This bias can be removed by stereological methods *if* the grains are of simple regular shapes (cubes, spheres, parallelepipeds, etc.). Otherwise stereological transformation is not pos-

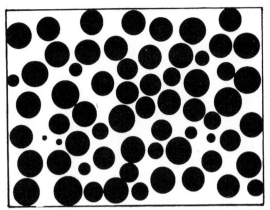

Fig. 2.1 A random section through a solid consisting of a framework of spheres of equal size.

sible and great care must be taken in using grain size measurements taken from sectioned specimens.

An *indication* of grain shape can be obtained by measuring a large number of intercepts (lengths of randomly chosen grain diameters) and analysing these, e.g. the presence of a substantial number of very small intercepts would indicate that the grains were angular and the opposite would show that the grains lacked edges and were smooth and convex (Jones 1987).

MODAL ANALYSIS

The most commonly used method is that of grid sampling. A real or imaginary grid is applied to a planar surface of a solid specimen and the minerals lying at intersections are identified and the number of intersections (points) per mineral recorded. These numbers, when converted to percentages, give us the areal proportions of the minerals and, since volumes in this situation are directly proportional to areas, these are also the volume percentages. There are few problems when coarse-grained unfoliated rocks or ores are measured, but care must be taken with foliated, banded rocks which should only be sampled at right angles to the banding, (Hutchison 1974). Very coarse-grained rocks such as pegmatites can be measured with a grid drawn on transparent material and placed on outcrop, joint or mine surfaces and a similar technique is used to assay ores where the sampler is sure that no ore mineral will be missed, e.g. tungsten deposits where the scheelite is the only ore mineral and can be picked out using ultra-violet light.

For work on thin or polished sections a point counter is attached to the microscope stage and the modal analysis carried out as in petrographic studies (Hutchison 1974, Tucker 1988). The size of the 'grid' and the number of points for a statistically valid analysis must be adjusted to take account of grain size and these points are comprehensively discussed by Chayes (1956).

The volume percentage of ore and/or deleterious minerals can be of crucial importance in mineral processing and a knowledge of whether a wanted metal is present in one or several minerals. If the latter is the case, their relative proportions may also be of great importance. Some examples of this case are: (1) gold ores — is all the gold present as native gold (free-milling gold) or is some in the form of tellurides (combined gold)? Native gold is readily leached from milled ores by cyanide solutions, but gold tellurides resist leaching and have to be roasted (after concentration) before cyaniding, thereby increasing the cost of the treatment and of course decreasing the value of the ore; (2) titanium ores in anorthosites will have significant amounts of titanium locked up in titaniferous magnetite, sphene and augite from which it is not recoverable; (3) the skarn iron orebody at Marmoraton, Ontario assayed on average 50% Fe, but only 37.5% (in magnetite) was recoverable, the rest was locked in silicates. These and similar devaluing features are readily detected and quantified by microscopic investigations. Valuations based on assays alone will be grossly exaggerated!

If the chemical compositions of the minerals are known, then by dividing volume percentages by mineral densities (and converting to per cent) we can obtain the weight percentages of the minerals and, by using Table 2.1 (or by calculation), of any wanted metals in the ore, which gives us an entirely independent way of obtaining assays that may be of value as checks on chemical or XRF assays.

Modal analyses on polished and thin sections together with grain size and intercept measurements can also be made using image analysers (Jones 1987, Sprigg 1987).

The mineral explorationist should train him or herself to make visual modal estimates in the field using hand specimens and natural exposures. By estimating the volume percentage of a metallic mineral such as chalcopyrite (often the only or principal copper mineral in an ore) and looking up the copper content in Table 2.1 a visual assay can be made. When the first laboratory assays become available for a prospect under investigation explorationists will have reference material with which they can compare their estimates and improve their accuracy for that particular ore type. The American Geological Institute Data Sheets (sheet 15.1) and various books (Spock 1953, Barnes 1981, Thorpe & Brown 1985, Tucker 1988 and others) have comparison charts to help the field geologist in estimating percentage compositions in hand specimens.

2.2.4 Economic significance of textures

MINERAL INTERLOCKING

Ores are crushed during milling to liberate the various minerals from each other (section 2.2.3) and for concentration a valuable mineral has to be reduced to less than its liberation size in order to separate it from its surrounding gangue. Crushing and grinding of rock is expensive and if the grain size of a mineral is below about 0.05 mm the cost may well be higher than the value of the liberated constituents. In addition there are lower limits to the degree of milling possible dictated by the separation processes to be employed because these are most effective over certain grain size ranges, e.g. magnetic separation—0.02–2.5 mm, froth flotation—0.01–0.3 mm and electrostatic separation—0.12–1.4 mm.

In Fig. 2.2 a number of intergrowth patterns are illustrated. Further crushing of the granular textured grains in (a) will give good separation of ore (black) from gangue—this is an ideal texture from the processing point of view. In (b) further crushing of the tiny pyrite grain veined by chalcopyrite (black) is out of the question and this copper will be lost to the tailings. The chalcopyrite (black) occurring as spheroids in sphalerite grains (c) is too small to be liberated and will go as a copper loss into the zinc concentrate. The grain of pyrite coated with supergene chalcocite (black) in (d) will, during froth flotation, carry the pyrite as a diluting impurity into the copper concentrate. The grain is too small for separation of the two minerals by crushing. It must be noted that the market price for a metal does not apply fully or directly to concentrates. The purchase terms quoted by a custom smelter are usually based on a nominal concentrate grade and lower concentrate grades are penalized according to the amount by which they fall below the contracted grade. Exsolution textures commonly devalue ores by locking up ore minerals and by introducing impurities. In (e) the tiny flame-shaped exsolution bodies of pentlandite (black) in the pyrrhotite grain will go with the pyrrhotite into the tailings and the ilmenite bodies (black) in magnetite (f) are likewise too small to be liberated by further grinding and will contaminate the magnetite concentrate. If this magnetite is from an ilmenite orebody then these interlocked ilmenite

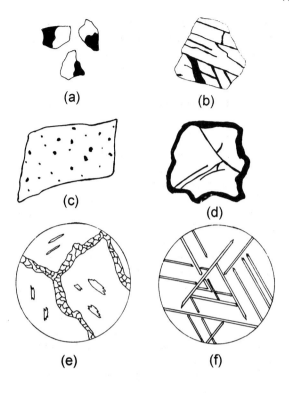

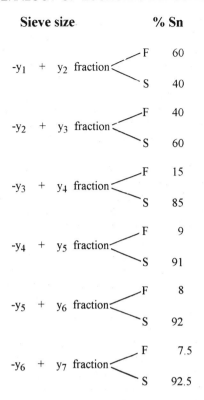

Fig 2.3 Mineral liberation size investigation of a possible tin ore. F=float, S=sink; y_1 to y_7=grain (sieve) sizes. For discussion see text.

Fig 2.2 (a)–(d): grains from a mineral dressing plant. (a) Granular texture; black represents an ore mineral, unornamented represents gangue ($\times 0.6$). (b) A pyrite grain veined by chalcopyrite (black) ($\times 177$). (c) Sphalerite grain containing small rounded inclusions of chalcopyrite (black) ($\times 133$). (d) Pyrite grain coated with supergene chalcocite (black) ($\times 233$). (e) Grains of pyrrhotite with exsolved granular pentlandite in the interstices and flame exsolution bodies within the pyrrhotite ($\times 57$). (f) Exsolution blades of ilmenite in a magnetite grain ($\times 163$).

bodies will be a titanium loss.

MINERAL LIBERATION SIZE

This is usually investigated as follows. Suppose we have a simple example such as a tin ore consisting of cassiterite ($\rho=6.99$) and quartz ($\rho=2.65$), then an ore sample is crushed, sieved and mineral separations, using heavy liquids such as di-iodomethane ($\rho=3.2$) and Clerici solution, carried out on the various size fractions. The float and sink fractions are analysed for each fraction and a possible result is shown in Fig. 2.3 which shows that little improvement in separation would be obtained by grinding beyond the y_5–y_6 fraction.

2.2.5 Deleterious substances

These have been discussed in section 2.2(b) with reference to arsenic, mercury, phosphorus, calcite and topaz in metallic ores. Among industrial minerals mention can be made of the presence of limonite coatings on quartz grains in sand required for glass making. Extra cost in exploitation will arise from the need for an acid leach, or other method, to remove the coatings. Coal fragments in gravel will render it valueless as gravel processing plants do not include equipment for eliminating the coal. Lastly it has become apparent recently that the alkaline-silica-reaction, which produces 'concrete cancer', is due in some cases to the presence of certain types of opaline silica in the aggregate.

2.2.6 Miscellaneous examples of the use of microscopy in ore evaluation and mineral processing

CHROMITE ORES

Chromite is never pure $FeCr_2O_4$ and iron(III) may substitute for chromium, particularly along grain boundaries and fractures, producing 'off colour grains' which can be detected in polished sections by the experienced observer. The magnetite rims will introduce mineral processing difficulties, as will badly fractured chromite grains present in some podiform deposits; these may disintegrate rapidly into a very fine-grained powder on grinding.

NICKEL SULPHIDE ORES

The normal opaque mineralogy is magnetite–pyrrhotite–pentlandite–chalcopyrite. The magnetite and pyrrhotite are separated magnetically and normally become waste. Nickel present in flame exsolution bodies, Fig 2.2(e), and in solid solution in the pyrrhotite will be lost. In an ore investigated by Stephens (1972) use of the electron probe microanalyser revealed that 20% of the nickel would be lost in this way. Minute grains of PGM minerals are often included in the pyrrhotite, again leading to losses if this has not been noted. Also in such ores the chalcopyrite may contain large exsolved bodies of cubanite, a magnetic mineral, which could mean a substantial copper loss in the pyrrhotite concentrate. In such a case the ore must be roasted (at extra expense) to destroy the magnetism of the cubanite.

TIN ORES

At mines where separate concentrates of cassiterite and copper–zinc–arsenic sulphides (often as a minor by-product) are produced, cassiterite coated with stannite (Cu_2FeSnS_4) will pass into the sulphide concentrate.

ZINC LOSS

Lead–zinc ore from a new orebody was tested by being processed in a mill at a nearby mine. Despite being apparently identical to the ore at the mine, substantial zinc losses into the tailings occurred. It was discovered with the use of an electron probe microanalyser that the siderite in the gangue carried 8–21% Zn in solid solution.

MERCURY IMPURITY

In 1976 Noranda Mines Ltd cut its copper–gold–silver concentrate purchases from Consolidated Rambler Mines Ltd, Newfoundland by nearly 50% because of 'relatively high impurities' (Anon. 1977). The major impurity was mercury. An electron probe investigation showed that this occurred in solid solution (1–2%) in the minor sphalerite in the ore. By depressing the zinc in the flotation circuit the mercury content of the concentrate was virtually eliminated.

Probably the first mine to recover mercury from copper concentrate was Rudnany in Czechoslovakia where it occurs in tetrahedrite. At the former Gortdrum Mine in Ireland the presence of cinnabar in the copper–silver ore, and of mercury in solid solution in the tennantite, remained unknown for several years and the smelting of concentrates from this mine at a custom smelter in Belgium presumably produced a marked mercury anomaly over a substantial part of western Europe! A mercury separation plant was then installed at Gortdrum and a record production of 1334 flasks was reached in 1973!

SULPHUR IN COAL

This is usually present as pyrite and/or marcasite. If it is coarse-grained then much can be removed during washing, but many coals, e.g. British ones, have such fine-grained pyrite that little can be done to reduce the sulphur content. Such coals are no longer easily marketable in this time of concern about acid rain.

2.3 FURTHER READING

General techniques in applied mineralogy are well discussed in Jones' *Applied Mineralogy—A Quantitative Approach* (1987). Hutchison's *Laboratory Handbook of Petrographic Techniques* (1974) is also an invaluable book as is Zussman (1977) *Physical Methods in Determinative Mineralogy*. Those working on polished sections should turn to *Ore Microscopy and Ore Petrography* by Craig and Vaughan (1981)—a new edition is in press. References to techniques not covered by these books are given in the text of this chapter.

3 Mineral Deposit Geology

ANTHONY M. EVANS

3.1 NATURE AND MORPHOLOGY OF OREBODIES

3.1.1 Size and shape of ore deposits

The size, shape and nature of ore deposits affects the workable grade. Large, low grade deposits which occur at the surface can be worked by cheap open pit methods, whilst thin tabular vein deposits will necessitate more expensive underground methods of extraction. Open pitting, aided by the savings from bulk handling of large daily tonnages (say >30 kt), has led to a trend towards the large scale mining of low grade orebodies. As far as shape is concerned, orebodies of regular shape can generally be mined more cheaply than those of irregular shape, particularly when they include barren zones. For an open pit mine the shape and attitude of an orebody will also determine how much waste has to be removed during mining. The waste will often include not only overburden (waste rock above the orebody) but waste rock around and in the orebody, which has to be cut back to maintain a safe overall slope to the sides of the pit (section 10.2.1). Before discussing the nature of ore bodies we must learn some of the terms used in describing them.

If an orebody viewed in plan is longer in one direction than the other we can designate this long dimension as its strike (Fig. 3.1). The inclination of the orebody perpendicular to the strike will be its dip and the longest dimension of the orebody its axis. The plunge of the axis is measured in the vertical plane ABC but its pitch or rake can be measured in any other plane, the usual choice being the plane containing the strike, although if the orebody is fault controlled then the pitch may be measured in the fault plane. The meanings of other terms are self-evident from the figure.

It is possible to classify orebodies in the same way as we divide up igneous intrusions according to whether they are discordant or concordant with the lithological banding (often bedding) in the enclosing rocks. Considering discordant orebodies first, this large class can be subdivided into those orebodies which have an approxi-

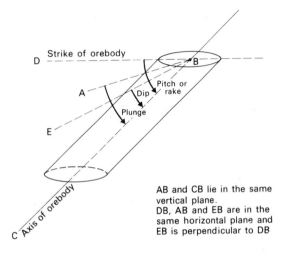

AB and CB lie in the same vertical plane.
DB, AB and EB are in the same horizontal plane and EB is perpendicular to DB

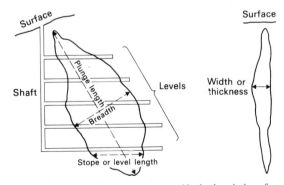

Fig. 3.1 Diagrams illustrating terms used in the description of orebodies.

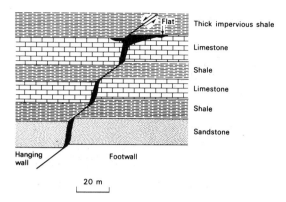

Fig. 3.2 Vein occupying a normal fault and exhibiting pinch-and-swell structure, giving rise to ribbon ore shoots. The development of a flat beneath impervious cover is shown also.

mately regular shape and those which are thoroughly irregular in their outlines.

Discordant orebodies

REGULARLY SHAPED BODIES

(a) *Tabular orebodies.* These bodies are extensive in two dimensions, but have a restricted development in their third dimension. In this class we have veins (sometimes called fissure-veins) and lodes (Fig. 3.2). These are essentially the same and only the term vein is now normally used. Veins are often inclined, and in such cases, as with faults, we can speak of the hanging wall and the footwall. Veins frequently pinch and swell out as they are followed up or down a stratigraphical sequence (Fig. 3.2). This pinch-and-swell structure can create difficulties during both exploration and mining often because only the swells are workable. If these are imagined in a section at right angles to that in Fig. 3.2, it can be seen that they form ribbon ore shoots. Veins are usually developed in fracture systems and therefore show regularities in their orientation throughout the orefield in which they occur.

The infilling of veins may consist of one mineral but more usually it consists of an intergrowth of ore and gangue minerals. The boundaries of vein orebodies may be the vein walls or they may be assay boundaries within the veins.

(b) *Tubular orebodies.* These bodies are relatively short in two dimensions but extensive in the third. When vertical or subvertical they are called pipes or chimneys, when horizontal or subhorizontal, 'mantos'. The Spanish word manto is inappropriate in this context for its literal translation is blanket; it is, however, firmly entrenched in the English geological literature. The word has been and is employed by some workers for flat-lying tabular bodies, but the perfectly acceptable word 'flat' (Fig. 3.2) is available for these; therefore the reader must look carefully at the context when he or she encounters the term 'manto'. Mantos and pipes may branch and anastomose and pipes frequently act as feeders to mantos.

In eastern Australia, along a 2400 km belt from Queensland to New South Wales, there are hundreds of pipes in and close to granite intrusions. Most have quartz fillings and some are mineralized with bismuth, molybdenum, tungsten and tin; an example is shown in Fig. 3.3. Pipes may be of various types and origins (Mitcham 1974). Infillings of mineralized breccia are particularly common, a good example being the copper-bearing breccia pipes of Messina in South Africa (Jacobsen & McCarthy 1976).

IRREGULARLY SHAPED BODIES

(a) *Disseminated deposits.* In these deposits, ore minerals are peppered throughout the body of the host rock in the same way as accessory minerals are disseminated through an igneous rock; in fact, they often *are* accessory minerals. A good example is that of diamonds in kimberlites. In other deposits, the disseminations may be wholly or mainly along close-spaced veinlets cutting the host rock and forming an interlacing network called a stockwork (Fig. 3.4) or the economic minerals may be disseminated through the host rock along veinlets. Whatever the mode of occurrence, mineralization of this type generally fades gradually outwards into subeconomic mineralization and the boundaries of the orebody are assay limits. They are, therefore, often irregular in form and may cut across geological boundaries. The overall shapes of some are cylindrical, others are caplike, whilst the mercury-bearing stockworks of Dubnik in Slovakia are sometimes pear-shaped.

Stockworks most commonly occur in porphyritic acid to intermediate plutonic igneous intrusions, but they may cut across the contact into the country rocks, and a few are wholly or mainly in the country rocks. Disseminated deposits produce most of the world's copper and

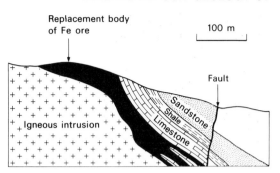

Fig. 3.5 Skarn deposit at Iron Springs, Utah. (After Gilluly *et al.* 1959.)

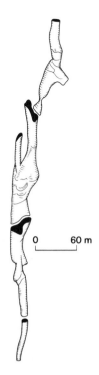

Fig. 3.3 Diagram of the Vulcan pipe, Herberton, Queensland. The average grade was 4.5% tin. (After Mason 1953.)

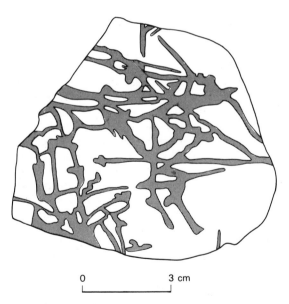

Fig. 3.4 Stockwork of molybdenite-bearing quartz veinlets in granite that has undergone phyllic alteration. Run of the mill ore, Climax, Colorado.

molybdenum (porphyry coppers and disseminated molybdenums) and they are also of some importance in the production of tin, gold, silver (Chapter 15), mercury and uranium. Porphyry coppers form some of the world's monster orebodies. Grades are generally 0.4–1.5% Cu and tonnages 50–5000 Mt.

(b) *Irregular replacement deposits.* Many ore deposits have been formed by the replacement of pre-existing rocks, particularly carbonate-rich sediments, e.g. magnesite deposits. These replacement processes often occurred at high temperatures, at contacts with medium-sized to large igneous intrusions. Such deposits have therefore been called contact metamorphic or pyrometasomatic; however, *skarn* is now the preferred and more popular term. The orebodies are characterized by the development of calc-silicate minerals such as diopside, wollastonite, andradite garnet and actinolite. These deposits are extremely irregular in shape (Fig. 3.5); tongues of ore may project along any available planar structure—bedding, joints, faults, etc. and the distribution within the contact aureole is often apparently capricious. Structural changes may cause abrupt termination of the orebodies. The principal materials produced from skarn deposits are: iron, copper, tungsten, graphite, zinc, lead, molybdenum, tin, uranium and talc.

Concordant orebodies

SEDIMENTARY HOST ROCKS

Concordant orebodies in sediments are very important producers of many different metals, being particularly important for base metals and iron, and are of course concordant with the bedding. They may be an integral part of the stratigraphical sequence, as is the case with

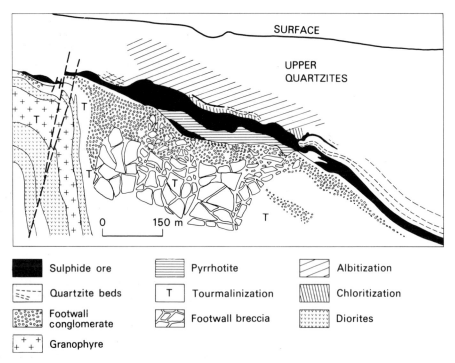

Fig. 3.6 Cross- section through the ore zone, Sullivan Mine, British Columbia. (After Sangster & Scott 1976.)

Phanerozoic ironstones—or they may be epigenetic infillings of pore spaces or replacement orebodies. Usually these orebodies show a considerable development in two dimensions, i.e. *parallel* to the bedding and a limited development *perpendicular* to it (Fig. 3.6) and for this reason such deposits are referred to as stratiform. This term must not be confused with strata-bound, which refers to any type or types of orebody, concordant or discordant, which are restricted to a particular part of the stratigraphical column. Thus the veins, pipes and flats of the Southern Pennine orefield of England can be designated as strata-bound, as they are virtually restricted to the Carboniferous limestone of that region. A number of examples of concordant deposits which occur in different types of sedimentary rocks will be considered.

(a) *Limestone hosts.* Limestones are very common host rocks for base metal sulphide deposits. In a dominantly carbonate sequence ore is often developed in a small number of preferred beds or at certain sedimentary interfaces. These are often zones in which the permeability has been increased by dolomitization or fracturing. When they form only a minor part of the stratigraphical succession, limestones, because of their

solubility and reactivity, can become favourable horizons for mineralization. For example the lead–zinc ores of Bingham, Utah, occur in limestones which make up 10% of a 2300 m succession mainly composed of quartzites.

(b) *Argillaceous hosts.* Shales, mudstones, argillites and slates are important host rocks for concordant orebodies which are often remarkably continuous and extensive. In Germany, the Kupferschiefer of the Upper Permian is a prime example. This is a copper-bearing shale a metre or so thick which, at Mansfeld, occurred in orebodies which had plan dimensions of 8, 16, 36 and 130 km^2. Mineralization occurs at exactly the same horizon in Poland, where it is being worked extensively, and across the North Sea in north-eastern England, where it is subeconomic.

The world's largest, single lead–zinc orebody occurs at Sullivan, British Columbia. The host rocks are late Precambrian argillites. Above the main orebody (Fig. 3.6) there are a number of other mineralized horizons with concordant mineralization. This deposit appears to be syngenetic and the lead, zinc and other metal sulphides form an integral part of the rocks in which they occur. The orebody occurs in a single, generally con-

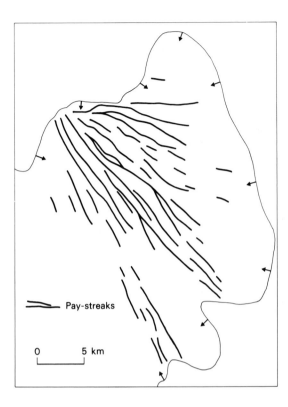

Fig. 3.7 Distribution of pay streaks (gold orebodies) in the Main Leader Reef in the East Rand Basin of the Witwatersrand Goldfield of South Africa. The arrows indicate the direction of dip at the outcrop or suboutcrop. For the location of this ore district see Fig. 13.2. (After Du Toit 1954.)

formable zone 60–90 m thick and runs 6.6% Pb and 5.9% Zn. Other metals recovered are silver, tin, cadmium, antimony, bismuth, copper and gold. This orebody originally contained at least 155 Mt of ore.

Other good examples of concordant deposits in argillaceous rocks, or slightly metamorphosed equivalents, are the lead–zinc deposits of Mount Isa, Queensland, many of the Zambian Copperbelt deposits and the copper shales of the White Pine Mine, Michigan.

(c) *Arenaceous hosts.* Not all the Zambian Copperbelt deposits occur in shales and metashales. Some bodies occur in altered feldspathic sandstones such as Mufulira, which consists of three extensive lenticular orebodies stacked one above the other and where the ore reserves in 1974 stood at 282 Mt assaying 3.47% Cu. The largest orebody has a strike length of 5.8 km and extends several kilometres down dip. Many other concordant sandstone-hosted orebodies occur around the world such as those in

desert sands (red bed coppers), which are very important in China where they make up nearly 21% of the stratiform copper reserves of that country (Chen 1988).

Many mechanical accumulations of high density minerals such as magnetite, ilmenite, rutile and zircon occur in arenaceous hosts, usually taking the form of layers rich in heavy minerals in Pleistocene and Holocene sands. As the sands are usually unlithified, the deposits are easily worked and no costly crushing of the ore is required. These orebodies belong to the group called placer deposits. Beach sand placers supply much of the world's titanium, zirconium, thorium, cerium and yttrium. They occur along present-day beaches or ancient beaches where longshore drift is well developed and frequent storms occur. Economic grades can be very low and sands running as little as 0.6% heavy minerals are worked along Australia's eastern coast.

(d) *Rudaceous hosts.* Alluvial gravels and conglomerates also form important recent and ancient placer deposits. Alluvial gold deposits are often marked by 'white runs' of vein quartz pebbles as in the White Channels of the Yukon, the White Bars of California and the White Leads of Australia. Such deposits form one of the few types of economic placer deposits in fully lithified rocks, and indeed the majority of the world's gold is won from Precambrian deposits of this type in South Africa (Chapter 13). Figure 3.7 shows the distribution of the gold orebodies in the East Rand Basin where the vein quartz pebble conglomerates occur in quartzites of the Witwatersrand Supergroup. Their fan-shaped distribution strongly suggests that they occupy distributary channels. Uranium is recovered as a by-product of the working of the Witwatersrand goldfields. In the very similar Blind River area of Ontario uranium is the only metal produced.

(e) *Chemical sediments.* Sedimentary iron and manganese formations and evaporites occur scattered through the stratigraphical column where they form very extensive beds conformable with the stratigraphy.

IGNEOUS HOST ROCKS

(a) *Volcanic hosts.* The most important deposit type in volcanic rocks is the volcanic-associated massive sulphide (Chapter 14) or oxide type. The sulphide variety often consists of over 90% iron sulphide usually as pyrite. They are generally stratiform bodies, lenticular to sheetlike (Fig. 3.8), developed at the interfaces between volcanic units or at volcanic–sedimentary interfaces. With increasing magnetite content, these sulphide ores grade into massive oxide ores of magnetite

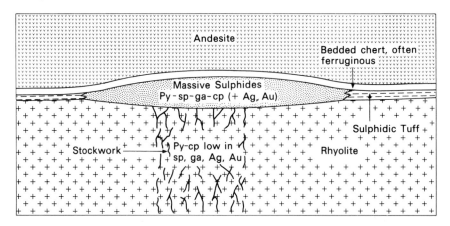

Fig. 3.8 Schematic section through an idealized volcanic-associated massive sulphide deposit showing the underlying feeder stockwork and typical mineralogy. Py = pyrite, sp = sphalerite, ga = galena, cp = chalcopyrite.

and/or hematite such as Savage River in Tasmania, Fosdalen in Norway and Kiruna in Sweden (Solomon 1976). They can be divided into three classes of deposit: (a) zinc–lead–copper, (b) zinc–copper, and (c) copper. Typical tonnages and copper grades are 0.5–60 Mt and 1–5%, but these are commonly polymetallic deposits often carrying other base metals and significant precious metal values which make them plum targets for exploration, e.g. Neves-Corvo (section 1.2.3, METAL AND MINERAL PRICES).

The most important host rock is rhyolite and lead-bearing ores are only associated with this rock-type. The copper class is usually, but not invariably, associated with mafic volcanics. Massive sulphide deposits commonly occur in groups and in any one area they are found at one or a restricted number of horizons within the succession (section 14.2.5). These horizons may represent changes in composition of the volcanic rocks, a change from volcanism to sedimentation, or simply a pause in volcanism. There is a close association with volcaniclastic rocks and many orebodies overlie the explosive products of rhyolite domes. These ore deposits are usually underlain by a stockwork that may itself be ore grade and which appears to have been the feeder channel up which mineralizing fluids penetrated to form the overlying massive sulphide deposit. All these relationships are of great importance in the search for this orebody type.

(b) *Plutonic hosts.* Many plutonic igneous intrusions possess rhythmic layering and this is particularly well developed in some basic intrusions. Usually the layering takes the form of alternating bands of mafic and felsic minerals, but sometimes minerals of economic interest

such as chromite, magnetite and ilmenite may form discrete mineable seams within such layered complexes. These seams are naturally stratiform and may extend over many kilometres as is the case with the chromite seams in the Bushveld Complex of South Africa and the Great Dyke of Zimbabwe.

Another form of orthomagmatic deposit is the nickel–copper sulphide orebody formed by the sinking of an immiscible sulphide liquid to the bottom of a magma chamber containing ultrabasic or basic magma. These are known as liquation deposits and they may be formed in the bottom of lava flows as well as in plutonic intrusions. The sulphide usually accumulates in hollows in the base of the igneous body and generally forms sheets or irregular lenses conformable with the overlying silicate rock. From the base upwards, massive sulphide gives way through disseminated sulphides in a silicate gangue to lightly mineralized and then barren rock (Fig. 3.9).

METAMORPHIC HOST ROCKS

Apart from some deposits of metamorphic origin such as the irregular replacement deposits already described and deposits generated in contact metamorphic aureoles; e.g. wollastonite, andalusite, garnet, graphite; metamorphic rocks are important for the metamorphosed equivalents of deposits that originated in sedimentary and igneous rocks and which have been discussed above.

RESIDUAL DEPOSITS

These are deposits formed by the removal of non-ore

Surface

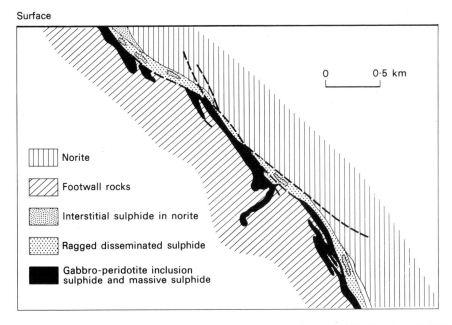

Norite

Footwall rocks

Interstitial sulphide in norite

Ragged disseminated sulphide

Gabbro-peridotite inclusion
sulphide and massive sulphide

0   0·5 km

Fig. 3.9 Generalized section through the Creighton ore zone, Sudbury, Ontario, looking west. (After Souch *et al.* 1969.)

material from protore (rock in which an initial but uneconomic concentration of minerals is present that may by further natural processes be upgraded to form ore). For example, the leaching of silica and alkalis from a nepheline–syenite may leave behind a surface capping of hydrous aluminium oxides (bauxite). Some residual bauxites occur at the present surface, others have been buried under younger sediments to which they form conformable basal beds. The weathering of feldspathic rocks (granites, arkoses) can produce important kaolin deposits which, in the Cornish granites of England, form funnel or trough-shaped bodies extending downwards from the surface for as much as 230 m.

Other examples of residual deposits include some laterites sufficiently high in iron to be worked and nickeliferous laterites formed by the weathering of peridotites.

3.2 WALL ROCK ALTERATION

Many ore deposits, particularly the epigenetic ones, may have beside or around them a zone or zones of wall rock alteration. This alteration of the host rock is marked by colour, textural, mineralogical or chemical changes or any combination of these. The areal extent of the alteration can vary considerably, sometimes being limited to a few centimetres on either side of a vein, at other times

forming a thick halo around an orebody and then, since it widens the drilling target, it may be of considerable exploration value. Hoeve (1984) estimated that the drilling targets in the uranium field of the Athabasca Basin in Saskatchewan are enlarged by a factor of ten to twenty times by the wall rock alteration.

3.3 PLATE TECTONICS AND THE GLOBAL DISTRIBUTION OF ORE DEPOSITS

3.3.1 Metallogenic provinces and epochs

It has long been recognized that specific regions of the world possess a notable concentration of deposits of a certain metal or metals and these regions are known as metallogenic provinces. Such provinces can be delineated by reference to a single metal (Fig. 3.10) or to several metals or metal associations. In the latter case, the metallogenic province may show a zonal distribution of the various metallic deposits. The recognition of metallogenic provinces has usually been by reference to epigenetic hydrothermal deposits, but there is no reason why the concept should not be used to show a geochemical similarity. For example, the volcanic-exhalative antimony–tungsten–mercury deposits in the Lower Palaeozoic inliers of the eastern Alps form a metallogenic province stretching from eastern Switzerland through

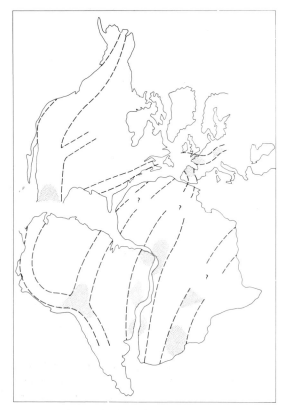

Fig. 3.10 Tin belts on continents around the Atlantic Ocean. Dotted areas indicate concentrations of workable deposits. (Modified from Schuiling 1967.)

Austria to the Hungarian border.

Within a metallogenic province there may have been periods of time during which the deposition of a metal or a certain group of metals was most pronounced. These periods are called metallogenic epochs. Some epochs are close in time to orogenic maxima, others may occur later.

Tin deposits are an excellent example of an element restricted from the economic point of view to a few metallogenic provinces and those around the Atlantic Ocean are shown in Fig. 3.10. Even more striking is the fact that most tin mineralization is post-Precambrian and confined to certain well-marked epochs. Equally striking is the strong association of these deposits with post-tectonic granites. Among tin deposits of the whole world, 63.1% are associated with Mesozoic granites, 18.1% with Hercynian (late Palaeozoic) granites, 6.6% with Caledonian (mid Palaeozoic) granites and 3.3% with Precambrian granites.

Many other metals and minerals are concentrated in space and time and this has obvious implications for mineral exploration programmes. Recent delineations of metallogenic provinces can be found in Jankovic & Petraschek (1987), Mead *et al.* (1988) and Feiss (1989).

3.3.2 **Mineral deposits in the principal plate tectonic regimes**

A vast deal of information has now been published on this subject including books by Mitchell & Garson (1981), Hutchison (1983) and Sawkins (1984), but space will allow only a bare outline to be given here. For this purpose the six tectonic settings discussed by Mitchell and Reading (1986) will be adopted, these are:

1 continental interior basins, intracontinental rifts and aulacogens;
2 oceanic basins and rises;
3 passive continental margins;
4 subduction-related settings;
5 strike-slip settings;
6 collision-related settings.

CONTINENTAL INTERIOR BASINS,
INTRA- CONTINENTAL RIFTS AND
AULACOGENS

There are two types of sedimentary basins within continental interiors: large basins often over 1 000 km across and relatively narrow, fault-bounded rift valleys. The large basins may contain entirely continental sediments, much of which may have been deposited in large lakes like those of the Chad Basin in Africa. Some interior basins, e.g. Hudson Bay, have been inundated by the sea and contain mainly marine sediments. These can contain important mineral deposits, an example being the Permian Zechstein Sea of northern Europe. Here the porous and permeable aeolian sandstones of the Lower Permian Rotliegendes, which provide the important gas reservoirs of north-western Germany, Holland and the southern North Sea, are succeeded by the marine Kupferschiefer, the world's best known copper shales, which were exploited in Germany for about one thousand years and whose mining in Poland makes that country Europe's leading copper producer. Less than 100 m above the Kupferschiefer the famous Zechstein cycles, with their valuable sulphate, sodium and potassium salts, commence. Other evaporites in this setting are those of the Devonian Elk Point Basin of western Canada and the Silurian evaporites of the Michigan Basin. The first continental basins were developed as soon as sufficient craton was present — about 3000 Ma ago in southern Africa (Windley 1984), where the Dominion Reef and Witwatersrand Supergroups with important gold and uranium mineralization were laid down. Similar Proterozoic basins contain uraniferous conglomerates at Elliot Lake in Ontario and the rich unconformity-associated uranium deposits of Athabasca, Canada and Alligator River, Australia. Following the development of stable lithospheric plates, banded iron formation was deposited synchronously over very wide areas; this took place possibly in continental basins and certainly on continental shelves.

Rifts and aulacogens (failed rifts) are initiated by the doming of continental areas over the mantle plumes (hot spots). Plumes may cause melting of the continental crust forming A-type granite intrusions with important tin mineralization as in Nigeria, Rondônia, Brazil and the Sudan. Other mineralization that may be associated with hot spot activity includes the chromium–PGM of the Bushveld Complex and the Great Dyke, southern Africa, the Palabora Complex, RSA with its copper-bearing carbonatites, phosphate rock and vermiculite and the alkaline complexes and carbonatites of the Kola Peninsula, Russia with their important production of phosphate rock, nepheline, iron, vermiculite and various by-products.

The initial stage of rift valley development is often marked by alkaline igneous activity with the development of carbonate lavas and intrusives and occasionally kimberlites. Erosion of the lavas may lead to the formation of soda deposits, e.g Lakes Natron and Magadi in East Africa, and the intrusive carbonatites may carry a number of metals of economic interest (Fe, Nb, Zr and REE) as well as fluorite, baryte and strontianite whilst some provide a source of lime. Carbonatites are highly susceptible to weathering and therefore may have residual, eluvial and alluvial placers associated with them. Some of the kimberlites are diamondiferous.

The sediments and volcanics within rift valleys and aulacogens may host important orebodies. The Central African Copperbelt rocks (Zambia and Zaïre) were probably formed within a rift or a continental basin as were those hosting the important Cu–Pb–Zn Mount Isa and Hilton orebodies of Queensland and the 2000 Mt Cu–U–Au deposit of Olympic Dam, South Australia (Lambert *et al.* 1987). Recently formed aulacogens may contain important oilfields (Ras Morgan, Egypt, North Sea) and many are important for evaporites. Burke and Dewey (1973) identified a number of important Proterozoic aulacogens in North America (Fig. 3.11). Those running perpendicular to the western margin of

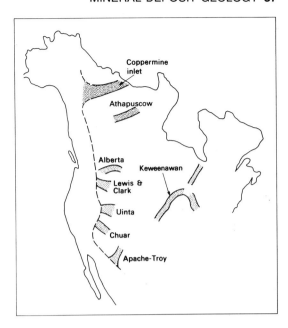

Fig. 3.11 Some of the Proterozoic aulacogens of North America. (Modified from Burke & Dewey 1973.)

the continent carry great thicknesses of rocks of the Belt Series. The northernmost trough contains the Coppermine River Group consisting of more than 3 km of basalt flows with native copper mineralization overlain by greater than 4 km of sediments with evaporites near the top. Epigenetic deposits associated with rifting include the Great Bear Lake Uranium Field. To the south is the Athapuscow Aulacogen with alkali syenite and granite with associated beryllium–REE deposits (Trueman *et al.* 1988), red beds, sedimentary uranium deposits and evidence of the former presence of evaporites (Stanworth & Badham 1984). Further south is the Alberta Rift which passes into British Columbiua and contains about 11 km of late Precambrian sediments in which important stratiform and epigenetic lead–zinc mineralization occur, including the famous Sullivan Mine at Kimberly, British Columbia. Phosphorite deposits are also present. To the east, the Keweenawan Aulacogen, which is the same age as the Coppermine Aulacogen, also carries a considerable thickness of basalt and clastic sediments, and again native copper mineralization is present (Elmore 1984).

OCEAN BASINS AND RISES

There is strong evidence that a number of mineral deposit types are formed during the birth of new oceanic

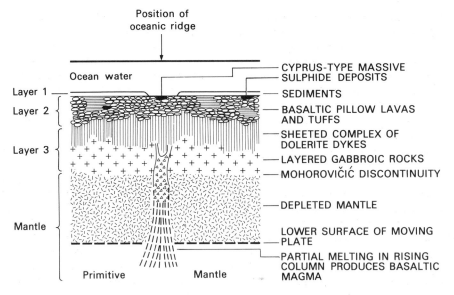

Fig. 3.12 Schematic representation of the development of oceanic crust along a spreading axis. The crustal layering and the possible locations of Cyprus-type massive sulphide deposits are shown. (After Cann 1970 and Sillitoe 1972a.)

crust along mid-ocean ridges, back-arc spreading ridges and the median zones of embryonic oceans like the Red Sea. The possible mode of development of oceanic crust along the median zone of the Red Sea is shown in Fig. 3.12. As this new crustal material moves away from the median zone, layer 1 of the oceanic crust, in the form of pelagic sediment, is added to it. There is strong evidence that many Cyprus-type sulphide deposits are formed during this process of crustal birth. If slices of the oceanic crust are thrust into mélanges at convergent junctures or preserved in some other way then sections similar to that shown in Fig. 3.12 may be expected. Similar situations to the oceanic spreading ridges are found in back-arc basins and indeed many workers hold that the Troodos Massif of Cyprus was developed in such a milieu. In this massif, non-economic nickel–copper mineralization and economic chromite deposits occur in the basic plutonic rocks.

Hydrothermal mineralization with the development of epigenetic copper, zinc, silver, tin, mercury, fluorite and baryte has been reported from a number of oceanic ridges but the most striking is the exhalative copper and zinc massive sulphide mineralization associated with black smokers. Ten deposits estimated to contain 3–5 Mt of sulphide have been found along an 8 km segment of the eastern axial valley at Southern Explorer Ridge, about 150 km west of British Columbia (McConachy & Scott 1987). Gold values up to 1.4 ppm and silver up to 640 ppm have been found in these deposits (Hannington & Scott 1988). Massive sulphide mineralization is also known from modern back-arc spreading centres, e.g. the Lau Basin, east of the Fiji Islands (von Stackelberg *et al*. 1988).

PASSIVE CONTINENTAL MARGINS

As ocean spreading gradually forces two continents apart, both sides of the original rift become passive margins with the development of a continental shelf, bounded oceanwards by a slope and landwards by a shoreline or an epicontinental sea. The type of sediment forming on shelves today depends on latitude and climate, on the facing of a shelf relative to the major wind belts and on tidal range (Mitchell & Reading 1986). In the past, shelves in low latitudes were often covered with substantial platform carbonate successions, and these can be hosts for base metal deposits of both epigenetic and syngenetic nature (so-called Mississippi Valley-type, Irish-type and Alpine-type deposits).

A number of small, stratiform, sandstone-hosted copper and lead–zinc deposits occur in the Cretaceous along the western edge of Africa from Nigeria to Namibia, and further north continue inland to the margin of the Ahaggar Massif. Similar deposits are known along the northern margin of the High Atlas Mountains on the margin of the Morocco Rift, and older ones, which are related to

Infracambrian rifting and the development of a Palaeotethys Ocean, are found along the northern edge of Gondwanaland (Olade 1980, Sillitoe 1980). The lead ores of Laisvall in Sweden and Largentière and Les Malines in France appear to have been developed in a similar tectonic setting (Ramboz & Charef 1988).

A number of the world's most important sedimentary manganese deposits occur just above an unconformity and were formed under shallow marine conditions on shelf areas e.g. Nikopol, Ukraine, Chiatura, Georgia and Groote Eylandt in Australia.

Passive continental margins that have suffered marine transgressions are also important for phosphorite deposits. There are a number of factors that control the development and distribution of phosphorites, including a low palaeolatitude and a broad shallow downwarp flanking a major seaway, such as may result from the complete rifting apart of a continent (Cook 1984). The Tertiary phosphorites of the Atlantic Coastal Plain of the USA appear to have formed under just these conditions. It should be noted that they are a source of by-product uranium.

Many workers now favour a continental shelf environment for the deposition of the Proterozoic Superior-type Banded Iron Formation (Windley 1984). McConchie (1984), in a comprehensive assessment of the evidence from the Hamersley Group, Western Australia, which contains the most extensive accumulation of sedimentary iron deposits known, makes a compelling case for a mid to outer shelf environment.

Finally, we must note the important placer deposits that are developed along the trailing edges of many continents, particularly those where trade winds blow in obliquely to the shoreline combining with ocean currents to give rise to marked longshore drift. Among these we can list the diamond placers of the Namibian coast, the rutile–zircon–monazite–ilmenite deposits of the eastern and western coasts of Australia, and the similar deposits of Florida and the eastern coasts of Africa and South America.

SUBDUCTION-RELATED SETTINGS

Oceanic lithosphere is subducted beneath a volcanic island arc or a continental margin arc on the overriding plate in these settings. In continental margin arcs the volcanic arc is situated landward of the oceanic crust–continental crust boundary as in the Andes. Behind volcanic arcs there is the back-arc area usually referred to when behind island arcs as the back-arc basin or marginal basin or when behind continental arcs as a foreland basin or molasse trough. It is convenient to divide the mineral deposits in island arcs according to whether they were formed outside the arc-trench environment and transported to it by plate motion (allochthonous deposits), or whether they originated within the arc (autochthonous deposits).

Allochthonous deposits. Rocks and mineral deposits formed during the development of oceanic crust may by various trains of circumstances arrive at the trench at the top of a subduction zone (Fig. 3.13). This material is largely subducted, when some of it may be recycled, whilst some is thrust into mélanges. There is, however, only one area so far discovered with what appears to be oceanic ridge-type massive sulphide orebodies thrust into a mélange and that is north-western California, where we find the Island Mountain deposit and some smaller occurrences in the Franciscan mélange. Massive sulphides are also likely to be present at the base of island arc sequences, whether these are the initial succession or a second or later one formed by the migration of the Benioff Zone, because in most cases the arc basement will have originated at an oceanic ridge.

Clearly, any other deposits formed in new oceanic crust at oceanic ridges may also eventually be mechanically incorporated into island arcs and the most likely victims will be the chromite deposits of Alpine-type peridotites and gabbros. Such podiform chromitites occur in the peridotites of the obducted, ophiolitic Papuan Ultramafic Belt. Small platinum metal deposits are known in some of the peridotites, but they are more important as the source rocks for placer deposits of these metals. Economic deposits of podiform chromitite occur in present day arcs in Cuba, where they are found in dunite pods surrounded by peridotite, and on Luzon in the Philippines, again in dunite in a layered ultramafic complex. Numerous occurrences are present in ancient island arc successions and both recent and ancient settings clearly form prime exploration targets for such deposits. A recent development of great importance in this context, and for exploration programmes, is the conclusion of Pearce *et al.* (1984) that all major podiform chromite deposits were formed in marginal basins and not at mid-ocean ridges.

The plutonic rocks of ophiolite suites have not as yet yielded much in the way of magmatic sulphides. Uneconomic pyrrhotite–pentlandite accumulations are known in the Troodos Massif, Cyprus but the only deposits of this type being exploited today are the nickel–platinum orebodies of the Acoje Mine in the Philippines.

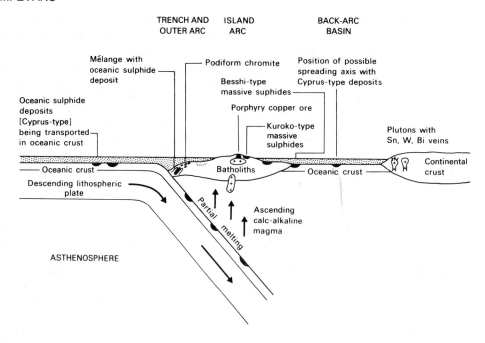

Fig. 3.13 Diagram showing the development and emplacement of some mineral deposits in an island arc and its adjacent regions. (Modified from Sillitoe 1972a,b.)

Autochthonous deposits. In considering these deposits it is convenient to divide the geosynclinal development into three stages: initial or tholeiitic, main calc–alkaline stage and waning calc–alkaline stage. The main calc–alkaline stage is the principal period of arc development with much plutonic activity and a transition from basaltic to andesitic and dacitic volcanics. Much more rock is exposed to subaerial erosion thus increasing the volume of sediment which reaches inter-island gaps, the arc-trench gap and the back-arc basin. Massive sulphide orebodies, usually yielding copper and zinc and known as Besshi-type (after the Japanese occurrences), occur in back-arc turbidite-volcanic (basalt to andesite) associations (Fig. 3.13).

Intrusions hosting porphyry copper deposits belong to this stage. The deposits are most commonly important for copper, gold and silver; molybdenum is not a usual by-product. Skarn deposits of tin, copper and other metals may be present at igneous–carbonate country rock contacts. However much exploration is now in progress in both plutonic and associated volcanic rocks for disseminated gold deposits and the arcs of the southwestern Pacific are proving to be well endowed with both porphyry and epithermal gold orebodies of this type.

During the waning stage of island arc formation the felsic volcanism may be accompanied by the development of massive sulphide orebodies containing economic amounts of copper, zinc and lead commonly accompanied by gold–silver values and known as Kuroko-type.

Continental margin arcs, like those of the Andes, are frequently regions of high relief exposing regularly arranged, regional metamorphic zones and monzonitic–granodioritic plutons and batholiths, as well as active volcanoes at the higher levels. The crust is of greater than normal thickness and although these are regions of plate convergence with its concomitant compression, many graben structures are present running parallel or obliquely to the arcs. Important magmatic activity and mineralization may be related to this rifting, e.g. the world's most important molybdenite deposits, Climax and Urad-Henderson in Colorado which are related to the Rio Grande Rift System (Sawkins 1984).

The Andes are immensely rich in orebodies of various types and metals, but on the whole these show a plutonic–epigenetic affiliation rather than a volcanic-syngenetic one which is to some extent a function of erosion. The Andes are characterized by linear belts of mineralization (Peterson 1970, Sillitoe 1976, Frutos

1982) that coincide with the morphotectonic belts which run down the mountain chain (Grant *et al*. 1980). The Coastal Belt, mainly consisting of Precambrian metamorphic rock, is important for its iron–apatite deposits of skarn and other types. The Western Cordillera (made up mainly of Andean igneous rocks) is a copper province particularly important for porphyry-type deposits carrying copper, molybdenum and gold; and the Altiplano, a Cretaceous–Tertiary intermontane basin filled with molasse sediments, is part of a larger province of vein and replacement Cu–Pb–Zn–Ag deposits that extends along the Eastern Cordillera. This belt consists of Palaeozoic sedimentary rocks and Palaeozoic and Andean igneous rocks, with the tin (+W–Ag–Bi) belt along its eastern margin.

Metallogenic variations are known from other parts of the continental margin arcs of the Americas and they have been tabulated by Windley (1984). The North American Cordillera is a vast storehouse of mineral treasures developed during a complicated tectonic history too involved for discussion here. A useful summary for the western USA can be found in Guild (1978). One point, however, should be made, and that is the important connexion between back-arc rifting and base and precious metal vein deposits (Ivosevic 1984).

STRIKE-SLIP SETTINGS

Two different types of fault are included here: the transform fault and the long known tear or wrench fault. Strike-slip faults vary in size from plate boundary faults such as the San Andreas Fault of California and the Alpine Fault of New Zealand, through microplate boundaries and intraplate faults, such as those of Asia north of the Himalayas, down to small scale fractures with only a few metres offset. Whether these structures occur as single faults or as fault zones (and where complex patterns with some extension are present in the resulting sedimentary basins), they may be important in locating economic mineral deposits.

Well-recognized transform faults in continental crust have little or no associated mineralization, although some potential is perhaps indicated by the location of the Salton Sea Geothermal System (which is a potential ore-forming fluid) within the San Adreas Fault Zone. However it must be noted that this system is not underlain by normal continental crust. Other possible examples of transform faults acting as structural controls of mineralization come from a study of the hypothetical continental continuations of transform faults. For example, many of the diamond-bearing kimberlites of West Africa lie

along such lines (Williams & Williams 1977) and some onshore base metal deposits of the Red Sea region show a similar relationship.

Modern strike-slip basins occur in oceanic, continental and continental margin settings (Mitchell & Reading 1986). They may also occur in back-arc basins. The classic onshore strike-slip basin is the Dead Sea with its important salt production, and the best known offshore basins are those of the Californian Continental Borderland, some of which host commercial oil pools. Ancient strike-slip basins are difficult to identify. However, two examples of economic importance are the Tertiary Bovey Basin of England with its economic deposits of sedimentary kaolin and minor lignite, and the late Mesozoic, coal-bearing, lacustrine basins of north-eastern China.

Fundamental faults and lineaments often have orebodies and even orefields spatially related to them. Among the first to point to the importance of this relationship were Billingsley & Locke (1941) in their consideration of orebody distribution in the Cordillera of the USA and Wilson (1949) dealing with the Ontario–Quebec region of Canada. Lineament analyses have made great advances in the last twenty years as the result of the availability of data from remote sensing and photogeological techniques (Chapter 6). Here we only have space to consider a few examples.

Heyl (1972) drew attention to the 38th Parallel Lineament of the central USA and its relationship to a number of orefields in the Mississippi Valley and elsewhere (Fig. 3.14). This lineament is a zone marked by wrench faults, lines of alkalic, gabbroic, ultramafic and kimberlitic intrusions, suggesting connexions into the mantle, and changes in stratigraphy across it. Gravity anomaly patterns suggest an 80 km, Precambrian dextral movement, with a Phanerozoic offset of 8–10 km. Important orefields occur at the intersections of this lineament and other structures, for example the Central Kentucky Orefield occurs where the West Hickman Fault Zone and the Cincinnati Arch cross it and the very important South-east Missouri Field where it is crossed by the Sainte Genevieve Fault System and the Ozark Dome.

Many workers have commented on the close connexion between gold deposits in the Archaean Abitibi Greenstone Belt of Ontario–Quebec and regional fault zones and a good recent discussion of this is in Kerrich (1986). About 18% of the world's production of antimony comes from the Archaean Murchison Schist Belt — a greenstone belt in the Kaapvaal Craton, RSA. The antimony-gold mines lie along the 35–40 km Antimony Line which is considered to be a zone of shearing hosting most of the mineralization (Vearncombe *et al*. 1988). Finally we can

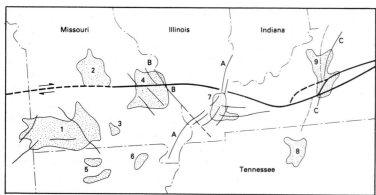

Fig. 3.14 Map showing the 38th Parallel Lineament as outlined by Heyl (1972) (broad dark lines), other faults (narrow lines) and some of the base metal–fluorite–baryte orefields of the Mississippi Valley. Key to the orefields: 1 = Tri-state, 2 = Central Missouri, 3 = Seymour, 4 = South-east Missouri, 5 = Northern Arkansas, 6 = North-east Arkansas, 7 = Illinois–Kentucky, 8 = Central Tennessee, 9 = Central Kentucky; faults: A = New Madrid, B = Ste Genevieve Fault Zone, C = West Hickman Fault Zone.

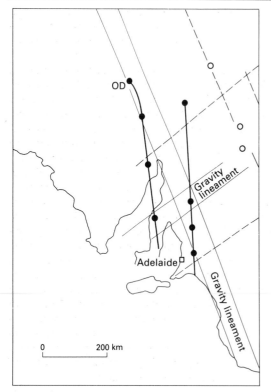

Fig. 3.15 Map showing the two alignments on which the largest of the copper deposits in this part of South Australia lie. The pecked lines represent geological lineaments, open circles are uranium deposits, OD = Olympic Dam. (After Lambert *et al.* 1987).

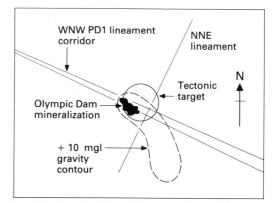

Fig. 3.16 Plan of the Olympic Dam orebody in relation to the west-north-west photolineament corridor PD1 and the north-north-west trending gravity anomaly, which has a sinistral flexure at its northern end bringing it into line with the photolineament. The location of Olympic Dam is shown on Fig. 3.15. (After O'Driscoll 1986.)

gion of South Australia lie on two alignments (Fig. 3.15). The giant among these deposits, Olympic Dam, is close to the intersection of two lineaments and a sinistral flexure in the NNW-trending gravity anomaly (Fig. 3.16).

COLLISION-RELATED SETTINGS

Collision can occur between two active arc systems, between an arc and an oceanic island chain or between an arc and a microcontinent, but the most extreme tectonic effects are produced when a continent on the subducting plate meets either a continental margin or island arc on the overriding plate. Subduction can then lead to melting in the heated continental slab and

glance at a small part of the important lineament analysis carried out in Australia by O'Driscoll which 'defined the Olympic Dam location as a priority drilling target' (Woodall, 1984, O'Driscoll 1986). The largest copper deposits of the Stuart Shelf–Adelaide Geosyncline re-the production of S-type granites (Pitcher 1983). In many continental collision belts, highly differentiated granites of this suite are accompanied by tin and tungsten deposits of greisen and vein type. The tin and tungsten deposits of the Hercynides of Europe (Cornwall, Erzegebirge, Portugal, etc.) are good examples. An excellent study of this activity is to be found in Beckinsale (1979) who discusses the generation of tin deposits in south-east Asia in the context of collision tectonics.

Uranium deposits, particularly of vein type, may be associated with S-type granites as in the Hercynides and Damarides (Rössing deposit). Unfortunately, not all S-type granites have associated mineralization, for example those found in a belt stretching from Idaho to Baja California (Miller & Bradfish 1980). Sawkins (1984) has suggested that the reason that some have associated tin mineralization, some uranium and others neither, presumably reflects the geochemistry of the protolith from which these granites were derived, and their subsequent magmatic history.

Of course, the various collision possibilities described above can result in deposits developed in arcs of various types being incorporated into collision belts. The other features, such as intermontane basins and foreland basins, which are present in arc orogens will have similar mineralization potential.

As a final point it must be emphasized that the type and style of mineralization has changed throughout geological time and this is most important in designing mineral exploration programmes. A short summary of these changes can be found in Evans (1993).

3.4 FURTHER READING

Longer discussions of the subjects forming the main sections of this chapter can be found in *Ores and Industrial Minerals — An Introduction* by Evans (1993). Sawkins' *Metal Deposits in Relation to Plate Tectonics* (second edition 1990, Springer Verlag) covers many of the topics discussed in sections 3.3 and 3.1.

3.5 SUMMARY OF CHAPTERS 1, 2 AND 3

The general nature of ore, industrial minerals and orebodies are discussed in sections 1.1 and 1.2 and it is emphasized that these deposits must contain valuable constituents that can be economically recovered using suitable treatment. (Chapter 10 is largely devoted to a detailed coverage of this principle.) In the past too much interest in many mining circles has been placed on the more glamorous metallic deposits and the general importance of industrial minerals neglected (section 1.2.2). The value of both is governed by demand and supply (section 1.2.3) and many factors including government action, recycling and substitution play a part in determining their market prices. In section 1.3 the principal steps involved in the exploration for, and development of, a mineral deposit are summarized and these are the subjects that are covered in more detail in the rest of this book. A sound knowledge of these is necessary in choosing exploration areas (section 1.4.) leading to the development of a rationale of mineral exploration (section 1.5.).

Economic mineral deposits are extremely variable in their mineralogy and grade and the geologist must be able to identify, or have identified for him or her, *every mineral* in a possible orebody (Chapter 2). This will help them to assess the full economic potential of any material to be mined from it and prevent them overlooking the presence of additional valuable constituents or deleterious substances that may render the deposit unworkable. There are many techniques now available for comprehensive mineralogical examination of mineral samples and these are discussed in section 2.2. However, such investigations must be quantitative as well as qualitative; grain size and shape, relative mineral amounts and the manner of interlocking must be determined.

The nature and morphology of mineral deposits are very varied and only a restricted coverage of these subjects can be given in this book (section 3.1.) The tyro is therefore strongly recommended to acquire a broad knowledge of these subjects from extended reading so that when he or she detects signs of mineralization they may soon develop a working hypothesis of the nature of the particular beast they have come upon.

In assessing the deposit potential of a region it is essential to consider its plate tectonic setting (section 3.3). These settings exert important controls on the nature and distribution of mineral deposits within them. Again this is a vast subject which can only be dealt with in outline in this book, so extensive further reading for the new recruit to the exploration ranks is a *sine qua non*.

4 Reconnaissance Exploration

CHARLES J. MOON AND MICHAEL K.G. WHATELEY

Exploration can be divided into a number of interlinked and sequential stages which involve increasing expenditure and decreasing risk (Fig. 4.1). The terminology used to describe these stages is highly varied. The widely accepted terms used for the early stages of exploration are planning and reconnaissance phases. These phases cover the stages leading to the selection of an area for detailed ground work; this is usually the point at which land is acquired. The planning stage covers the selection of commodity, type of deposit, exploration methods and the setting up of an exploration organization. The process of selecting drill targets within licence blocks we term target selection and that of drilling, target testing (discussed in Chapter 5). The deposit is then at the stage of pre-development followed by a feasibility study (Chapter 10). Before we consider how exploration is planned we should discuss who explores.

THE EXPLORATION PLAYERS

Private sector. Most mineral exploration in developed countries is conducted by companies with a substantial capital base generated either from existing mineral production or from investors on stock markets. The size of the company can range from major multinational mining companies, such as RTZ plc or Anglo American Corporation with operations on several continents to small venture capital companies with one or two geologists. Exploration by individual prospectors is an important factor in countries with large unexplored areas and liberal land tenure laws, for example, Canada, Australia and Brazil. Here you can still meet the grizzled prospector or garimpeiro. Although they usually lack the sophisticated training of the corporate geologist, this can be compensated for by a keen eye and the willingness to expend a little boot leather.

State organizations. In more centrally directed economies, most exploration is carried out by state run companies, geological surveys and in addition in the case of developing countries, international aid organizations. Examples of state companies are ADARO in

Spain and Mindeco in Zambia, which compete with private companies. The role of a geological survey usually includes some provision of information on mineral exploration to government and the private sector. Usually this takes the form of reconnaissance work but the Bureau de Recherches Géologiques et Minières (BRGM) in France is actively involved in detailed exploration, both inside and outside France. The United Nations finances exploration in developing countries under two schemes, the United Nations Development Programme and the Revolving Fund for Mineral Exploration. In contrast, the Soviet Ministry of Geology had until 1991 exclusive prospecting rights in the former USSR, although exploration was carried out by a wide variety of organizations at the Union (federal) and republic levels whereas in China exploration is undertaken by a number of state groups, including the army.

These two groups essentially explore in similar ways although state enterprises are more constrained by political considerations and need not necessarily make a profit. The remainder of the chapter will be devoted to private sector organizations although much may be applicable to state organizations.

4.1 EXPLORATION PLANNING

Mineral exploration is a long-term commitment and there must be careful planning of a company's long-term objectives. This should take particular regard of the company's resources and the changing environment in which it operates (Riddler 1989). The key factors are.

1 *Location of demand for products*. This will depend on the areas of growth in demand. For metals the most obvious areas are the newly industrializing countries of the Pacific Rim, whose Gross Domestic Products increased at around 7% between 1970 and 1985 and who are resource deficient.

2 *Metal prices*. Price cycles should be estimated as far as possible and supply and demand forecast (section 1.2.3).

3 *Host country factors*. The choice of country for

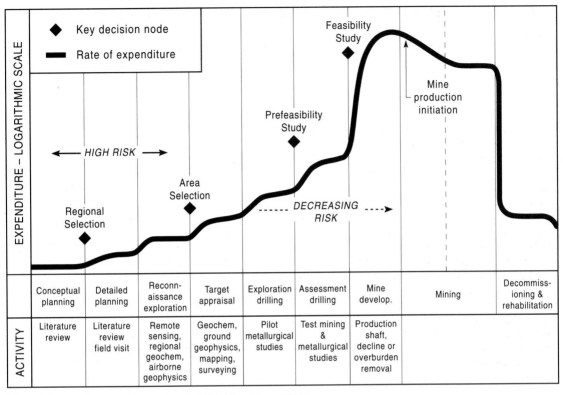

Fig. 4.1 Stages of an exploration project. (Modified from Eimon 1988.)

operation is important in an industry which has seen substantial nationalization, such as copper in Zaïre and Zambia and coal in the UK. The standing of foreign investment, the degree of control permitted, percentage of profits remittable to the home country and most of all government stability and attitude are important. Other factors are availability of land, security of tenure and supply of services and skilled labour (section 1.4).

4 *The structure of the mining industry.* Barriers to the entry of new producers are competition from existing producers and the need for capital to achieve economies of scale.

These factors have combined to encourage exploration companies in the 1980s to concentrate on precious metals in Australia, Canada and the USA (section 1.2.3 METAL AND MINERAL PRICES). Generally low base metal prices during that decade discouraged all but existing metal producers from exploring for other metals. Some existing producers have undertaken limited exploration to maintain market share and smelter output, two legitimate strategies.

After the initial corporate planning, usually by senior executives, an exploration strategy must be chosen, a budget allocated and desirable deposit type(s) defined.

The choice of exploration strategy varies considerably between companies and depends on the objects of the company and its willingness to take risks. For new entrants into a country the choice is between exploration by acquisition of existing prospects or grass roots (i.e. from scratch) exploration. Acquisition requires the larger outlay of capital but carries lower risk and has, potentially, a shorter lead time to production. Acquisitions of potential small producers are particularly attractive to the smaller company with limited cash flow from existing production. Potential large producers interest larger companies which have the capital necessary to finance a large project. Larger companies tend to explore both by acquisition and by grass roots methods and often find that exploration presence in an area will bring offers of properties ('submittals'). Existing producers have the additional choice of exploring in the immediate vicinity of their mines, where it is likely that

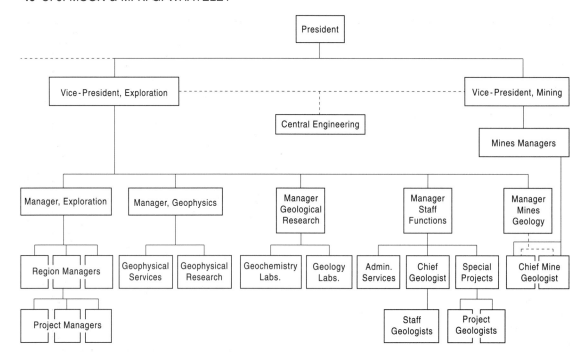

Fig. 4.2 Organization of exploration in a large mining group with producing mines.

they will have substantial advantages in cost saving by using existing facilities. Most of the following section refers to grass roots exploration, although evaluation of potential acquisitions could run in parallel.

4.1.1 Organization

The key to exploration organization is to have the best available staff and adequate finance in order to create confidence throughout the organization (Woodall 1984). A number of factors that characterize a successful exploration team have been recognized by Snow & Mackenzie (1981) and Regan (1971).

1 *High quality staff and orientation towards people.* Successful organizations tend to provide more in-house training.

2 *Sound basis of operations.* The organization works within corporate guidelines towards objectives.

3 *Creative and productive atmosphere.* The group encourages independent creative and innovative thinking in an environment free from bureaucratic disruption.

4 *High standard of performance, integrity and ethics.*

5 *Entrepreneurial acumen.* Innovation is fostered in a high risk, high reward environment.

6 *Morale and team spirit.* High morale, enthusiasm and a 'can do' attitude.

7 *The quality of communication is high.* The 'top brass' are aware of the ideas of geologists.

8 *Pre-development group.* Successful organizations are more likely to have a specialist group responsible for the transition of a deposit from exploration to development.

All these points make it clear that the management must consist of flexible individuals with considerable experience of exploration.

If these are the optimum characteristics of an exploration group, is there an optimum size and what structure should it have? Studies such as those of Holmes (1977) show that the most effective size is in the range seven to ten geologists; larger organizations tend to become too formalized and bureaucratic, leading to inefficiency whereas small groups lack the budgets and the manpower to mount a successful programme. For the large mining group which wishes to remain competitive while spending a large budget, the solution is to

Table 4.1 Corporate Exploration Spending in 1991. (From Metals Economics Group 1992.)

Company	US$ million
Anglo American	102
De Beers	60
Johannesburg Consolidated Investments	25
Minorco	22
Gold Fields of South Africa	12
RTZ	56
CRA	86
Rio Algom	17.5
Noranda + Hemlo Gold	66
Falconbridge	62
Mount Isa Mines	30
Asarco + Asarco Australia	16.9
Cominco	27
Teck	14
Placer Dome	60
Anglo Vaal	48
Broken Hill Pty.	54
Newmont + Newmont Gold	50
Western Mining	44
INCO	38.5
Gencor	36
Cogema	34.6
Battle Montain	31
Phelps Dodge	30
Lac Minerals	30
Billiton	30
Homestake	29
Newcrest	29
BRGM	28
Outukumpu	23
Renison Goldfields	21
Cyprus Minerals	19
Sante Fe Pacific	17
Gold Fields Mining	17
Companhia Vale do Rio Doce	15
Total spending (including others 519.5)	1800

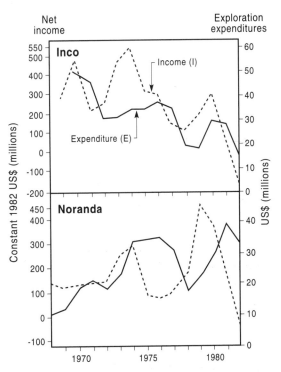

Fig. 4.3 Relation between exploration expenditure and income for two Canadian mining companies. (After Eggert 1988.)

shown in Fig. 4.2. In a company with existing production there may be close liaison with mines and engineering divisions.

4.1.2 **Budgets**

Exploration costs are considered in two ways, (1) as an expenditure within an organization and (2) within the context of a project. It is usually the exploration manager that considers the former, but it is as a geologist on a specific exploration project that one becomes involved in the latter.

Corporate exploration expenditure. Finance for corporate exploration is derived from two main sources, revenue from existing production and by selling shares on the stock market. In the first case the company sets aside a percentage of its before tax profits for exploration. The decision as to the percentage set aside is based upon how much the company wishes to keep as capital, for their running costs, as dividend for their shareholders and for taxes. Exploration costs may range from 1% to 20%, of the annual cash flow. This may be anything between US$0.5M to $100 million per annum depend-

divide its explorationists into semi-autonomous groups.

Exploration groups can be organized on the basis of geographical location or of deposit type. The advantage of having deposit specialists is that in depth expertise is accumulated; however, the more usual arrangement is to organize by location with geologists in each region forming specialist subgroups. At the reconnaissance stage most work will be carried out in offices located in the national head office or state office but as exploration focuses in more detail smaller district offices can be opened. A typical arrangement for a large company is

Table 4.2 Example of a budget worksheet used to estimate the annual budget requirement for a project

Project Name:
Project No:

Month	Jan	Feb	Mar	Apr	May	Jun	Jul	Aug	Sep	Oct	Nov	Dec	Total
Salaries													
Wages													
Drilling													
Transport													
Office Lease													
Administration													
Analytical Costs													
Field Expenses													
Options – New													
Options – Renewals													
Surveys													
Other Contracts													
MONTH TOTALS													

ing upon the size of the company and the size of the deposit for which they are searching (Table 4.1). For producing companies most exploration can be written off against tax. The smaller company is at a disadvantage in that money must be raised from shareholders. This is easy in times of buoyant share prices. For example about $C2 billion were raised in the late 1980s for gold exploration in Canada, largely on the Vancouver Stock Exchange. This was aided by flow-through schemes which enabled share purchasers to offset this against tax liabilities. In a similar way, but without tax breaks, about US$100 million was raised on the Irish Stock Exchange from 1983–1989 (Gardiner 1989). Although major mining groups should be able to maintain consistent budgets and avoid business cyclicity, there seems little evidence of this. In a study of major groups Eggert (1988) demonstrated that spending was linked to income, and therefore metal prices, but lagged about 18 months behind the changes in income (Fig. 4.3).

The overall budget is then subdivided. For example a multinational corporation may have several regional exploration centres, each of which may have anything from one or two persons to a fully equipped office with up to 50 people employed. The budget for the latter must include the salaries, equipment, office rentals and vehicular leases before a single geologist sets foot in the field. Each project would be given a percentage of the regional office's budget, e.g. exploration for coal 20%, base metals 20%, uranium 15%, gold 25% and indus-

trial minerals 20%. Particular exploration projects then have to compete for funds. If a particular project is successful then it will attract additional spending. This distribution of funds is carried out on a regular basis and is usually coupled with a technical review of exploration projects in which the geologist in charge of a project has to account for money spent and put forward a bid for further funding.

Careful control of funds is essential and usually involves the nomination of budget holders and authorization levels. For example, the exploration manager may be able to authorize expenditure of $100 000 but a project·geologist only $2000, so all drilling accounts will be sent to head office whereas the geologist will deal with vehicle hire and field expenses. In this case written authorization from the exploration manager would also be required before drilling starts. Summary accounts will be kept in head office under a qualified accountant or bookkeeper and will be subject to regular audit by external accountants. For most companies global expenditure on exploration will be included in annual accounts and will be written off against the year's income.

Project basis. In order to contain costs and provide a basis for future budgeting, costs will be calculated for each exploration project. Firstly to monitor spending accurately and secondly so that the expenses can be written off against production, if exploration is successful, or if the project is sold to another company. Budgets

Table 4.3 Examples of budget forcasting for a low cost project and a project requiring drilling. Land costs are excluded

Low Cost Project Cost in US$	
Time: 6 days @ US$200 per day	1200
Hotels and meals for 2 nights	350
Transport (vehicle hire)	200
Assays	200
Office staff time	100
TOTAL	2050

Medium Cost Project	
20 Boreholes @ 200m each=4000m	
4000m at US$50 per metre	200 000
Overhead costs	200 000
TOTAL	400 000

are normally calculated on a yearly basis within which projects can be allocated funds on a monthly basis. At the desk study stage the main costs are salaries of staff and these can be calculated from a nominal staff cost per month, the average of the salaries is usually multiplied by two to cover pensions, housing costs and secretarial support. Other support costs must be included such as vehicles and helicopter transport in remote areas. Direct costs of exploration are easier to estimate and include geochemistry, remote sensing and geophysical costs. These can be easily obtained by asking for quotations from contractors. Perhaps a more contentious matter is the allocation of overhead costs to each project; these cover management time and will include the time of head office staff who review projects for presentation to board level management. Often these can be frighteningly large, for example, the costs of keeping a senior geologist in a metropolitan centre will be much larger than a small town. Table 4.2 shows a worksheet for the calculation of budget costs and Table 4.3 gives an example of a completed budget and typical costs for reconnaissance programmes.

4.1.3 Geological models

One of the aims of the planning stage is to identify areas for reconnaissance and to do this we must have some idea of how the materials sought relate to geological factors including geophysics and geochemistry. This is best achieved by setting up a model or models of the type of deposit sought. But what is a model? The term has been defined in various ways but a useful one is that

of 'a functional idealization of a real world situation used to aid in the analysis of a problem'. As such it is a synthesis of available data and should include the most informative and reliable characteristics of a deposit type, identified on a variety of scales and including definition of the average and range of each characteristic (Adams 1985). It is therefore subject to uncertainty and change; each new discovery of an example of a deposit type should be added to the data base.

In mineral deposit models there are two main types which are often combined; the empirical model based on deposit descriptions and a genetic model which explains deposits in term of causative geological processes. The genetic model is necessarily more subjective but can be more powerful, as it can predict deposits not contained in the descriptive data base. Another type of model which is extremely useful for preliminary economic evaluations is a grade–tonnage model. This accumulates grade and tonnage data for known deposits and from this it is possible to estimate the size and grade of an average or large deposit and the cashflow if one were found. Examples of this type of modelling are given by Gorman (1994) for South American gold and copper deposits.

To understand how models are constructed we will examine the deposit model for unconformity related uranium deposits. These deposits are currently the main source of high grade uranium in the Western World and mainly occur in two basins; the Athabasca Basin of Northern Saskatchewan and the Alligator River area of the Northern Territory in Australia. Good accounts of most deposit models can be found in two publications by North American Geological Surveys, Cox & Singer (1986) and Eckstrand (1984). Cox & Singer adopt a strictly descriptive approach but also give grade–tonnage curves and geophysical signatures whereas Eckstrand is more succinct but gives brief genetic models. Some major deposit types are also dealt with in depth by Roberts & Sheahan (1988). Table 4.4 shows the elements that were used by Eckstrand and Cox & Singer in their unconformity related deposit models.

Both accounts agree that the deposits occur at or near the unconformable contact between regionally metamorphosed Archaean to Lower Proterozoic basement and Lower to Middle Proterozoic (1900–1200 Ma) continental clastic sediments, as shown in Fig. 4.4. The details of the controls on mineralization are more subjective; the Canadian deposits occur both in the overlying sandstones and basement whereas the Australian deposits are almost exclusively in the basement. Most deposits in Canada are spatially associated very

Table 4.4 Summary of elements used in the models for unconformity related uranium deposits by Grauch & Moiser in Cox & Singer (1986) and Tremblay & Ruzicka in Eckstrand (1984)

Element	Cox & Singer	Eckstrand
Synonym: Commodities (Eckstrand)	Vein-like type U	U (Ni, Co, As, Se, Ag, Au, Mo)
Description	U mineralization in fractures and breccia fills in metapelites, metapsammites and quartz arenites below, across and above an unconformity separating early and middle Proterozoic rocks	
Geological environment Rock types	Regionally metamorphosed carbonaceous pelites, psammites and carbonates. Younger unmetamorphosed quartz arenites	Ores occur in clay sericite and chlorite masses at the unconformity and along intersecting faults and in clay altered basement rocks and kaolinized cover rocks
Textures	Metamorphic foliation and later brecciation	
Age range	Early and middle Proterozoic affected by Proterozoic regional metamorphism	Most basement rocks are Aphebian, some may be Archaean. Helikian cover rocks. Ore: Helikian (1.28 ± 0.11 Ga)
Depositional environment: Associated rocks (Eckstrand)	Host rocks are shelf deposits and overlying continental sandstone	Basement: graphitic schist and gneiss, coarse-grained granitoids, calc silicate metasediments. Cover rocks: sandstone and shale
Tectonic setting (Cox & Singer) Geological setting (Eckstrand)	Intracratonic sedimentary basins on the flanks of Archaean domes. Tectonically stable since the Proterozoic	Relatively undeformed, intracratonic, Helikian sedimentary basin resting unconformably on intensely deformed Archaean and Aphebian basement. Deposits are associated with unconformity where intersected by faults. Palaeoregolith present in Saskatchewan
Associated deposit types	Gold and nickel-rich deposits may occur	
Form of deposit		Flattened cigar shaped, high grade bodies oriented and distributed along the unconformity and faults, especially localized at unconformity–fault intersections. High grade bodies grade outwards into lower grade
Deposit description Mineralogy	Pitchblende + uraninite ± coffinite ± pyrite ± galena ± sphalerite ± arsenopyrite ± niccolite. Chlorite + quartz + calcite + dolomite + hematite + siderite + sericite. Late veins contain native gold, uranium and tellurides. Latest quartz–calcite veins contain pyrite, chalcopyrite and bituminous matter	Pitchblende coffinite, minor uranium oxides. Ni and Co arsenides and sulphides, native selenium and selenides, native gold and gold tellurides, galena, minor molybdenite, Cu and Fe sulphides, clay minerals, chlorite, quartz, graphite, carbonate

Table 4.4 *(continued)*

Element	Cox and Singer	Eckstrand
Texture & structure	Breccias, veins and disseminations. Uranium minerals coarse and colloform. Latest veins have open space filling textures	
Alteration	Multistage chloritzation dominant. Local sercitization, hematization, kaolinization and dolomitization. Vein silicification in alteration envelope. Alteration enriched in Mg, F, REE and various metals. Alkalis depleted	
Ore controls	Fracture porosity controlled ore distribution in metamorphics. Unconformity acted as disruption in fluid flow but not necessarily ore locus	1 At or near unconformity between Helikian sandstone and Archaean basement. 2 Intersection of unconformity with reactivated basement faults. 3 Associated with graphitic basement rocks and grey and/or multicoloured shale and sandstone cover. 4 Intense clay, chlorite and sericite alteration of basement and cover rocks. 5 Basement rocks with higher than average U content
Weathering	Various secondary U minerals	
Geochemical and geophysical Signature (Cox & Singer)	Increase in U, Mg, P and locally Ni, Cu, Pb, Zn, Co, As; decrease in SiO_2. Locally Au with Ag, Te, Ni, Pd, Re, Mo, Hg, REE and Rb. Anomalous radioactivity . Graphitic schists in some deposits are strong e.m. conductors	
Genetic model		Combinations of 1 preconcentration of U during deposition of Aphebian sediments and their anatexis, 2 concentration in lateritic regolith (Helikian), 3 mobilization by heated oxidized solutions and precipitation in reducing environment at unconformity & fault locus, 4 additional cycles of mobilization and precipitation leading to redistribution of uranium
Examples	Rabbit Lake, Saskatchewan Cluff Lake Saskatchewan	Key Lake, Rabbit Lake, Cluff Lake, Canada. Jabiluka I and II, Ranger, N.T., Australia
Importance		Canada: 35% of current U production but 50% of reserves. World: 15% of reserves.
Typical grade & tonnage	see Fig. 4.5	Canada: small to 5 Mt of 0.3-3% U. Australia: maximum 200 000 t of contained U but grade lower than Canada

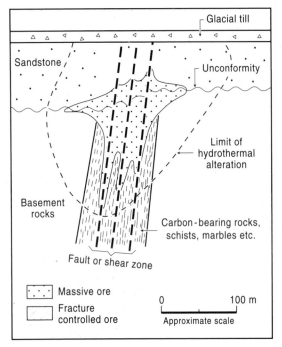

Fig. 4.4 Generalized diagram of an unconformity associated uranium deposit. (After Clark *et al*. 1982.)

closely with graphitic schists whereas the Australian deposits are often in carbonates, although many are carbonaceous. Both models agree that the key alteration is chloritization together with sericitization, kaolinization and hematitization along the intersection of faults and the unconformity. Generalization on the mineralogy of the deposits is difficult but the key mineral is pitchblende with lesser amounts of the uranium silicate, coffinite. The more contentious issue is the associated minerals; in Canada high grade nickel arsenides are common, but by no means ubiquitous, whereas the mineralogy of the Australian deposits is simpler. However, gold is more common and may be sufficiently abundant to change the deposit model to unconformity related uranium–gold. Selenides and tellurides are present in some deposits, although Cox & Singer omit the former but mention enrichment in palladium.

The genetic model for these deposits includes elements of the following.
1 Preconcentration of uranium and associated elements in basement sedimentary rocks.
2 Concentration during the weathering of the basement prior to the deposition of the overlying sediments.

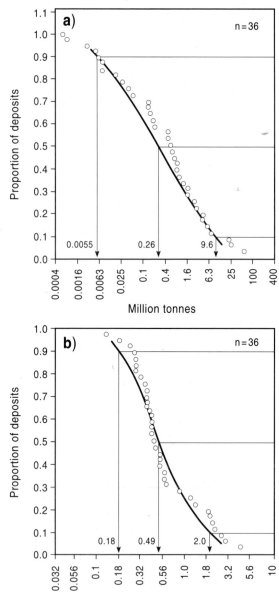

Fig. 4.5 Grade–tonnage curves for unconformity associated deposits. (From Cox & Singer 1986.)

3 Mobilization of uranium and the associated elements by oxidizing fluids and precipitation in a reducing environment at fault–unconformity intersections.
4 Additional cycles of oxidation and mobilization.

As an example of the differences between the descriptive and genetic models, the descriptive model would suggest exploration for graphitic conductors whereas the genetic model would broaden the possible depositional sites to any reducing environment. Descriptive and genetic models can be combined with the grade–tonnage curves of Cox & Singer (Fig. 4.5) to suggest the probable economic benefits. It should be noted that high tonnage does not necessarily mean low grade and large deposits such as Cigar Lake (1.47 Mt of 11.9% U_3O_8) in Saskatchewan have grade and tonnage that rank in the top ten per cent of all deposits.

The history of exploration in the two areas clearly illustrates the uses and limitations of models. Unconformity vein deposits were first discovered as a result of exploration in known uraniferous areas of Northern Canada (Fig. 4.6) and Australia in the late 1960s. The initial Canadian discoveries were made by a French company which was exploring for uranium veins around the Beaverlodge deposit. These have no obvious relation to the mid Proterozoic sandstones and it was only the company's previous experience in Gabon and minor known occurrences near the unconformity which suggested that the sandstones might be mineralized (Tona *et al.* 1985). On this premise the company (Amok) conducted an airborne radiometric survey which led them to boulders of mineralized sandstone and massive pitchblende close to the source deposit (Cluff Lake A). At roughly the same time a small Canadian company explored the eastern edge of the basin knowing that they had an aircraft available and that the sandstones might be prospective by comparison with the much younger Western US deposits (Reeves & Beck 1982). They also discovered glacial boulders which led them back to the deposit (Rabbit Lake). At roughly the same time (1969) in Australia airborne surveys in the East Alligator Valley detected major anomalies at Ranger, Koongarra and Narbalek. The area selection was initially based on the similarity of the geology with known producing areas in South Alligator Valley, with acid volcanics as the postulated source rock, and by a re-interpretation of granites, that were previously thought to be intrusive, as basement gneiss domes (Dunn *et al.* 1990b). When drilling started it became clear that the mineralization was closely associated with the unconformity. Thus the overlying sandstones, rather than being responsible for the erosion of the deposits or being a barren cover, became a target. In Canada the Key Lake and Midwest Lake deposits were discovered as a result of this change in model (Gatzweiler *et al.* 1981, Scott 1981). The exploration model has been further developed in Canada, to the point at which deposits with no surface expression

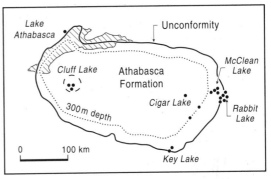

Fig. 4.6 Outline of the Athabasca Basin and distribution of the associated uranium deposits. (Modified after Clark *et al.* 1982.)

(blind deposits) have been found by deep drilling of electromagnetic conductors caused by graphitic schists (McMullan *et al.* 1989). Similar developments could have been anticipated in Australia but for government control on uranium exports, which inhibited uranium exploration. The importance of recognizing the elements associated with this type of deposit became obvious in the late 1980s with the discovery of the Coronation Hill Au–Pd deposit in the Northern Territory and similar prospects in Saskatchewan. Unfortunately for the company concerned Coronation Hill is on the edge of the Kakadu National Park and will not be mined in the foreseeable future.

The application of a particular deposit model will depend on the quality of the database and should not be regarded as a panacea for the exploration geologist. Some deposit types such as placer gold are easy to understand and have well-developed models whereas others such as Besshi style massive sulphides or the Olympic Dam models are not well developed and may be represented by a single deposit on which information is difficult to obtain.

The major pitfalls of using models are ably illustrated in cartoon form by Hodgson (1989). He identifies a number of problems which he likens to religious cults.
(a) *The cult of the fad or fashion.* An obsession with being up-to-date and in possession of the newest model.
(b) *The cult of the panacea.* The attitude that one model is the ultimate and will end all controversy.
(c) *The cult of the classicists.* All new ideas are rejected as they have been generated in the hot house research environment.
(d) *The cult of the corporate iconoclasts.* Only models generated within an organization are valid, all outside models are wrong.

(e) *The cult of the specialist*. In which only one aspect of the model is tested and usually not in the field.

The onus is thus on the exploration geologist to ensure that a model is correctly applied and not to exclude deviations from the norm. Other examples of the application of models are given in Chapters 11 to 16.

4.2 DESK STUDIES AND RECONNAISSANCE

4.2.1 Desk studies

Once the exploration organization is in place, initial finance budgeted and target deposit type selected then desk studies can start and areas be selected for reconnaissance—but on what basis?

The first stage in a totally new programme is to acquire information about the areas selected. Besides background information on geology, data on the occurrence of currently producing mines and prospects and their economic status are essential. This information will normally be based largely on published material but could also include open-file material from geological surveys and departments of mines, data from colleagues and from consultants with particular expertise in the area concerned.

Background geological information is available for most areas in the world although its scale and quality vary considerably. In some parts of Europe geological maps are published at 1:50 000 and manuscript field sheets at 1:10 000 are available. In less populated parts of the world, such as Canada or Australia, the base cover is 1:250 000 with more detailed areas at 1:100 000. For the mineral exploration geologist the published geological data will only be a beginning and he or she will interpret the geology using the geological features defined in the deposit model. This is best achieved by starting with a synoptic view of the geology from satellite imagery. Unless the area is extensively vegetated this is likely to be from Landsat or SPOT imagery (Chapter 6). Landsat has the added advantage of highlighting areas of hydrothermal alteration within arid areas if processed correctly. In areas of dense vegetation side looking radar can be used. For example, most of Amazonian Brazil has been mapped in this way. For smaller areas air photography provides better resolution often at considerably less cost, although air photographs are still regarded as top secret in some less developed countries, in spite of the availability of satellite information. Besides surface mapping familiar to most geologists, a number of other sources of information on regional geology are widely used in exploration.

An invaluable addition to surface regional geology is the use of regional geophysics (Chapter 7). Airborne magnetics, radiometrics and regional gravity data are available for much of the developed world and help in refining geological interpretation and, particularly, in mapping deep structures. For example, a belt of the Superior Province in northern Manitoba which hosts a number of major nickel deposits, including the large deposit at Thompson, can be clearly followed under Palaeozoic cover to the south (Fig. 4.7). Regional seismic data are helpful but are usually only available if oil companies donate them to the public domain. Specially commissioned seismic surveys have greatly helped in deciphering the subsurface geology of the Witwatersrand Basin (section 13.5.4). Subsurface interpretations of geophysical information can be checked by linking them with information from any available deep drill hole logs.

Regional geochemical surveys (section 8.4) also provide much information in areas of poor outcrop and have defined major lithological provinces covered by boulder clay in Finland.

The sources of information for mineral occurrence localities are similar to those for regional geology. Geological surveys usually have the most comprehensive data base within a country and much of this is normally published (for example the summary of UK mineral potential of Colman 1990). Many surveys have collated all the mineral occurrences within their country and the results are available as maps, reports or even on computerized databases. Two useful types of maps are mineral occurrence maps and metallogenic maps. The former type merely shows the location of the occurrences whereas the latter attempts to show the form of the deposit and associated elements overlain on background geology. Overlays of the mineral prospects with geology will provide clues to regional controls on mineralization. At this point some economic input is required as it is often the case that significant prospects have different controls from weakly mineralized occurrences. The sort of economic information that is useful are the grades and tonnages mined, recovery methods and the reasons for stopping mining. This can be obtained from journals or from company reports.

An example of a desk study is the recognition of target areas for epithermal gold deposits in western Turkey. Here mineral exploration has increased significantly since 1985 due to the reform of Turkish mining laws and the ability of non-Turkish mining companies

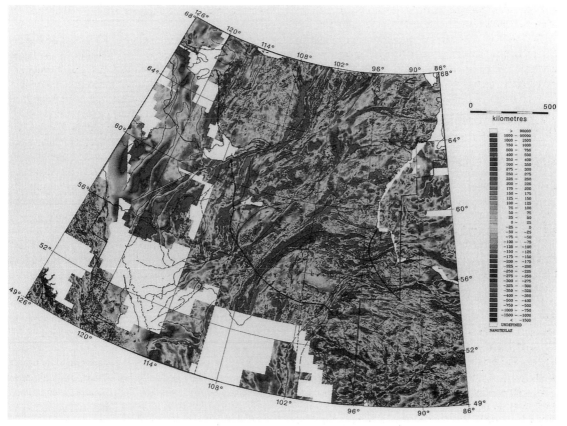

Fig. 4.7 Aeromagnetic map of western Canada showing the continuity of anomalies underneath sedimentary cover. The edge of the sedimentary cover is the bold dashed line striking NW–SE . The belt containing the Thompson nickel deposit runs NNE from 102°E, 52°N. Hudson Bay forms the north-east corner of the map. (Reproduced with permission from Ross *et al.* 1991.)

to obtain a majority shareholding in any discovery. Western Turkey is of interest as its geological setting is similar to eastern Nevada (section 15.4), with extension and graben formation in the Miocene accompanied by extensive volcanism and much current hot spring activity. In addition a cursory glance at the Metallogenic Map of Europe shows that a number of gold occurrences, as well as other elements, such as arsenic, antimony and mercury which are often associated with epithermal activity, are present (Fig. 4.8). The mineral map of Turkey (Erseçen 1989) provides further information on reserves at several of these localities. The compiler must test this information against the epithermal model in which gold is likely to be associated with graben bounding faults or volcanics. Most of the gold occurrences shown in western Turkey are within high grade metamorphic rocks and represent

metamorphosed quartz veins of little interest. However the gold occurrences within volcanics to the south of Çanakkale and to the NW of Izmir (points A and B, Figs 4.8, 4.9) remain of interest, as do the mercury and antimony occurrences along the graben bounding faults (points C, D, E). The area south of Çanakkale was mined by a British company until the outbreak of the 1914–18 World War and has lain nearly dormant since then. The first company to appreciate the potential of this area has defined a resource of 9 Mt at 1.25 g t^{-1} Au within one of the brecciated volcanics and a number of other targets have been identified in the vicinity. The area near Izmir has been mined since Roman times but the gold mineralization is largely confined to a vein structure. Recent exploration by the Turkish geological survey (MTA) who hold the licence has defined a further resource of 1.4 Mt at 1.3 gt^{-1} Au. The mercury and antimony are

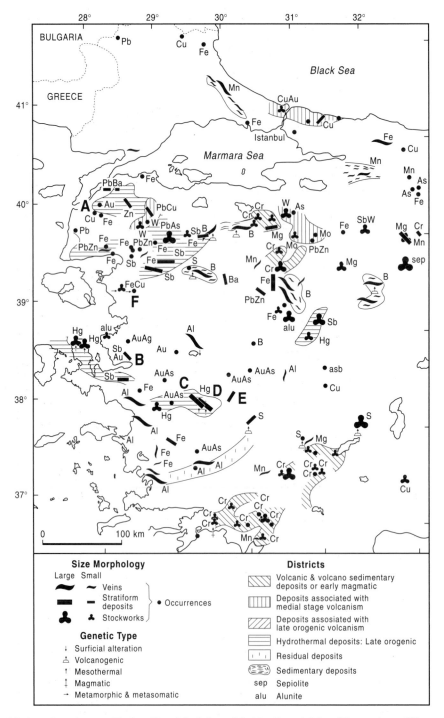

Fig. 4.8 Simplified metallogenic map of Turkey. The original sheet of the Metallogenic Map of Europe shows different deposit types and commodities in different colours as well as underlying geology. (From Lafitte 1970, Sheet 8.)

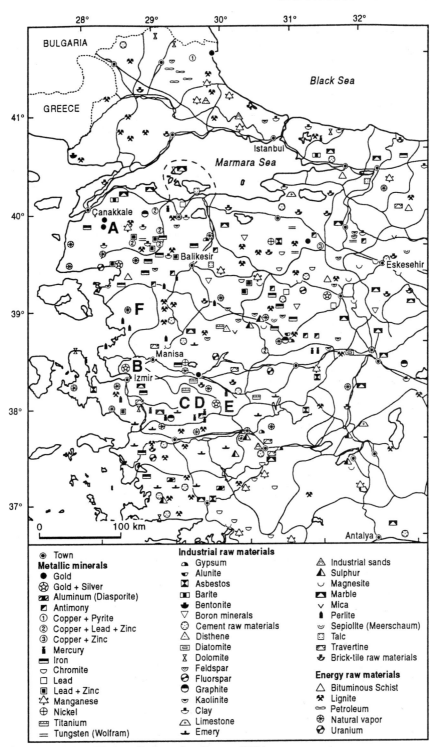

Fig. 4.9 Part of the Mineral Map of Turkey. (Redrawn from Erseçen 1989.)

The legend of the map:

Metallic minerals
- ⊙ Town

- ● Gold
- ⊛ Gold + Silver
- ◪ Aluminum (Diasporite)
- ◪ Antimony
- ① Copper + Pyrite
- ② Copper + Lead + Zinc
- ③ Copper + Zinc
- ◗ Mercury
- ▬ Iron
- ▽ Chromite
- ▢ Lead
- ▣ Lead + Zinc
- ⟁ Manganese
- ⊕ Nickel
- ▱ Titanium
- ═ Tungsten (Wolfram)

Industrial raw materials
- ◖ Gypsum
- ◗ Alunite
- ◪ Asbestos
- ◪ Barite
- ♣ Bentonite
- ▽ Boron minerals
- ☉ Cement raw materials
- △ Disthene
- ▭ Diatomite
- ✕ Dolomite
- ▽ Feldspar
- ∅ Fluorspar
- ● Graphite
- ▽ Kaolinite
- ♣ Clay
- △ Limestone
- ↯ Emery

- △ Industrial sands
- ▲ Sulphur
- ∪ Magnesite
- ◼ Barite
- ∨ Mica
- ◗ Perlite
- ◡ Sepiolite (Meerschaum)
- ▣ Talc
- ▱ Travertine
- ♣ Brick-tile raw materials

Energy raw materials
- △ Bituminous Schist
- ✹ Lignite
- ◡ Petroleum
- ⊛ Natural vapor
- ◔ Uranium

largely the concessions of the state mining group, Etibank, but at least one exploration company was able to agree a joint venture and found significant, although sub-economic, gold grades at the Emirli Antimony Mine (point C, Figs 4.8, 4.9). The most significant gold discovery in the area, Ovaçik (point F), is not shown on either map, although a number of veins in the area had, at least in part, been mined by ancient workers. The Ovaçik deposit was discovered during a regional reconnaissance of favourable volcanic areas using stream sediment geochemistry and prospecting. Larson (1989) provides a further overview of the potential of the area including much on logistics.

4.2.2 Reconnaissance

The aim of reconnaissance is to evaluate areas of interest highlighted in the desk study rapidly and to generate other, previously unknown, targets preferably without taking out licences.

In areas of reasonable outcrop the first stage will be to check the geology by driving along roads or from the air if ground access is impossible. A considerable number of disseminated gold prospects have been found by helicopter follow up of Landsat data processed to highlight argillic alteration and siliceous caps in arid areas, particularly in Chile. Airborne geology can be extremely effective in solving major questions quickly but is expensive, US$500 per hour, and is best undertaken by experienced geologists with specific problems to solve. In any case a preliminary visit to the field should be made as soon as practicable to check the accessibility of the area, examine on the ground some of the data emerging from the desk study and check on competitor activity.

Reconnaissance techniques can be divided into those which enable a geologist to locate mineral prospects directly and those which provide background information to reduce the search area.

Airborne geophysics and stream or lake sediment geochemistry are the principal tools for directly detecting mineralization. Perhaps the most successful of these has been the use of airborne electromagnetic techniques in the search for massive sulphide deposits within the glaciated areas of the Canadian Shield. This has led to the discovery of a number of deposits, such as Kidd Creek (section 14.4.2), by drilling the airborne anomalies after very little ground work. However, each method has drawbacks. In the case of airborne e.m. it is the inability to distinguish base metal rich sulphides from pyrite and graphite and interferences from con-

ductive overburden. Airborne radiometrics have been very successful in finding uranium deposits in nonglaciated areas and led to the discovery of deposits such as Ranger 1 and Yeelirrie in Australia (Dunn *et al.* 1990a,b). In glaciated areas radiometric surveys have found a number of boulder trains which have led indirectly to deposits. In areas of residual overburden and active weathering stream sediment sampling has directly located a number of deposits, such as Bougainville in Papua New Guinea but it is more likely to highlight areas for ground follow up and licensing.

An example of an exploration strategy can be seen in the search for sandstone hosted, uranium deposits in the Karoo Basin of South Africa during the 1970s and early 1980s. The initial reconnaissance was conducted by Union Carbide in the late 1960s. They became interested in the basin because of the similarity of the Karoo sediments to those of the Uravan area in Colorado, where the company had uranium mines, and because of the discovery of uranium anomalies in Karoo sediments in Botswana and Zimbabwe. As little of the basin, which is about 600 000 km^2, had been mapped, the company chose a carborne radiometric survey as its principal technique. In this technique a scintillometer is carried in a vehicle and gamma radiation for the area surrounding the road is measured. Although a very limited area around the road is measured it has the advantages that the driver can note the geology through which he is passing, can stop immediately if an anomaly is detected and it is cheap. In this case a driver was chosen with experience in Uravan geology and he was able to use that experience to compare the geology traversed with favourable rocks in the western USA. After a good deal of traversing a significant anomaly (Fig. 4.11) within sandstones was discovered to the west of the small town of Beaufort West (Fig. 4.10). Once the initial discovery was made the higher cost reconnaissance technique of fixed wing airborne radiometrics was used as this provided much fuller coverage of the area, about 30% at 30 m flying height. This immediately indicated to Union Carbide a number of areas for follow up work and drilling.

It became rapidly apparent that the deposits are similar to those of the Uravan area. In both the uranium is in the fluviatile sandstones in an alternating sequence of sandstone, siltstone and mudstone; in the South African case, in the Upper Permian Beaufort Group of the Karoo Supergroup (Turner 1985). These sediments were deposited by braided to meandering streams which resulted in interfingering and overlapping of sandstone–mudstone transitions and made stratigraphical

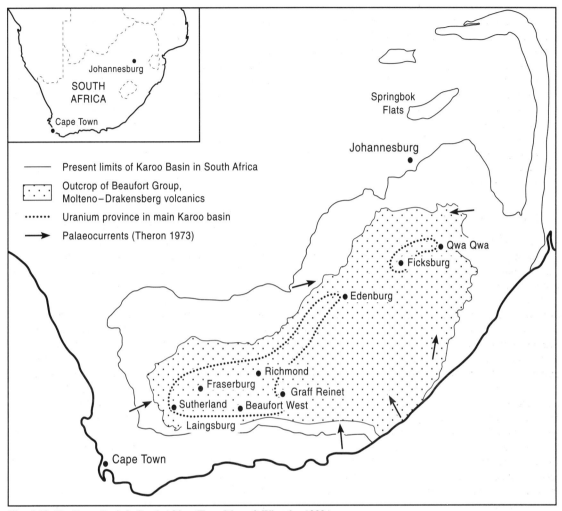

Fig. 4.10 The Karoo Basin in South Africa. (From Moon & Whateley 1989.)

correlation difficult. Significant mineralization is confined to thicker multi-storey sandstones, usually >14 m thick, and is epigenetic (Gableman & Conel 1985). The source of the uranium is unknown, possibly it was leached from intrabasinal volcanic detritus but certainly mobilized as uranyl carbonate complexes by oxygen-rich, alkaline ground waters which then migrated into the porous sandstone (Turner 1985, Sabins 1987). The migrating ground water then encountered reducing conditions in the presence of organic material, bacterially generated hydrogen sulphide and pyrite. This change from oxic to anoxic conditions caused the uranium to precipitate as minerals (coffinite and uraninite) that coated sand grains and filled pore spaces.

Diagenesis rendered the deposits impermeable to further remobilization. The outcrop characteristics of the deposits are either calcareous, manganese-rich sandstone or bleached alteration stained with limonite and hematite, that locally impart yellow, red and brown colouration. Typical sizes of deposits, which occur as a number of pods, are 500 by 100 m. Average grades are about 0.1% U_3O_8 and thicknesses vary from 0.1 to 6 m but are typically 1 m.

From the airborne radiometrics and follow up drilling Union Carbide defined one area of potential economic interest and inevitably word of discovery leaked out. In addition the price of uranium rose following the 1973 Arab–Israeli war leading to interest on the part of

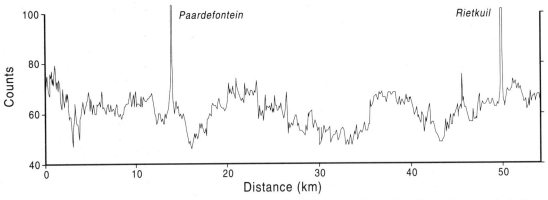

Fig. 4.11 Trace of a carborne radiometric survey similar to that which discovered the first major radiometric anomaly in the Karoo Basin. (After Moon 1973.)

a large number of other companies. The reconnaissance approach of competitor companies was however markedly different in that they knew that the basin hosted significant mineralization and that Union Carbide had important areas under option to purchase. In general these companies selected at least one area around the known Union Carbide discovery and a number of areas elsewhere within the basin. These were chosen on the basis of comparison with deposits in the western USA where the deposits are preferentially hosted at specific horizons and within thicker fluvial or lacustrine sandstones. Mosaics of the, then (1973) newly available, Landsat MSS images, as well as government flown air photography, were used to derive a regional stratigraphy and to highlight thicker sandstone packages. Landsat imagery offered a hitherto unavailable synoptic view and the ability to use the spectral characteristics of the different bands to discriminate alteration and rock types. On false colour images (sections 6.2.3 & 6.2.6) sandstone and mudstone rocks could be differentiated, but it was not possible to distinguish the altered sandstone from sandstone that had a natural desert varnish of limonite. To enhance the appearance of the altered rocks, colour ratio images were produced by combining the ratios of bands 4/7, 6/4 and 7/4 in blue, green and red respectively. The altered sandstones were identified by light yellow patches, which were used by field geologists as targets for ground follow-up surveys. This was less successful than flying radiometric surveys over areas of favourable sandstones. In more rugged areas helicopters were used to trace the sandstones and check for radiometric anomalies (Fig 4.12).

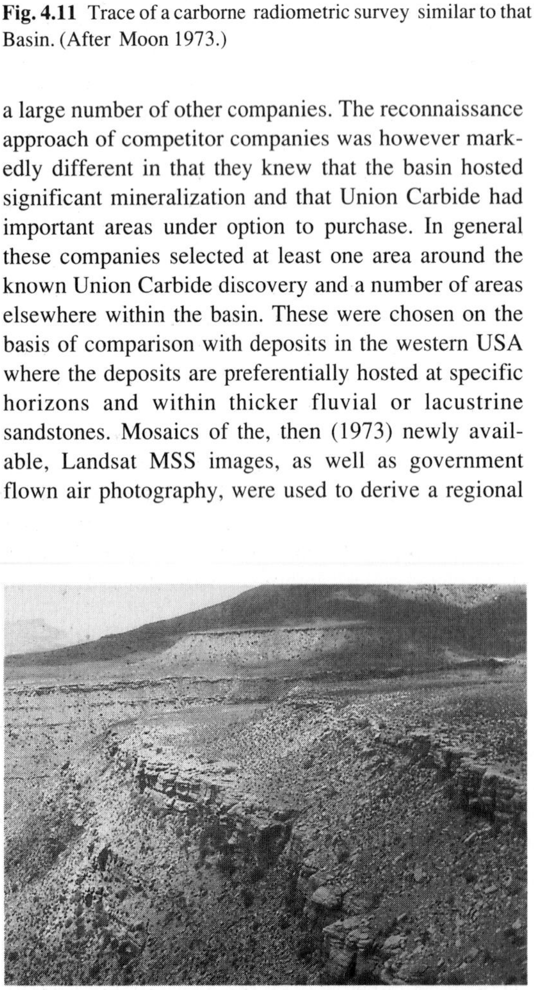

Although airborne radiometric surveys were extremely successful in areas of good outcrop, they were ineffective in locating deposits concealed under inclined sedimentary rock cover. The most successful exploration methods in the search for subsurface mineralization were detailed surface mapping coupled with hand-held surface radiometric surveys and down the hole logging (section 7.13) of all available boreholes drilled for farm water supplies (Moon & Whateley 1989).

4.2.3 Land acquisition

The term land is deliberately used as the actual legal requirements for exploration and mining varies from

Fig. 4.12 View from a helicopter traversing in the central Karoo Basin, north-west of Beaufort West.

country to country. What the explorer needs to acquire is the right (preferably exclusive) to explore and to mine a deposit if the exploration is successful. Normally a company will obtain the right to explore the property for a particular period of time and the option to convert this into a right to mine, if desired, in return for an annual payment and in some cases the agreement to expend a minimum amount on exploration and to report all results. Unfortunately most legal systems are extremely complicated and the explorer may not be able to obtain the exact right that he or she requires, for example gold or energy minerals may be excluded and the right to the surface of the land (surface rights) may be separate from the right to mine (mineral rights). Two end member legal systems can be distinguished as far as mineral rights are concerned: the first in which all mineral rights are owned by the state and the explorer can mine with no regard to the current occupier of the surface rights and the second in which all mineral and surface rights are privately owned. The first normally results from governmental decree or revolution whereas the second is typical of many former British colonies.

In the private situation, for example in Britain, the company's lawyer will negotiate with a private mineral rights owner and obtain an exploration or option agreement under which the company will be able to explore for a minimum period, normally three years, and then be able to renew the option; or to buy the mineral rights for a fixed sum, normally in excess of the free market value. In exchange, the mineral rights owner will receive a fixed annual sum option payment and compensation for any damage to the surface if he or she owns the surface rights. If the surface rights are separately owned then an agreement must be made with that owner. In the case of Britain the rights to gold are owned by the Crown and must be covered in a further separate agreement. There is no legal obligation to report the results to government although summary drill hole results must be reported to the British Geological Survey. Most physical exploration of any significance and drilling of greater than 28 days duration requires consent from the local planning authority.

In the case of state ownership the state will normally own the mineral rights and be able to grant access to the surface. In this case application for an exploration licence will usually be made to the department of mines. The size of the exploration area may be fixed. For example the law in Western Australia limits a two year prospecting licence to 200 ha although a five year exploration licence can cover between 10 and 200 km^2. Annual work commitments on the former are $A40 per

ha and $A300 km^{-2} on the latter. Mining leases are granted for a renewable period of 21 years with a maximum area of 1000 ha. Normally in areas with state ownership a full report of exploration results must be filed to the mines department every year and at the termination of the lease. Such reports often provide information for future exploration as well as data for government decision making.

An example of a new mineral law designed to encourage exploration is the 1985 law of Turkey. Under this exploration licences are granted for 30 months, these can be converted into pre-operation licences for three years and then into an operating or mining licence. The cost of an exploration licence in 1992 was 4000 Turkish Lira (US$0.10) ha^{-1}, which is refundable when the licence is relinquished. The rights of small miners are protected by a right of denunciation under which any Turkish citizen who can demonstrate a previous discovery in the area can claim 3% of the gross profit. The whole operation is policed by a unit in the geological survey, which uses a computerized system to monitor licence areas.

Legal advice is a virtual necessity in land acquisition and most exploration groups have a lawyer or land person on the staff or on a retainer. If legal tenure is insecure then all exploration effort may be wasted and all revenue from the property go elsewhere. The lawyer will also be responsible for checking the laws that need to be observed, for example, information from drill holes may need to be reported to government. Corporate administration should be checked so that there is no insider dealing or any activities that could be considered unethical. A famous recent case was that in which Lac Minerals lost control of part of the large and profitable Hemlo deposit in Ontario. It was alleged that Lac Minerals had bought claims from the widow of a prospector after receiving confidential information from another company, Corona Corp., and in spite of a verbal agreement not to stake that ground. Besides losing control of the deposit, Lac was faced with a large legal bill.

A more recent example was the dispute between Western Mining Corporation and Savage Exploration Pty Ltd over control of the Ernest Henry deposit in northern Queensland (Western Mining 1993). Leases in the area, including a tenement named ML2671, had originally been pegged in 1974 by Savage Exploration for iron ore on the basis of the results of a government aeromagnetic survey. In 1989 Western Mining decided to explore the area for base metals and obtained information in the public domain including the government airborne geophysics. When Western Mining selected areas for further work they found that much of the land was held by Hunter Resources Ltd and therefore entered into a joint venture with Hunter but operated by Western Mining.

One of the areas targeted for further work by the joint venture was in the general area of ML2671 as described in the Mines Department files. A baseline was pegged in July 1990 at 100 m intervals and crossed the described position of ML2671 although no pegs were seen on the ground to indicate the position of the lease. A magnetic survey was undertaken including readings within the described area of ML2671 and an initial TEM survey in August 1990. Savage Exploration was approached in March 1991 and agreement made over the terms of an option by which the Western Mining–Hunter joint venture could acquire a number of leases, including ML2671. In May 1991 further TEM surveys were made and a formal option signed in October 1991. In late October 1991 the first hole was drilled and this intersected strong mineralization.

After further drilling the discovery of the Ernest Henry deposit was announced and Savage Exploration were advised in June 1992 that the joint venture wished to exercise its option and buy the lease of ML2671. Savage Exploration then commenced court proceedings alleging Western Mining had misrepresented the situation when agreeing an option on ML2671 and that Western Mining had trespassed on ML2671 during the ground surveys. In the ensuing court case it transpired that the pegs for ML2671 were not in the place indicated on the Mines Dept files and were 850 m north of the stated location and the lease was rotated several degrees anticlockwise from its plotted position. The case was settled out of court with Western Mining agreeing to give up any claim to ML 2671 and paying Hunter $A17 million and certain legal costs.

4.3 SUMMARY

Mineral exploration is conducted by both the private and public sectors in both market and most centrally planned economies. Groups from both sectors must be clear about what commodity and type of deposit they are seeking before setting up an exploration organization. Mineral exploration is a long-term commitment and there must be careful planning of the participant's long-term objectives to ensure viable budgeting. With the objective decided upon an exploration organization must be set up and its success will be enhanced by developing all the factors listed in section 4.1.1.

Budgets must be carefully evaluated and not be figures drawn out of the blue. An underfunded project will in most cases be a failure. Careful control of funds is essential and it provides a basis for future budgeting.

Having recognized the plate tectonic setting of an exploration region, models of the deposits likely to be present, and particularly of those being sought, should be set up. Both empirical and genetic models are used, often in combination, to show how the materials sought relate to geological factors including geophysics and geochemistry. To understand how models are constructed and used, unconformity related uranium deposits are discussed in some detail (section 4.1.3).

With the organization in place and the target deposit type selected, desk studies can start in earnest and areas for reconnaissance be chosen. Relevant information must be acquired, assessed and selected using published works and open-file material from government institutions (section 4.2.1). The aim of reconnaissance is to evaluate rapidly areas highlighted in the desk study and to identify targets for follow up work and drilling. If this is successful then the exclusive rights to explore and to mine any deposits found in the target area must be acquired (section 4.2.3).

4.4 FURTHER READING

Literature on the design and execution of exploration programmes is very scattered. Peters (1987) provides a good introduction. Mineral deposit models are discussed in Roberts & Sheahan (1988) as well as the summary volumes of Cox & Singer (1986) and Eckstrand (1984). The link between mineral deposit studies and exploration is made in the introductory text of Edwards & Atkinson (1986). Papers edited by Huchinson & Grauch (1991) discuss the eveolution of genetic concepts as well as giving some case histories of major discoveries.

An enthuiastic view from industry, tempered by economic reality, is provided by Woodall (1984 & 1992). Much data on the economics of exploration can be found in the volume edited by Tilton *et al.* (1987).

5 From Prospect to Predevelopment

CHARLES J. MOON AND MICHAEL K.G. WHATELEY

Once land has been acquired the geologist must direct his or her efforts to proving whether or not a mineral prospect is worthy of commercial evaluation. Proving a discovery to be of sufficient size and quality inevitably involves a subsurface investigation and the geologist usually faces the task of generating a target for drilling. In exceptional cases, such as very shallow mineralization, a resource may be proved by digging pits or trenches or, in mountainous areas by driving adits into the mountainside.

Whatever the method used the key requirement is to explore the area at the lowest cost without missing significant targets. Fulfilling this requirement is not easy and the mineral deposit models, such as those discussed in Chapter 4, must be modified to include economic considerations. There should be a clear idea of the size of the deposit sought, the maximum depth of interest and whether underground mining is acceptable. Finding a drilling target normally involves the commissioning of a number of different surveys, such as a geophysical survey, to locate the target and indicate its probable subsurface extension. The role of the geologist and the exploration management is to decide which are necessary and to integrate the surveys to maximum effect.

5.1 FINDING A DRILLING TARGET

5.1.1 Organization and budgeting

Once land has been acquired, an organization and budget must be set up to explore it and bring the exploration to a successful conclusion. The scale and speed of exploration will depend on the land acquisition agreements (section 4.2.3) and the overall budget. If a purchase decision is required in three years, then the budget and organization must be geared to this. Usually the exploration of a particular piece of land is given project status and allocated a separate budget under the responsibility of a geologist, normally called the project geologist. This geologist is then allocated a support team and he or she proceeds to plan the various surveys, usually in collaboration with in-house experts, and reports his or her recommendations to the exploration management. Typically, reporting of progress is carried out in monthly reports by all staff and in six-monthly reviews of progress with senior exploration management. These reviews are often also oral presentations ('show and tell') sessions and linked to budget proposals for the next financial period.

Reconnaissance projects are normally directed from an existing exploration office but once a project has been established serious consideration should be given (in inhabited areas) to setting up an office nearer the project location. Initially this may be an abandoned farmhouse or caravan but for large projects it will be a formal office in the nearest town with good communications and supplies. A small office of this type is becoming much easier to organize following the improvement of communication and computing facilities during the last decade. Exploration data can be transferred by modem, fax and satellite communications to even the remotest location. If the project grows into one with a major drilling commitment then it is probable that married staff will work more effectively if their families are moved to this town. In general, the town should be less than one hour's commuting time so that a visit to a drilling rig is not a chore. In the remotest areas staff will be housed in field camps and will commute by air to their home base. If possible exploration staff should not be expected to stay for long periods in field camps as this is bad for morale and efficiency declines.

Budgeting for the project is more detailed than at the reconnaissance stage and will take account of the more expensive aspects of exploration, notably drilling. A typical budget sheet is shown in Table 4.2. Usually the major expenditure will be on labour and on the various surveys. If the area is remote then transport costs, particularly helicopter charter, can become significant. Labour costs are normally calculated on the basis of man months allocated and should include an overhead component to cover office rental, secretarial and drafting support. Commonly overheads equal salary costs. For geophysical, and sometimes geochemical surveys, contractors are often hired as their costs can be accurately estimated and companies are not then faced with the possibility of having to generate work for staff. Estimates are normally made on the basis of cost per line-kilometre or sample. In remote areas careful consideration should be given to contractor availability. For example in northern Canada the field seasons are often very short and a number of companies will be competing for contractors. For other types of survey and for drilling, estimates can be obtained from reputable operators.

5.1.2 Topographical surveys

The accurate location of exploration surveys relative to each other is of crucial importance and requires that the explorationist knows the basics of topographical surveying. The effort put into the survey varies with the success and importance of the project. At an early stage in a remote area a rough survey with the accuracy of a few metres will be adequate. This can be achieved using aerial photographs and photogrammetry (Ritchie *et al.* 1977, Sabins 1987) where only a few survey points are needed. For a major drilling programme surveying to a few millimetres will be required for accurate borehole location.

The usual practice when starting work on a prospect is to define a local grid for the prospect. Some convenient point such as a wind pump or large rock is normally taken as the origin. The orientation of the grid will be parallel to the regional strike if steeply dipping mineralization is suspected (from geophysics) but otherwise should be N–S or E–W. In better mapped areas the local grid will be at the intersection of national grid lines. Elsewhere the prospect grid can be linked to the national grid, if necessary.

SURVEYING METHODS

Geologists generally use simple and cheap techniques in contrast to those used by professional surveyors. Further details can be found in Ritchie *et al.* (1977).

One of the simplest and most widely used techniques is the tape and compass survey. This type of survey starts from a fixed point with directions measured with a prismatic or Brunton compass and distances measured with a tape or chain. Closed traverses, i.e. traverses returning to the initial point, are often used to minimize the errors in this method. Errors of distance and orientation may be distributed through the traverse (section 5.1.4). Grids are often laid out using this technique with base lines measured along a compass bearing. Distances

are best measured with a chain which is less vulnerable to wear and more accurate than a tape. Longer baselines in flat country can be measured using a bicycle wheel with a cyclometer attached. Lightweight hip mounted chains are commonly used in remote areas. Straight grid lines are usually best laid out on flat ground by back and forward sighting along lines of pegs or sticks. Sturdy wood or metal pegs should be used, the grid locations marked with metallic tags and flagged with coloured tape. The tape should be animal proof; goats have a particular fondness for coloured tape.

A very simple method of determining elevation in relatively uncharted areas is with the use of an aneroid barometer. It can be calibrated at a known datum (base station) and the elevation of all points read during the day are relative to this datum. Ideally, a second barometer should be read at regular intervals at the base station so corrections for diurnal changes in atmospheric pressure can be made.

Levelling is the most accurate method of obtaining height differences between stations and is used for example to obtain the elevation of each station when undertaking gravity surveys. This method measures the height difference between a pair of stations using a surveyor's level. Levelling is required in an underground mine to determine the minimum slope required to drain an adit or drive and in a surface mine the maximum gradient up which load, haul, dump (LHD) trucks can climb when fully laden (usually <10%).

A major recent development has been the advent of cheap (<$1000) hand held instruments that use the world wide network of global positioning satellites (GPS) installed by the US Department of Defense. These allow the exploration geologist to get an immediate fix of his or her position on the ground to within 100 m and this is invaluable in sparsely inhabited or poorly mapped areas.

More exact ground surveys use a theodolite as a substitute for a compass. It is simple enough for geologists to learn to use. Numerical triangulation is carried out using angles measured by theodolite to calculate x and y coordinates. By also recording the vertical angles, the heights of points can be computed (Ritchie *et al.*. 1977). Professional surveyors are readily available in most parts of the world and they should be contracted for more exact surveys. They will use a theodolite and electronic distance measuring (EDM) equipment, often combined in one instrument, and can produce an immediate printout of the grid location of points measured.

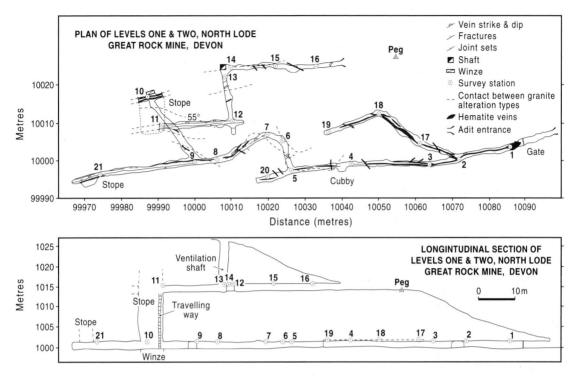

Fig. 5.1 Typical underground survey of a small vein mine using tape and compass.

5.1.3 **Geological mapping**

One of the key elements during the exploration of prospect is the preparation of a geological map. Its quality and scale will vary with the importance of the programme and the finance available. Initial investigation of a prospect may only require sketch mapping on an aerial photograph whereas detailed investigations prior to drilling may necessitate mapping every exposure. Mapping at the prospect scale is generally undertaken at 1:10 000 to 1:2 500. For detailed, accurate mapping a telescopic alidade and plane table may be used. The principle of the alidade is the same as for a theodolite, except that the vertical and horizontal distances between each point are calculated in the field. The base of the alidade is used to plot the position of the next point on the waterproof drafting film covering the plane table.

The process of geological mapping of mineral prospects is similar to that of general geological mapping but is more focused. Although the regional environment of the prospect is important, particular attention will fall on known mineralization or any discovered during the survey. The geological relationship of the mineralization must be assessed in detail, in particular whether it

has any of the features of the geological model sought. For example if the target is a volcanic-associated massive sulphide deposit then any sulphides should be carefully mapped to determine if they are concordant or cross cutting. If the sulphides are cross cutting it should be established whether the sulphides are in a stockwork or a vein. Particular attention should be paid to mapping hydrothermal alteration, which is described in detail by Pirajno (1992).

Detailed guidance on geological mapping is beyond the scope of this book but can be found in a series of handbooks published by the Geological Society of London (Fry 1993, Thorpe & Brown 1993, Tucker 1993), particularly in the summary volume of Barnes (1993).

One of the key elements of mapping is its final presentation. Conventionally this is in the form of a map drafted by Indian ink pen on to transparent film. From film, multiple copies either on film or on paper can be made using the dye line process and the map can be overlaid on other maps of the same scale, allowing easy comparison of features and selection of targets.

The conventional pen and paper approach is rapidly being superseded by computerized drafting that allows the storage of information in digital form. Computer

Table 5.1 An example of balancing a closed traverse. (Modified from Reedman 1979.)

Line	Bearing $\emptyset°$	Quadrant	Distance (D) m	$\pm \sin\emptyset$	$\pm \cos\emptyset$	ΔE $\pm D.\sin\emptyset$	ΔN $\pm D.\cos\emptyset$	Unadjusted Coords E	N	Final Coords E	N	
AB	56	1	82	+0.8290	+0.5592	+67.9811	+45.8538	10000	10000	10000	10000	A
BC	331	4	74	−0.4848	+0.8746	−35.8759	+64.7219	10067.98	10045.76	10069.76	10045.53	B
CD	282	4	91	−0.9781	+0.2079	−89.0114	+18.9200	10032.10	10110.58	10035.49	10109.97	C
DE	172	2	100	+0.1392	−0.9903	+13.9173	−99.0268	9943.09	10129.50	9948.46	10128.53	D
EA	130	2	45	+0.7660	−0.6428	+34.4720	−28.9254	9957.01	10030.47	9964.54	10029.11	E
								9991.48	10001.54	10000.00	10000.00	A'
Total			392					-8.52	+1.54			

Corrections for Eastings

Correction for B	$82/392 \times 8.52 =$		1.78
Correction for C	$74/392 \times 8.52 + B =$	1.61+1.78=	3.39
Correction for D	$91/392 \times 8.52 + C =$	1.98 +3.39=	5.37
Correction for E	$100/392 \times 8.52 + D =$	2.17 +5.37 =	7.54
Correction for A	$45/392 \times 8.52 + E =$	0.98 +7.54=	8.52

Quadrants		N		
4			1	
sin	−ve		sin	+ve
cos	+ve		cos	+ve
sin	−ve		sin	+ve
cos	−ve		cos	−ve
3			2	

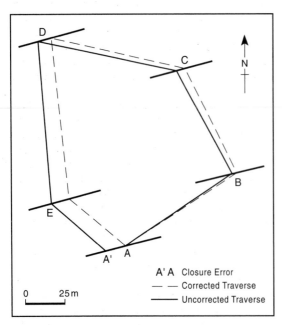

Fig. 5.2 Plot of a typical compass traverse. The errors in closure of the traverse can be corrected using the calculations shown in Table 5.1. (Modified from Reedman 1979.)

packages, such as AUTOCAD, a computer aided drafting package, are widely used in industry. Maps and plans can be produced to scale and different features of the overall data set, held on different layers in the computer, can be selected for viewing on the computer screen or printed as a hard copy. The data can also be transmitted to more sophisticated Geographical Information Systems (GIS), for example ARC/INFO, that allow the inquisition of data.

5.1.4 Mapping and sampling of old mines

Many prospects contain or are based around old mines. They may become attractive as exploration targets because of rising commodity prices, cheaper mining and processing costs, the development of new technology which may improve recovery or the development of a new geological model which could lead to undiscovered mineralization. The presence of a mining district indicates mineral potential, which must reduce the exploration risk. However, there will be a premium to pay, as the property will probably already be under option to, or owned by, a rival company.

The type of examination warranted by an old mine will depend on its antiquity, size and known history. In

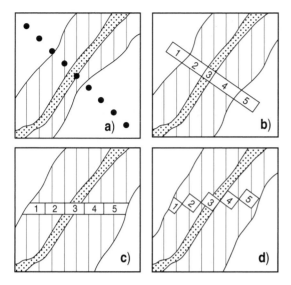

Fig. 5.3 Sampling strategies across a mineralized vein. (a) grab samples across the vein, (b) channel samples normal to the vein, (c) channel samples at the same height above the floor and (d) channel samples at the same height but with individual samples oriented normal to the strike of the vein. In all cases the vein is shown in vertical section.

Europe and west Asia old mines may be over 2 000 years in age and be the result of Roman or earlier activity. In this case there are few, if any, records and the target commodity can only be guessed at. In such cases small areas of disturbed ground, indicating the presence of old prospecting pits or trenches, can best be found from aerial photography. Field checking and grab sampling will confirm the presence and indicate the possible type of mineralization.

Nineteenth or twentieth century mines are likely to be larger and have more extensive records. Available records should be obtained but should be treated with some caution, as many reports are unreliable and plans likely to be incomplete, as discussed in sections 16.5.3 and 16.5.4. An aim of this type of investigation is to check any records carefully by using systematic underground sampling above the water level, as old mines are frequently flooded. Evaluation of extensive underground workings requires considerable planning and will be more expensive than surface exploration because equipment and labour for development and securing old underground workings are costly. A key consideration is safety and access to the old workings must be made safe before any sampling programme is established. It may be necessary to undertake trenching and pitting in areas

adjacent to the old workings, and eventually drilling may be used to examine the deeper parts of the inaccessible mineralization. Guidelines on safe working practices in old workings can be found in Peters (1987) and Berkman (1989) and in the UK in publications of the National Association of Mining History Organizations (NAMHO 1985) and the Institution of Geologists (now the Geological Society of London) (IG 1989).

Before a sampling programme can be put into operation, a map of the old workings will be required, if none is available from archives. The exploration geologist is often the first person at the site and it is up to him or her to produce a plan and section of the old workings. This would be done using a tape and compass survey (Ritchie *et al.* 1977, Reedman 1979, Peters 1987) (Table 5.1, Figs. 5.1 & 5.2). Once the layout of the old workings is known, the mapping and complementary sampling programme can begin. The survey pegs established during the surveying will be used to locate the sample points and guide mapping. With the tape held between the pegs, the sample points are marked on the drive or crosscut walls and the distance from one peg to the sample point recorded in the field note book along with the sample number. The same number is written on a sample ticket and included with the sample in the sample bag. The samples are normally collected at regular intervals from channels cut normal to the dip of the mineralized rock (Fig. 5.3). The sample interval varies depending upon the type of mineralization. A vein gold deposit may well be sampled at 1 m intervals along every drive, while a copper deposit may only be sampled every 5 or 10 m.

If old records are available and reliable, then their data should be evaluated in conjunction with new sample data acquired during the remapping and resampling exercises. Geological controls on mineralization should be established using isopach and structure contour maps as discussed in section 5.2.2. It may be necessary to apply a cut-off value below which the mineralization is not considered mineable. In Table 5.2 two cut-off parameters (Lane 1988) are used, in one a direct cut-off and in the second a weighted average value is used.

In the first case the upper sample cut-off is taken where the individual grade falls below 1.5%. Some assays within this sample are also below 1.5%, but they are surrounded by higher values, which, when averaged out locally have a mean >1.5%. The samples between the two lines are then averaged using the sample thickness as the weighting function, giving an average of 2.26% Zn over 2.10 m. In the second case, the samples are averaged from the base upwards using thickness as

Table 5.2 Example of the use of a grade cut-off to establish the mining width of a Pb–Zn vein deposit

Assay width	Zn%	Average grade at a 1.5% cut-off	Grade using a 1.5% weighted average cut-off
0.30	0.01		
0.30	0.05		
0.30	0.13		
0.20	0.17		
0.20	0.25		
0.20	0.64		
0.20	0.92		
0.20	1.10		
0.20	2.30		
0.20	1.20		
0.20	2.10	2.26 %	1.59%
0.20	5.30	over	over
0.15	2.00	2.10 m	3.40 m
0.23	1.40		
0.20	1.90		
0.20	1.80		
0.20	1.30		
0.20	2.60		
0.12	3.50		
4.00 m	1.35% (overall average)		

the weighting function. The weighted average is calculated until the average falls above the cut-off, which in this case makes 1.59% Zn over 3.40 m. The inclusion of just one more sample would bring the average down below 1.5%. The overall weighted average is 1.35% Zn over 4.00 m.

5.1.5 Laboratory programme

Once the samples have been collected, bagged and labelled, they must be sent to the laboratory for analysis. Not only must the elements of interest be specified, but the type of analytical procedure should be discussed with the laboratory. Cost should not be the overriding factor when choosing a laboratory. Accuracy, precision and an efficient procedure are also needed. An efficient

Table 5.3 An example of a sample control form used to ensure correct analyses are requested

Submitted by: Submitted to:
Sample Control File: Date submitted:
Returned to: Job File:

Job No. /Area: Date returned:
Personnel File:

Channel or borehole No	Sample No	From (m)	To (m)	Thick -ness (m)	Description, analyses and elements requested

procedure within one's own office is also required. Sample numbers and the analyses requested should be noted down rigorously on a sample control form (Table 5.3).

An alternative practice is to state clearly the instructions for test work on the sample sheet (Table 5.4). In coal analytical work there are several ways in which the proximate analyses (moisture, volatile, ash and fixed carbon contents) can be reported; e.g. on an as received, moisture free or dry, ash free basis (Stach 1982, Speight 1983, Ward 1984, Thomas 1992).

A similar form may well be utilized for sand and gravel or crushed rock analyses where the geologist requires special test work on his or her samples, such as size analysis, aggregate crushing value (ACV) or polished stone value (PSV) to name but a few.

In today's computerized era, results are often returned to the company either on a floppy disk or direct to the company's computer from the laboratory's computer via a modem and a telephone link. Care must be taken when entering the results thus obtained into the company's data base, that the columns of data in the laboratory's results correspond with the columns in your own data base. Gold values of several per cent and copper values

Table 5.4 An example of a form used to give instructions to the laboratory for detailed coal test work

To:	Date:
	Order No:
Please analyse the following samples as per the attached instructions	

SAMPLE IDENTIFICATION	DETERMINATION REQUIRED
	On individual samples marked B or M undertake the following:
	1. Specific Gravity of sample
	2. Crush to 4.75 mm and screen out the <0.5 mm fraction
	3. Cut out a representative sample of the raw coal and undertake:
	4. Proximate analysis
	5. Ultimate analysis (H, N, C, & O) on a moisture free basis
	6. Analyse for inorganic and organic forms of sulphur, and Cl
	7. Calorific Value on a moisture free basis
	8. Free silica %
	9. Combined silica %

Submitted by:	Signature:

in the ppb range should alert even the most unsuspecting operator to an error!

ELEMENT SELECTION

Before samples are submitted to the laboratory, discussions between the project manager, chief geologist and the field staff should take place to ensure that all the elements that may be associated with the mineral deposit in question are included on the analytical request sheet and that analysis includes possible pathfinder elements. Typical elemental associations are discussed in detail in Chapter 8.

ANALYTICAL TECHNIQUES

There are a wide variety of analytical techniques available to the exploration geologist. The method selected depends upon the element which is being analysed and upon the amount expected. Amongst the instumental methods available are atomic absorption spectrometry (AAS), X-ray fluorescence (XRF), X-ray diffraction (XRD), neutron activation analysis (NAA) and inductively coupled plasma emission or mass spectrometry (ICP-ES/MS). These analytical methods and their application to different mineral deposit types are discussed in Chapter 8. AAS is a relatively inexpensive method of analysis and some exploration camps now have an instrument in the field, this ensures rapid analysis of the samples for immediate follow up. The other methods involve purchasing expensive equipment and this is usually left to specialist commercial laboratories.

Detailed identification of individual minerals is usually undertaken using a scanning electron microscope (SEM) or an electron microprobe, and these and other techniques are discussed in section 2.2.2.

5.1.6 **Prospecting**

One of the key exploration activities is the location of surface mineralization and any old workings. Although this often results from the follow-up of geochemical and geophysical anomalies or is part of routine geological mapping, it can also be the province of less formally trained persons. These prospectors compensate for their lack of formal training with a detailed knowledge of the countryside and an acute eye — an 'eye for ore'. This eye for ore is the result of experience of the recognition of weathered outcrops and the use of a number of simple field tests.

Many deposits which crop out have been recognized because they have a very different appearance from the surrounding rocks and form distinct hills or depressions. A classic example of this is the Ertsberg copper deposit in Irian Jaya, Indonesia. It was recognized because its green stained top stood out through the surrounding jungle. Its presence was first noted by two oil exploration geologists on a mountaineering holiday in 1936 and reported in a Dutch university geological journal. A literature search by geologists working for a Freeport Sulphur–East Borneo company joint venture found the report and investigated the discovery resulting in one of the largest gold deposits outside South Africa. The Carajas iron deposits in the Brazilian Amazon were also recognized because they protruded through rain forest (Machamer *et al.* 1991). On a smaller scale silicification is characteristic of many hydrothermal deposits, resulting in slight topographical ridges. It is said that the silicification of disseminated gold deposits is so characteristic that one deposit in Nevada was discovered by the crunchy sound of a geologist's boots walking across an altered zone after dark. Many kimberlite pipes form

Fig. 5.4 Gossan overlying massive sulphides exposed in the pit wall of the Filon Sur deposit, Rio Tinto, southern Spain.

Table 5.5 Colours of minerals in outcrop

Mineral or metal	Outcrop colour	Mineral/compound in outcrop
Iron sulphides	Yellows, browns, chestnuts, reds	Goethite, hematite, limonite, sulphates
Manganese	Blacks	Mn oxides, wad
Antimony	White	Antimony bloom
Arsenic	Greenish, greens, yellowish	Iron arsenate
Bismuth	Light yellow	Bismuth ochres
Cadmium	Light yellow	Cadmium sulphide
Cobalt	Black, pink, sometimes violet	Oxides, erythrite
Copper	Greens, blues	Carbonates, silicates, sulphates, oxides, native Cu
Lead	White, yellow	Cerussite, anglesite, pyromorphite
Mercury	Red	Cinnabar
Molybdenum	Bright yellow	Molydenum oxides, iron molybdate
Nickel	Green	Annabergite, garnierite
Silver	Waxy green, yellow	Chlorides, native Ag
Uranium	Bright green, yellow	Torbernite, autunite
Vanadium	Green, yellow	Vanadates
Zinc	White	Smithsonite

distinct depressions as do karst-hosted deposits in limestone terrain.

Besides forming topographical features most outcropping mineral deposits have a characteristic colour anomaly at the surface. The most common of these is the development of a red, yellow or black colour over iron rich rocks, particularly those containing sulphides. These altered iron rich rocks are known generically as ironstones, and iron oxides overlying metallic sulphide deposits as gossans or iron hats (an example is shown in Fig. 5.4). These relic ironstones can be found in most areas of the world, with the exception of alpine mountains and polar regions, and result from the instability of iron sulphides, particularly pyrite. Weathering releases SO_4^{2-} ions, leaving relic red iron oxides (hematite) or yellow–brown oxy-hydroxides (limonite) that are easily recognized in the field. Other metallic sulphides weather to form even more distinctively coloured oxides or secondary minerals. For example copper sulphides oxidize to secondary minerals that have distinctive green or blue colours such as malachite and azurite depending on the concentration of carbonate or sulphate ions. Metals such as lead and zinc normally form white secondary minerals in carbonate areas that are not easily distinguished on colour grounds from the host carbonates. A fuller list is given in Table 5.5. Besides providing ions to form secondary minerals, sulphides often leave recognizable traces of their presence in the form of the spaces that they occupied. These spaces are relic textures, often known as boxworks from their distinctive shapes and are frequently infilled with limonite and goethite. Ironstones overlying sulphides have a varied appearance with much

alternation of colour and texture giving rise to the term 'live' in contrast to ironstones of non-sulphide origin that show little variation and are known as 'dead' ironstones. Ironstones also preserve chemical characteristics of their parent rock although these can be considerably modified due to the leaching of mobile elements.

The recognition of weathered sulphides overlying base metal or gold deposits and the prediction of subsurface grade is therefore of extreme importance. This particularly applies in areas of laterite development. These areas, such as much of Western Australia, have been stable for long periods of geological time, all rocks are deeply weathered, and the percentage of ironstones that overlie non-base metal sulphides (known rather loosely as false gossans) large. In Western Australia the main techniques for investigating these are (Butt & Zeegers 1992):

1 visual description of weathered rocks,
2 examination for relic textures and
3 chemical analysis.

Visual recognition requires experience and a good knowledge of primary rock textures. Visual recognition

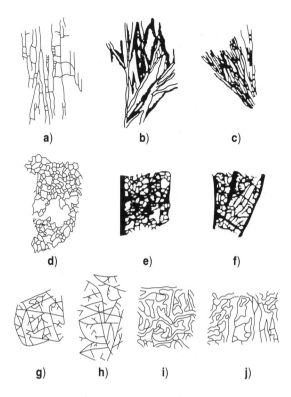

Fig. 5.5 Typical boxwork structures. Primary ore minerals were a,b,c - galena (a - cleavage, b - mesh, c - radiate); d,e - sphalerite (d - sponge structure, e - cellular boxwork); f - chalcopyrite; g,h - bornite (triangular cellular structure); i,j - tetrahedrite (contour boxwork). Approximately 4× magnification. (After Blanchard and Boswell 1934.)

can also be supplemented by chemical analysis for trace and major elements. Hallberg (1984) showed that Zr–TiO$_2$ plots are extremely useful in confirming rock types. Both elements are immobile and easy and cheap to determine at the levels required. They may be supplemented by Cr determinations in ultrabasic areas, although Cr is more mobile than the other two elements.

Relic textures can sometimes be observed in hand specimen but this is usually supplemented by microscopic examination using either a binocular microscope or a petrographic microscope and an impregnated thin section. In a classic study of relic textures, Blanchard (1968) described the textures resulting from the weathering of sulphides. Major types are shown in Figure 5.5. Although boxwork textures are diagnostic they are, unfortunately, not present in every gossan overlying base metal sulphides and textural examination must be supplemented by chemical analysis.

The choice of material to be sampled *and* the interpretation need to be carefully scrutinized. Andrew (1978) recommends taking 20 samples as representative of each gossan and using a multi-element analysis, either XRF or ICP–ES following a total attack to dissolve silica. The interpretation needs to be treated with care. A large amount of money was wasted in Western Australia at the height of the nickel boom in the early 1970s because ironstones were evaluated for potential on the basis of their nickel content. While most ironstones with very high nickel contents do overlie nickel sulphide deposits, a number have scavenged the relatively mobile nickel from circulating ground waters or overlie silicate sources of nickel and a number of deposits have a weak nickel expression at surface. More careful research demonstrated that it is better to consider the ratio of nickel to more immobile elements, such as copper, or even better immobile iridium that is present at the sub-ppm level in the deposits. These elements can be combined into a discriminant index (Travis *et al.* 1976, Moeskops 1977). Similar considerations also apply to base metal deposits; barium and lead have been shown to be useful immobile tracers in these (summarized in Butt & Zeegers 1992).

FIELD TESTS

Although a large number of field tests have been proposed in the literature only a few of the simplest are in routine use. At present only two geophysical instruments can be routinely carried, a scintillometer to detect gamma radiation and a magnetic susceptibility meter to determine rock type. The application of the former is discussed in section 7.5. Ultraviolet lamps to detect fluorescence of minerals at night are commonly used, particularly for scheelite detection although a number of other minerals glow (Chaussier & Morer 1987). In more detailed prospecting darkness can be created using a tarpaulin.

A number of simple field chemical tests aid in the recognition of metal enrichments in the field, usually by staining. Particularly useful tests aid the recognition of secondary lead and zinc minerals in carbonate areas. Lead minerals can be identified in outcrops by reaction with potassium iodide following acidification with hydrochloric acid. Lead minerals form a bright yellow lead iodide. A bright red precipitate results when zinc reacts with potassium ferrocyanide in oxalic acid with diethylaniline. Grey copper minerals can be detected with an acidified mixture of ammonium pyrophosphate and molybdate and nickel sulphides with dimethyl

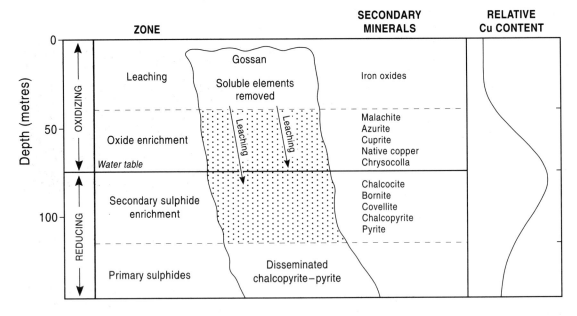

Fig. 5.6 Schematic diagram showing the gossan and copper minerals formed by weathering of a copper sulphide vein. (After Levinson 1980.)

glyoxine. Full details are given in Chaussier & Morer (1987).

SUPERGENE ENRICHMENT

As metallic ions are leached at the surface they are moved in solution and deposited at changes in environmental conditions, usually at the water table, where active oxidation and generation of electrical charge takes place. This leaching and deposition has very important economic implications that are particularly significant in the case of copper deposits as illustrated in Fig. 5.6. Copper is leached from the surface, leaving relic oxidized minerals concentrated at the water table in the form of secondary sulphides. A simple Eh–pH diagram enables the prediction of the occurrence of the different copper minerals. From the surface these are malachite, cuprite, covellite, chalcocite, bornite and chalcopyrite (the primary phase). The changes in mineralogy have important implications for the recovery of the minerals and the overall grade of the rock. Oxide minerals cannot be recovered by the flotation process used for sulphides and their recovery needs a separate circuit. The overall grade of the surface rock is much lower than the primary ore and sampling of the surface outcrops does not reflect that of the primary ore. Near the water

table the copper grade increases, as the result of the conversion of chalcopyrite (35% Cu) to chalcocite (80% Cu). This enrichment, known as supergene enrichment, often provides high grade zones in disseminated copper deposits, such as those of the porphyry type. These high grade zones provide much extra revenue and may even be the basis of mines with low primary grades. In contrast to the behaviour of copper, gold is less mobile and is usually concentrated in oxide zones. If these oxide zones represent a significant amount of leaching over a long period then the gold grades may be become economic. An example of this is the porphyry deposit of Ok Tedi discussed in section 10.7.3. These near surface, high grade zones are especially important as they provide high revenue during the early years of a mine and the opportunity to repay loans at an early stage. These gold rich gossans can also be mechanically transported for considerable distances e.g. the Rio Tinto Mine in southern Spain.

FLOAT MAPPING

The skill of tracing mineralized boulders or rock fragments is extremely valuable in areas of poor exposure or in mountainous areas. In mountainous areas the rock fragments have moved downslope under gravity and the

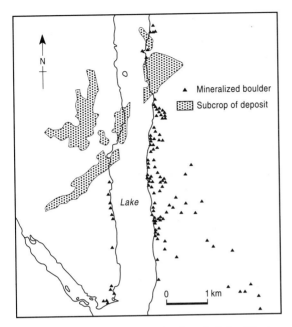

Fig. 5.7 Train of mineralized boulders from the Laisvall lead deposit, Sweden.

Fig. 5.8 Typical exploration trench, in this case on the South African Highveld.

lithology hosting the mineralization can be matched with a probable source in a nearby cliff and a climb attempted. Float mapping and sampling is often combined with stream sediment sampling and a number of successful surveys have been reported from Papua New Guinea (Lindley 1987). In lowland areas mineralized boulders are often disturbed during cultivation and may

be moved to nearby walls. In this case it is often difficult to establish a source for the boulders and a soil survey and a subsequent trench may be necessary. Burrowing animals may also be of help. Moles or rabbits and termites in tropical areas, often bring small fragments to the surface.

In glaciated areas boulders may be moved up to tens of kilometres and distinct boulder trains can be mapped. Such trains have been followed back to deposits, notably in central Canada, Ireland and Scandinavia (Fig. 5.7). In Finland the technique proved so successful that the government offered monetary rewards for finding mineralized boulders. Dogs are also trained to sniff out the

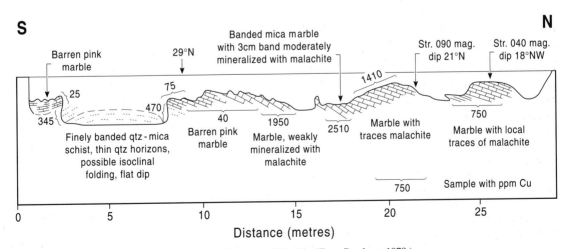

Fig. 5.9 Sketch of a trench testing a copper anomaly in central Zambia. (From Reedman 1979.)

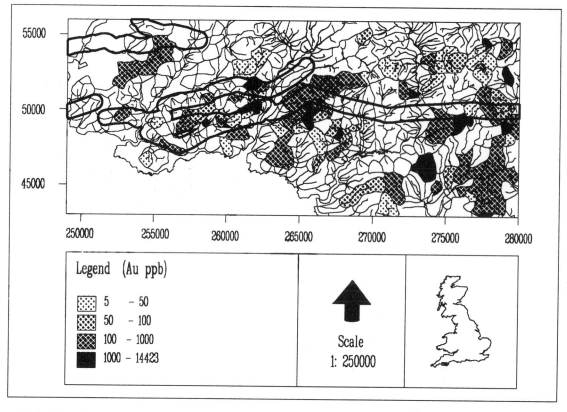

Fig. 5.10 Typical output from a vector based GIS system, in this case ARC/INFO. The plot shows concentrations of panned concentrate gold combined with a 1 km buffer around mapped thrusts. This enables the selection of catchments that contain possible thrust-related gold prospects. (Courtesy C. L. Wang, Leicester University.)

sulphide boulders as their sense of smell is more acute than that of the exploration geologist!

5.1.7 Physical exploration: pitting and trenching

In areas of poor to moderate outcrop a trench (Fig. 5.8) or pit is invaluable in confirming the bedrock source of an anomaly, be it geological, geochemical or geophysical. The geology of a trench or pit wall should be described and illustrated in detail (Fig. 5.9). For further details see section 9.2. Trenches and pits also provide large samples for more accurate grade estimates as well as for undertaking pilot processing plant test work to determine likely recoveries.

Some operators in remote areas, particularly in central Canada, strip relatively large areas of the overburden to enable systematic mapping of bedrock. However,

this would now (probably) be regarded as environmentally unfriendly.

An alternative to disturbing the environment by trenching is to use a hand held drill for shallow drilling. This type of drill is lightweight and can be transported by two people. It produces a small core, usually around 25–30 mm in diameter. Penetration is usually limited, but varies from around 5 m to as much as 45 m depending on the rig, rock type and skill of the operators!

5.1.8 Merging the data

The key skill in generating a drilling target is integrating the information from the various surveys. Not all the information obtained will be useful, indeed some may be misleading, and distinction should be clearly made between measured and interpreted data. Geophysical and geochemical surveys can indicate the surface or sub-

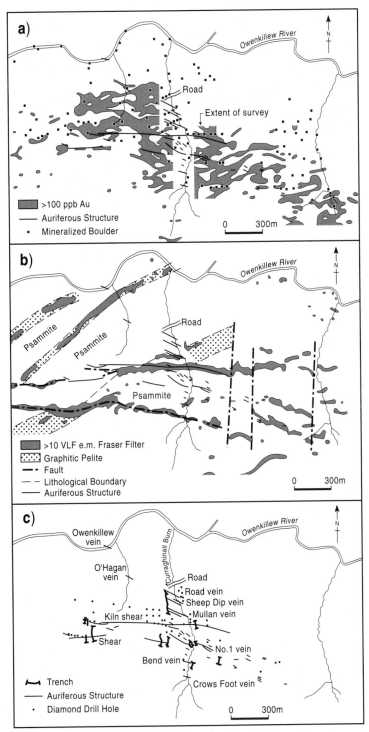

Fig. 5.11 Drilling based on a combination of (a) geochemistry, (b) geophysics and (c) trenching and prospecting, Sperrin Mountains, Northern Ireland. (After Clifford *et al*. 1992.)

surface geology of the potential host rocks or more directly the presence of mineralization as discussed in Chapters 7 and 8. Traditionally the results of these surveys have been combined by overlaying coloured transparent copies of the data on a topographical or geological paper base. It is then possible to determine the inter-relation between the various data sets. More recent developments have allowed the use of computerized methods (GIS) which allow rapid integration and interrogation of the databases, either in vector form or on a grid base (Fig. 5.10). These systems run well on portable computers and images or raw data can be examined in the field. This is a rapidly developing field and details can be found in Bonham-Carter (1994) and Maguire *et al.* (1991).

A straightforward example of the overlaying of different data types is shown in Figure 5.11 taken from the work of Ennex plc in Northern Ireland as detailed in Clifford *et al.* (1990, 1992). The area was chosen for investigation because of the presence of favourable geology, placer gold in gravels and anomalous arsenic geochemistry discovered during government surveys. A licence was applied for in 1980. Ennex geologists investigated the anomalous areas and by careful panning and float mapping identified the source of the gold as thin quartz veins within schist. Although the veins identified assayed about 1 g t-1 Au and were not of economic grade, they were an indication that the area was mineralized and prospecting was continued over a wider area. In late 1983 economically significant gold in bedrock was found and subsequently three other gold-bearing quartz veins were found where a small stream has eroded to bedrock. More than 1 300 mineralized boulders were mapped over a strike of 2 700 m. Channel samples of the first identified vein gave assays of 14 g t-1 Au over 3.8 m with selected samples assaying over 150 g t-1 Au.

Detailed follow-up concentrated on building up a systematic picture of the area where there is no outcrop. This focused on boulder mapping and sampling, deep overburden geochemistry and geophysical mapping using Very Low Frequency Electromagnetic Resistivity (VLF–e.m./R) surveys. Deep overburden geochemistry was undertaken using a small core overburden drill to extract 250 g samples on a 50 × 20 m grid. This spacing was decided on following an orientation study over a trenched vein. Geochemical sampling defined 33 anomalies (>100 ppb Au) (Fig. 5.11a). A variety of geophysical methods were tested including induced polarization and VLF–e.m./R. No method located the quartz veins but the VLF–e.m./R method did define previously unidentified gold bearing shears (Fig. 5.11b) and the subcrop of

graphitic pelites. The nature of the geophysical and geochemical anomalies was defined by a 2 850 m trenching programme (Fig. 5.11c). Besides increasing the number of gold bearing structures to 16, this trenching was also instrumental in demonstrating the mineralogy of the mineralized structures. The quartz veins contain significant sulphides, mainly pyrite but with lesser chalcopyrite, galena, sphalerite, native copper and tetrahedrite–tennantite. In contrast the shears are richer in arsenopyrite with chloritization often observed. Thus geochemistry defined the approximate location of the quartz veins and auriferous shears, geophysics the location of the shears and trenching the subcrop of the quartz veins.

Once the gold bearing structures had been defined their depth potential was tested by diamond drilling (Fig. 5.11c). The diamond drilling (6 490 m in 63 holes) took place during 1985-6 and defined a resource of 900 000 t at 9.6 g t^{-1}. Subsequent underground exploration confirmed the grades and tonnages indicated by drilling but the deposit is undeveloped because of a ban on the regular use of explosives in Northern Ireland by the security authorities.

5.1.9 Environmental aspects

Obtaining an environmental permit to operate a mine has become a vital part of the feasibility process. This involves the preparation of an environmental impact statement (EIS) describing the problems that mining will cause and the rehabilitation programme that will be followed once mining is complete (Hinde 1993). Such an impact statement requires that the condition of the environment in the potential mining area before development began is recorded. Thus it is useful to collect data during the exploration stage for use in these EISs. Initial data might include surface descriptions and photographs, and geochemical analyses indicating background levels of metals and acidity. It is of course essential to minimize damage during exploration and to set a high standard for environmental management during any exploitation. Trenches and pits should be filled and any damage by tracked vehicles should be minimized and if possible made good.

Another aspect of environmental studies is public relations, particularly keeping the local population informed of progress. Most locals are curious about outsiders and appreciate non-technical information. Most exploration companies organize presentations for locals and liaise closely with elected officials, such as councillors or mayors.

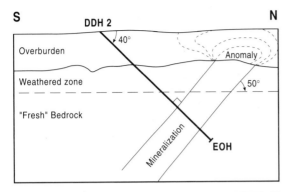

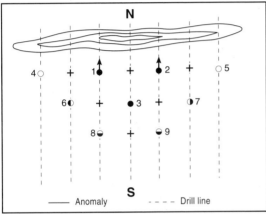

Fig. 5.12 Idealized initial drill grid. EOH = End of hole. (After Annels 1991.)

5.2 WHEN TO DRILL AND WHEN TO STOP

One of the hardest decisions in exploration is to decide when to start drilling, and an even harder one is when to stop.

The pressure to drill will be evident when the programme has identified surface mineralization. Management will naturally be keen to test this and gain an idea of subsurface mineralization as soon as possible. However, this pressure should be resisted until there is a reasonable idea of the overall surface geology and the inferences that can be made from this knowledge concerning mineralization in depth.

5.2.1 Setting up a drilling programme

The geologist in charge of a drilling programme is faced with a number of problems, both logistical and geological. There must be a decision on the type of drilling required, the drill hole spacing (section 9.4.4), the

timing of drilling and the contractor to be used. The logistics of drilling should be considered carefully as the drill will need drill crews, consumables and spare parts; this will require helicopter support in remote areas and vehicle access in more populated areas. Many drills require vehicle access and access roads must be made and pads for drilling constructed so that the drill rigs can be set on an almost horizontal surface.

The pattern of drilling used is dependent on the assumed attitude and thickness of the drilling target. This depends on the available information which may, of course, be inaccurate. Drilling often causes reconsideration of geological ideas and prejudices. Vertical boreholes are the easiest and cheapest to drill and widely used for mineralization with a shallow dip or for disseminated deposits. However, inclined holes are usually preferred for targets with steep dips. The aim will be to cut the mineralization at 90° with the initial hole cutting immediately below the zone of oxidation (weathered zone) (Fig. 5.12).

Drilling is used to define the outlines of any deposit and also the continuity of mineralization for purposes of resource estimation. The initial pattern of drilling will depend on surface access, which may be very limited in mountainous areas. In areas without access problems, typical drill hole patterns are square with a regular pattern or with rows of holes offset from adjacent holes. The first hole normally aimed at the down dip projection of surface anomalies or the interpreted centre of subsurface geophysics. Most programmes are planned on the basis of a few test holes per target with a review of results while drilling. The spacing between holes will be based on anticipated target size, previous company experience with deposits of a similar type and any information on previous competitor drilling in the district (Whateley 1992). The subsequent drilling location and orientation of the second and third holes will depend on the success of the first hole. Success will prompt step outs from the first hole whereas a barren and geologically uninteresting first hole will suggest that another target should be tested.

Once a deposit has been at least partly defined then the continuity of mineralization must be assessed. The spacing between holes depends on the type of mineralization and its anticipated continuity. In an extreme case, e.g. some vein deposits, boreholes are mainly of use in indicating structure and not much use in defining grade, which can only be accurately determined by underground sampling (see Chapter 9). Typical borehole spacing for a vein deposit is 25–50 m and for stratiform deposits anything from 100 m to several hundreds of

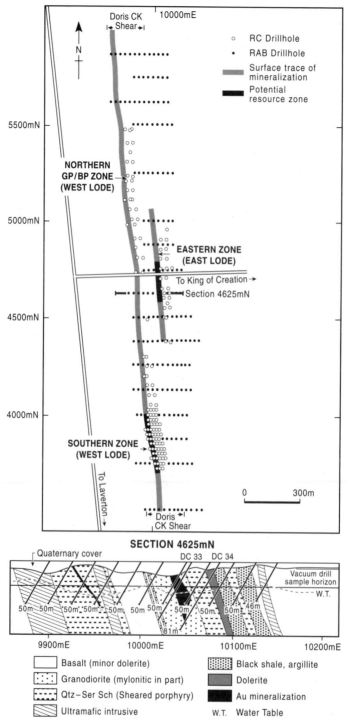

Fig. 5.13 (a) Typical combination of shallow and deeper drilling over the Doris Creek vein gold deposit in Western Australia. The shallow (rotary air blast, RAB) drilling defines the subcrop of the vein whereas the deeper (reverse circulation, RC) drilling defines grades. (After Jeffery *et al.* 1991.)

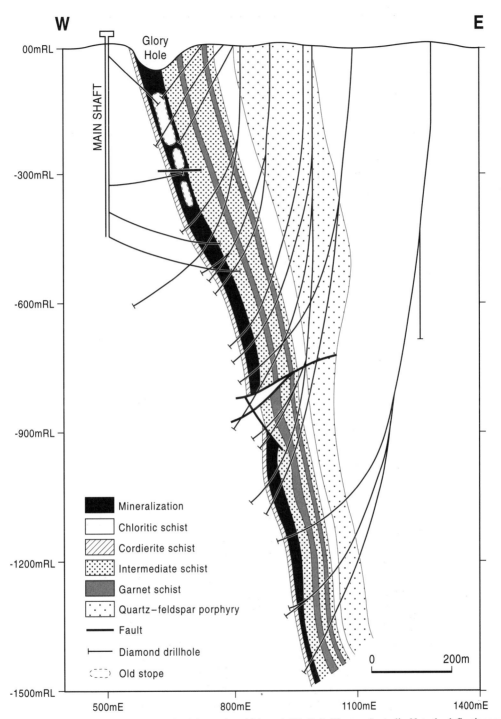

Fig. 5.13(b) Typical drilling of a previously mined deep vein gold deposit, Big Bell, Western Australia. Note the deflections on the deeper holes, the abandoned holes and combination of drilling both from surface and underground. (After Handley & Carrey 1990.)

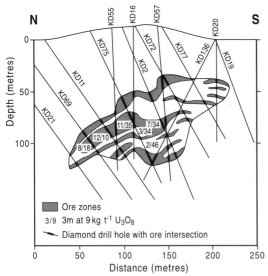

Fig. 5.13(c) Typical drill section of an ill defined mass, Kintyre uranium deposit, Western Australia. (After Jackson & Andrew 1990.)

metres. Examples of drill spacing and orientation for a variety of deposits is shown in Figure 5.13(a–c).

5.2.2 Monitoring drilling programmes

Monitoring the geology and mineralization intersected during a drilling programme is vital in controlling costs. In the initial phases of drilling this may involve the geologist staying beside the rig if it is making rapid progress, for example when using percussion drilling, and logging material as it comes out of the drill hole. In the case of diamond drilling twice daily visits to examine core, make initial logs and decide on the location of the next drill holes are usually sufficient, although longer visits will be needed when cutting potentially mineralized zones or nearing the scheduled end of the hole. Often the geologist will be required by contractors to sign for progress or the use of casing. Besides detailed core logs, down hole geophysical logging (section 7.13) is often used and arrangements should be made with contractors for timely logging as holes can become rapidly blocked and cleaning of holes is an expensive undertaking.

Data on mineralization, the lithologies and structures hosting it should be recorded and plotted on to a graphic log as soon as the information becomes available. Initially, strip drill logs (Fig. 5.14) can be used and sections incorporating the known surface geology. Assay infor-

mation will usually be delayed for a few days but estimates of the location and importance of mineralization can be plotted alongside the lithological log (Fig. 5.16e). As further drilling proceeds the structural and stratigraphical controls on mineralization should become clearer. Ambiguities concerned with the interpretation of drilling are common and often cannot be resolved until there has been underground development. A typical example is shown in Figure 5.15 in which three different interpretations are possible from the information available.

Once a stratigraphy has been established and several boreholes are available, more sophisticated plotting techniques can be used. Typical methods used to plot drill hole information are structure contour plans, isopach, grade (quality), thickness and grade multiplied by thickness, known as the grade–thickness product or accumulation maps. Grade and grade–thickness product maps are extremely useful in helping to decide on the areal location of oreshoots and of helping direct drilling towards these shoots (Fig. 5.16 a–e).

One of the key issues in any drilling programme is continuity of mineralization. This determines the spacing of drill holes and the accuracy of any resource estimation (as discussed in Chapter 9). In most exploration programmes the continuity can be guessed at by comparison with deposits of a similar type in the same district. However, it is usual to drill holes to test continuity once a reasonable sized body has been defined. Typical tests of continuity are to drill holes immediately adjacent to others and to test a small part of the drilled area with a closer spacing of new holes.

5.2.3 Deciding when to stop

Usually the hardest decision when directing a drilling programme is to decide when to stop. The main situations are.

1 No mineralization has been encountered.

2 Mineralization has been intersected, but it is not of economic grade or width.

3 Drill intercepts have some mineralization of economic grade but there is limited continuity of grade or rough estimates show that the size is too small to be of interest.

4 A body of potentially economic grade and size has been established.

5 The budget is exhausted.

The decision in the first case is easy, although the cause of the surface or sub-surface anomalies that determined the siting of the drill hole should be established. The

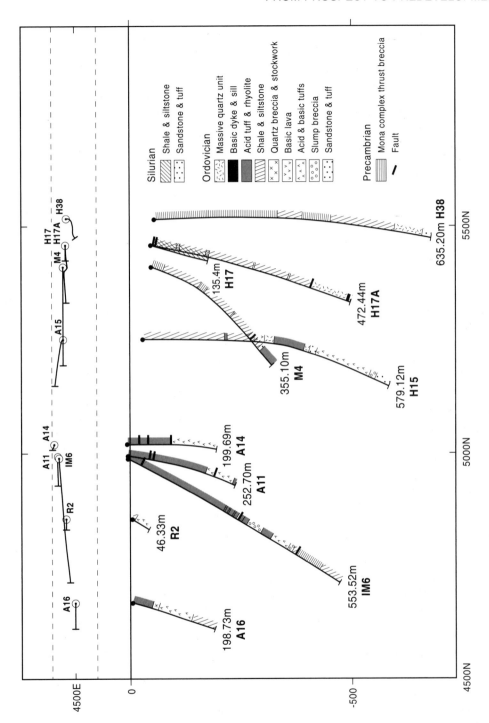

Fig. 5.14 Geological strip logs and plot of drill holes, Parys Mountain zinc prospect, Anglesey, Wales. (Data courtesy Anglesey Mining, plot by L. Agnew.)

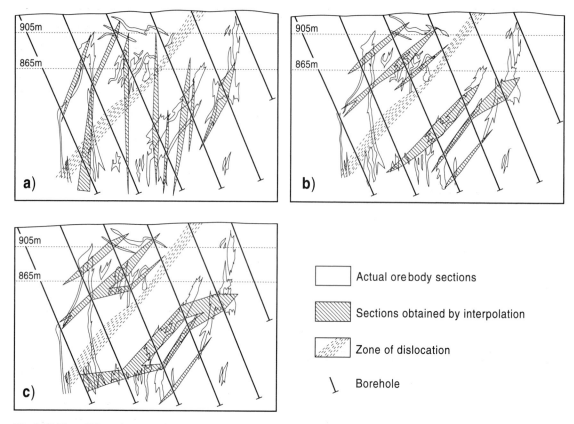

Fig. 5.15 Three different interpretations of drill intersections in an irregular deposit compared with outlines defined by mining. (After Kreiter 1968.)

second case is more difficult and the possibility of higher grade mineralization should have been eliminated. The presence of some interesting intersections in a prospect suggests that the processes forming economic grade mineralization were operative but not at a sufficient scale to produce a large deposit, at least not in the area drilled. The encouragement of high grade mineralization will often result in several phases of drilling to test all hypotheses and surrounding areas to check whether the same processes may have taken place on a larger scale. This type of prospect is often recycled (i.e. sold) and investigated by several different companies before being abandoned or a deposit discovered. Often it is only some factor such as the impending deadline on a mineral rights option that will result in a decision to exercise the option or drop this type of prospect.

The definition of a potentially economic deposit is an altogether more agreeable problem, and the exact·ending of the exploration phase will depend on corporate policy. If a small company is having to raise money for the evaluation phase, it will probably leave the deposit partly open (i.e. certain boundaries undetermined), as a large potential will encourage investors. Larger companies normally require a better idea of the size, grade and continuity of a deposit before handing over the deposit to their pre-development group.

It is usual for a successful programme to extend over several years, resulting in high costs. Exhausting a budget is common; however overspending may result in a sharp reprimand from accountants and the painful search for new employment!

5.3 RECYCLING PROSPECTS

The potential of many prospects is not resolved by the first exploration programme and they are recycled to another company or entrepreneur. This recycling has many advantages for the explorer. If for some reason the

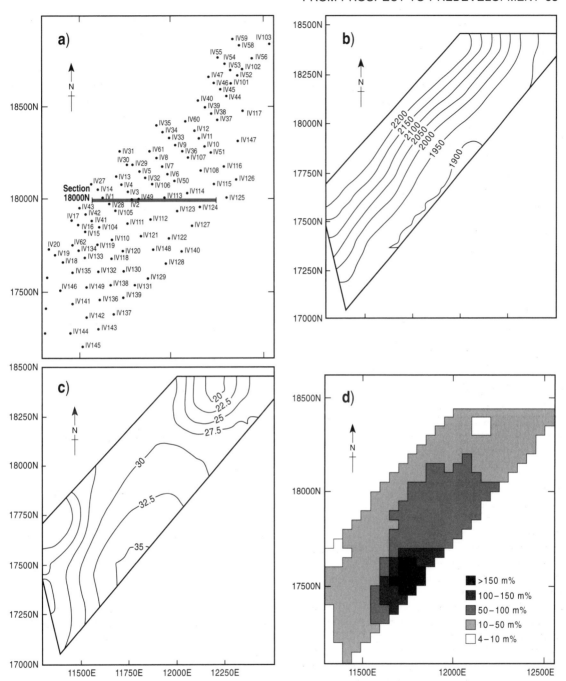

Fig. 5.16 (a–d) Typical plots used in interpretation and planning of drilling. The example is a stratiform copper deposit of the Zambian type — (a) drill hole plan and numbers, (b) structure contour of the top of the mineralized interval (m above sea level), (c) isopach of the mineralized interval (m), (d) grade thickness product (accumulation) of the mineralized interval (m%).

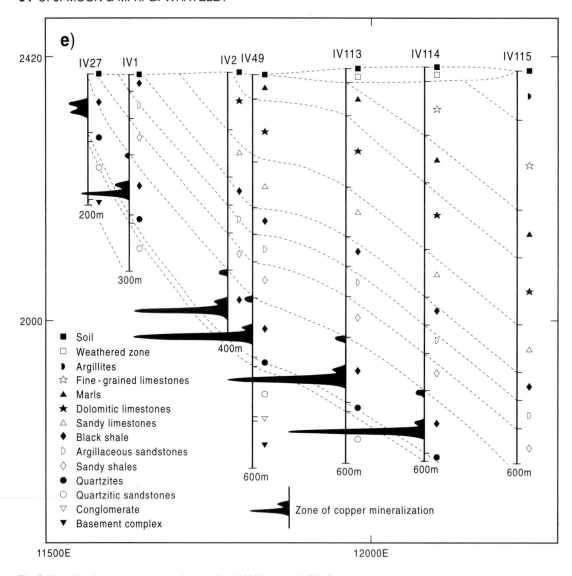

Fig. 5.16 (e) Section through the deposit along line 18000N shown in Fig. 5.16(a).

prospect does not meet their requirements, they are often able to recover some of the investment in the prospect with the sale of the option or the data to a new investor. Often it is not an outright sale but one in which the seller retains some interest. This is termed a joint venture or farm out. Typical agreements are ones in which the buyer agrees to spend a fixed amount of money over a fixed period of time in order to earn a pre-agreed equity stake in the prospect. The period over which the funding takes place is known as the earn-in phase. Normally the seller retains an equity stake in the project and the right to regain complete control if the buyer does not meet its contractual obligations. The type of equity retained by the seller varies; it may include a duty to fund its share of development costs or it may have no further obligations—a free or carried interest. The seller may as an alternative choose merely to have a royalty if the project is eventually brought to production.

Joint venturing is normally carried out by circulating prospective partners with brief details of the project, but this may exclude location and other sensitive details. Seriously interested parties will then be provided with full details and the possibility of visiting the prospect to gain a more accurate impression before committing any money.

Buying or re-examining a prospect that a previous party has discarded is tempting, but the buyer must be satisfied that they can be more successful than the other explorer. This success may result from different exploration models or improved techniques.

MULBERRY PROSPECT

A good example of this is the exploration of the area on the north side of the St. Austell Granite in south-western England, here termed Mulberry after the most productive old mine (Dines 1956). It was examined by three different companies from 1963–1982, as reported in British Geological Survey (BGS) open file reports. The Mulberry area is of obvious interest to companies as it is one of the few in Cornwall that supported open pit mining of disseminated tin deposits in the nineteenth century and might therefore be more efficiently mined using advances in mining technology (see section 16.3.1 for further details of the regional geology). The deposits worked (Fig. 5.17) were the Mulberry open pit with a N–S strike and a series of open pits on the Wheal Prosper structure which runs E–W. All the open pits were based on a sheeted vein complex or stockwork of thin veins containing cassiterite, tourmaline and quartz. In addition there were a number of small underground workings based on E–W copper- and N–S iron-bearing veins. All the deposits are within the metamorphic aureole of the St Austell Granite, hosted in slate and another metamorphic rock known locally as calc-flinta. This consists of garnet, diopside, actinolite, quartz and calcite and was probably originally an impure tuff or limestone.

The initial examination in 1964 resulted from a regional appraisal discussed in detail in sections 16.5.1–3 in which Consolidated Gold Fields (CGF) sought bulk mining targets that were amenable to improved mining and processing techniques. CGF immediately identified Mulberry as being of interest and confirmed this by sampling the old open pit and the underground accesses. In addition the company investigated the potential of the area around the open pit and particularly its strike extensions by taking soil samples for geochemical analysis along E–W lines (Hosking 1971). These suggested that the deposit continued to the south into the calc flintas. CGF then undertook a drilling programme to test the potential of the deposit at depth (Fig. 5.17). Drilling

over the period 1964–66 confirmed that the mineralization continued at roughly the width of the open pit but that the cassiterite was contained in thin erratic stringers. Drilling to the north of the open pit showed a sharp cut-off whereas drilling to the south showed that tin values continued from the slate host into a calc flinta host where mineralization became much more diffuse. Although the grades intersected were similar to those in the open pit, the size and grade were not considered economic, especially when compared with Wheal Jane (Chapter 16) which CGF was also investigating, and the lease was terminated.

The next phase of exploration was from 1971–2 when the area was investigated by Noranda–Kerr Ltd, a subsidiary of the Canadian mining group Noranda Inc. This exploration built on the work of CGF and investigated the possibility that the diffuse mineralization within the calc silicates might be stratabound and more extensive than previously suspected. Noranda conducted another soil survey (Fig. 5.18) that showed strong Sn and Cu anomalies over the calc silicates. Four drill holes were then sited to test these anomalies after pits were dug to confirm mineralization in bedrock. The drill holes intersected the calc silicates but the structure was unclear and not as predicted from the surface. In addition although extensive (>20 m thick) mineralization was intersected it was weak (mainly 0.2–0.5% Sn) with few high grade (>1% Sn) intersections. The programme therefore confirmed the exploration model but was terminated.

As part of an effort to encourage mineral exploration in the UK, the British Geological Survey investigated the southern part of the area (Bennett *et al.* 1981). Their surveys also defined the southern geochemical anomaly to the SW of Wheal Prosper, as well as an area of alluvial tin in the west of the area. The extension of the Mulberry trend was not as clearly detected, probably because of the use of a wider sampling spacing. BGS also drilled two holes (Fig. 5.20) to check the subsurface extension of the western end of the Wheal Prosper system as well as a copper anomaly adjacent to the old pit. Limited assays gave similar results (up to 0.3% Sn) to surface samples.

The third commercial programme was a more exhaustive attempt from 1979–82 by Central Mining and Finance Ltd (CMF), a subsidiary of Charter Consolidated Ltd, themselves an associate company of the Anglo American group. Charter Consolidated had substantial interests in tin and wolfram mining, including an interest in the South Crofty Mine and an interest in Tehidy Minerals, the main mineral rights owners in the area. CMF's approach included a re-evaluation of the

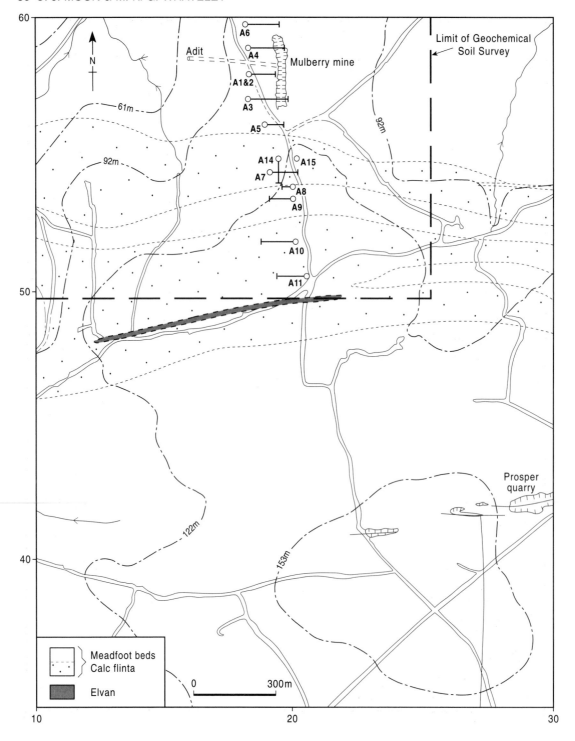

Fig. 5.17 Consolidated Gold Fields Programme 1964–66: drilling along strike from the Mulberry open pit.

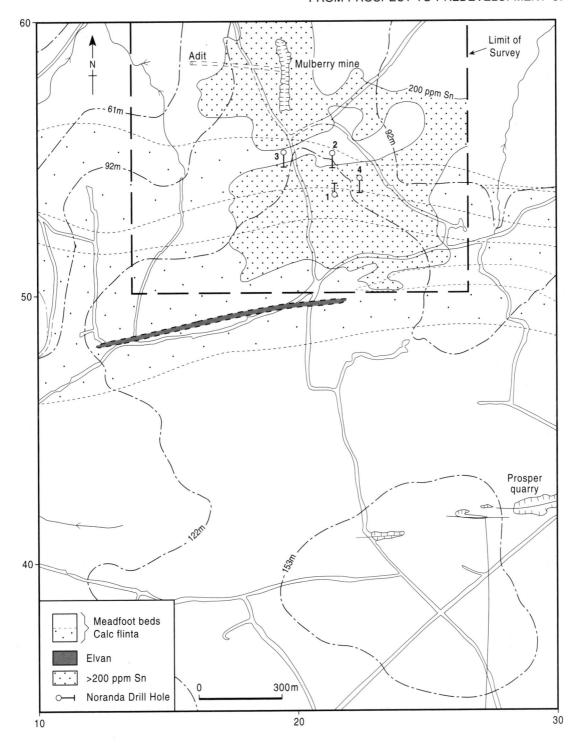

Fig. 5.18 Noranda Kerr Programme 1971–2: soil surveys and drilling of calc silicates.

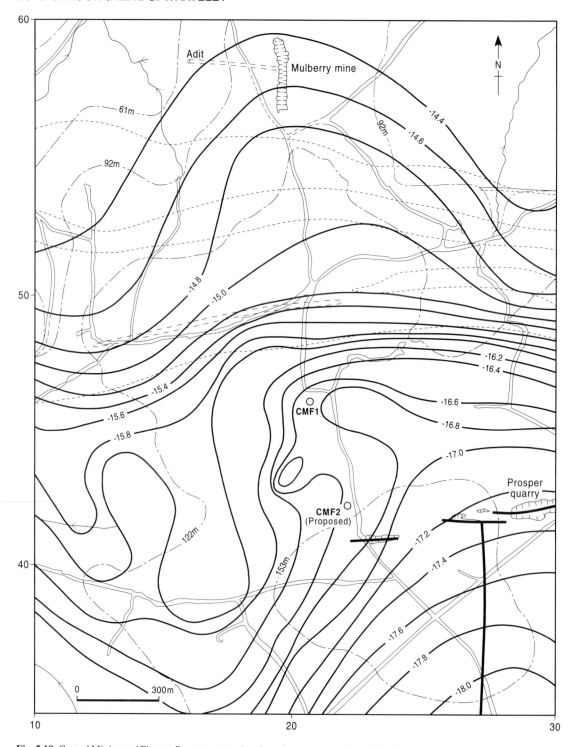

Fig. 5.19 Central Mining and Finance Programme: regional gravity survey and deep drill holes.

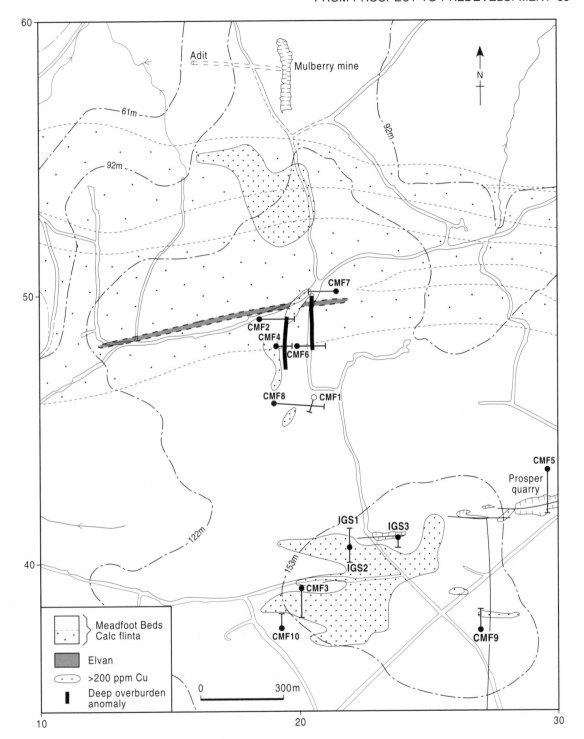

Fig. 5.20 Central Mining and Finance Programme 1980–82: drilling to test deep overburden anomalies. The British Geological Survey drill holes to test the Wheal Prosper system (IGS 1–3) are also shown.

Mulberry pit and a search for extensions, as well as a search for extensions of the Wheal Prosper system and an attempt to identify any buried granite cupolas with their associated mineralization. Their programme contained four phases. In the first phase the Mulberry pit was resampled and regional soil geochemistry and gravity surveys were undertaken. The gravity survey identified a perturbation in the regional gravity field along the Mulberry trend (Fig. 5.19). Regional shallow soil geochemistry samples were taken on 100 m centres and analysed for Sn, Cu, As, Mo and W. They showed an ill-defined southern continuation of the Mulberry system in Cu data and a large Cu and As anomaly to the SW of the Wheal Prosper system. Resampling of the Mulberry pit led to a re-estimation of the resource to 1.1 Mt at 0.44% Sn using a 0.2% Sn cut off. This would form part of the resource for any mine.

The Phase 2 programme in 1980 consisted of a detailed geochemical soil survey follow up of anomalous areas identified in phase 1, as well as the planning of two holes to test the geophysical anomaly. The first hole (CMF1, Fig. 5.20) was a 540 m deep diamond cored borehole that failed to find any indication of granite or associated contact metamorphism although it did cut vein mineralization of 0.51% Sn over 3.1 m at 206.5 m depth. Because of the lack of granite the planned second hole was cancelled.

The results of the soil geochemistry survey encouraged CMF to progress to phase 3 with a deep overburden sampling programme to the south of the Mulberry trend. This defined two parallel mineralized zones with encouraging assays in bedrock. Their trends were tested with five holes (CMF 2,4,6–8, Fig. 5.20) which demonstrated that the mineralized zones are steeply dipping veins. Although some higher grade intersections were made the overall tenor did not appear to be viable, especially when the mineralogy was carefully examined. Laboratory mineral processing trials of the higher grade intersections failed to recover sufficient cassiterite to account for the total tin content determined. It therefore seemed likely that some tin was contained in silicates, probably partly in the tin garnet, malayaite. This tin is not recoverable during conventional mineral processing and considerably devalued these intersections. Two holes were also drilled to test the subsurface extension of the Wheal Prosper pit (CMF 5) and the major copper anomaly (CMF 3). Hole CMF 5 confirmed that surface grades extended to depth but the grade was too low to be of further interest. In contrast CMF 3 was more encouraging with several intersections, including 1.5 m at 0.5% Sn at 95 m.

The encouragement of Phase 3 in the south of the area led to a Phase 4 (1982) programme which further examined the large copper anomaly at the intersection of the Wheal Prosper and Mulberry trends. Two holes were drilled (CMF9 and CMF10) and although one of these cut high grade mineralization (0.85 m of 7.0% Sn at 15.9 m) the programme was terminated.

5.4 SUMMARY

Following the identification of areas with possible mineralization and the acquisition of land, the next step is to generate targets for drilling or other physical examination such as pitting and trenching or, in mountainous places, by driving adits into the mountainside. Budgeting and the organization of suitable staff are again involved.

If the area is remote then suitable topographical maps may not be available and the ground must be surveyed to provide a base map for geological, geochemical and geophysical work. Some of the geological investigations may include the mapping and sampling of old mines and the dispatch of samples for assaying. Any of this work may locate areas of mineralization which may be recognized by topographical effects, colouring or wall rock alteration. Different ore minerals may give rise to characteristic colour stains and relic textures in weathered rocks and gossans and below these supergene enrichment may have occurred. All these investigations together with pitting and trenching may indicate some promising drilling targets.

Geologists should never rush into drilling which is a most expensive undertaking! Very careful planning is necessary (section 5.2.1) and drilling programmes require meticulous monitoring. Once the drilling is in full swing deciding when to stop may be very difficult and various possibilities are reviewed in section 5.2.3.

If an exploration programme does not resolve the potential of some mineralized ground then the operating company may sell the option, with or without the data obtained on it, to a new investor. This is often nowadays termed recycling and it enables the original operator to recover some of its investment. Such recycling is exemplified by the Mulberry Prospect in Cornwall (section 5.3).

5.5 FURTHER READING

Basic field techniques are covered in Reedman (1979), Peters (1987) and Chaussier & Morer (1987). Reviews of geochemical, geophysical and remote sensing techniques, covered in Chapters 6–8, as well as some case histories of their integration, can be found in the volume edited by Garland (1989). The textbook by Bonham-Carter (1994) provides a comprehensive introduction to the use of GIS in mineral exploration.

Case histories of mineral exploration are scattered through the literature but the volume of Glasson & Rattigan (1990) compiled much Australian experience. Evaluation techniques from exploration to production are well covered by Annels (1991).

6 Remote Sensing and Photogeology

MICHAEL K. G. WHATELEY

6.1 INTRODUCTION

Geologists have been using aerial photography to help their exploration efforts for over four decades. Since the advent of high resolution satellite imagery with the launch of the Landsat 1 satellite in 1972, exploration geologists are increasingly required to interpret digital images (computerized data) of the terrain. More recently, geologists involved in research and commercial exploration have been seeking out the more illusive potential mineral deposits; those hidden by vegetation or by Quaternary cover. In these instances geochemical, geophysical and other map data are available. It is now possible to express these map data as digital images, allowing the geologist to manipulate and combine them using digital image processing techniques and geographical information systems (GIS).

As image interpretation and photogeology are commonly used in exploration programmes today, it is the intention in this chapter to describe a typical satellite system and explain how the digital images can be processed, interpreted and used in an exploration programme to select targets. More detailed photogeological studies using aerial photographs are then carried out on the target areas.

6.1.1 Data collection

Remote sensing is the collection of information about an object or area without being in physical contact with it. Data gathering systems used in remote sensing are:
1 photographs obtained from manned space flights or airborne cameras, and
2 electronic scanners or sensors such as multispectral scanners in satellites or aeroplanes and TV cameras, all of which record data digitally.

Most people are familiar with the weather satellite data shown on national and regional television. For most geologists and other earth scientists, multispectral satellite imagery is synonymous with the Landsat series. It is images from these satellites that are most readily available to exploration geologists and they are discussed below. The use of imagery from the French satellite, SPOT, is becoming more widespread. Remote sensing data gathering systems are divided into two fundamental types, i.e. those with passive or active sensors.

PASSIVE SENSORS

These sensors gather data using available reflected or transmitted parts of the electromagnetic (e.m.) spectrum, i.e. they rely on solar illumination of the ground or natural thermal radiation for their source of energy respectively. Some examples are:
1 Landsat Multispectral Scanner (MSS);
2 Landsat Thematic Mapper (TM) which utilizes additional wavelengths, and has superior spectral and spatial resolution compared with MSS images;
3 airborne scanning systems which have even greater resolution and can look at more and narrower wavebands;
4 SPOT, a French commercial satellite with stereoscopic capabilities;
5 Space Shuttle.

Different satellites collect different data with various degrees of resolution (Fig. 6.1) depending on their application e.g. Landsat cf. TIROS-N, Seasat or NOAA weather satellites.

ACTIVE SENSORS

These sensors use their own source of energy. They emit energy and measure the intensity of energy reflected by a target. Some examples are (1) Radar and (2) Lasers.

6.2 THE LANDSAT SYSTEM

The first Landsat satellite, originally called Earth Resources Technology Satellite (ERTS), was launched in July 1972. So far five satellites in the series have been launched (Table 6.1).

Each satellite is solar powered and has a data collecting system which transmits data to the home station (Fig. 6.2). On Landsats 1 to 3 data were recorded on tape when out of range of the home station for later transmission. Tracking and data relay system satellites (TDRSS) were operational from March 1984 (Fig. 6.2).

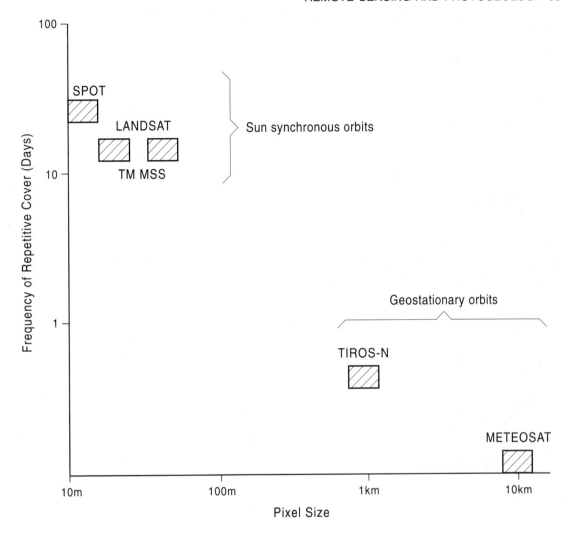

Fig. 6.1 Different resolution and frequencies of some satellite imagery.

The Landsat series was designed initially to provide multispectral imagery for the study of renewable and non-renewable resources, particularly land use resources using MSS and Return Beam Vidicon (RBV) cameras. RBV images are not commonly used today. Geologists immediately recognized the geological potential of the Landsat images and in July 1982 when Landsat 4 was launched it housed, in addition to MSS, a Thematic Mapper which scans in seven bands, two of which (5 and 7) were chosen specifically for their geological worth (Table 6.1 & Fig. 6.6). Landsat 5 also has TM.

Not all wavelengths are available for remote sensing (Fig. 6.3). Even on an apparently clear day ultraviolet (UV) light is considerably absorbed by ozone in the upper atmosphere. UV, violet and blue light are particularly affected by scattering within the atmosphere and are of little use in Landsat work. Blue light is only usefully captured using airborne systems. The blanks in the spectrum are due mainly to absorption by water vapour, CO_2 and O_3 (Fig. 6.3). Only wavelengths >3 μm, e.g. microwave wavelengths (or radar) can penetrate cloud (Fig. 6.3), so radar imagery is particularly useful in areas of considerable cloud cover (e.g. equatorial regions).

Table 6.1 Characteristics of several remote sensing systems. (* The MSS bands on Landsats 1–3 were numbered 4, 5, 6 & 7 because of a 3 band RBV sensor on Landsats 1 & 2. P+ = panchromatic mode). (After Schowengerdt 1983.)

	Landsat MSS	Thematic mapper	SPOT/HVR
Launch	1 July 1972 –Jan 1978 2 Jan 1975–July 1983 3 Mar 1978–Sept 1983 4 July 1982–present 5 Mar 1984–present	4 July 1982–Feb 1983 5 Mar 1984–present	1 Feb 1983–present 2 Jan 1990–present
Altitude (km)	918 (1–3)	700	822
Spectral bands (µm)		1 0.45–0.53 2 0.52–0.60 3 0.63–0.69	1 0.50–0.59 2 0.61–0.68 3 0.79–0.89
	4* 0.50–0.60 5 0.60–0.70 6 0.70–0.80 7 0.80–1.10	4 0.76–0.90 5 1.55–1.75 6 10.40–12.5 7 2.08–2.35	P+ 0.51–0.73
IFOV (m)	79 × 79 (1–3) 82 × 82 (4–5)	30 × 30 (bands 1–5, 7) 120 × 120 (band 6)	20 × 20 (bands 1–3) 10 × 10 (P)
Pixel interval (m)	57 × 82 (1–3) 57 × 80 (4) 60 × 60 (6)	30 ×30 (bands 1–5, 7) 120 × 120 (band 6)	20 × 20 (bands 1–3) 10 × 10 (P)
Scene dimensions (km)	185 × 185	185 × 185	60 × 60
Pixels per scene (× 10^6)	28	231	27 (bands 1-3)

6.2.1 Characteristics of digital images

MSS data are recorded by a set of six detectors for each spectral band (Table 6.1). Six scan lines are simultaneously generated for each of the four spectral bands (Fig. 6.3). Data are recorded in a series of sweeps and recorded only on the east bound sweep (Fig. 6.2). TM data are recorded on both east and west bound sweeps and each TM band uses an array of 16 detectors (band 6 uses only four detectors). There is continuous data collection and the data are transmitted to earth for storage on Computer Compatible Tapes (CCTs). These tapes are then processed on a computer to produce images. The scene dimensions on the surface under the satellite (its swath) is 185 km and the data are (for convenience) divided into sets which equal 185 km along its path.

The satellites have a polar, sunsynchronous orbit, and the scanners only record on the southbound path (Fig. 6.2) because it is night-time on the northbound path. The paths of Landsats 1 to 3 shifted west by 160 km at the equator each day so that every eighteen days the paths repeat resulting in repetitive, worldwide, MSS coverage. Landsats 4 and 5 have a slightly different orbit resulting in a revisit frequency of 16 days. Images are collected at the same local time on each pass—generally between 9.30 and 10.00 a.m. to ensure similar illumination conditions on adjacent tracks. Successive paths overlap by 34% at 40°N, and 14% at the equator, but they do not give stereoscopic coverage, although it can be added digitally.

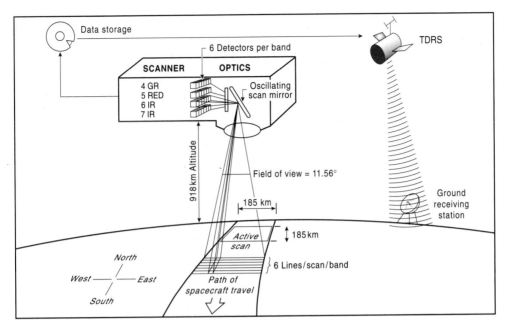

Fig. 6.2 Schema of a satellite system. MSS data are collected on Landsats 1 to 3 in four bands and transmitted via the TDRS (relay satellite) to the ground receiving station. (Modified after Sabins 1987 and Lo 1986.)

6.2.2 **Pixel parameters**

Digital images consist of discrete picture elements called pixels. Associated with each pixel is a number that is the average radiance, or brightness, of that very small area within the scene. In satellite remote sensing we are looking through the atmosphere at the earth's surface, and the sensor therefore measures the reflected radiation from the surface and radiation scattered by the atmosphere. The pixel value represents both. Fortunately, the level of scattered radiation is nearly constant, so changes in pixel value are essentially caused by changes in the radiance of the surface.

The Instantaneous Field of View (IFOV) is the distance between consecutive measurements of pixel radiance (79 m × 79 m in MSS), which is commonly the same as the pixel size (Table 6.1). However, MSS scan lines overlap, hence the pixel interval is less than the IFOV (56 m × 79 m). The TM detectors have a greatly improved spatial resolution, giving an IFOV of 30 m × 30 m.

The radiance for each pixel is quantified into discrete grey levels and a finite number of bits are used to represent these data (Table 6.2). This table shows how decimal numbers in the range 0–255 can be coded using eight individual bits (0 or 1) of data. Each is grouped as an 8 bit 'byte' of information, with each bit used to

indicate ascending powers of 2 from 2^0 (=1) to 2^7 (=128). This scheme enables us to code just 256 (0–255) values. On the display terminal 256 different brightness levels are used corresponding to different shades of grey ranging from black (0) to white (255), giving us a grey scale. Commonly 6 bits per pixel (64 grey levels) are used for MSS and 8 bits per pixel (256 grey levels) for TM. A pixel can therefore be located in the image by an x and a y coordinate, while the z value defines the grey-scale value between 0 and 255. Pixel size determines spatial resolution whilst the radiance quantization effects radiometric resolution (Schowengerdt 1983).

6.2.3 **Image parameters**

An image is built up of a series of rows and columns of pixels. Rows of pixels multiplied by the number of columns in a typical MSS image (3240 × 2340 respectively) gives approximately 7M pixels per image. The improved resolution and larger number of channels scanned by the TM results in nine times as many pixels per scene. Hence the need for computers to handle the data. Remote sensing images are commonly multispectral, e.g. the same scene is imaged simultaneously in several spectral bands. Landsat MSS records four spectral bands (Table 6.3) and TM records seven spectral bands. Visible

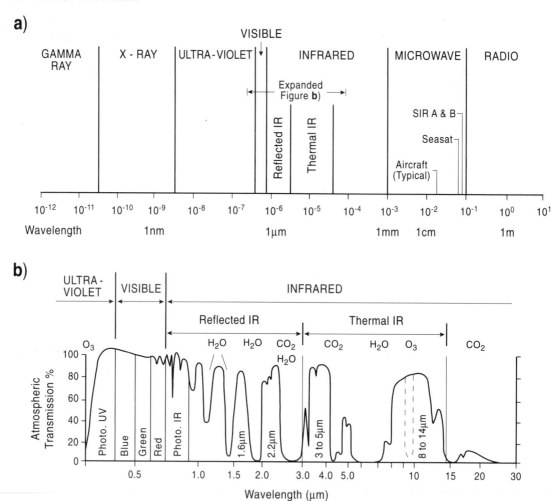

Fig. 6.3 (a). The electromagnetic spectrum. (b). Expanded portion of the spectrum showing the majority of the data gathering wavelengths. Additional data are acquired in the microwave part of the spectrum for use by aircraft and to measure the surface roughness of either wave action or ice accumulation. Certain parts of the spectrum are wholly or partly absorbed by atmospheric gases. Atmospheric windows where transmission occurs are shown and the imaging systems that use these wavelengths are indicated. (Modified after Sabins 1987.)

Table 6.2 The principle of binary coding using 8 bit binary digits (bits)

	Place values									
Binary	2^7	2^6	2^5	2^4	2^3	2^2	2^1	2^0		
Decimal	128	64	32	16	8	4	2	1		
	0	0	0	0	0	0	0	0	=	0
	0	0	0	0	0	0	0	1	=	1
	0	0	0	0	0	0	1	0	=	2
	0	0	0	0	0	1	0	0	=	4
	0	0	0	0	1	0	0	0	=	8
	0	0	0	1	0	0	0	0	=	16
	0	0	1	0	0	0	0	0	=	32
	0	1	0	0	0	0	0	0	=	64
	1	0	0	0	0	0	0	0	=	128
	1	0	0	0	0	1	0	1	=	128+4+1
									=	133
	1	1	1	1	1	1	1	1	=	128+64+32+16+8+4+2+1
									=	255

Table 6.3 False colours assigned to the Landsat MSS band

MSS band	Wavelength (μm)	Natural colour recorded	False colour assigned
4	0.5-0.6	Green	Blue
5	0.6-0.7	Red	Green
6	0.7-0.8	IR	Red
7	0.8-1.1	IR	Red

Table 6.4 Some sources of remotely sensed data

US Geological Survey 1981. Worldwide directory of national earth science agencies and related international organizations. Geological Survey Circular 834, USGS, 507 National Center, Reston, Virginia, 22092, USA

SPOT-Image, 18 avenue Eduoard Belin, F31055, Toulouse Cedex, France

Meteorological satellite imagery
US Department of Commerce, NOAA/NESDIS/NSDC, Satellite Data Services Division (E/CCGI), World Weather Building, Room 10, Washington DC, 20233, USA

Landsat (pre-September 1985)
US Geological Survey, EROS Data Center (EDC), Mundt Federal Building, Sioux Falls, South Dakota, 57198, USA

Landsat (post September 1985)
Earth Observation Satellite Company (EOSAT), 1901 North Moore Street, Arlington, Virginia, 22209, USA

Landsat imagery is also available from local and national agencies

National Remote Sensing Centre, R190 Building, Royal Aircraft Establishment, Farnborough, Hampshire, GU14 6TD, UK

blue light is not recorded by Landsat MSS, but it is recorded in TM band 1. By combining three spectral bands and assigning a colour to each band (so that the grey scale becomes a red, green or blue scale instead), a multispectral false colour composite image is produced.

The image intensity level histogram is a useful indicator of image quality (Fig. 6.4). The histogram describes the statistical distribution of intensity levels or grey levels in an image in terms of the number of pixels (or percentage of the total number of pixels) having each grey level. Figure 6.4 shows the general characteristics of histograms for a variety of images.

A photographic print of a geometrically corrected image (Fig. 6.5) is a parallelogram. The earth rotates during the time it takes the satellite to scan a 185 km length of its swath, resulting in the skewed image (Fig. 6.11f).

6.2.4 Availability of Landsat data

1 Landsat CCTs for MSS or TM imagery are available for computer processing (Table 6.4).
2 Black and white, single band prints in a standard 23 cm × 23 cm format at a scale of 1:1 000 000 are available from the above sources.
3 Colour composite prints in a similar format are also available. It is possible to enlarge Landsat MSS images up to 1:100 000, before the picture quality degrades severely.

6.2.5 Mosaics

Each image has a uniform scale and relative lack of distortion at the edges. Black and white or colour prints of each scene can be cut and spliced into mosaics. With the repetitive cover most of the world has cloud-free Landsat imagery available. The biggest difficulty is in matching the grey tone during printing, and the images from different seasons.

6.2.6 Digital image processing

Data from the satellites are collected at the ground station on magnetic tape, i.e. in a digital form. The CCTs are used in conjunction with image processing computer systems. The main functions of these systems in a geological exploration programme are: image restoration, image enhancement and data extraction (Sabins 1987). Drury (1993) explains the methods of digital image processing in more detail and shows how they can be applied to geological remote sensing.

IMAGE RESTORATION

Image restoration is the process of correcting inherent defects in the image caused during data collection. Some of the routines used to correct these defects are:
(1) replacing lost data, i.e. dropped scan lines or bad pixels, (2) filtering out atmospheric noise and (3) geometrical corrections. The last named correct the data for cartographic projection, which is particularly important if the imagery is to be integrated with geophysical, topographical or other map-based data.

IMAGE ENHANCEMENT

Image enhancement transforms the original data to improve the information content. Some of the routines used to enhance the images are as follows.
1 *Contrast enhancement.* An image histogram has already been described in section 6.2.3 (Fig. 6.4). A histogram of a typical untransformed image has low contrast (Fig. 6.4a) and in this case the input grey level is equivalent to the transformed grey level (Schowengerdt 1983, Drury 1993). A simple linear transformation, commonly called a contrast stretch, is routinely used to increase the contrast of a displayed image by expanding the original grey level range to fill the dynamic range of the display device (Fig. 6.4b).
2 *Spatial filtering* is a technique used to enhance naturally occurring straight features such as fractures, faults, joints, etc. It is described in more detail by Drury (1986, 1993) and Sabins (1987).
3 *Density slicing* converts the continuous grey tone range into a series of density intervals (slices), each corresponding to a specific digital range. Each slice may be given a separate colour or line printer symbol. Density slicing has been successfully used in mapping bathymetric variations in shallow water and in mapping temperature variations in the cooling water of thermal power stations (Sabins 1986).
4 *False colour composite images* of three bands, e.g. MSS bands 4, 5, & 7, increase the amount of information available for interpretation.

INFORMATION EXTRACTION

Routines employed for information extraction use the speed and decision making capability of the computers to classify pixels on some predetermined grey level. This is carrried out interactively on computers by band ratioing, multispectral classification and principal component analysis. These can all be used to enhance specific

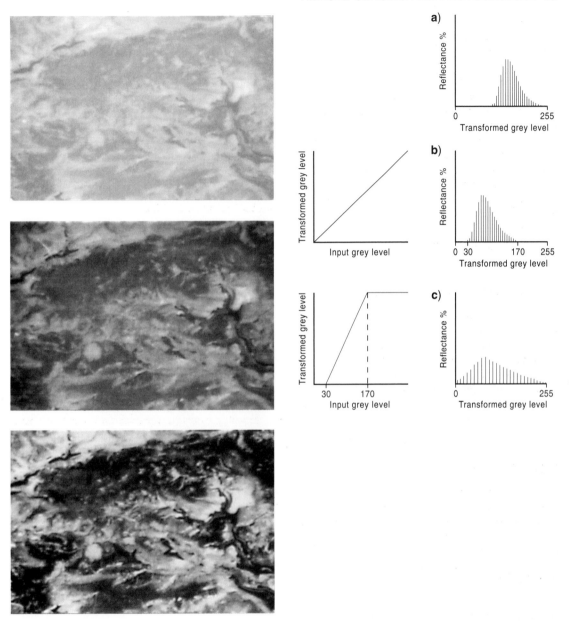

Fig. 6.4 Histograms of the grey levels (intensity levels) of four different types of image. The image is part of Landsat 5, TM band 3, taken on 26 April 1984 of the High Peak District of northern England. The area is approximately 15 × 15 km. The dark areas are water bodies; those in the the upper left corner are the Longdale reservoirs while those in the right centre and lower right are the Derwent Valley reservoirs. The two E–W trending bright areas in the lower right of the images are the lowland valleys of Edale and Castleton. (From Mather 1987.)

(a) A bright, low intensity image histogram with low spectral resolution. (b) A dark, low intensity, low spectral resolution, image histogram. (c) A histogram with high spectral resolution but low reflectance values. Increased spectral resolution has been achieved with a simple histogram transformation (also known as a histogram stretch). (Modified after Schowengerdt 1983.)

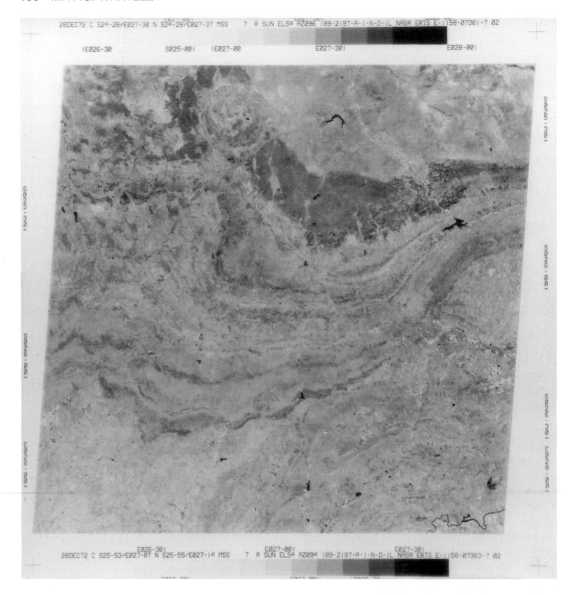

Fig. 6.5 A black and white Landsat 1 (ERTS), band 7 photograph of part of the Bushveld Igneous Complex in South Africa. This Summer image (wet season) illustrates the strong geological control over vegetation and variations in soil moisture. The circular Pilansberg Alkali Complex is seen in the north with the layered basic and ultra-basic intrusion to the south-east of it (dark tone). The banding in the Transvaal Supergroup dolomites, in the south centre (medium grey tone) was unsuspected until viewed on Landsat photographs.

geological features.

1 *Ratios* are prepared by dividing the grey level of a pixel in one band by that in another band. Ratios are important in helping to recognize ferruginous and limonitic cappings (gossans) (Fig. 6.6). Rocks and soils rich in iron oxides and hydroxides absorb wavelengths less than 0.55 µm and this is responsible for their strong red colouration (Drury 1993). These iron oxides are often mixed with other minerals which masks this colouration. A ratio of MSS band 4 over band 5 will enhance the small contribu-

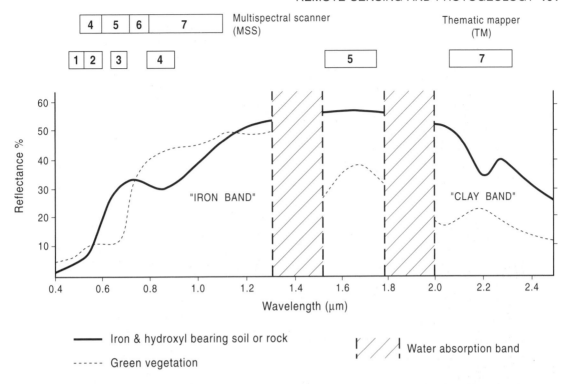

Fig. 6.6 A reflectance profile of the visible and IR parts of the EM spectrum, showing the changing reflectance profile of soil associated with 'typical' hydrothermal alteration compared with that of green vegetation. (Modified after Settle *et al.* 1984.)

tion of iron minerals (Fig. 6.6). Similarly a ratio of MSS band 6 over band 7 will discriminate areas of limonite alteration. Unfortunately limonite also occurs in unaltered sedimentary and volcanic rocks. However, by ratioing TM bands, mineralogical spectral characteristics related to alteration can be detected. The Landsat TM scanner has two critical, additional bands (bands 5 & 7) which are important in being used to identify hydrothermally altered rocks. Absorption in the 'clay band' causes low reflectance at 2.2 μm (TM band 7), but altered rocks have a high reflectance at 1.6 μm (TM band 5). A ratio of bands 5/7 will result in enhancement of altered rocks (Fig. 6.6).
2 *Multispectral classification.* Using this routine a symbol or colour is given to a pixel or small group of pixels which have a high probability of representing the same kind of surface material e.g. alteration. This is a decision, best made by computers. Multispectral classification is particularly useful for a large area or a region covered by dense vegetation and differences in the vegetation reflect the underlying geology.
3 *Principal component analysis* is used to enhance or distinguish lithological differences, because spectral dif-

ferences between rock types may be more apparent in principal component images than in single bands. The reflectances in different bands in MSS or TM images have a high degree of correlation. Principal component analysis is a commonly used method to improve the spread of data by redistributing them about another set of axes. This effectively exaggerates differences in the data.

6.2.7 Interpretation

Two approaches are used to extract geological information from satellite imagery.
1 *Spectral approach*—spectral properties are used to separate units in image data based on spectral reflectance. This is done interactively on computers using multispectral data in areas with or without dense vegetation.
2 *Photogeological approach*—known weathering and erosional characteristics (topographical expression) are used, on black and white or colour prints, to imply the presence of geological structure or lithology. Photogeological elements include topography, erosion,

tone, texture, drainage pattern, vegetation and land use. These elements are discussed in section 6.4.

6.2.8 Applications of Landsat satellite imagery

The use of satellite imagery is now a standard technique in mineral exploration (Nash *et al*. 1980, Goetz & Rowan 1981, Peters 1983, Drury 1993). It has also been used in structural investigations (Drury 1986) and in hydrogeology (Deutsch *et al*. 1981). In mineral exploration Landsat imagery has been used to provide basic geological maps, to detect hydrothermal alteration associated with mineral deposits and to produce maps of regional and local fracture patterns, which may have controlled mineralization or hydrocarbon accumulations.

STRUCTURAL MAPS

The synoptic view afforded by MSS and TM imagery is ideal for regional structural analyses, especially if a scene illuminated with a low sun angle is chosen (e.g. Autumn to Spring) which emphasizes topographical features. Manual lineament analysis on overlays of photographic prints is then carried out (usually after a spatial filtering technique has been employed) and lineaments are digitized. The original apparently random pattern can be quantified by computer processing, thus providing an objective method for evaluating lineaments. A rose diagram is prepared to depict the strike–frequency distribution. At the Helvetia porphyry copper test site in Arizona, the three trends identified by these methods correspond to deformation events in the Precambrian, the Palaeozoic and the Laramide Orgeny (Abrams *et al*. 1984).

The most common structures seen on Landsat images are faults, fractures, lineaments of uncertain nature and circular features. In conjunction with the statistical approach above, deductions regarding stress patterns in an area can be made. These structural studies may yield clues to the location of concealed mineral deposits. Linear zones, such as the 38th Parallel Lineament of the USA (Fig. 3.14) (Heyl 1972), consist of weak crustal areas that have provided favourable sites for upwelling intrusives and mineralizing fluids. These areas contain faults or fractures associated with deep-seated basement structures and the intersections of these features with other faults can provide favourable loci for mineralization.

LITHOLOGY AND ALTERATION MAPS

Landsat imagery has made significant contributions to the advancement of geological mapping both in known and unmapped areas of the world. Many third world countries now have geological maps which were previously too expensive to produce by conventional field mapping. Today regional geological mapping often starts with Landsat imagery, followed by rapid field reconnaissance for verification where possible. The maps thus produced are used to select exploration target zones on the basis of favourable geology and structure. Detailed follow-up photogeology and field work then takes place.

Ferruginous residual deposits (gossans) which overlie mineralized ground can often be identified from colour anomalies on enhanced false colour ratio composites, whereas these may not be detected on standard imagery, e.g. at the Silver Bell porphyry copper site (Abrams *et al*. 1984).

6.2.9 Advantages of Landsat imagery

1 This imagery provides a synoptic view of large areas of the earth's crust (185 km × 185 km), revealing structures previously unrecognized because of their great extent.
2 The evolution of rapidly developing dynamic geological phenomena can be examined through the use of successive images of the same area produced at 16 to 26 day intervals, e.g. delta growth or glaciation.
3 Some geological features are only intermittently visible, e.g. under certain conditions of climate or vegetation cover. Landsat offers 'revisit capability' and multitemporal coverage (Viljoen *et al*. 1975).
4 Digital images can be displayed in colour. This is useful in mapping rock types and alteration products, either direct from rocks or from vegetation changes (Abrams *et al*. 1984).
5 Landsat is valuable in providing a tool for rapid mapping of regional and local fracture systems. These systems may have controlled ore deposit location.
6 It is very cost effective, with an outlay of only a few pence km^{-2} for map production.
7 Computer processing enables discrimination and detection of specified rocks or areas.

6.3 SPOT

The French satellite, SPOT, was launched in February 1986 and SPOT 2 in January 1990. They are more sophisticated than Landsat, having a high resolution visible (HRV) imaging system, with a 20 m ground resolution in MS mode and a 10 m resolution in panchromatic mode (Table 6.1). SPOT also has off-nadir viewing capabilities which means that an area at 45° N can be imaged eleven times in the 26 day orbital cycle. Repeated off-nadir viewing introduces parallax from which stereoscopic image pairs can be produced. The commercial nature of SPOT means that images are slightly more expensive, but the increased resolution and stereoscopic capability mean that the additional cost may be warranted in an exploration programme, although SPOT does not have as good a spectral range as TM.

6.4 **PHOTOGEOLOGY**

Photogeology is the name given to the use of aerial photographs for geological studies. To get the best out of photographs we must plan the photogeological work in the office and in the field. A typical scheme is

1 annotation of aerial photographs,
2 compilation of photogeology on to topographical base maps,
3 field checking,
4 re-annotation and
5 re-compilation for production of a final photogeological map.

The common types of aerial photos are: panchromatic black and white photographs (B&W), B&W infrared (IR) sensitive film, colour film and colour IR.

These films make use of different parts of the spectrum, e.g. visible light (0.4–0.7 μm) and photographic near infrared (0.7–0.9 μm). Panchromatic film produces a print of grey tones between black and white in the visible part of the spectrum. These are by far the most common and cheapest form of aerial photography. Colour film produces a print of the visible part of the spectrum but in full natural colour. It is expensive but very useful in certain terrains. By contrast, colour infrared film records the green, red, and near infrared (to about 0.9 μm) parts of the spectrum. The dyes developed in each of these layers are yellow, magenta, and cyan. The result is a 'false colour' film in which blue areas in the image result from objects reflecting primarily green energy, green areas in the image result from objects reflecting primarily red energy, and red areas in the image result from objects reflecting primarily in the photographic near infrared portion of the spectrum. Vegetation reflects IR particularly well (Fig. 6.6) and so IR is used extensively where differences in vegetation may help in exploration. Aerial photographs are generally classified as either oblique or vertical.

OBLIQUE PHOTOGRAPHS

Oblique photographs can be either high angle oblique photographs which include a horizon or low angle oblique photographs which do not include a horizon. Oblique photographs are useful for obtaining a permanent record of cliffs and similar features which are difficult of access, and these photographs can be studied at leisure in the office. Similarly studies can be made of quarry faces to detect structural problems, to plan potential dam sites, etc.

VERTICAL PHOTOGRAPHS

Vertical photographs are those taken by a camera pointing vertically downward. A typical aerial photograph with tilting strip is shown in Fig. 6.7. The principal point is the point on the photograph that lies on the optical axis of the camera. It is found on a photograph by joining the fiducial marks. The title strip includes bubble balance, flight number, photograph number, date and time of the exposure, sun elevation, flight height and camera focal length.

Photographs are taken along flight lines (Fig. 6.8). Successive exposures are taken so that adjacent photographs overlap by about 60%. This is essential for stereoscopic coverage. Adjacent pairs of overlapping photographs are called stereopairs. To photograph a large block of ground a number of parallel flight lines are flown. These must overlap laterally to ensure that no area is left unphotographed. This area of sidelap is usually about 30%. Less sidelap is needed in flat areas than in mountainous terrain. Flight lines are recorded on a topographical map and drawn as a flight plan. Each principal point is shown and labelled on the flight plan. These are used for ordering photographs, from local and national Government agencies. At the start of a photogeological exercise, a photographic overview of the area is achieved by taking every alternate photograph from successive flight lines and roughly joining them to make a print laydown. The photographs are not cut, but simply trimmed to overlap and then laid down.

A slightly more sophisticated approach is to make an uncontrolled mosaic. For this the best fit approach is used and photographs are cut and matched but of course they are then unusable for other purposes. The advantage is that apparent topographical mis-matches and tonal changes are reduced to a minimum and one achieves a continuous photographic coverage. Unfortunately the scale varies across the mosaic and there are occasional areas of loss, duplication or breaks in topographical detail.

RC8-DM/22

Fig. 6.7 A typical B&W, vertical, aerial photograph with a titling strip showing date and time of the photograph, the level and sun elevation.

These imperfections can be corrected by making a controlled mosaic using scale-corrected, rectified, matched (tonal) prints. However this can be very expensive. When photographs are scale corrected, matched, and produced as a series with contours on them, they are known as orthophotographs.

6.4.1 Scale of aerial photographs

The amount of detail in an aerial photograph (resolution) is dependent upon the scale of the photograph. The simplest way to determine a photograph's scale is to compare the distance between any two points on the ground and on the photographs.

In vertical photographs taken over flat terrain, scale (S) is a function of the focal length (f) of the camera and the

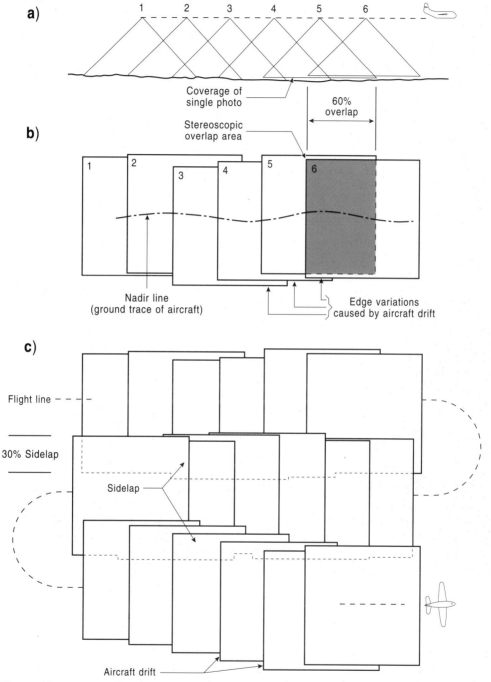

Fig. 6.8 Photographic coverage. (a) Showing the overlap of photographs during exposure. (b) A print laydown of the photographs from a single strip. (c) A print laydown of photographs from adjacent flight lines illustrating sidelap and drift. (Modified after Lillesand & Kiefer 1979.)

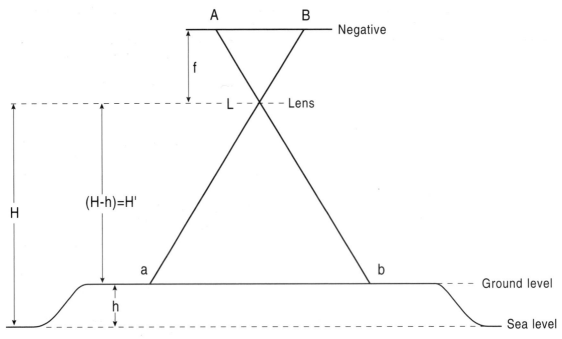

Fig. 6.9 The factors used to calculate the scale of aerial photographs. (Modified after Allum 1966.)

flying height above the ground (H') of the aircraft (Fig. 6.9).

$$\text{Scale (S)} = \frac{f}{H'}$$

H' is obtained by subtracting the terrain elevation (h) from the height of the aircraft above a datum (H), usually sea level which is the value given by the aircraft's altimeter. The important principle to understand is that photograph scale is a function of terrain elevation. A plane flies at a constant (or nearly constant) height. When a plane flies over varying terrain elevation, such as in mountainous areas, then the scale will vary rapidly across the photographs.

6.4.2 Parallax

All points on a topographical map are shown in their true relative horizontal positions, but points on a photograph taken over terrain of varying height are displaced from their actual relative position. This apparent displacement is known as parallax. Objects at a higher elevation lie closer to the camera and appear larger than similar objects at a lower elevation. The tops of objects are always displaced relative to their bases (Fig.6.10). This

distortion on aerial photographs is known as relief displacement and results in objects appearing to lean away radially from the principal point of a photograph (Fig. 6.7).

The effect of relief displacement is illustrated in Fig. 6.10 where the radial displacement differs on two adjacent photographs. It can be seen that the parallax of the base at the object is less than the parallax of the top of the object. Thus, a difference in elevation, in addition to producing radial displacement, also produces a difference in parallax. It is this difference in parallax which gives a three-dimensional effect when stereopairs are viewed stereoscopically. Allum (1966) gives a good description of relief and parallax.

6.4.3 Photographic resolution

The resolution of a photographic film is influenced by the following factors.

1 *Scale*, which has already been discussed.
2 *Resolving power of the film*. A 25 ISO film (a slow film) has a large number of silver halide grains per unit area and needs a long exposure time to obtain an image. A 400 ISO film (a fast film) has relatively fewer grains per unit area but requires less exposure time. The 25 ISO film has

Photograph 1

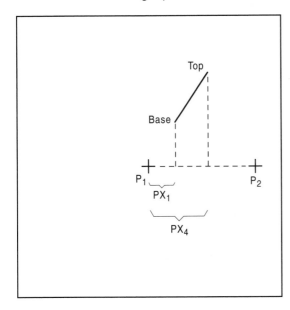

Photograph 2

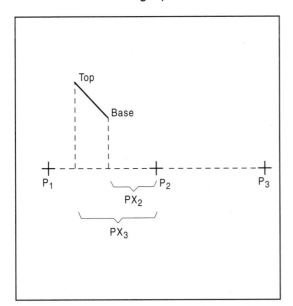

better resolution, while the 400 ISO film tends to be grainy.

3 *Resolving power of the camera lens.* Compare the results from a Nikon lens with those of a cheap camera lens.

4 *Uncompensated camera motion* during exposure.

5 *Atmospheric conditions.*

6 *Conditions of film processing.*

The effect of scale and resolution is expressed as ground resolution distance (GRD). Typically GRDs for panchromatic film vary from centimetres for low flown photography to about a metre for high flown photography.

6.4.4 **Problems with aerial photographs and flying**

1 *Drift*—edges of the photograph are parallel to flight line but the plane drifts off course (Fig. 6.8c).

2 *Scale*—not uniform because of parallax, terrain elevation, aircraft elevation (Fig. 6.11a) and the factors described below.

3 *Tilt*—front to back or side to side, causing distortions of the photograph (Figs. 6.11b & c)

4 *Yaw* (or crab)—the plane turns into wind to keep to the flight line, resulting in photographs whose edges are not parallel to the flight line (Fig. 6.11d).

5 *Platform velocity*—changing speed of the aircraft during film exposure (Fig. 6.11e). This applies particularly to satellite imagery.

6 *Earth rotation*—see section 6.2.3 (Fig. 6.11f).

6.4.5 **Photointerpretation equipment**

The two main pieces of photointerpretation equipment that exploration geologists use are (1) field stereoscopes capable of being taken into the field for quick field

---

Fig. 6.10 *(left)* Distortion on aerial photographs, where objects appear to lean away radially from the principal point (P_1 on photograph 1 and P_2 on photograph 2) of the photograph, is known as relief displacement. The principal points are joined to construct the base line of the photograph and a perpendicular line is dropped from the base of the object to the base line on each photograph. The distance from the principal point on photograph 1 to this intersection point is measured (PX_1). The same exercise is carried out on photograph 2 (PX_2). The parallax of the base of the object is given by the sum of PX_3 and PX_4. It is obvious that the parallax of the base is less than at the top. The top has greater parallax. (Modified after Allum 1966.)

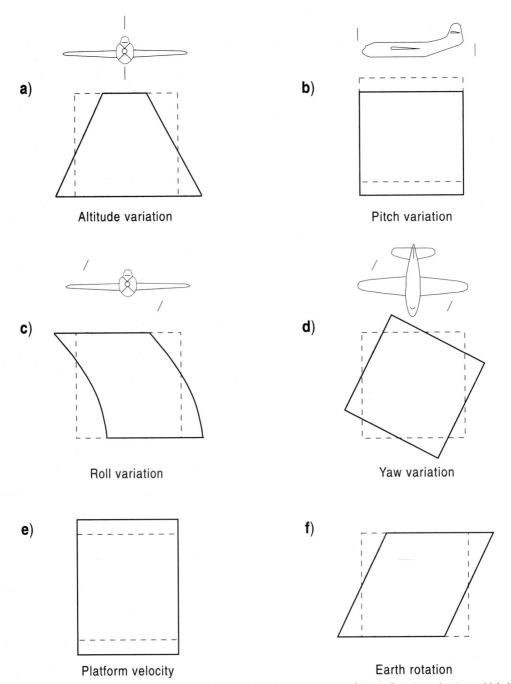

Fig. 6.11 Some of the problems that can arise as a result of variations in the movement of the platform (aeroplane) on which the imaging device is mounted. (Modified after Drury 1987.)

checking, and (2) mirror stereoscopes which are used mainly in the office and can view full 23 cm × 23 cm photographs without overlap. These pieces of equipment and their use are fully described by Moseley (1981) and Drury (1993).

Extra equipment which can be utilized in an exploration programme are a colour additive viewer and an electronic image analyzer. (1) A colour additive viewer colour codes and superimposes three multispectral photographs to generate a false colour composite. Multispectral photographs need three or four cameras taking simultaneous images in narrow spectral bands, e.g. 0.4–0.5, 0.5–0.6, 0.6–0.7 µm. Colour is far easier to interpret because the human eye can differentiate more colours than grey tones. (2) The electronic image analyser consists of a closed circuit TV scanner which scans a B&W image and produces a video digital image. The tape is then processed by computer and the results are shown on a TV monitor. This brings us back to the digital image processing described in section 6.2.6.

6.4.6 Elements of aerial photograph interpretation

Aerial photograph interpretation is based on a systematic observation and evaluation of key elements, and involves the identification from pairs of stereophotographs of relief, tonal and textural variations, drainage patterns and texture, erosion forms, vegetation and land use (Lillesand & Kieffer 1979). Drury (1993) includes additional characteristics of the surface such as shape (of familiar geological features), context and scale of a feature.

TOPOGRAPHY (Relief)

Each rock type generally has its own characteristic topographical form. There is often a distinct topographical change between two rock types, e.g. sandstone and shale. However, it is not an absolute quantity because a dyke may form a positive feature in certain climatic conditions but a negative feature elsewhere.

Vertical photographs with their 60% overlap result in an exaggeration of relief by three or four times. Consequently slopes and bedding dips appear steeper than they actually are. It is necessary to judge dips and group them into estimated categories, e.g. 0–5°, 6–10°, 11–25°, 26–45°, 46–85°, vertical. The vertical exaggeration works to the advantage of the photogeologist, as it may lead to the identification of subtle changes in slope in otherwise rounded and subdued topographical features.

TONE

Tone refers to the brightness at any point on a panchromatic photograph. Tone is affected by many factors such as: nature of the rock (sandstone is light, but shale is dark), light conditions at time of photographing (cloud, haze, sun angle), film, filters and film processing.

These effects mean that we are interpreting relative tone values. In general terms basic extrusive and intrusive igneous rocks (lava, dolerite) have a darker tone, while bedded sandstone, limestone, quartz schists, quartzite and acid igneous rocks are generally lighter. Mudstone, shale and slate have intermediate tones.

Subtle differences in rock colours are more readily detected using colour photographs, but these are more expensive. Subtle differences in soil moisture and vegetation vigour can be more readily detected using colour IR, but even these change with the time of the year.

TEXTURE

There is a large variation in apparent texture of the ground surface as seen on aerial photographs. Texture is often relative and subjective, but some examples are limestone areas which may be mottled or speckled, whilst shale is generally smooth, sandstone is blocky and granite is rounded.

DRAINAGE PATTERN

This indicates the bedrock type which in turn influences soil characteristics and site drainage conditions. The six most common drainage patterns are: dendritic, rectangular, trellis, radial, centripetal, deranged (Fig. 6.12).

Dendritic drainage (Fig. 6.12a) occurs on relatively homogenous material such as flat lying sedimentary rock and granite. Rectangular drainage (Fig. 6.12b) is a dendritic pattern modified by structure, such as a well jointed, flat lying, massive sandstone. Trellis drainage (Fig. 6.12c) has one dominant direction with subsidiary drainage at right angles. It occurs in areas of folded sedimentary rocks. Radial drainage (Fig. 6.12d) radiates outwards from a central area, typical of domes and volcanoes. Centripetal drainage (Fig. 6.12e) is the reverse of radial drainage where drainage is directed inwards. It occurs in limestone sinkholes, glacial kettle holes, volcanic craters and interior basins (e.g. Lake Eyre, Australia and Lake Chad). Deranged drainage (Fig. 6.12f) consists of disordered, short, aimlessly wandering streams typical of ablation till areas.

These are all destructional drainage patterns. There

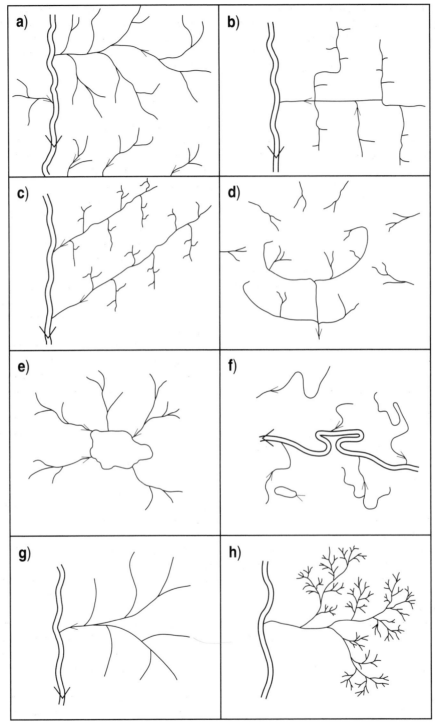

Fig. 6.12 Typical drainage patterns which can be used to interpret the underlying geology (a–f), see text for details. Drainage texture can be coarse (g) or fine (h). (Modified after Lillesand & Kiefer 1979.)

are numerous constructional landforms such as alluvial fans, deltas, glacial outwash plains and other superficial deposits. These are only indirectly of value in exploration studies (except for placer and sand and gravel exploration) and are described in detail by Siegal & Gillespie (1980) and Drury (1993).

DRAINAGE TEXTURE

Drainage texture is described as either coarse or fine (Fig. 6.12 g & h). The coarse texture develops on well drained soil and rock with little surface run off, e.g. limestone, chalk. Fine texture develops where soils and rocks have poor internal drainage and high run off, e.g. lava and shale.

EROSION

Erosion is a direct extension of the description of drainage above, but gullies etc. often follow lines of weakness and thus exaggerate features such as joints, fractures and faults.

VEGETATION AND LAND USE

The distribution of natural and cultivated vegetation often indicates differences in rock type, e.g. sandstone and shale may be cultivated, while dolerite is left as rough pasture. On the other hand forests may well obscure differences so great care must be taken to draw meaningful conclusions when annotating areas which have a dense vegetation cover. If an area is heavily forested, such as tropical jungles, then the use of conventional photography becomes limited and we have to look at alternative imagery such as reflected IR and side-looking airborne radar (SLAR) (Sabins 1987, Drury 1993).

Scanners with narrow apertures viewing the reflectance from vegetation (thematic mappers) are now being used (Goetz *et al.* 1983). One way in which all imagery can be used is by taking pictures of the same area at different times of the year (multitemporal photography). This may emphasize certain features at a given time of the year. Lines of vegetation, e.g. bushes, trees, etc., are a good indicator of fractures, faults, veins and joints. The joints are more porous and have more available ground water, producing lines of more vigorous vegetation.

LINEAMENTS

This is a word used in photogeology to describe any line on an aerial photograph that is structurally controlled by joints, fractures, faults, mineral veins, lithological horizons, rock boundaries, etc. These have already been discussed above, e.g. streams, gullies, lines of vegetation. It is important to differentiate these from fence lines, roads, rails, etc.

6.4.7 **Interpretation**

By using the elements of aerial photograph interpretation described above we identify landforms, man-made lines *vs.* photogeological lineaments, vegetation, rock outcrop boundaries, rock types and rock structures (folds, faults, fractures). Once identified, the geological units and structure are transferred on to a topographical base map.

Most near surface mineral deposits in accessible regions have been discovered. Emphasis is now on remote regions or deep-seated deposits. The advantages of using imagery in remote, inaccessible regions is obvious. Much information about potential areas with deep-seated mineralization can be provided by interpretation of surface features, e.g. possible deep-seated faults and/or fractures which may have been pathways for rising mineralizing solutions.

Aerial photographs are normally taken between midmorning and mid-afternoon when the sun is high and shadows have a minimal effect. Low sun angle photographs contain shadow areas which, in areas of low relief, can reveal subtle relief and textural patterns not normally visible in high sun elevation photographs. There are disadvantages in using low sun angles, especially in high relief areas, where loss of detail in shadow and rapid light change conditions occur very early and very late in the day.

Most exploration studies involve multi-image interpretation. Often Landsat images are examined first (at scales from 1:1 M, up to 1:250 000) for the regional view. If high altitude photography is available (1:120 000 to 1:60 000) then this may also be examined. Target areas are selected from the above studies and, in conjunction with literature reviews, it is then possible to select areas for detailed photogeological studies.

Aerial photographs are mainly used by various ordnance surveys for the production of topographic maps using photogrammetry. Photogrammetry is described in detail by Allum (1966). Some additional uses of aerial photography are regional geological mapping (1:36 000–1:70 000), detailed geological mapping (1:5 000–1:20 000), civil engineering for road sites, dam sites, etc., open pit management for monthly calculations of volumes extracted during mining, soil mapping, land use and land cover mapping, agricultural applications, forestry applications, water resources applications, e.g. irrigation, pollution, power, drinking water and flood damage, manufacturing and recreation, urban and regional planning, wetland mapping, wildlife ecology, archaeology and environmental impact assessment.

6.5 **SUMMARY**

In remote sensing we are concerned with the collection of information about an area without being in contact with it and this can be achieved using satellites or aeroplanes carrying electronic scanners or sensors or photographic and TV cameras.

There are two satellite systems producing data useful for mineral exploration. The American Landsat and the French SPOT. Each have their advantages (section 6.3). The first Landsat satellite was launched in 1972 and SPOT 1 in 1986. Satellites are solar powered and transmit data to their home station in digital form which enables the geologist to manipulate, combine and compare this data with geological, geochemical and geophysical data which has itself been expressed as digital images. Satellite imagery is used in structural investigations, in hydrogeology, to provide basic geological maps, to detect hydrothermal alteration and to produce maps of regional and local fracture patterns that may have con-

trolled mineralization or hydrocarbon accumulation.

Aerial photography has been used for much longer than satellite imagery and is important to the geologist for the production of topographical maps (photogrammetry) as well as in the making of geological maps (photogeology). In areas of good exposure aerial photographs yield much valuable geological information and even in areas of only 5% outcrop the amount of information they provide is often invaluable to the explorationist and some of this is discussed in section 6.4.6.

6.6 **FURTHER READING**

Good text books which cover most aspects of remote sensing in geology are Siegal & Gillespie (1980) and Drury (1993). The classic book about photogeology is Allum (1966), which despite its age is still worth reading. Detailed information on all aspects of remote sensing are covered by Colwell (1983). Recent developments in remote sensing appear monthly in the *International Journal of Remote Sensing* published by Taylor & Francis. A good introduction to digital image processing is given by Niblak (1986) and techniques for image processing and classification are explained well by Schowengerdt (1983).

7 Geophysical Methods

JOHN MILSOM

7.1 INTRODUCTION

Geophysical methods allow conclusions to be drawn about concealed geology on the basis of measurements of physical quantities made at or above the ground surface or, more rarely, in boreholes. Lines may have to be surveyed and cleared, heavy equipment may have to be brought on site and cables may have to be laid out, so geophysical work on the ground is normally rather slow. Airborne methods, on the other hand, provide the quickest, and often the most cost-effective, ways of obtaining information about large areas. In some cases, as at Elura in central New South Wales, the airborne indications have been so clear and definitive that the ground follow-up work could virtually be confined to defining drill sites (Emerson 1980).

For a geophysical technique to be useful, there must be a physical contrast between the various rocks in an area which is related, either directly or indirectly, to the presence of economically useful minerals. It will then be possible to record geophysical anomalies, which are departures seen on profiles or contours of the quantity being measured from a constant or slowly varying background value. Anomalies may take many different forms and need not necessarily be centred over their sources (Fig. 7.1). Ideally they will be produced by the actual ores but the existence of even a strong physical contrast between ore minerals and the surrounding rocks does not guarantee a significant anomaly. For example, the effect of gold in deposits suitable for large scale mining is negligible because of the very low concentrations and in such cases geophysicists have to rely on detecting associated minerals or, as in the use of seismic reflection to locate offshore placers, on defining favourable environments.

Geophysical methods can be divided into those which involve only measurements of naturally existing fields and those in which the response of the earth to some stimulus is observed. The first group includes measurements of magnetic and gravity fields, naturally occurring alpha and gamma radiation and natural electrical fields (static SP and magnetotellurics), the second includes all other electrical and electromagnetic methods, seismic methods and some downhole methods which use artificial radioactive sources. Before discussing any of these in detail, the general principles of airborne reconnaissance are considered. The final sections of this chapter are concerned with practicalities and particularly with the role of non-specialist geologists in geophysical work.

7.2 AIRBORNE SURVEYS

Magnetic, electromagnetic and gamma-ray measurements, which do not require physical contact with the ground, can be made from aircraft. Inevitably, there is some loss of sensitivity, since detectors are further from sources, but this may have the advantage of filtering out local effects. The main virtue of airborne work, however, is the speed of coverage of large areas. Surveys are often flown with several sensors and fixed or rotary wing aircraft may be used, flying either at constant altitude or at (nominally) constant terrain clearance. The latter is much the more common in mineral exploration. Aerial surveys require good navigational control, both at the time of the survey and later, when flight paths have to be 'recovered'. Traditionally, the pilot is guided by a navigator equipped with maps or photo-mosaics showing the planned locations of the lines. Low level navigation is not easy since even good landmarks may be visible for only a few seconds when flying only a few hundred metres above the ground. Course changes should not be made too often, since, although reflying is necessary if lines diverge too much, a line which is slightly out of position is preferable to one which continually twists and turns and is therefore difficult to plot precisely. The navigator's main job is to ensure that each line is at least begun in the right place.

A navigator's opinion of where he or she had been would be a very inadequate basis for a geophysical map. The tracking camera which provides the incontrovertible evidence records images either continuously or as overlapping frames on 35 mm film. Because of the generally small terrain clearance, a very wide-angle or fish-eye lens is used to give a sufficient, although distorted, field of view. Plotting is done directly from the negatives and even the most experienced plotters are

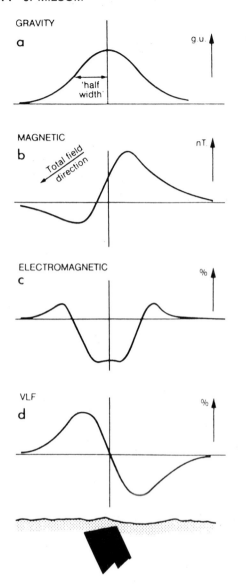

Fig. 7.1 Geophysical anomalies over a massive sulphide orebody. The electromagnetic anomaly (c) is the ratio of secondary to primary field, the VLF anomaly (d) is the ratio of vertical to horizontal magnetic field.

likely to misidentify one or two points in every hundred, so rigorous checking is needed. Fiducial numbers and marks printed on the film provide the essential cross reference between the flight path and the geophysical data, which are now usually recorded on magnetic tape.

Navigation in savannah areas may be very difficult, since one clump of trees looks much like another when seen sideways on, but plotting is often easy because tree patterns are quite distinctive when viewed from above. Visual navigation and flight-path recovery are difficult over jungles or deserts and are, of course, impossible over water. If there are no existing radio navigational aids, temporary beacons may be set up, but since several will be required and each will have to be manned and supplied, survey costs increase considerably. Self-contained Doppler or inertial systems which can be carried entirely within the aircraft have obvious advantages and were formerly widely used, but have now been almost entirely replaced by systems which utilize global positioning satellites (GPS). GPS navigation has also largely eliminated the need for tracking cameras, although these are still sometimes carried for back-up and verification, if not for primary flight-path recovery. The high accuracies with which velocities can now be estimated is also greatly improving the accuracy of airborne gravimetry, but GPS positioning may still not be sufficiently accurate for absolute positioning of a rapidly moving aircraft which is carrying out a detailed survey. Problems may also arise because of deliberate GPS signal degradation in response to military requirements.

Survey lines must not only be flown in the right places but at the right heights. Ground clearance is usually measured by radioaltimeter and is often displayed using limit lights on a head-up display. The extent to which the nominal clearance can be maintained depends on the terrain. Areas too rugged for fixed-wing aircraft may be flown by helicopter, but even then there may be lines that can only be flown downhill or are too dangerous to be attempted. Survey contracts define allowable deviations from specified heights and line separations, and lines normally have to be reflown if limits are exceeded for more than a specified distance but contracts also contain clauses emphasizing the paramount importance of safety.

The practical problems of survey navigation provide one very good argument for carrying magnetometers on all surveys. Electromagnetic or radiometric anomalies are often either rare or poorly defined and can easily be misplotted without the fact being noticed. Errors in plotting magnetic field anomalies are usually much more obvious. Parallax errors due to faulty synchronization between the navigation system and the geophysical records will also be apparent on magnetic maps as herring-bone patterns, provided that, as is usually the case, adjacent lines are flown in opposite directions (Fig. 7.2).

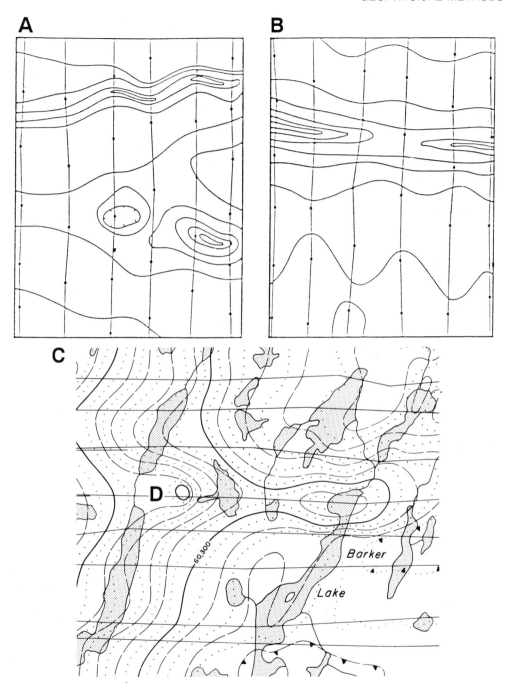

Fig. 7.2 Herringbone on magnetic maps. Adjacent lines were flown in opposite directions in all three cases. In (A) the effect is due to a time displacement (parallax) between the navigational and geophysical systems and is most obvious where magnetic gradients are steep. In (B) there is an uncompensated heading error which causes a difference in the base levels of lines flown in opposite directions. This is most obvious in areas of low magnetic relief. In (C), the dramatic elongation of contours along a single flight line at D is almost certainly an artefact, due either to a major error in baselining or an error in flight path recovery.

7.3 **MAGNETIC SURVEYS**

Magnetic surveys are the quickest, and often the cheapest, form of geophysics which can provide useful exploration information. Certain minerals, of which magnetite is by far the most common, produce easily detectable anomalies in the earth's magnetic field because the rocks containing them become magnetized. The magnetization is either temporary (induced) and in the same direction as the earth's field, or permanent (remanent) and fixed in direction with respect to the rock, regardless of folding or rotation (cf. Parkinson 1983).

Since magnetite is a very minor constituent of sediments, a magnetic map generally records the distribution of magnetic material in the crystalline basement beneath. Even sediments which contain magnetite have little effect on airborne sensors, partly because the fields from randomly oriented magnetite grains tend to cancel each other and partly because the fields due to thin, flat-lying sources decrease rapidly with height. Because the small pieces of iron scrap which are ubiquitous in populated areas strongly affect ground magnetometers, and also because ground coverage is slow, most magnetic work is done from the air. However there have been some significant triumphs using ground magnetic measurements—e.g. see section 13.5.2. Flying heights and line-spacings vary widely, but lines in mineral exploration may be as little as 200 m apart; a constant ground-clearance, which may be less than 200 m, is usually specified.

The earth's magnetic field is approximately that of a dipole located at the earth's centre and inclined at $10°$ to the spin axis. Distortions covering areas several hundreds of kilometres across can be regarded as due to a small number of subsidiary dipoles located at the core–mantle boundary. At the earth's magnetic poles the field is about 60 000 nT and vertical, whereas at the equator it is about 30 000 nT and horizontal (Fig. 7.3). The unit of magnetic field, the nanoTesla or nT, is commonly known to geophysicists as the gamma.

The slow variations in magnitude and direction of the earth's field which must be taken into account when comparing surveys made more than a few months apart are described, together with large-scale variations with latitude and longitude, by a complicated experimentally determined formula, the International Geomagnetic Reference Field or IGRF (Regan 1983). This is generally a reasonable approximation to the regional field in well surveyed areas where the control on its formulation was good, but may be unsatisfactory in remote areas. In addition to the long-term variations, there are daily cyclical (diurnal) changes, normally of the order of 20–60 nT in amplitude. Variation is low at night but the field begins to increase at about dawn, reaches a peak at about 10 : 00, declines rapidly to a minimum at about 16 : 00, then rises more slowly to the overnight value (Riddihough 1971). Although this pattern is repeated from day to day, the changes are not predictable in detail and must be determined by actual measurement. Diurnal effects are especially important in airborne surveys, which typically use precisions of 1 nT and contour intervals of 5 nT, and are monitored during such surveys by fixed ground magnetometers which provide permanent records on chart or magnetic tape. Corrections are made using these records directly or by adjustments based on systems of tie lines at right angles to the main lines of the survey, or by a combination of the two. Tie lines can be misleading if the intersections occur in regions of steep magnetic gradients.

During periods of sunspot activity, streams of high-energy protons strike the upper atmosphere and enormously increase the ionospheric currents, causing changes of sometimes hundreds of nT over times as short as twenty minutes. Corrections cannot be satisfactorily made for these magnetic storms (which are not related to meteorological storms and may occur when the weather is fine and clear) and work must be delayed until they are over.

For geophysical interpreters, regional changes in dip of the earth's main field are more important than changes in absolute magnitude, since bodies magnetized in the direction of the earth's field produce very different anomalies in areas of different dip (Fig. 7.4). Because of the bipolar nature of magnetic sources, all magnetic anomalies have both positive and negative parts and the net total flux will be zero. This is true even of an inductively magnetized source at one of the earth's magnetic poles, although the negative flux is then likely to be widely dispersed. Additional complications occur if, as is usual with strongly magnetic materials, there is significant permanent magnetization.

Magnetic anomalies are caused by maghemite (a form of hematite with the crystal form of magnetite), pyrrhotite and magnetite, magnetite being by far the commonest. Ordinary hematite, the most plentiful ore of iron, does not produce anomalies detectable in aeromagnetic surveys. Because all geologically important magnetic minerals lose their magnetic properties at about 600°C, a temperature reached near the base of the continental crust, local features on magnetic maps are virtually all of crustal origin. Magnetic field variations over sedimentary basins are often only a few nT in amplitude, but

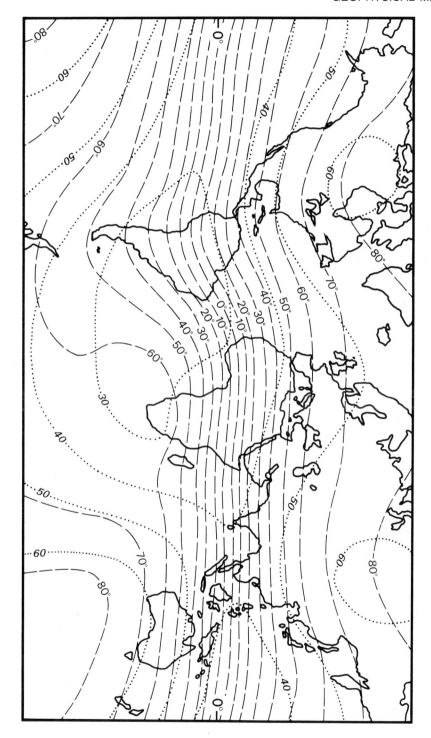

Fig. 7.3 Variations in dip and intensity of the earth's main magnetic field. Pecked lines are contours of dip angle, in degrees, dotted lines are contours of total magnetic field, in thousands of nanoTesla.

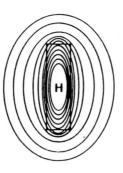

INCLINATION 90° (MAGNETIC POLE)

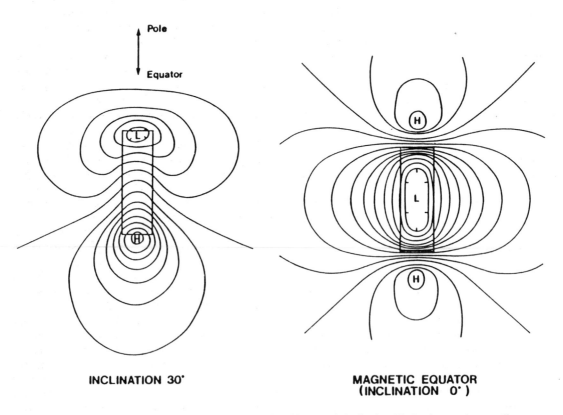

INCLINATION 30°

MAGNETIC EQUATOR
(INCLINATION 0°)

Fig. 7.4 Variations in the shapes of total magnetic field anomalies with magnetic inclination. The local magnetic anomalies are produced by a prism source magnetized by induction only.

changes of hundreds and even thousands of nT are common in areas of exposed basement. The largest anomalies known reach to more than 150 000 nT, several times the strength of the earth's normal field.

Magnetic maps are often interpreted qualitatively since they contain a wealth of information about rock types and structural trends which can be appreciated without mathematical analysis (cf. Fig 7.19), and automatic (e.g. Euler deconvolution) and image processing methods are increasingly being used in analysing magnetic data (cf. Fraser and Fogaresi 1993). Depths, shapes and magnetization intensities of sources of individual anomalies can be estimated, but it would seldom be useful, or even practicable, to do this for all the anomalies in an area of basement outcrop. In particular, magnetization intensities are not directly related to any economically important parameter, even for magnetite ores, and are therefore rarely of interest.

Magnetic surveys are not only amongst the most useful types of airborne work but are also, because of the low weight and simplicity of the equipment, the cheapest. The standard instrument, the proton magnetometer, is self-orienting and can be mounted either on the aircraft or in a towed bird. Sensors on aircraft are usually mounted in specially constructed non-magnetic booms or stingers placed as far from the main aircraft sources of magnetic field as possible (Fig. 7.5). The aircraft field will vary slightly with heading and is usually compensated by a system of coils and permanent magnets. The magnitude of any heading error will need to be checked on a regular basis.

7.4 GRAVITY METHOD

The gravity field at the surface of the earth is influenced, to a very minor extent, by density variations in the rocks near the surface. Densities range from less than 2.0 $Mg\ m^{-3}$ for soft sediments and coals to more than 3.0 $Mg\ m^{-3}$ for mafic and ultramafic rocks. Many ore minerals, particularly metal sulphides and oxides, are very much denser than the minerals which make up the bulk of most rocks, and orebodies are thus likely to be denser than their surroundings. The actual effects on the gravity field are minute, amounting to generally less than one part per million (one milligal) even in the case of large massive sulphide deposits and being often concealed by variations in regional background (Fig. 7.6). Gravity meters must therefore be extremely sensitive, a requirement which to some extent conflicts with the need for them also to be rugged and field-worthy. They measure only gravity differences and are subject to drift, so that

surveys involve repeated references to base stations. Airborne gravity is still in its infancy and much less accurate than ground survey but is already being used for some regional studies. Surface gravity methods have played an important part in unravelling the stratigraphy and structure of mineralized areas—e.g. see section 13.5.3.

The observed gravity field is affected by the distance of the measuring point from the earth's centre, and so by its latitude (because of polar flattening) and elevation. The effect of a height change of only 5 cm is detectable and corrections for height must be made on all surveys. Applying the free-air correction, which allows for height differences, and the Bouguer correction, which approximates the effect of the rock mass between the station and the reference surface, to the latitude-corrected gravity value produces a quantity known as the Bouguer anomaly. Provided the density of the topography has been accurately estimated, a Bouguer anomaly map is usually a good guide to the effect of subsurface geology on the gravity field but additional corrections are needed in rugged terrain. In some cases these are so large, and subject to such large errors, that there is little chance of detecting the small anomalies produced by local exploration targets.

Geophysical interpretations are notoriously ambiguous but the gravity method does at least provide, in theory, a unique and unambiguous answer to one exploration question. If an anomaly is fully defined over the ground surface, the total gravitational flux is proportional to the total excess mass of the source body. Any errors or uncertainties are due to the difficulty, under most circumstances, of defining background, and are not implicit in the calculations. The true total mass of the body is obtained by adding in the mass of an equivalent volume of country rock, so some knowledge of source and country-rock density is needed.

Very much less can be deduced about either the shape of the source or its depth and all estimates are subject to a fundamental ambiguity which no amount of more detailed survey or increased precision can remove. The peak of the anomaly is generally located over the centre of the source (which is not the case with many other geophysical methods), but the depth below surface is not easily determined. Rule of thumb methods rely on the rough general relationship between the lateral extent of an anomaly and the depth of its source (cf. Milsom, 1989). Since field lines diverge from any source, a deeper source will, other things being equal, give rise to a broader (and flatter) anomaly. Strictly speaking, only homogeneous spherical bodies can be approximated by

Fig. 7.5 Airborne survey aircraft tail member, showing attached stinger within which a magnetometer sensor may be mounted.

point sources, and the width of a real anomaly obviously depends on the width of its source, but the rule does give a rough guide to the depth of the centre of the body in many cases.

Quantitive interpretations are usually made by entering geological models into a computer and modifying them until there is an acceptable degree of fit between the observed and calculated fields. Several microcomputer program packages based on algorithms published by Talwani *et al.* (1959) and Cady (1980) are now available for modelling fields due to 2D bodies (of constant cross-section and infinite strike extent) and to the so-called 2½D bodies which are limited in strike length.

7.5 RADIOMETRICS

Alpha particles, which consist of two neutrons and two protons bound together, beta particles, which are high-energy electrons, and gamma rays (very high-frequency electromagnetic waves which quantum theory allows us to treat as high-energy particles) are all produced in natural radioactive decay. Particulate radiations are screened out by one or two centimetres of solid rock, so a little transported soil may conceal the effects of mineralization. Gamma rays are the most useful in exploration because of their relatively long range (1–2 m) in solid matter.

The traditional picture of a radiometric survey is of the bearded prospector, with or without donkey, plodding through the desert and listening hopefully to the clicking noises coming from a box on his hip. The Geiger

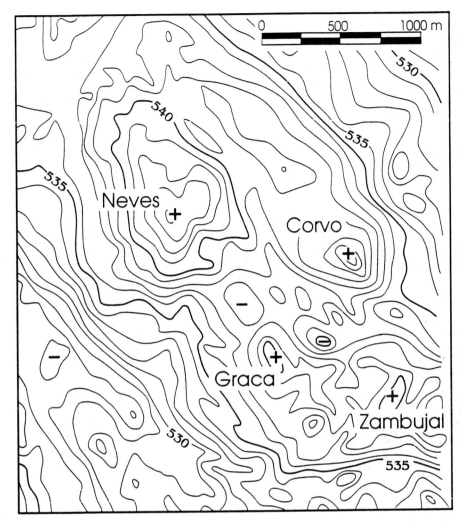

Fig. 7.6 Bouguer anomalies of massive pyrite orebodies. A massive sulphide orebody was discovered on the basis of gravity data in 1973 in the Neves region in the Portuguese section of the Iberian Pyrite Belt. The exploration programme was guided by gravity maps produced by the Société Fomento Mineiro and later extended by the exploration consortium. The anomalies are superimposed on a rapidly varying regional background field and very careful analysis of the Bouguer anomaly map, contoured here at 0.1 mGal intervals, is required in order to identify features likely to be associated with mineralization. (Map drawn after Leca 1990.)

counter he used was sensitive only to alpha particles and is now obsolete, but tradition dies hard. Modern scintillometers, which detect gamma rays, may have dials or digital readouts but ground instruments can usually also be set to click in areas of high radioactivity. Radiometric methods are geophysical oddities because measurements are of count rates which are subject to statistical rules rather than of fields with definite, even if diurnally variable, values. Survey procedures must strike a balance between speed of coverage, dictated by

economics, and the high precision obtainable by observing for long periods at each station. Statistics can be improved by using very large detector crystals; these are essential for airborne work but their cost rises dramatically with size. The slow speeds needed to obtain statistically valid counts from the air generally require helicopter installations (Fig. 7.7).

Sources of radiation can be distinguished if a gamma ray spectrometer is used to estimate the energy of each detected photon. Terrestrial radiation may come from

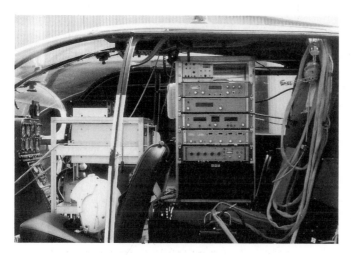

Fig. 7.7 Helicopter installation for radiometric survey, from back and front. The instrumentation includes scintillometer electronics, magnetometer and magnetic tape recorder.

the decay of ^{40}K, which makes up about 0.01% of natural potassium, or from thorium and uranium isotopes or their daughters, the latter being isotopes of lighter elements formed during decay but themselves unstable and producing further offspring. Photons with energies above 2.7 MeV must be of extraterrestrial (usually solar) origin but a 2.615 MeV peak is produced by a thorium daughter. The most prominent uranium (actually ^{214}Bi) peak is at 1.76 MeV and the single potassium peak is at 1.46 MeV. These and other peaks are superimposed on a background of scattered radiation but the relative importance of the three main radioelements in any source can be estimated by analysing the spectrum using rather simple techniques. Discrimination is important, as was shown by the discovery of the Yeeleerie uranium deposit in Western Australia in a salt-lake environment where the radiometric anomaly had initially been attributed to potash in the evaporites. It is normal to use three windows centred on the energy peaks and also to record total count. Corrections can be made for the effects of thorium radiation in the uranium window and for the effect of both thorium and uranium in the potassium window. The main radiometric exploration target is uranium but unfortunately the strongest and most diagnostic radiation is produced by the decay of the late stage daughter, ^{214}Bi, which lies below radon

in the decay chain. Radon is a gas and the isotope involved has a half-life of several days, allowing considerable dispersion from the primary source and giving rise to interpretational difficulties.

The current unpopularity of nuclear power has made exploration for uranium much less attractive and the importance of radiometric methods has declined accordingly. There are applications in general geological mapping, in the search for alteration zones in granite and in the search for, and evaluation of, phosphate and placer deposits, but these are all of rather minor importance. Most radiometric work being undertaken at present is for public health purposes and in such surveys techniques using alpha cups and alpha cards, which monitor alpha particles from radon gas, may be as useful as those involving detection of gamma rays.

7.6 RESISTIVITY

Electrical properties can be characterized by resistivity, the resistance of a metre cube. Most rock-forming minerals are insulators and rock resistivities, which vary widely but are normally within the range from 0.1 to 1000 ohm-metres, are generally determined by porosity and by the salinity of the pore waters. Clay minerals are electrically polarized and enhance the conductivity of

(a) Wenner

$$\rho_a = 2\pi a \frac{V}{I}$$

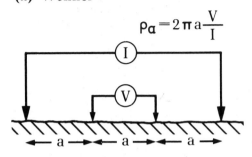

(b) Dipole-dipole

$$\rho_a = \pi n(n+1)(n+2)\, a \frac{V}{I}$$

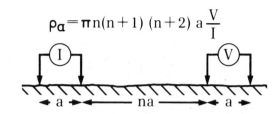

(c) Schlumberger

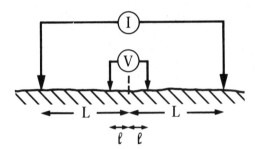

(d) Gradient

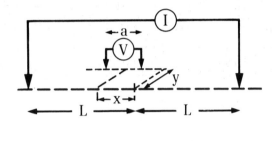

Fig. 7.8 Common electrode arrays. The geometric factors required to convert Wenner and dipole–dipole readings to apparent resistivity are shown. The Schlumberger factor is $\rho L^2/2l$, provided that L is much larger than l, and the gradient array factor is a complicated function of L, x, y and a.

wet rocks. A few minerals, notably graphite and the base metal sulphides (except sphalerite), conduct by electron flow and can reduce rock resistivity to very low values if present in significant amounts, but straightforward measurement of d.c. resistivity is seldom used alone in the search for base metals. Resistivity measurements can also indicate overburden thickness and the extent of deposits of various bulk minerals. Clean, and therefore resistive, gravels are easily located in clays and, conversely, china clays in granites or deposited as ball clays in bedrock depressions can be investigated directly.

The two requirements for any resistivity survey are the introduction of current and the detection of voltage but, because of contact effects at electrodes, subsurface resistivity cannot be estimated using the same electrodes for both purposes. Two current and two voltage electrodes are used in arrays which are usually, but not necessarily, linear. For convenience, current is usually supplied through the outer electrodes but the calculations would be identical were the two inner electrodes to be used for this purpose. Power may be provided by

generators or re-chargeable batteries. Separate generators and voltmeters may be used or a single unit which produces a resistance value directly. To reduce polarization effects, and also to compensate for any natural currents and voltages, the current is usually reversed at intervals of the order of a second. Polarization effects may be further reduced by using non-polarizing electrodes consisting of porous pots filled with saturated copper sulphate solution and containing copper electrodes. Contact with the ground is made by solution leaking through the base of the pot. Such electrodes are messy and inconvenient and are used only when, as in induced polarization surveys, they are absolutely essential.

For studying lateral variations in resistivity, the Wenner and gradient arrays (Fig. 7.8a,d) are the most convenient. In Wenner traversing, the leading electrode is moved one electrode space along the line for each new station and each following electrode is moved into the place vacated by its neighbour. With the gradient array, which requires very powerful current generators, the outer electrodes are fixed and far apart and the inner electrodes, separated by only a few metres, are moved together. The bulk apparent resistivity of the ground is obtained by dividing the voltage reading by the current reading and multiplying by a geometrical factor.

Resistivity surveys are also used to investigate interfaces, such as the water table or a bedrock surface, which are approximately horizontal. Current can be sent progressively deeper into the ground by moving electrodes further apart. The Wenner array is often used, as is the Schlumberger array, which is symmetrical but has outer electrodes very much further apart than the inner ones (Fig. 7.8c). On expansion, all four Wenner electrodes must be moved but with the Schlumberger array, within limits, the inner electrodes can be left in the same positions. To interpret the results, from either array, apparent resistivity is plotted against array expansion on log–log paper and compared with type curves. A single sheet of curves is sufficient if only two layers are present but a book is needed for three layers and a library for four! Multiple layers are usually interpreted either by matching successive curve segments using two-layer curves and auxiliary curves, which constrain the two-layer curve positions, or by computer modelling.

7.7 SPONTANEOUS POLARIZATION (SP)

Natural currents and natural potentials exist and have exploration significance. In particular, sulphide orebodies may produce negative anomalies of several hundred millivolts. It was originally thought that these potentials were maintained by oxidation of the ore itself, but it is now generally agreed that the ore acts as a passive conductor, focussing currents associated with oxidation–reduction reactions which occur at the water table. If a potential is to be produced, the orebody must straddle the water table and the method, although quick and simple, requiring only a high-impedance voltmeter, some cables and a pair of non-polarizing electrodes, is often rejected because economically exploitable bodies do not necessarily produce anomalies.

7.8 INDUCED POLARIZATION (IP)

Flow of electric current in the ground causes parts of the rock mass to become electrically polarized. The effect is almost negligible in sandstones, quite marked in clays, in which small pore spaces and electrically active surfaces impede ionic flow, and can be very strong at the surfaces of electronic conductors such as graphite and metallic sulphides (Sumner 1976). If current flow ceases,

the polarization cells discharge, causing a brief flow in the reverse direction. The ways in which this effect can be measured can be illustrated by considering a simple square wave.

A square-wave voltage applied to the ground will cause currents to flow, but the current waveform will not be perfectly square and will, moreover, be different in different parts of the subsurface. Its shape in any given area can be monitored by observing the potential between two appropriately placed non-polarizing electrodes. Polarization delays attainment of the steady voltage, V_o, for some time after current begins to flow. When the applied voltage is cut off the voltage measured drops rapidly to a much lower value, V_p, and then decays slowly to zero (Fig. 7.9A). If, as in Fig 7.9B, the voltage is applied for a very short time only, V_o is never reached and, since resistance is proportional to voltage in any of the array formulae, a low apparent resistivity is calculated. Induced polarization (IP) effects can thus be observed either in time, by measuring decay voltages, or in frequency, by measuring resistivity at different frequencies. The asymmetry of the observed voltage curve also implies a phase difference between it and the primary voltage, and advances in solid-state circuitry and precise crystal-controlled clocks now allow this difference to be measured and plotted as a function of frequency (cf. Hallof 1993).

Time and frequency techniques both have their supporters. Time–domain work tends to require dangerously high voltage and current levels, and correspondingly powerful and bulky generators. Frequency–domain equipment can be lighter and more portable but the measurements are more vulnerable to electromagnetic noise. Phase measurements, which span a range of frequencies, allow electromagnetic coupling to be estimated and eliminated, but their use adds complexity to field operations. Coupling can be significant in IP work of any type and will be especially severe if cables carrying current pass close to cables connected to voltage electrodes. Keeping cables apart is easy with the dipole–dipole array (Fig. 7.8b), which is therefore very popular.

Time–domain IP effects are defined in terms of chargeability, theoretically defined as the ratio of V_p to V_o. In frequency work, the per cent frequency effect (PFE) is obtained by dividing the difference between the d.c. and high frequency resistivities by the high frequency resistivity. Electromagnetic noise and practical difficulties prevent either of these theoretical quantities being measured and time–domain chargeability is defined in terms of voltages observed after one or more preset delay

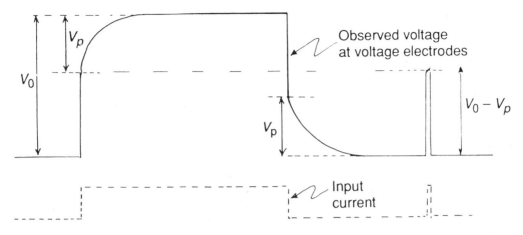

Fig. 7.9 Effects of ground polarization on long period (A) and short period (B) square waves. Note that in the case of the short period wave, the observed voltage can rise only a little way above its initial level before the applied voltage is reversed.

times. It is expressed in millivolts per volt or, if areas under the decay curve are measured, in milliseconds. PFEs are calculated from resistivities measured at two frequencies which are generally a few Hz and a few tenths of a Hz. Because the details of measurement are arbitrary, results obtained with different makes of equipment can usually not be quantitatively related and even different instruments of the same make may give slightly different results.

Low frequency or d.c. resistivity is necessarily obtained in frequency IP and can be calculated in time–domain work provided that the input current is recorded. IP surveys are thus also resistivity surveys. In the frequency–domain a metal factor is often calculated by dividing the PFE by the resistivity and, since this would be a rather small quantity, multiplying by a factor of the order of ten thousand. Metal factors emphasize regions which are both chargeable and conductive and can thus be useful in detecting massive sulphides, but may very effectively conceal orebodies of other types. A clear chargeability anomaly produced by disseminated copper ore can be completely masked on a metal factor plot by the effects of an adjacent conductive alteration zone. Resistivity and IP data should each be respected as providing useful information and should not be confused into a single composite quantity.

Gradient arrays are often used for IP reconnaissance, with results presented as profiles. Dipole–dipole results may also be presented in this way, with separate profiles for different values of n, the ratio of the inter-dipole to intra-dipole distance. Increasing n increases the penetration, and an alternative form of presentation, the

pseudo-section is often used (Fig. 7.10). Pseudo-sections give some indication of the depths of chargeable or conductive bodies but can be very misleading. For example, the classic 'pant's leg' anomaly, apparently due to a body dipping both ways at 45° from the surface, may actually be caused by a small source near to one specific dipole position which makes every measurement which uses that position anomalous. Indications of dip on pseudo-sections are notoriously ambiguous and it is quite possible for the apparent dip of a body on the section to be in the opposite direction to its actual dip. Model studies can provide guides to the location of a chargeable mass but it may still require several drill holes to actually locate it.

IP is important in mineral exploration because it depends on the surface area presented by the conductive mineral grains rather than on their connectivity and is especially sensitive to disseminated mineralization which may produce no resistivity anomaly. It suffers from the same drawbacks as other electrical methods, in responding to barren pyrite (IP has, rather unkindly, been said to stand for 'indicator of pyrite') but not to sphalerite. Galena also may produce little in the way of an anomaly, especially in very fine-grained deposits. There is now considerable interest in the possibility of using phase/ frequency plots to discriminate between different types of chargeable materials. It seems that the curve shapes are dictated by grain size rather than actual mineral type but even this may allow sulphide mineralization to be distinguished from graphite because of its generally smaller grains.

In theory, a solid mass of conductive sulphides will

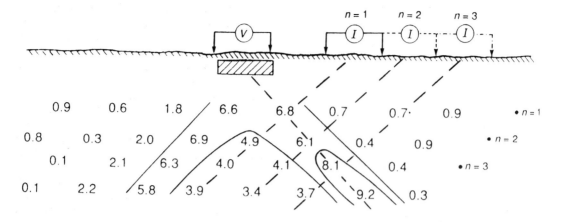

Fig. 7.10 Dipole–dipole "Pant's-leg" anomaly due to a small shallow source. All readings with either dipole immediately above the body will record high chargeability. The pecked lines at 45° illustrate the principles involved in plotting values on the pseudo-section.

give a negligible IP anomaly but real massive conductive ores are always sufficiently complex to respond well. Since both massive and disseminated deposits can be detected, IP is very widely used, even though it is rather slow, requires moderately large field crews and is consequently rather expensive.

7.9 ELECTROMAGNETIC METHODS

Electrical current flowing in a circuit produces a magnetic field. Conversely, voltages will be induced in any conductor exposed to a changing magnetic field, and current will flow in any closed circuit. This interaction between adjacent circuits is termed electromagnetic induction. The self-inductance of a circuit determines the extent to which changes in current flow will be resisted. Figure 7.11 shows, in schematic form, an electromagnetic (e.m.) prospecting system as used in exploration for massive sulphide orebodies. Because the magnetic field from the transmitter cuts the plane of the receiver coil at right angles, the system is said to be maximum-coupled.

Eddy currents induced in geological conductors are rather unpredictable, but the basic principles involved can be illustrated by the currents induced in a simple conducting loop, a rough analogue of a narrow, steeply dipping conductor. The response is determined by a quantity known as the response parameter, which for a loop is equal to the frequency multiplied by the self

inductance and divided by the resistance. Similar but more complex relationships apply to real solid bodies and magnetic permeability also appears in the equations (Grant and West 1965).

The total response of an e.m. system is determined by the response function, shown plotted against response parameter in Fig. 7.12, which demonstrates how an anomaly can be expected to vary either with frequency or with the conductivity of the target body. There is a phase difference between the primary and secondary fields at the receiver, the out-of-phase (quadrature) signal being small at high resistivities, because eddy currents are small, and over good conductors, where eddy currents are almost in phase with the transmitted field. At or near the point where the quadrature and in-phase responses are equal, the quadrature signal reaches a maximum; the in-phase/quadrature ratio provides crucial information on conductivity. Readings at a single frequency would be sufficient in the absence of noise, but several frequencies may be used to improve the chances of distinguishing signal from noise. The depth of penetration of an electromagnetic system is determined in most circumstances by the source–receiver spacing but there is also an effect associated with the frequency-dependent attenuation of the electromagnetic field in a conducting medium. This is conveniently expressed in terms of the 'skin depth', in which the field strength falls to about one-third of its surface value.

The eddy currents circulating in each part of a conduc-

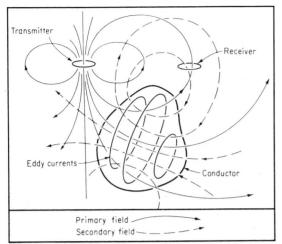

Transmitter

Receiver

Eddy currents

Conductor

Primary field
Secondary field

Fig. 7.11 Schematic of a horizontal-loop electromagnetic prospecting system. The eddy-currents induced in the conductor produce an alternating magnetic field which opposes the primary (transmitted) field. (Reproduced from F.S. Grant and G.F. West, Interpretation Theory in Applied Geophysics, McGraw-Hill 1965, with the permission of McGraw-Hill Inc.)

tor produce fields which act on every other part and computed solutions, except in very simple cases, have to be obtained by successive approximations. Considerable computer power is needed, even for simple two-dimensional bodies, and three-dimensional studies are only rarely attempted. An alternative approach uses physical scale models. A valid model can be achieved if the skin depths in the model and in the real geology are in the same ratio as the lateral dimensions. In principle the permeabilities and dielectric constants could be scaled but if only the conductivities and lateral dimensions are varied, field instruments can be used in the laboratory, coupled to miniature coils and transmitting loops. Aluminium sheets often model conductive overburden and copper sheets mimic sulphide orebodies. Physical models are convenient for three-dimensional studies, being easily configured to represent the complicated combinations of orebodies and formational conductors found in the real environment.

Electromagnetic surveys use either continuous, usually sinusoidal, waves or transients. Sources are usually small coils through which alternating electric current is passed at frequencies somewhere in the range from 200

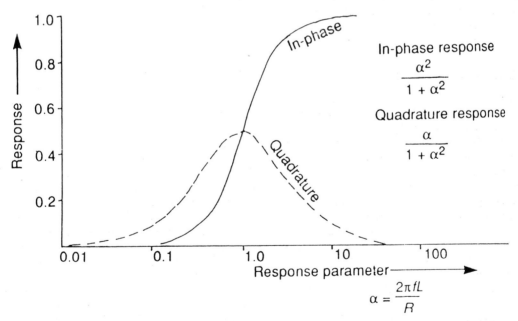

In-phase response

$$\frac{\alpha^2}{1 + \alpha^2}$$

Quadrature response

$$\frac{\alpha}{1 + \alpha^2}$$

$$\alpha = \frac{2\pi fL}{R}$$

Fig. 7.12 Response recorded by a horizontal-loop e.m. prospecting system over a vertical conducting loop. The size of the anomaly is determined by the response parameter which is a function of the frequency, f, the self-inductance, L, and the resistance, R, of the loop. Natural conductors respond in more complicated but generally similar ways.

to 4000 Hz. Field strength from such sources falls off as the inverse cube of distance. Other methods use sources in the form of long, grounded, current-carrying wires or extended current loops with dimensions comparable to the areas being surveyed. Receivers are almost always small coils. Physical contact with the ground is not required and the e.m. method can therefore be used from aircraft. In early systems, transmitter coils were mounted on the aircraft and the receiver coils were towed behind in aerodynamic birds, avoiding the problems due to variation in primary field at the receiver by recording only the out-of-phase signals. There was considerable argument as to whether conductors exist in nature which are so good that quadrature anomalies are negligible. The question never seems to have been formally resolved, there being commercial vested interests on both sides, but it does seem to be generally agreed that some good targets might produce only very small phase shifts. In any case, the in-phase to quadrature ratio is an important interpretation parameter.

Transmitter and receiver coils are now usually rigidly mounted either on the aircraft frame or in a single large towed bird. Birds are now used almost entirely from helicopters (Fig 7.13), as civil aviation authorites enforce increasingly stringent safety regulations on fixed-wing aircraft. Traditional airborne systems used the vertical-loop coaxial maximum-coupled coil configuration, but multi-coil systems have become more popular in recent years. Because the source–receiver separation in airborne systems is inevitably small compared to the distance to the conducting bodies, anomalies generally amount to only a few parts per million of the primary field. However rigidly the coils are mounted, some degree of flexure always remains and produces noise of similar amplitudes. Flexure in some systems is monitored electronically and corrections are made automatically.

7.10 **TRANSIENT ELECTROMAGNETICS (TEM)**

One solution to the problem of variable source–receiver coupling is to work in the time-domain. A magnetic field can be suddenly collapsed by cutting off the current producing it, and the magnetic effects of the currents induced in ground conductors by this abrupt change can be observed at a time when no primary field exists. This method was first patented for airborne use by Barringer Research of Toronto as the INduced PUlse Transient system (INPUT). The transmitter coil in this system is usually rigidly attached, and may extend, in the case of

Fig. 7.13 Helicopter electromagnetic system. Transmitter and receiver coils are mounted at opposite ends of the towed bird. The probe at the front of the aircraft is a VLF sensor (see section 7.11) and a scintillometer crystal (section 7.5) is mounted on the external tray.

a fixed-wing aircraft, from the tailplane to the nose via the wingtips. The receiver coil is normally mounted in a towed bird (Fig. 7.14); the variability in coupling to the primary field does not matter since this is zero when measurements are being made. The magnetic fields associated with the eddy currents were originally observed at four delay-times but in later systems, and in successor systems such as GEOTEM, the number of channels has been increased. Eddy currents in the relatively poor ionic conductors die away quite rapidly but persist in massive sulphide orebodies (and often in graphitic shales) for much longer periods, giving five- or six-channel anomalies (Fig. 7.15).

The success of INPUT delayed development of transient systems for use on the ground for some years except

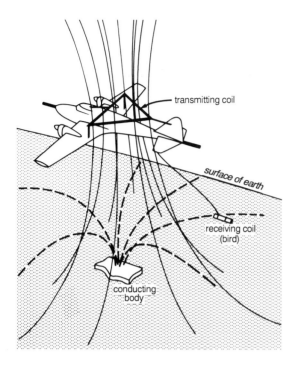

Fig. 7.14 Airborne transient electromagnetic system, showing one method of mounting the transmitter coil. Solid lines indicate primary field and pecked lines indicate secondary field. Note that with this system the secondary field is measured only after the primary field has been turned off.

in the USSR, but in 1968 a Russian instrument was imported into Australia by the Western Mining Corporation for surveys in the regions of high-conductivity overburden around Kalgoorlie. This was followed by the development of the SIROTEM by an Australian Government research agency. At about the same time the Crone PEM system was developed independently in North America and another system, the EM37, was produced in Canada by Geonics. A different approach to transients is found in UTEM systems (West *et al.* 1984), which use transmitter currents with precisely triangular waveforms. In the absence of ground conductivity, the received signal, proportional to the time-derivative of the magnetic field, is a square-wave. Distortions are observed by sampling at eight separate time intervals.

Since transient measurements are made after current has ceased to flow in the transmitter loop, this loop can also be used to receive the signal. Even if separate loops are used, the absence of primary field at measurement time allows the receiver to be placed within the transmitter loop, which would be impractical using continu-

ous waves because of the very strong coupling to the primary field. The effects on transients of inhomogeneities in ground conductivity, and particularly of highly conductive ores, are understandably complex, and physical models have so far proved more useful aids to interpretation than computers. However, published model studies of any kind are still rather scarce.

The Fourier theorem states that any transient can be regarded as the sum of a large number of sine and cosine waves, and transient methods can be viewed as multi-frequency continuous wave surveys. Although first developed almost entirely for the detection of massive sulphide ores, their multi-frequency character allows preparation of conductivity–depth sections which can be used in studies of porphyry systems, in detection of stratiform orebodies and even in the development of water resources. Transient methods are also superficially similar to time–domain IP, and in IP also there is a time–frequency duality. The most obvious difference between the e.m. and IP methods is that currents are applied directly to the ground in IP surveys whereas in most e.m. work they are induced. More fundamentally, time–domain IP systems sample at delay times of from 0.1 to 2 sec, the initial delay being introduced specifically to avoid TEM effects. Although some modern IP units are designed to work over a wide range of frequencies to obtain conductivity spectra, it is quite possible to avoid working in regions, of either frequency or time–delay, where e.m. and IP effects are both significant.

7.11 VLF AND MAGNETOTELLURICS

Life would be much easier for e.m. field crews if they had only to measure field strength and did not also have to transmit primary signals. Attempts have therefore been made to use virtually all the various forms of background electromagnetic radiations for geophysical purposes. The most successful instruments use military transmissions in the 15–25 kHz range (termed Very Low Frequency or VLF by radio engineers, although very high for geophysicists) or natural radiation in the 10 Hz to 20 kHz range produced by thunderstorms (audio-frequency magnetotellurics or AMT).

In both AMT and VLF work, the electromagnetic wavefront is considered to be essentially planar, and in neither is it possible to measure phase differences between primary and secondary radiation. However, differences between the phases of various magnetic and electrical components of the waves can be measured and used diagnostically. The magnetic components of the

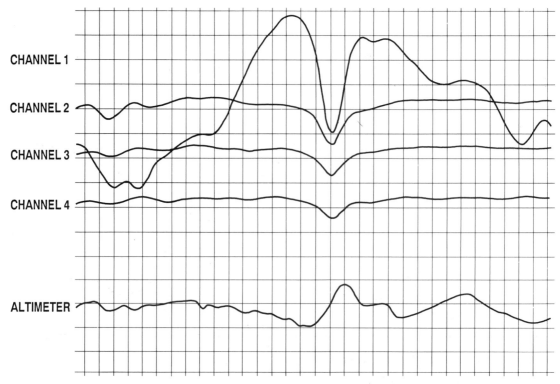

Fig. 7.15 INPUT Mark II anomaly over the Timmins massive sulphide deposit. Note that the anomaly on the right, which is due to the orebody, persists through all four channels, whereas the large anomaly on the left, due to overburden conductors, is virtually confined to Channel 1.

VLF field are generally most important if the conductors are relatively good and dip steeply (Fig. 7.16), the electrical components provide information on flat-lying and, particularly, surficial conductors. Because of the high frequencies used, even moderate conductors respond strongly provided they are large enough and in directions which ensure good coupling to the primary field. The VLF method is especially useful in locating steeply dipping shear zones.

Magnetotellurics offer a wide frequency range but poor consistency in source direction and power, and there has been increased interest in using local artificial sources which cover the same frequency range. In controlled-source magnetotellurics (CSAMT), the source is usually a grounded wire several kilometres long. It should be located far enough from the survey area for plane-wave approximations to apply and complex corrections must be made where this is not the case. Unfortunately, because distances are expressed in the equations in terms of wavelengths, near-field corrections have to be applied to at least the low frequency data

in almost all surveys.

7.12 SEISMIC METHODS

Seismic methods dominate oil industry geophysics but are comparatively little used in mineral exploration, partly because of their high cost but more especially because orebodies are often found in igneous or metamorphic rocks which lack coherent layering. Where ores occur in sedimentary rocks which have been only gently folded or faulted, seismic surveys may be useful (section 13.5.4) and another obvious application is in the search for offshore placers. Sub-bottom profiling using a sparker or boomer source and perhaps only a single hydrophone detector can allow bedrock depressions, and hence areas of possible heavy mineral accumulation, to be identified. Subsea resources of bulk minerals such as sands and gravels may be similarly evaluated. The images produced can be very striking but, being scaled vertically in two-way reflection time rather than depth (Fig. 7.17), need some form of velocity control before being

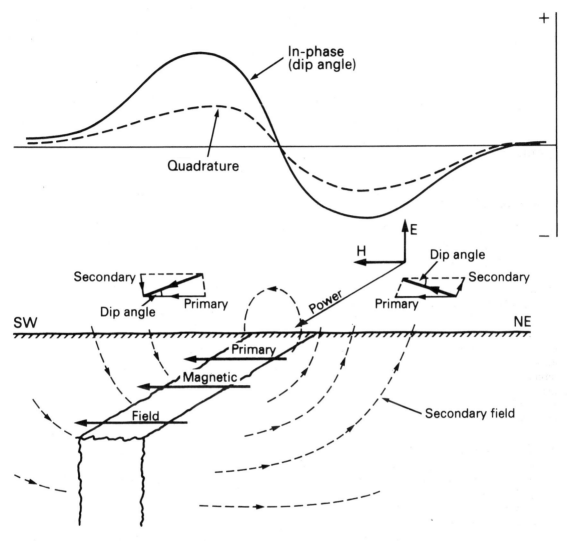

Fig. 7.16 Dip angle anomaly in the VLF magnetic field. Over homogeneous ground the VLF magnetic vector is horizontal. The largest anomalies are produced when, as in this case, the conductor is elongated in the direction of the VLF transmitter (which may be thousands of kilometres away).

used quantitatively. Onshore, seismic reflection work is more difficult because detectors and sources have to be positioned manually and may need to be buried. Also, because of the variability of the soil and weathered layers, results require sophisticated processing and are less easy to interpret. The use of reflection methods in onshore exploration for solid minerals other than coal is rare.

Refraction methods use simpler equipment and need less processing but can succeed only if the subsurface is regularly layered and velocity increases with depth at each interface. Generally, only the first arrivals of energy at each detector (geophone) are used, minimizing processing but limiting applications to areas with a maximum of four significant interfaces at any one locality. Depths to interfaces can be estimated because, except close to the shot point, where the direct path is the quickest, the first arrivals will have been critically refracted. Such rays can be pictured as travelling quite steeply from the surface to the refractor and back again but also as travelling along the refractor at the velocity of the underlying layer (Fig. 7.18). Horizontal velocities estimated from plots of distance against time are used in calculations as if they were vertical velocities. Calibration against borehole or other subsurface data is therefore desired, the more so since an interface may be hidden by others if the velocity contrast or layer thickness beneath it is small, and will be completely undetectable if it involves a velocity reversal. Despite these limitations, there are many cases, especially in the assessment of bulk mineral resources, in which refraction surveys can be useful.

7.13 **BOREHOLE GEOPHYSICS**

Most geophysical techniques can be modified for use in boreholes but gravity and magnetic logs are rare and radiometric and seismic velocity logs, both very important in the oil industry, are little used in mineral exploration. Electrical logs, however, do have a place in exploration programmes. As in surface electrical work, four electrodes must be used but usually at least one is outside the hole. In the 'normal' logs, one current and one voltage electrode, a few tens of centimetres apart, are lowered downhole, with the other two electrodes 'at infinity' on the surface. The limitations of d.c. resistivity in the search for base metal deposits apply downhole as much as on the surface and more useful results may be obtained with IP or e.m. logging. For IP, electrode polarization can be reduced by using lead (Pb) electrodes and eliminated, with some practical difficulty, by using porous pots. In e.m. work, small coils are used for both transmitting and detecting the signal, which may be a continuous wave or a transient.

The 'mise-à-la-masse' method can be used when conductive ore has been intersected. One current electrode is positioned at the downhole intersection and the other is placed on the surface well beyond the area of interest. If the earth were electrically homogeneous (and the ground surface flat), equipotentials mapped at the surface would be circles centred above the downhole electrode. Departures from this pattern indicate the strike direction and strike extent of continuous mineralization. Equipotentials are actually diverted away from good conductors which are not in electrical contact with the electrode, providing a powerful method for investigating the continuity of mineralization between intersections in a number of boreholes.

Because mineral exploration holes are normally cored completely, the mineral industry has been slower than the oil industry to recognize the advantages of geophysi-

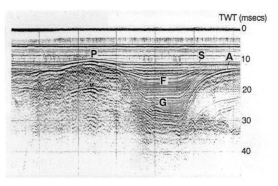

TWT (msecs)

Fig. 7.17 Seismic reflection profile across a buried submarine channel. Vertical scale is two-way reflection time, in milliseconds. Ten milliseconds is approximately equivalent to 7.5 m of water or 10 m of soft sediments. (Reproduced with the permission of Dr. D.E. Searle and the Geological Survey of Queensland.)

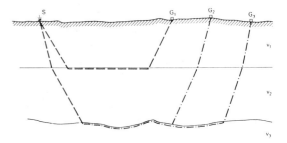

Fig. 7.18 Principles of seismic refraction. If $v_1 < v_2 < v_3$, critical refraction can occur at both interfaces. Near to the shot point the direct wave (not shown) will arrive first, but it will soon be overtaken by the wave that travels via the first interface and, eventually, by the wave that travels via the second.

cal logging. However, the high cost of drilling makes it essential to obtain the maximum possible information from each hole and the bulk electrical characteristics of the wall rocks can be a better guide to the significance of an ore intersection than the information contained in a few centimetres of core.

7.14 **EXPLORATION PROGRAMMES**

A logically designed exploration programme progresses through a number of stages, from regional reconnaissance to semi-detailed follow-up and thence to detailed evaluation. Its geophysical component will tend to pass through the same stages, reconnaissance often being airborne and the final evaluation perhaps involving downhole techniques. The exact route followed will depend on the nature of the target, but a typical base metal exploration programme might have, as its first stage, the assembling of all the geophysical data already available. Lease conditions may in some cases require the integration of certain types of survey data into national databases. Most countries have at least partial magnetic and gravity coverage and there may have been local surveys of these or other types. With gravity and magnetics it is usually worthwhile going beyond the published maps, which often contain errors, to obtaining the original data.

For sulphide ores, the next stage may well be to carry out an aerial combined magnetic and electromagnetic survey, using the smallest possible terrain clearance and very closely spaced lines. Semi-regional magnetic maps would be produced and, one hopes, a number of interesting electromagnetic anomalies would be located. Some sources may be obvious (e.g. power lines, pipelines, corrugated-iron barns) but all other anomalies will have to be checked on the ground, with geological mapping to determine rock type, magnetics and electromagnetics to locate the anomaly precisely, and perhaps a gravity survey if density differences are probable between ore and country rock. Other electrical methods, such as IP and even d.c. resistivity, might also be used to define drilling targets. Ground geophysics might also be employed in areas which had not been shown to be anomalous from the air but which geology or geochemistry had defined as promising. Once a drilling programme has begun, information can be fed back into the geophysical interpretation. The holes can also be used for electrical or radiometric logging or, if there are ore intersections, for 'mise-à-la-masse', and work of this sort may well continue into any development stage.

Not all geophysical exploration follows this pattern. Geophysicists are often called in at rather late stages to solve specific problems, and sometimes only one or two methods have any chance of success. Seismic techniques, whether reflection or refraction, tend to be used alone, being applicable where, as in the search for placer deposits, other methods fail, and useless in dealing with the complex geologies enountered in hard rock mining. Thought should always be given to the possibility of unusual and specialized uses for geophysical methods; the estimation, in some deposits, of phosphate percentages in uncored boreholes by radiometric logging is a case in point.

7.15 INTEGRATION OF GEOLOGICAL AND GEOPHYSICAL DATA

Making good use of the data from a variety of very different sources is the most testing part of an exploration geologist's work, requiring theoretical knowledge, practical understanding and a degree of flair and imagination. All these will be useless if the information is not to hand in convenient and easily usable form. The most effective way of recognizing correlations is to overlay one set of data on another (section 5.1.8.), so decisions on map scales and map boundaries must be made at an early stage and from then on all data should be compiled in accordance with these decisions, although the advent of computer-based GIS (geographical information systems) has made second thoughts easier to implement.

The proven usefulness of overlaying and the decreasing cost of computer processing have produced something of a vogue for presenting different types of data together on a single map. This is generally useful if they are shown as separate entities which can still be clearly distinguished, but sometimes data of different types are combined before being plotted. Such approaches have long been familiar in exploration geochemistry, in the form of factor analysis, but their extension to geophysics has not been notably successful. Displays of mathematically combined gravity and magnetic data have generally succeeded only in disguising important features while producing no new insights. Allowing computers to cross-correlate geophysics with geology or geochemistry is even less likely to give meaningful results. Correlation is necessary, but should take place in the brain of the interpreter.

Of the various possible correlations, the ones most likely to be useful in the early stages of an exploration programme are those between geology, photo-interpretation and the magnetic map. Structural dislocations such as faults and shears are likely to be indicated on magnetic maps and the intersections of magnetically defined lineaments have proved fruitful exploration targets in many areas (Fig. 7.19). Since the aim is to obtain the best possible understanding of the geology, gravity or radiometric data may also help, along with infra-red or radar imagery and soils geochemistry. Geophysics provides independent information on the third dimension and may be especially useful in preparing regional cross-sections.

Detailed integration of geological and geophysical data is achieved largely through computer modelling. For there to be even a chance of a model being correct, it must satisfy all the known geological and geophysical constraints. The inherent ambiguity of all geophysical interpretation makes it essential to use all possible additional information and the final integration stage occurs during a drilling programme. If a worthwhile target has been defined but the hole has proved unsuccessful, it is not enough to blame the geophysicists. There is a reason for any failure. It may be obvious, as in the case of an IP anomaly produced by barren pyrite, but until the reason has been understood, a valid target remains to be tested.

7.16 THE ROLE OF THE GEOLOGIST IN GEOPHYSICAL EXPLORATION

A mineral exploration programme in a given area will usually be the responsibility of a small team and whether this should include a geophysicist is itself a decision which requires geophysical advice. Factors to be considered are the types of targets sought (and therefore the sort of exploration methods appropriate), the duration of the project and, of course, the availability of personnel. If a full-time geophysicist cannot be justified but some geophysical techniques are to be used, there may be company geophysicists who can allocate a percentage of their time to the project. Alternatively, consultants can be retained or one or more of the geologists in the team can be given responsibility for the geophysical aspects, which may include contract preparation, supervision of contractors and even the conduct of some surveys. It is usually best if one (and preferably only one) geologist is responsible for routine geophysical matters. He or she must be able to call on more geophysically experienced help when needed and his or her freedom to seek such help must not be unduly restricted. Geophysical surveys are often among the most expensive parts of an exploration programme and much expenditure can be wasted by a few apparently trivial but wrong decisions.

Project geologists should not be expected to decide, unaided, on methods, instrument purchases and overall geophysical exploration strategy. In companies with their own geophysicists it would be unusual for them to have to do so, but it is all too common for consultants to be called in only when things have already begun to go irrevocably wrong. Instruments may have been purchased, and used, contracts may have been let and contractors may have been selected which are all totally unsuitable for the work required.

What, then, can reasonably be expected of a project geologist whose only experience of geophysics may be a few dimly remembered lectures on a BSc or MSc course several years in the past, possibly accompanied by some miserable days in the field pressing the keys on a proton magnetometer or trying to level a gravity meter? First, and perhaps most important, a degree of humility and a recognition that help is likely to be needed, and that even simple geophysical surveys can go badly wrong if mishandled. Second, a variety of geophysical texts and a willingness to read them all (beware of relying on any single authority). Third, a realistic attitude. Exploration programmes have limited and often pre-defined budgets, and geophysical instruments are neither magic wands, which reveal everything about an area, nor pieces of useless electronic circuitry inflicted on hard-working geologists specifically to complicate their lives. In considering geophysics, a geologist must ask him- or herself a series of questions. What information do they hope to get? Is there a realistic expectation that they will actually get that information? What information will they not get? Can they do without such information or, indeed, can they tolerate any incompleteness? Can they afford geophysics? Would the money be better spent on something else? They may not be in a position to answer all these questions themselves but they need to ask them, and to get the answers. General approaches are best illustrated by specific examples.

(1) An exploration area some hundreds of square kilometres in extent is covered by regional aeromagnetic maps. Are these adequate or should a new, detailed survey be planned? (Seek geophysical advice, but not from a contractor who may be hoping for work.) Is it reasonable to suppose that the targets will be magnetic or that there will be critical structural or lithological information in the magnetic data? Will such data only be obtained by a high sensitivity or gradient survey or will a 1 nT magnetometer do? Given that airborne work is to

---

Fig. 7.19 *(Opposite)* The 'Aeromagnetic Ridge' at Tennant Creek in northern Australia. The contours are of total magnetic field intensity, measured in nanoTesla (contour interval 100 nT). The 'ridge', which, despite the strong magnetic expression, is not marked by any topographical feature, is host to a number of gold and copper orebodies. Mines and prospects known at the time of publication of the map are indicated by the crossed pickaxe symbol. Although the original regional airborne survey shown here has since been superseded by more detailed work, it was a primary guide to exploration in the area for many years. Attention has been focussed on the intersections of magnetic lineations, two of which, intersecting near the subsequently discovered Ivanhoe deposit, are indicated on the map. The map also shows areas with gamma radiation levels significantly higher than the regional background levels, enclosed by hachured dashed lines. The map is reproduced with the permission of the Director, Bureau of Mineral Resources, Canberra.

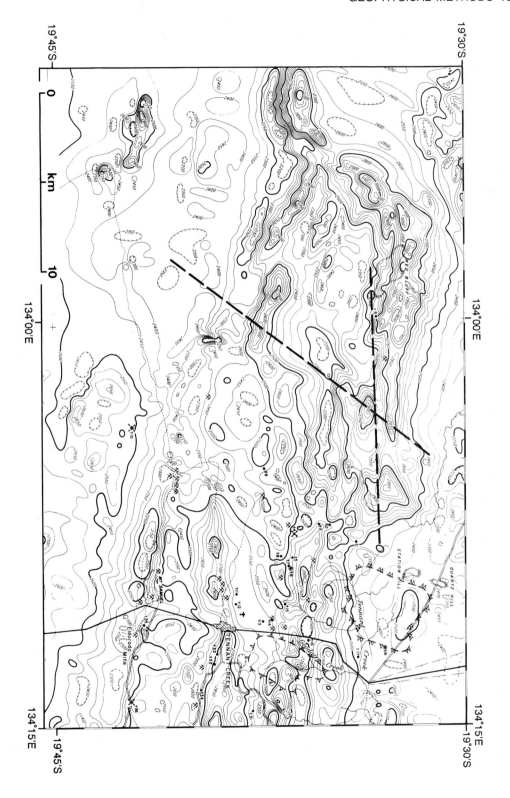

be done, is there a case for adding other sensors (electro-magnetic, radiometric), at a possible tripling in survey cost? Is the area topographically suitable for fixed-wing survey or will a helicopter have to be used? How much help will have to be bought-in for contract supervision and interpretation, and can this be accommodated within the existing budget?

(2) An area of a few square kilometres is to be explored for massive sulphides. SP surveys have located some orebodies in similar environments but have missed others. Bearing in mind that the method is cheap but that other, more expensive methods will also have to be used if the area is to be fully evaluated, is SP worthwhile? Should it be kept in reserve (the equipment is cheap) in case there are field-hands spare for a few days? Or, having gone to the trouble of gridding and pegging the area, should not every possible method be used every-where?

Any ground work of the type considered in the second example will involve surveying, and may require clear-ing and cutting as well. At this point the geologist responsible for geophysics may well find him- or herself in conflict with the geologist responsible for geochemistry (who might be themself, complicating their life even further). Geochemists, once beyond the reconnaissance, stream sediment stage, also take samples along pegged lines, but they then like to be able to plot locations at regular intervals on planimetric base maps. Planimetrically regular spacing (secant-chaining) is also desirable for geophysical 'point data' (gravity, magnetic and SP), but in resistivity, IP and electromagnetic sur-veys it is the actual distances between coils or electrodes that must be constant and these may differ significantly from plan distances. Whether lines are secant- or slope-chained will depend on how much, of what, is being measured, but this apparently trivial problem has some-times been a major cause of friction between team members.

The final requirements of exploration geologists, and not least of those who involve themselves in geophysics, are serendipity and awareness of the possible. Geo-physical surveys always provide some information, even though it may not seem directly relevant at the time, and many mineral deposits have been discovered by what may be considered luck but at least required geologists with their wits about them. Airborne radiometric sur-veys in the Solomon Islands in the 1960s were seen primarily as a means of detecting phosphates. A ground follow-up team sent to Rennel Island expected to find phosphatic rock, of which there was very little, but did

recognize that the source of the anomaly was uraniferous bauxite, in mineable quantities.

7.17 NEGOTIATIONS WITH CONTRACTORS

Field geologists may be involved with geophysical con-tractors at many different stages, and will need the support and advice of an experienced geophysicist who is committed to the exploration project and not to the contractor. The most important stage comes when the contract is first being drafted. Geophysical consultants find it frustrating, but also often very profitable, to be asked to supervise contracts, already written and agreed, which are either so loose that data of almost any quality have to be accepted or which specify procedures which actually prevent useful data being acquired. Project geologists generally share in the frustration but not in the profit!

Survey parameters such as line spacing and line orientation, and all specifications relating to final re-ports and data displays, should be decided between the project geologist and his or her geophysical adviser. More technical questions, such as tolerances (naviga-tional and instrumental), instrument settings, flying heights in airborne surveys and data control and reduc-tion procedures, can be left to the geophysicist. A geologist who sets specifications without this sort of help, perhaps relying on the contractor's advice and good will, is taking grave risks. Goodwill is usually abundant during initial negotiations and contractors will, after all, only stay in business if they keep their clients reasonably happy, but as deadlines draw closer and profit margins come to look tighter, the temptation to rely on the letter of the contract will increase.

Once a contract has been signed, the work has to be done. The day-to-day supervision can usually be left to the geologist on the spot, who will find a comprehensive and tightly-written contract invaluable. The more detail in which procedures are specified, the easier it is to see if they are being followed and to monitor actual progress. Random and unannounced visits to the contractor's field crew, especially early in the morning (are they getting up at a reasonable hour, without hangovers, and do they have precise and sensible plans for the day's work?) or at the end of the day (are they working a full day and are their field notes for that day clear and intelligible?) are likely to be more useful than any number of formal meetings with the party chief. Evening visits are also useful, revealing whether data reductions and paperwork are keeping pace with the field operations and whether the evenings are being passed in an alcoholic haze.

A geophysical contractor can reasonably be expected to carry out field work competently, reduce data correctly and present results clearly. Many also offer interpretational services, but these may be of limited value. Only very specialized contractors are likely to have anything approaching the knowledge of an area and its geological problems possessed by the members of an exploration team, and companies are not often willing to provide contractors, who may next week be working for a competitor, with recently acquired data about an area. Since such information may, however, be vital for good interpretation, contractors' efforts, are often, through no fault of their own, bland and superficial. They may only be worth having if the method used is so new or arcane that only their own geophysicists actually understand it.

7.18 **FURTHER READING**

A number of textbooks cover the range of geophysical techniques used in mineral exploration in some detail. Of these, perhaps the most comprehensive is *Applied Geophysics* (Telford *et al.* 1976), while *Practical Geophysics II* (van Blaricom 1993), which omits much of the theory, provides a wealth of case histories and practical discussions. A briefer but still excellent coverage is provided by *An Introduction to Geophysical Exploration* by Kearey and Brooks (1991). The theoretical bases of many of the geophysical methods are described by Grant and West (1965) in their *Interpretation Theory in Applied Geophysics*, while the practical aspects of small scale geophysical surveys and some elementary interpretational rules are discussed in *Field Geophysics* (Milsom 1989). Finally, two Australian compilations, on orebodies at Elura (Emerson 1980) and Woodlawn (Whitely 1981), provide case histories which cover the application of ranges of geophysical techniques to two specific base metal deposits.

8 Exploration Geochemistry

CHARLES J. MOON

Geochemistry is now used in virtually every exploration programme, if only to determine the grade of the material to be mined. However exploration geochemistry has evolved from its early origins in assaying, to using the chemistry of the environment surrounding a deposit in order to locate it. This particularly applies to the use of surficial material, such as soil, till or vegetation, that can be used in areas where there is little outcrop. The object is to define a geochemical anomaly which distinguishes the deposit from enhancements in background and non-significant deposits. This chapter explains how geochemistry may be employed in the search for mineral deposits. Further details of the theory behind exploration geochemistry are given in Rose *et al.* (1979) and Levinson (1980).

The literature on exploration geochemistry is large but reasonably accessible. Reviews of most geochemical techniques are given in Garland (1989) and full bibliographies are provided in Hawkes (1982, 1985, 1988). These techniques are updated in *Explore,* the newsletter of the Association of Exploration Geochemists which also publishes the more formal *Journal of Geochemical Exploration.* Articles of interest will also be found in *Applied Geochemistry, Transactions of the Institution of Mining & Metallurgy Section B* and *Transactions of the Canadian Institution of Mining & Metallurgy.*

Exploration geologists are likely to be more directly involved in geochemistry than with geophysics which is usually conducted by contractors and supervised by specialist geophysicists. A geochemical programme can be divided into the following phases.

1 Planning
2 Sampling
3 Chemical analysis
4 Interpretation
5 Follow up

The field geologist will probably carry out phases 1, 2, 4 and 5 while analysis is normally performed by a commercial laboratory.

8.1 PLANNING

The choice of the field survey technique and the analyti-cal methods depends on the commodity sought and its location. In the same way as geological and grade–tonnage models are generated (section 4.1.3) modelling can be extended to include geochemical factors. Thus the geologist will start with a knowledge of the elements associated with a particular deposit type, an idea of the economic size of the deposit to be sought, the minera-logical form of the elements and the probable size of the elemental anomalies around it. The outline of a deposit is defined by economic criteria and the mineable mate-rial is surrounded by lower concentrations of the mined elements which are however substantially enriched com-pared with unmineralized rock. This area of enrichment is known as the primary halo, by analogy with the light surrounding the outline of the moon and the process of enrichment as *primary dispersion*. In addition ore form-ing processes concentrate or deplete elements other than those mined. For example, massive sulphide deposits often contain substantial arsenic and gold in addition to the copper, lead and zinc for which they are mined. A summary of typical elemental associations is shown in Table 8.1.

The geologist's problem is then to adapt this knowl-edge of primary concentration to the exploration area. The geochemical response at the surface depends on the type of terrain and especially on the type of material covering the deposit as shown in Fig. 8.1. The response in an area of two metres of overburden is very different from that of an area with 100 m deep cover. Also elements behave differently in the near surface environ-ment from that in which the deposit formed. For example of copper, lead and zinc associated together in volcanic-associated massive sulphide deposits, zinc is normally more mobile in the surface environment than copper and much more so than lead. Lead is more likely to be concentrated immediately over the deposit, as it is relatively insoluble, whereas zinc will move or disperse from the deposit. This process of movement away from the primary source is termed *secondary dispersion* and it can also be effected by mechanical movement of fragments under gravity, movement as a gas, or diffusion of the elements in the form of ions as well as movement in solution.

Table 8.1 Elemental associations and associated elements (pathfinders) useful in exploration. (Largely from Rose *et al.* 1979 with some data from Beus and Grigorian 1977, p 232 and Boyle 1974.)

Type of deposit	Major components	Associated elements
Magmatic deposits		
Chromite ores (Bushveld)	Cr	Ni, Fe, Mg
Layered magnetite (Bushveld)	Fe	V, Ti, P
Immiscible Cu–Ni–sulphide (Sudbury)	Cu, Ni, S	Pt, Co, As, Au
Pt–Ni–Cu in layered intrusion (Bushveld)	Pt, Ni, Cu	Sr, Co, S
Immiscible Fe–Ti–oxide (Allard Lake)	Fe, Ti	P
Nb–Ta carbonatite (Oka)	Nb, Ta	Na, Zr, P
Rare–metal pegmatite	Be, Li, Cs, Rb	B, U, Th, rare earths
Hydrothermal deposits		
Porphyry copper (Bingham)	Cu, S	Mo, Au, Ag, Re, As, Pb, Zn, K
Porphyry molybdenum (Climax)	Mo, S	W, Sn, F, Cu
Skarn–magnetite (Iron Springs)	Fe	Cu, Co, S
Skarn–Cu (Yerington)	Cu, Fe, S	Au, Ag
Skarn–Pb–Zn (Hanover)	Pb, Zn, S	Cu, Co
Skarn–W–Mo–Sn (Bishop)	W, Mo, Sn	F, S, Cu, Be, Bi
Base metal veins	Pb, Zn, Cu, S	Ag, Au, As, Sb, Mn
Sn–W greisens	Sn, W	Cu, Mo, Bi, Li, Rb, Si, Cs, Re, F, B
Sn–sulphide veins	Sn, S	Cu, Pb, Zn, Ag, Sb
Co–Ni–Ag veins (Cobalt)	Co, Ni, Ag, S	As, Sb, Bi, U
Epithermal precious metal	Au, Ag	Sb, As, Hg, Te, Se, S, Cu
Sediment hosted precious metal (Carlin)	Au, Ag	As, Sb, Hg, W
Vein gold (Archaean)	Au	As, Sb, W
Mercury	Hg, S	Sb, As
Uranium vein in granite	U	Mo, Pb, F
Unconformity associated uranium	U	Ni, Se, Au, Pd, As
Copper in basalt (L. Superior type)	Cu	Ag, As, S
Volcanic-associated massive sulphide Cu	Cu, S	Zn, Au
Volcanic-associated massive sulphide Zn–Cu–Pb	Zn, Pb, Cu, S	Ag, Ba, Au, As
Au–As rich Fe formation	Au, As, S	Sb
Mississippi Valley Pb–Zn	Zn, Pb, S	Ba, F, Cd, Cu, Ni, Co, Hg
Mississippi Valley fluorite	F	Ba, Pb, Zn
Sandstone-type U	U	Se, Mo, V, Cu, Pb
Red bed Cu	Cu, S	Ag, Pb
Sedimentary types		
Copper shale (Kupferschiefer)	Cu, S	Ag, Zn, Pb, Co, Ni, Cd, Hg
Copper sandstone	Cu, S	Ag, Co, Ni
Calcrete U	U	V

The background levels of an element in rocks and soils also have to be considered when trying to find secondary dispersion from deposits. All elements are present in every rock and soil sample; the concentration will depend on the mode of formation of the rock and the process forming the soil. Indications of background levels of elements in soils are given in Table 8.2, as are known lithologies with elevated concentrations which might provide spurious or non-significant anomalies during a survey. These background levels can be of use in preparing geochemical maps which can be used to infer lithology in areas of poor outcrop. The reader is advised to get some idea of the background variation over ordinary rock formations from a geochemical atlas,

Residual

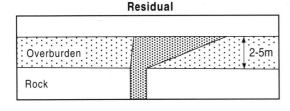

Glacial

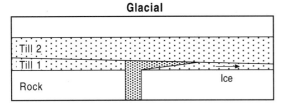

Deep Tropical

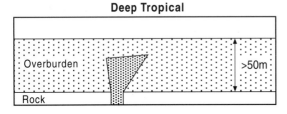

Fig. 8.1 Sketch showing dispersion through major types of overburden.

such as those for England and Wales (Webb *et al.* 1978), Alaska (Weaver *et al.* 1983) and the former West Germany (Fauth *et al.* 1985).

The basis of a geochemical programme is a systematic sampling programme (Thomson 1987) and thus decisions must be made in a cost-effective manner as to the material to be sampled, the density of sampling and the analytical method to be employed. Cost/benefit ratios should be considered carefully as it may be that a slightly more expensive method will be the only effective technique. The material to be sampled will largely be determined by the overburden conditions but the density and exact nature of the sample will be based either on previous experience in the area or, if possible, on an orientation survey.

8.1.1 **Orientation surveys**

One of the key aspects of planning is to evaluate which techniques are effective for the commodity sought and in the area of search. This is known as an orientation survey. The best orientation survey is that in which a variety of sampling methods is tested over a prospect or

deposit of similar geology to the target and in similar topographical conditions to determine the method which yields the best results. A check list for an orientation study is given below (Closs & Nichol 1989).

1 Clear understanding of target deposit type.
2 Understanding of surficial environment of the search area.
3 Nature of primary and secondary dispersion from the mineralization.
4 Sample types available.
5 Sample collection procedures.
6 Sample size requirements.
7 Sample interval, orientation and areal density.
8 Field observations required.
9 Sample preparation procedures.
10 Sample fraction for analysis.
11 Analytical method required.
12 Elemental suite to be analysed.
13 Data format for interpretation.

The relatively small cost involved in undertaking an orientation survey compared with that of a major geochemical survey is always justified and whenever possible this approach should be used. However, if a physical orientation survey is not possible, then a thorough review of the available literature and discussion with geochemical experts is a reasonable second best. Of particular use in orientation studies are the models of dispersion produced for various parts of the world; Canadian Shield and Canadian Cordillera (Bradshaw 1975), Western USA (Lovering & McCarthy 1977) and Australia (Butt & Smith, 1980).

Particularly useful discussion on orientation procedures is provided by Thomson (1987) and Closs & Nichol (1989).

After the orientation survey has been conducted the logistics of the major survey need to be planned. The reader should devise a check list similar to that of Thomson (1987).

1 Hire field crew with appropriate experience and training.
2 Obtain base maps and devise simple sample numbering scheme.
3 Designate personnel for communication with the laboratory.
4 Arrange quality control of laboratory.
5 Arrange for data handling and interpretation.
6 Organize archive of samples and data.
7 Liaise with other project staff (e.g. geophysicists) and arrange reporting to management.

Reporting of geochemical surveys is important and readers should consult the short booklet by Hoffman

Table 8.2 Background concentration of trace elements and utility in geochemical exploration. (Largely after Rose *et al.* 1979, Levinson 1980.)

Element	Typical Concentration in Soils (ppm)	Enriched Lithologies	Surficial Mobility	Use in Exploration
Antimony	1		Low	Pathfinder
Arsenic	10	Ironstones	Mobile; Fe scavenged	Pathfinder especially for Au
Barium	300	Sandstones	Low often baryte	Panned concs in VMS search
Beryllium	1	Granites	High if not in beryl	Occasional use
Boron	30	Granites	Moderate	For borates
Bismuth	1	Granites	Low	Difficult to determine
Cadmium	100 ppb	Black shales	High	Zn deposits
Chromium	45	Ultramafics	Low	Chromite in panned concs.
Cobalt	10	Ultramafics	Moderate Mn scavenged	Widely used
Copper	15	Basic igneous	pH>5 low, else moderate	Most surveys
Fluorine	300	Alkaline igneous	High	F deposits
Gold	1 ppb	Black shales	Low	Gold deposits
Lead	15	Sandstones	Low	Wide use
Lithium	20	Granites	Moderate	Tin deposits
Manganese	300		Moderate, high at acid pH	Scavenges Co, Zn, Ag
Mercury	50 ppb		High	Wide use as pathfinder
Molydenum	3	Black shale	Moderate to high pH>10	Wide use
Nickel	17	Ultramafics	Low, scavenged	Wide use
Platinum	1 ppb	Ultramafics	Very low	Difficult to determine
Rare Earths	La 30	Beach sands	Very low	
Selenium	300 ppb	Black shales	High	Little use
Silver	100 ppb?		High; Mn scavenged	Difficult to use
Tellurium	10 ppb	Acid intrusives	Low	Difficult to use
Thallium	200 ppb?		Low	Epithermal Au
Tin	10	Granites	Very low	Panned concentrates
Tungsten	1	Granites	Very low	Scheelite fluoresces in UV
Uranium	1	Phosphorites	Very high; organic scavenged	Determine in water
Vanadium	55	Ultramafics	Moderate?	Little use
Zinc	35	Black shales	High; scavenged by Mn	Most surveys
Zirconium	270	Alkaline igneous	Very Low	Little use

(1986). This provides a wealth of information as well as the basis for fulfilling legal requirements in Canada.

8.2 ANALYSIS

As the geologist generally sees little of the process of analysis, which is usually done at some distance from the exploration project, analytical data tend to be used uncritically. While most laboratories provide good quality data they are usually in business to make a profit and it is up to the geologist to monitor the quality of data produced and investigate the appropriateness of the analytical methods used.

8.2.1 Accuracy and precision

The critical question for the geologist is how reproducible the analysis is and how representative of the 'correct' concentration the concentration is, as shown in Fig 8.2. The reproducibility of an analysis is termed the precision and its relation to the expected or consensus value the accuracy. For most purposes in exploration geochemistry it is vitally important that an analysis is precise but the accuracy is generally not so crucial, although some indication of the accuracy is needed. At the evaluation stage the analyses must be precise and accurate. The measurement of accuracy and precision requires careful planning and an understanding of the theory involved.

A number of schemes have been devised but the most comprehensive is that of Thompson (1982). Precision is measured by analysing samples in duplicate whereas accuracy requires the analysis of a sample of known composition, a reference material. The use of duplicate samples means that precision is monitored across the whole range of sample compositions. Reference materials can be acquired commercially but they are exceedingly expensive (>US$100 per 250 g) and the usual practice is to develop in-house materials which are then calibrated against international reference materials. The in-house reference materials can be made by thoroughly mixing and grinding weakly anomalous soils from a variety of sites. The contents should be high enough to give some indication of accuracy in the anticipated range but not so high as to require special treatment (for copper, materials in the range 30–100 ppm are recommended). If commercial laboratories are used then the reference materials and duplicates should be included at random; suggested frequencies are 10% for duplicates and 4% for reference materials. If the analyses are in-house then checks can be made on the purity of reagents

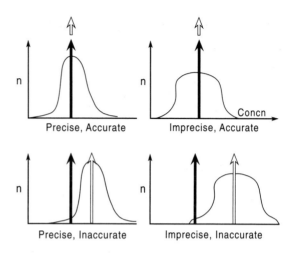

Fig. 8.2 Schematic representation of precision and accuracy assuming normal distribution of analytical error.

by running blank samples, i.e. chemicals with no sample. Samples should if possible be run in a different and random order to that in which they are collected. This enables the monitoring of systematic drift. In practice this is often not easy and it is not desirable to operate two numbering systems, so some arrangement should be made with the laboratories to randomize the samples within an analytical batch. The results can then easily be re-sorted by computer.

When laboratory results become available the data should be plotted batch by batch to examine the within- and between-batch effects. The effects observable within a batch are precision and systematic instrumental drift. Precision should be monitored by calculating the average and difference between duplicates on a chart such as is detailed by Thompson (1982). The precision for most applications should be less than about 20%, although older methods may reach 50%. If precision is worse than anticipated then the batch should be re-analysed. Systematic drift can be monitored from the reference materials within a batch and the reagent blanks. Between-batch effects represent the deviation from the expected value of the reference material and should be monitored by control charts. Any batch with results from reference materials outside the mean ±2 standard deviations should be re-analysed; commercial laboratories normally do this at no further charge. Long-term monitoring of drift can show some interesting effects. Coope (1991) was able to demonstrate the disruption caused by the moving of laboratories in a study of the gold reference materials used by Newmont Gold Inc.

8.2.2 Sample collection and preparation

Samples should be collected in non-metallic containers to avoid any contamination. Kraft paper bags are best suited for sampling soils and stream sediments because the bags retain their strength if the samples are wet and the samples can be oven dried without removing them from their bags. Thick gauge plastic or cloth are preferred for rock samples. All samples should be clearly labelled by pens containing non-metallic ink.

Most sample preparation is carried out in the field, particularly when it involves the collection of soils and stream sediments. The aim of the sample preparation is to reduce the bulk of the samples and prepare them for shipment. Soils and stream sediments are generally dried either in the sun, in low temperature ovens or freeze dried; the temperature should be below 65°C so that volatile elements such as mercury are not lost. Drying is generally followed by gentle disaggregation and sieving to obtain the desired size fraction. Care should be taken to avoid the use of metallic materials and to avoid carry over from highly mineralized to background samples. Preparation of rocks and vegetation is usually carried out in the laboratory and care should be taken in the selection of crushing materials. For example, in a rock geochemical programme a company searching for volcanic-associated massive sulphides found manganese anomalies associated with a hard amphibolite. They were encouraged by this and took it as a sign of exhalative activity. Unfortunately further work showed that the manganese highs were related to pieces of manganese steel breaking off the jaw crusher and contaminating the amphibolite samples. Other less systematic variation can be caused by carry-over from high grade samples, for example not cleaning small grains of mineralized vein material (e.g. 100 000 ppb Au) will cause significant anomalies when mixed with background (1 ppb Au) rock. Contamination can be eliminated by cleaning crushing equipment thoroughly between samples and by checking this by analysing materials such as silica sand.

8.2.3 Analytical methods

Most analysis is aimed at the determination of the elemental concentrations in a sample and usually of trace metals. At present it is impossible to analyse all elements simultaneously at the required levels, so some compromises have to be made (Fig. 8.3). In exploration for base metals it is usual to analyse for the elements sought, for example copper in the case of a copper

deposit, and as many useful elements as possible at a limited extra cost. With modern techniques it is often possible to get 20–30 extra elements, including some that provide little extra information but a lot of extra data for interpretation. The major methods are as shown in Table 8.4 but for detail on the methods used the reader should consult Fletcher (1981, 1987) and Thompson & Walsh (1989).

The differences between the methods shown are cost, the detection limits of analysis, speed of analysis and the need to take material into solution. Most general analysis in developed countries is carried out by Inductively Coupled Plasma Emission Spectrometry (ICP–ES) or X-ray Fluorescence (XRF); both require highly sophisticated laboratories, pure chemicals, continuous, non-fluctuating power supplies and readily available service personnel, features not always present in developing countries. In less sophisticated environments, high quality analysis can be provided by atomic absorption spectrophotometry (AAS), which was the most commonly used method in developed countries until about 1980. Another method which is widely used in industry is neutron activation analysis (NAA) but its use is restricted to countries with cheap nuclear reactor time, mainly Canada. Inductively coupled plasma sourced mass spectrometry (ICP–MS) provides much reduced detection limits for a wide range of metals and is widely used in water analysis but has yet to gain wide acceptance in the analysis of soils and rocks.

Precious metals (gold and platinum group elements) have been extremely difficult to determine accurately at background levels. The boom in precious metal exploration has, however, changed this and commercial laboratories are able to offer cheap gold analysis at geochemical levels (5 ppb–1 ppm) using solvent extraction and AAS, ICP–ES or alternatively NAA on solid samples. For evaluation the method of fire assay is still without equal: in this the precious metals are extracted into a small button which is then separated from the slag and determined by AAS, ICP–ES or ICP–MS. The analysis of precious metals is different from most major elements and base metals in that large sub-samples are preferred to overcome the occurrence of gold as discrete grains. Typically 30 or 50 g are taken in contrast to 0.25–1 g for base metals; in Australia, 8 kg are often leached with cyanide to provide better sampling statistics (further details in section 15.1.2)

Elements which occur as anionic species are generally difficult to measure, especially the chloride, bromide and iodide ions which serve as some of the ore transporting ligands. Although some of these elements can be

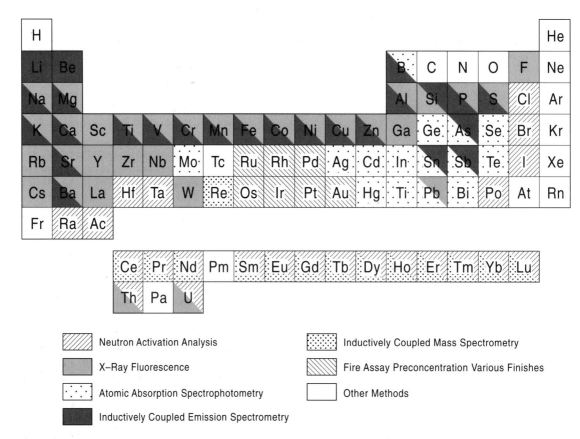

Fig. 8.3 Summary periodic table showing cost-effective methods of elemental analysis at levels encountered in background exploration samples.

determined by ICP–ES or XRF the most useful method is ion chromatography.

Isotopic analysis is not yet widely used in exploration although some pilot studies, such as that of Gulson (1986), have been carried out. The main reason for this is the difficulty and cost of analysis. Although ICP sourced mass spectrometers are finding their way into commercial laboratories and are the main hope for cheap analysis, they are not yet capable of resolving isotopic ratios sufficiently for the application of methods such as the discrimination of vein from large stratiform ore deposits. This still has to be carried out by thermal source mass spectrometers and is extremely expensive (US$200 per sample).

The choice of analytical method will aim at optimizing contrast of the main target element. For example, it is little use determining the total amount of nickel in an ultramafic rock when the majority of nickel is in olivine and the target sought is nickel sulphides. It would be better in this case to choose a reagent which will mainly extract nickel from sulphides and little from olivine. In soils and stream sediments, optimum contrast for base metals is normally obtained by a strong acid (e.g nitric + hydrochloric acids) attack, that does not dissolve all silicates, and an ICP–ES or AAS finish. Most rock analysis uses total analysis by XRF or by ICP–ES following a fusion or nitric–perchloric–hydrofluoric acid attack.

8.3 INTERPRETATION

Once the analytical data have been received from the laboratory and checked for precision and accuracy, there is the question of how the data should be treated and

Table 8.3 Summary of the main methods used in exploration geochemistry.

Method	Capital Cost $	Multi-element	Precis-ion	Sample type	Cost per analysis $	Comments
Colorimetry	8 000	No	Low	Solution	2–10	Good for W, Mo and field use
Atomic absorption spectrophotometry	60 000	No	Good	Solution	1–5	Cheap and precise
X–ray fluorescence	200 000	Yes	Good	Solid	20	Good for refractory elements
ICP–emission spectrometry	150 000	Yes	Good	Solution	10	Good for trace metals
ICP–mass spectrometry	300 000	Yes	Good	Solution	35	Good for heavy metals

interpreted. As the data are likely to be multi-element and there are likely to be a large number of samples this will involve the use of statistical analysis on a computer. It is recommended that the data are received from the analyst either in the form of a floppy disk or over a telephone line using a modem. Re-entering data into a computer from a paper copy is expensive and almost certain to introduce major errors. Normally the data will be transferred into an electronic database to allow easy access or, in the case of small data sets held in a spreadsheet, such as LOTUS 1-2-3 or EXCEL.

8.3.1 Statistics

The object of geochemical exploration is to define significant anomalies. In the simplest case these are the highest values of the element sought but they could be an elemental association reflecting hydrothermal altera-tion or even element depletion. Anomalies are defined by statistically grouping data and comparing these with geology and sampling information as shown in the flow chart (Fig. 8.4). Normally this grouping will be under-taken by computer and a wide variety of statistical packages are available; for microcomputers, one of MINITAB, SYSTAT and STATGRAPHICS is recom-mended at a cost of US $200–500 each.

The best means of statistically grouping data is graphi-cal examination using histograms and box plots (Howarth 1984, Garrett 1989). This is coupled with description

using measures of central tendency (mean or median) and of statistical dispersion (usually standard devia-tion). It would be expected that if data are homogeneous then they will form a continuous normal or, more likely, log-normal distribution but if the data fall into several groups then they will be multi-modal as shown in the example in Fig. 8.5. These are a set of copper determinations of soil samples from the Daisy Creek area of Montana and have been described in detail by Sinclair (1991). The histogram shows a break (dip) in the highly skewed data at 90 and 210 ppm. This grouping can be well seen by plotting the data with a probability scale on the x axis (Fig. 8.6); log-normal distributions form straight lines and multi-modal groups straight lines separated by curves (for full details see Sinclair 1976). Separation of data into sub-groups is best achieved using a computer program available from the Association of Exploration Geochemists, which also includes the Daisy Creek data on the diskette. The Daisy Creek data show thresholds (taken at the upper limit, 99%, of the lower sub-groups) which can be set at 100 ppm to divide the data into two groups and at 71 and 128 ppm if three groups are used. The relationship of the groups to other elemental data should then be examined by plotting and calculating the correlation matrix for the data set. In the Daisy Creek example there is a strong correlation (0.765) between copper and silver (Fig. 8.7) which reflects the close primary association between the two elements, whereas the correlation between copper and lead is low

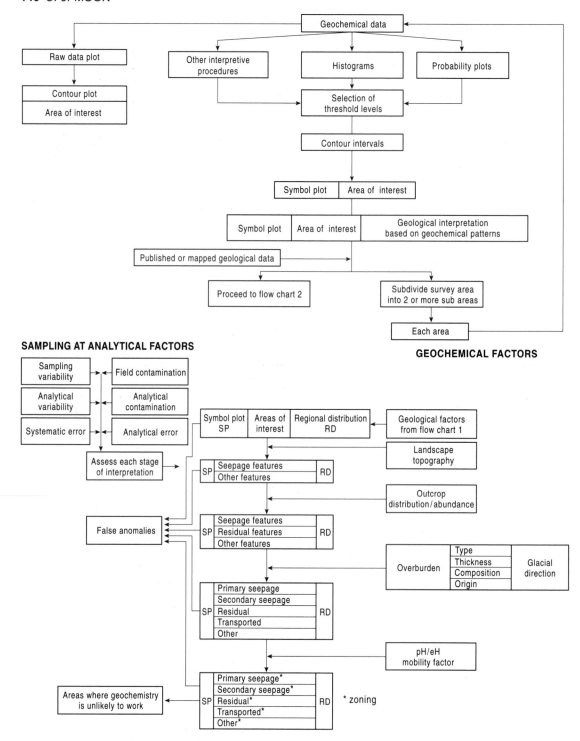

Fig. 8.4 *(Above and opposite)* Flow charts for the interpretation of soil geochemical data. (From Hoffman 1987.)

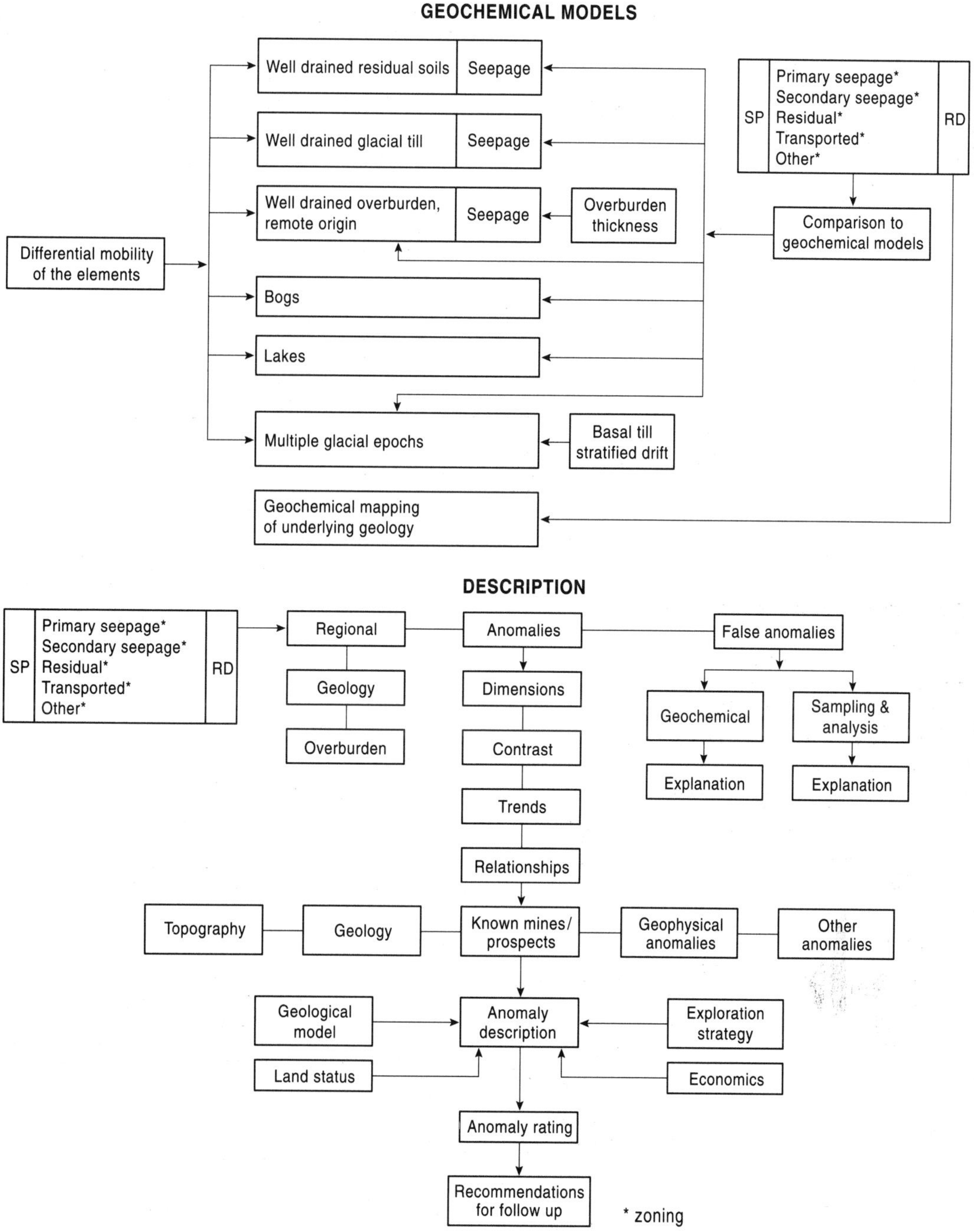

Fig. 8.4 *(Continued)*

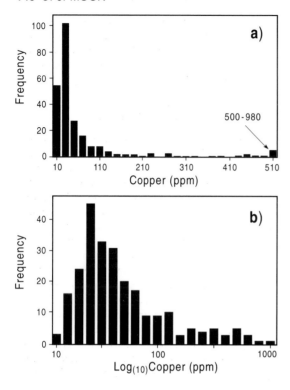

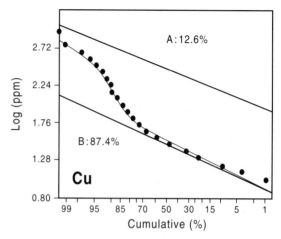

Fig. 8.6 Probability plot of the Daisy Creek copper data. The data have been sub-divided into two log normal groups.

Fig. 8.5 Arithmetic and $\log_{10}$ transformed histograms of Daisy Creek copper data. Note the positive skew of the arithmetic histogram. For further discussion see text and Sinclair 1991.

(0.12) reflecting the spatial displacement of galena veins from chalcopyrite. If geological or sampling data are coded it will be possible to compare these groups with that information.

The spatial distribution of the data must then be examined by plotting elemental data using the intervals that delineate the groups or by plotting the data by percentiles (percentage of data sorted into ascending order) if the data are normally distributed; 50,75,90,95 are recommended for elements that are enriched.

Geologists are used to thinking in terms of maps and the most useful end product to compare geochemical data with geology and geophysics is to summarize the data in map form. However, some care should be taken with the preparation of these maps as it is extremely easy to prejudice interpretation and many maps in the literature do not reflect the true meaning of the data but merely the easiest way of representing it. If the data reflect the chemistry of an area, such as a catchment, it is best if the whole area is shaded with an appropriate

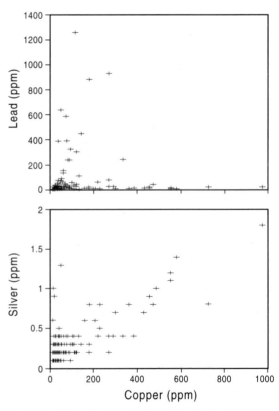

Fig. 8.7 Scatter plots of Daisy Creek data showing the strong correlation of copper and silver and lack of correlation of lead and copper.

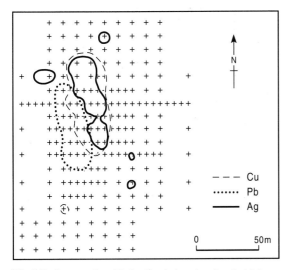

Fig. 8.8 Contour plot of Daisy Creek data showing the highest anomalous populations for copper (>128 ppm), lead (>69 ppm) and silver (>0.55 ppm). The relative size and relative positions of the anomalous areas reflect the primary zoning in the underlying prospect. (From Sinclair 1991.)

colour or tone, (e.g. Fig. 8.9). Data which essentially represent a point can be plotted by posting the value at that point or if there are a lot of data by representing them by symbols, which are easier to interpret. Typical symbols are given by Howarth (1982). A more usual method of presentation if the samples were collected on a grid is to present the data as a contour plot using the intervals from a histogram (Figs 8.8, 8.5). Plots are best coloured from blue to green to yellow to red—cold to hot, the colours used on seismic sections.

The combined statistical analysis and element mapping is used to define areas of interest which need to be investigated in detail, using detailed geology, topography sampling information and, probably, a return to the site. These areas can be classified as to their suggested origin, seepage and precipitation from local groundwaters, from transported material, or probable residual anomalies. The aim must be to provide a rational explanation for the chemistry of all the areas of interest and not merely to say that they are of unknown origin.

8.4 RECONNAISSANCE TECHNIQUES

The application of these techniques has been discussed in section 4.2.2 but the actual method to be used depends on the origin of the overburden. If it is residual, i.e.

derived from the rock underneath, the problem is relatively straightforward but if the overburden is exotic, i.e. transported, then different methods must be used. A summary of the application of various geochemical techniques is given in Table 8.4.

8.4.1 Stream sediment sampling

The most widely used reconnaissance technique in residual areas undergoing active weathering is stream sediment sampling, the object of which is to obtain a sample that is representative of the catchment area of the stream sampled. The active sediment in the bed of a river forms as a result of the passage of elements in solution and in particulate form past the sampling point. Thus a sample can be regarded as an integral of the element fluxes. The simplicity of the method allows the rapid evaluation of areas at relatively low cost. Interpretation of stream sediment data is carried out on an empirical basis by comparing catchments, as there is only a poorly defined relationship between the chemistry of the stream sediment and the catchment from which it is derived. The simple mass balance method put forward by Hawkes (1976) works in tropical areas of intense weathering but requires modification by topographical and elemental factors in other areas (Solovov 1987).

In stream sediment sampling the whole stream sediment or a particular grain size or mineralogical fraction of the sediment, such as a heavy mineral concentrate, can be collected. In temperate terrains maximum anomaly/background contrast for trace metals is obtained in the fine fraction of the sediments as this contains the majority of the organic material, clays and iron and manganese oxides. The coarser fractions including pebbles are generally of more local origin and depleted in trace elements. Usually the size cut-off is taken at 80 mesh (80 holes per inch or <190 μm). However, the grain size giving best contrast should be determined by an orientation survey. For base metal analysis and geochemical mapping a 0.5 kg sample is sufficient but a much larger sample is required for gold analysis due to the very erratic distribution of gold particles. A number of authors (Gunn 1989, Hawkins 1991) have collected 10 kg of -2 mm material and carefully sub-sampled this.

The most common sample collection method is to take a grab sample of active stream sediment at the chosen location. This is best achieved by taking a number of sub-samples over 20–30 m along the stream and at depths of 10–15 cm to avoid excessive iron and manganese oxides. Care must be taken, if a fine fraction is

Table 8.4 Application of the main exploration geochemical methods

Exploration Phase	Plotting Scale	Drainage Sediments	Lake Sediment	Water	Soil	Deep Overburden	Rock	Vegetation	Gas
Reconnaissance	1:10 000 to 1:100 000	[commonly used]	Glacial	Arid	[commonly used]	Laterites Glacial	Good Outcrop	Forests	
Detailed	1:2 500 to 1:10 000	[commonly used]			[commonly used]				Arid
Drilling	>1:1 000						[commonly used]		

[shaded] Commonly used   [light shaded] Useful in areas noted

required, that sufficient sample is collected. In addition contamination should be avoided by sampling upstream from roads, farms, factories and galvanized fences. Any old mine workings or adits should be carefully noted as the signature from these will mask natural anomalies. Most surveys use quite a dense sample spacing such as 1 km^{-2} but this should be determined by an orientation survey. The main mistake to avoid is sampling large rivers in which the signature from a deposit is diluted.

An alternative approach used by geological surveys, for example the British Geological Survey, is to sieve the samples in the field using a minimum of water. The advantages of this are that it is certain that enough fine fraction is collected and samples are easier to carry in remote areas. It also seems that the samples are more representative but the cost efficiency of this in small surveys remains to be demonstrated. A full review of the use of regional geochemical surveys is given in Plant *et al.* (1988).

An example of a stream sediment survey is shown in Fig. 8.9. The object was to define further mineralization in an area of Besshi style massive sulphide mineraliza-tion and the copper data is shown in the figure. The area sampled consists of interbedded units of amphibolite and mica schist, within which a deposit had been drilled at point A. The area suffers very high rainfall and much precipitation of iron and manganese; as a result dispersion trains are very short, of the order of 200 m, as shown by an orientation survey. Thus the regional survey was conducted at very close-spaced intervals. However, the deposit is clearly detected and other prospects were defined along strike from the deposit (point B) and disseminated chalcopyrite was found at C.

Heavy mineral concentrates have been widely used as a means of combating excessive dilution and of enhancing weak signals. The method is essentially a quantification of the gold panning method, which separates grains on the basis of density differences. Panning in water usually separates discrete minerals with a density of greater than 3. Besides precious metals, panning will detect gossanous fragments enriched in metals, secondary ore minerals such as anglesite and insoluble minerals such as cassiterite, zircon, cinnabar, baryte and most gemstones, including diamond. The

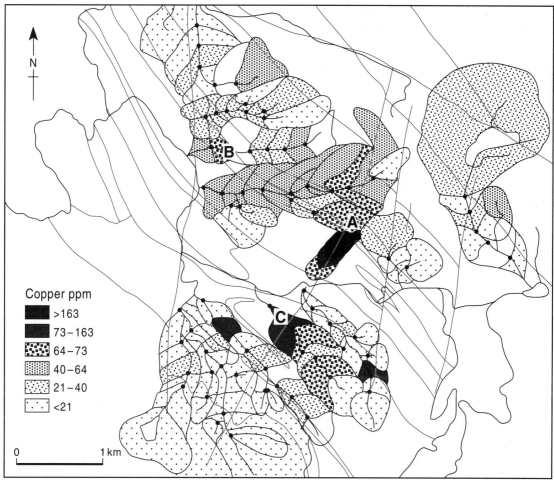

Fig. 8.9 Stream sediment survey of part of the Gairloch area, Wester Ross, NW Scotland. The survey, which was designed to follow up known mineralization (A), detected further anomalies along strike (B) and disseminated mineralization at C. The analytical method was ICP-ES following a $HNO_3–HClO_4$ digestion.

mobility of each heavy mineral will depend on its stability in water, for example sulphides can only be panned close to their source in temperate environments whereas diamonds will survive transport for thousands of kilometres.

The samples collected are usually analysed or the number of heavy mineral grains are counted. The examination of the concentrate can prove very useful in remote areas where laboratory turnaround is slow and the cost of revisiting the area high, as it will be possible to locate areas for immediate follow-up. The major problem is that panning is still something of an art and must be practised for a few days before the sampler is proficient. The diagram and photograph in Figs 8.10 and 8.11 show

panning procedures. Useful checks on panning technique can be made by trial runs in areas of known gold or by adding a known number of lead shot to the pan and checking their recovery. It is usual to start from a known sample size and finish with a concentrate of known mass. However, the differences in panning technique mean that comparisons between different surveys are usually impossible.

8.4.2 **Lake sediments**

In the glaciated areas of northern Canada and Scandinavia access to rivers is difficult on foot but the numerous small lakes provide an ideal reconnaissance sampling

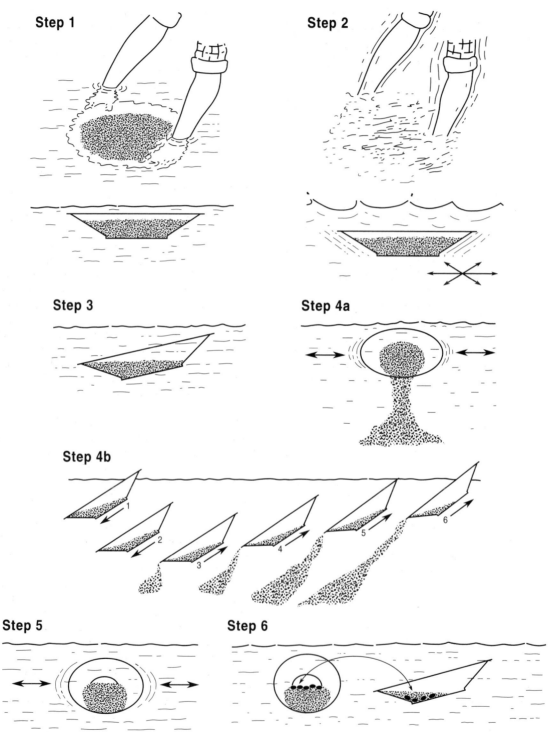

Fig. 8.10 The stages of panning a concentrate from sediment. Often the sediment is sieved to -2 mm although this may eliminate any large nugget. Note the importance of mixing the sediment well so that the heavy minerals stay at the bottom of the pan. For an experienced panner the operation will take about 20 minutes. (Modified from Goldspear 1987.)

Fig. 8.11 Panned concentrate sampling in northern Pakistan. The water is glacial so edge of river sampling is essential.

medium as they are accessible from the air. A sample is taken by dropping a heavy sampler into the lake sediment and retrieving it. The sampling density is highly varied and similar to stream sediments. Productivity is of the order of ten samples per flying hour. Essentially data are interpreted in a similar way to that from stream sediments and corrections should be made for organic matter. Lake sediment is only useful if glacial material is locally derived and is ineffective in areas of glacio-lacustrine material such as parts of Manitoba. Further details are contained in the reviews by Coker *et al.* (1979) and Hornbrook (1989).

8.4.3 Overburden geochemistry

In areas of residual overburden, overburden sampling is generally employed as a follow-up to stream sediment surveys but may be used as a primary survey for small licence blocks or more particularly in areas of exotic overburden. Perhaps the key feature is that it is only employed when land has been acquired. The results are generally plotted at scales from 1:10 000 to 1:1 000.

The method of sampling depends on the nature of the overburden; if the chemistry of the near surface soils

reflects that at depth then it is safe to use the cheap option of sampling soil. If not, then samples of deep overburden must be taken. The main areas where surface soils do not reflect the chemistry at depth are glaciated areas, where the overburden has been transported from another area, in areas of windblown sand and in areas of lateritic weathering where most trace metals have been removed from the near surface layers.

SURFACE SOIL SAMPLING

The simplest sampling scheme is to take near surface soil samples. The major problem is to decide which layer of the soil to sample, as the differences between the layers are often greater than that between sites. The type of soil reflects the surface processes but in general the most effective samples are from a zone at around 30 cm depth formed by the downward movement of clays, organic material and iron oxides, the B horizon indicated in Fig. 8.12, or the near surface organic material (A horizon). This downward movement of material is responsible for the depletion and concentration of trace elements causing variations that may be greater than that over mineralization. It is essential that the characteristics of the soil sampled are recorded and that an attempt is made to sample a consistent horizon. If different horizons are sampled anomalies will reflect this as shown in Fig. 8.13.

The usual method of soil collection in temperate terrains is to use a soil hand auger, as shown in Fig. 8.14. This allows sampling to a depth of the order of 1 m, although normally samples are taken from around 30 cm and masses of around 100–200 g collected for base metal exploration. In other climatic terrains, particularly where the surface is hard or where large samples (500 g–2 kg) are required for gold analysis, then small pits can be dug. The area of influence of a soil sample is relatively small and should be determined during an orientation survey. The spacing is dependent on the size of the primary halo expected to occur across the target and the type of overburden, but a rule of thumb is to have at least two anomalous samples per line if a target is cut. Spacing in the search for veins may be as little as 5 m between samples but 300 m between lines (see section 16.4.2) but for more regularly distributed disseminated deposits may be as much as 100 m by 100 m. Sample spacing may also be dictated by topography: in flat areas or where the topography is subdued then rectilinear grids are the ideal choice, but in mountainous areas ridge and spur sampling may be the only reasonable choice.

A typical example of overburden sampling is shown in

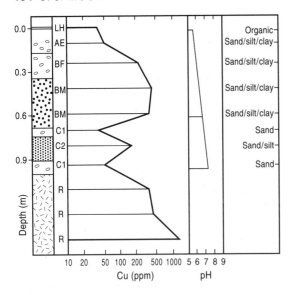

Fig. 8.12 Soil profile showing the effect of sample texture and soil horizons on copper content. Note the bedrock copper content of 0.15% Cu. Letters refer to soil horizons and sub-divisions, from the top L, A, B, C and R (rock). (From Hoffman 1987.)

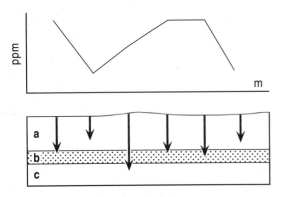

Fig. 8.13 Effect of sampling differing soil horizons with very different metals contents during a soil traverse.

Figs 8.15a–c. The area is on the eastern side of the Leinster Granite in Ireland and the target was tin–tungsten mineralization associated with microgranite dykes, which are also significantly enriched in arsenic and lesser copper and bismuth. Regional base of over-burden sampling (usually 2–5 m depth), which was undertaken as the area has been glaciated, clearly delineates the mineralized dykes (Fig. 8.15a). Surface soil sampling was also effective in delineating the dykes using arsenic and to some extent copper and Fig. 8.15b

Fig. 8.14 Typical surface soil sampling in temperate terrains using hand augers over the Coed-y-Brenin copper prospect, Wales.

demonstrates the movement of elements through the overburden. Tungsten is very immobile and merely moves down slope under the influence of gravity whereas copper and arsenic are more mobile and move further down slope. The density of soil samples on the surface traverse is clearly greater than needed. The reader may care to consider the maximum sample spacing necessary to detect two anomalous samples in the traverse.

A special form of residual overburden is found in lateritic areas such as in Amazonian Brazil or Western Australia where deep weathering has removed trace elements from the near surface environment and geochemical signatures are often very weak. One technique that has been widely used is the examination of gossans. As discussed in section 5.1.6 gossans have relic textures which allow the geologist to predict the primary sulphide textures present at depth. In addition the trace element signature of the primary mineralization is preserved by immobile elements. For example, it is difficult to discriminate gossans overlying nickel deposits from other iron rich rocks on the basis of their nickel contents, as nickel is mobile, but possible if their multi-element signatures are used (e.g Ni, Cr, Cu, Zn, Mo, Mn) or the

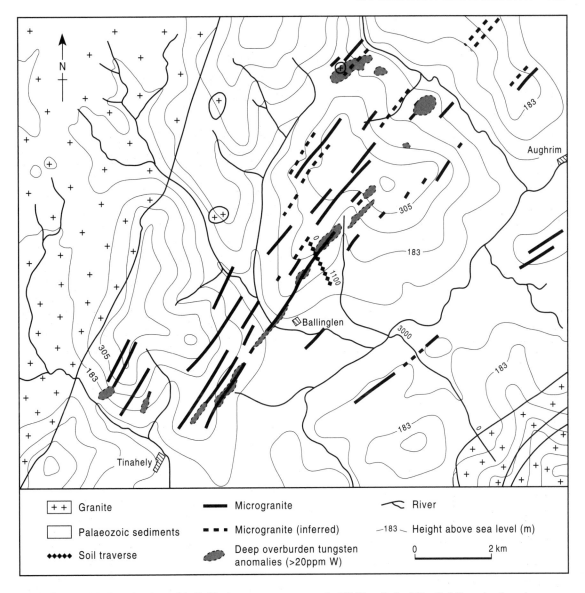

Fig. 8.15(a) Regional geochemistry of the Ballinglen tungsten prospects, Co. Wicklow, Ireland. Detailed dispersion through overburden and soil response are shown in Figs 15 (b) and (c). (From Steiger & Bowden 1982.)

Pt, Pd, Ir content (Travis *et al.* 1976, Moeskops 1977, Smith 1977).

Drilling to fresh rock, often 50–60 m, has been widely used to obtain a reasonable signature despite the costs involved. Mazzucchelli (1989) suggests that costs are of the order of A\$10 m⁻¹ and could easily reach A\$1Mkm⁻² for detailed grids. He recommends surface sampling to outline areas for follow-up work.

TRANSPORTED OVERBURDEN

In areas of transported overburden, such as glaciated terrains of the Canadian Shield or sandy deserts, sampling problems are severe and solutions to them expensive.

In glaciated terrains overburden rarely reflects the underlying bedrock and seepages of elements are only present where the overburden is less than about 5 m

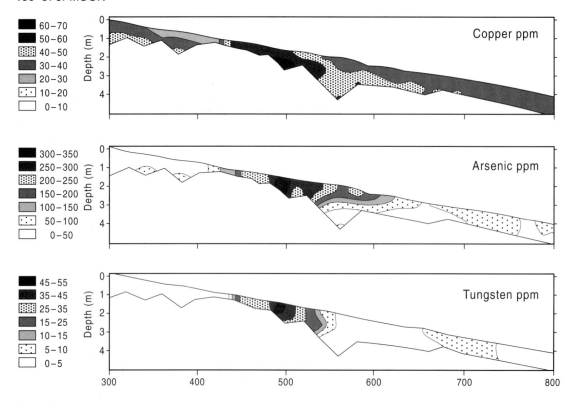

Fig. 8.15(b) Dispersion through overburden along the soil traverse shown in Fig. 8.15(a). Note the different mobilities of tungsten, copper and arsenic.

thick. In addition the overburden can be stratified with material of differing origins at different depths. If the mineralization is distinctive, it is often possible to use boulder tracing to follow the boulders back to the apex of the boulder fan, as in the case of boulders with visible gold, sulphides or radioactive material. Generally however the chemistry of the tills must be examined and basal tills which are usually of local origin sampled. Fig. 8.16 shows a typical glacial fan in Nova Scotia. Usually basal till sampling can only be accomplished by drilling; the most common methods are light percussion drills with flow-through samplers or heavier reverse circulation or sonic drills. Lightweight drills are cheaper and easier to operate but results are often ambiguous, as it is not easy to differentiate the base of overburden from striking a boulder. The use of heavy equipment in most glaciated areas is restricted to the Winter when the ground is frozen.

In sandy deserts water is scarce, most movement is mechanical and most fine material is windblown. Thus the -80 mesh fraction of overburden is enriched in wind blown material and is of little use. In such areas either

a coarse fraction (e.g. 2–6 mm), reflecting locally derived material or the clay fraction reflecting elements moved in solution is used (Carver *et al.* 1987). One of the most successful uses of this approach has been in the exploration for kimberlites in central Botswana. Fig. 8.17 shows the result of a regional sampling programme which discovered kimberlite pipes in the Jwaneng area. Samples were taken on a 0.5 km grid, heavy minerals separated from the +0.42 mm fraction and the number of kimberlite indicator minerals, such as picro-ilmenite counted. The anomalies shown are displaced from the sub-outcrop probably due to transport by the prevailing north-easterly winds.

8.4.4 Hydrogeochemistry

Hydrogeochemistry uses water as a sampling medium. Although water is the most widely available material for geochemistry, its use is restricted to very specific circumstances. The reasons for this are that not all elements show equal dissolution rates, indeed many are insoluble, contents of trace elements are very low and

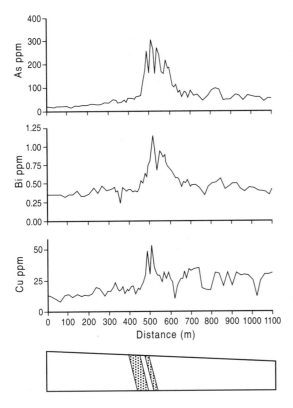

Fig. 8.15(c) Soil traverse across the mineralized dyke. Note the sharp cut-off upslope and dispersion downslope.

difficult to measure, being highly dependent on the weather and easily contaminated by human activity. In general where surface waters are present it is far more reliable and easier to take a stream sediment sample. The possible exception is when exploring for fluorite as fluoride is easily measured in the field using a portable single ion electrode. Where hydrogeochemistry becomes useful is in exploration of arid areas with poor outcrop. In this terrain water wells are often drilled for irrigation and these wells tap deep aquifers that can be used to explore in the subsurface. This approach has been used in uranium exploration although it has not been very successful as the uranium concentration is dependent on the age of groundwater, the amount of evaporation and its source, which are extremely difficult to determine.

8.4.5 Gases

Gases are potentially an attractive medium to sample as gases can diffuse through thick overburden; in practice

most surveys have met with discouraging results. A number of gases have been used of which mercury has been the most successful. Mercury is the only metallic element which forms a vapour at room temperature and it is widely present in sulphide deposits, particularly volcanic-associated base metal deposits. The gas radon is generated during the decay of uranium and has been widely used with some success. More recently the enrichment of carbon dioxide and depletion of oxygen caused by weathering of sulphide deposits has been tested, particularly in the western USA (Lovell & Reid 1989).

In general results have been disappointing because of the large variations in gas concentration (partial pressure) caused by changes in environmental conditions, particularly changes in barometric pressure and rainfall. The more successful radon and mercury surveys have sought to overcome these variations by integrating measurements over weeks or months. Gaseous methods work best in arid areas as diffusion in temperate climates tends to be overshadowed by movement in groundwater.

8.4.6 Vegetation

Vegetation is used in two ways in exploration geochemistry. Firstly the presence, absence or condition of a particular plant or species can indicate the presence of mineralization or a particular rock type, and is known as *geobotany*. Secondly the elemental content of a particular plant has been measured, this is known as *biogeochemistry*. Biogeochemistry has been used more widely than geobotany and has found particular application in the forest regions of northern Canada and Siberia where surface sampling is difficult, but it should only be used with caution. A comprehensive treatment of the subject can be found in Brooks (1983) and the biogeochemical part has been revised in Dunn (1989).

GEOBOTANY

One of the pleasurable, although all too infrequent, parts of exploration is to identify plants, in particular flowers associated with mineralization. The most famous of these is the small mauve copper flower of the Zambian Copper belt, *Beccium Homblei*. In general this plant requires a soil copper content of 50–1600 ppm Cu to thrive, conditions that are poisonous to most plants (Reedman 1979). This type of plant is known as an indicator plant. Unfortunately recent research has demonstrated that most indicator plants will flourish under other conditions and are not very reliable.

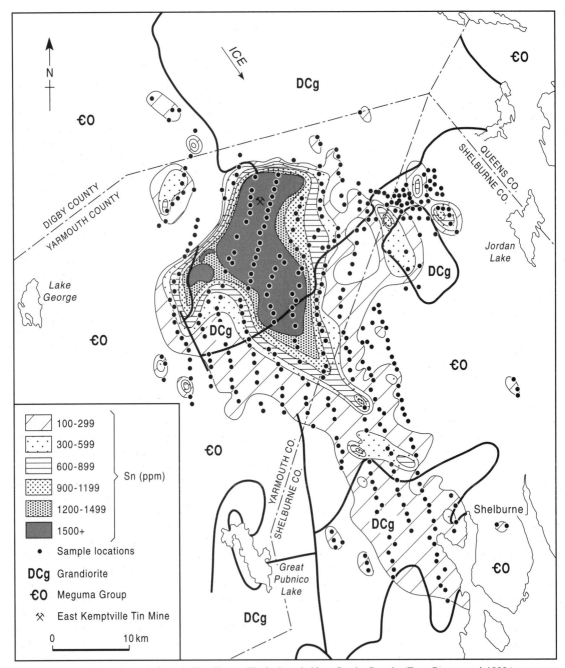

Fig. 8.16 Glacial dispersion train from the East Kemptville tin deposit, Nova Scotia, Canada. (From Rogers *et al.* 1990.)

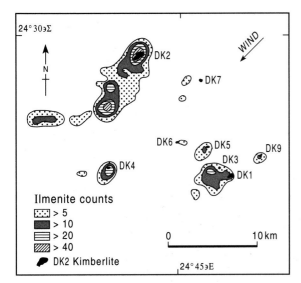

Fig. 8.17 Dispersion of ilmenite around the Jwaneng kimberlites, Botswana. Ilmenite is one of the heavy minerals used to locate kimberlites. (From Lock 1985.)

A more reliable indicator of metal in soils is the stunting of plants or yellowing (chlorosis). This condition is particularly amenable to remote sensing from satellites. Most gardeners appreciate that plants have favourite soil conditions and it is possible to make an estimate of bedrock based on the distribution of plant species. Unfortunately this method tends to be rather expensive and is little used.

BIOGEOCHEMISTRY

Plants require most trace elements for their survival and take these through their roots, transpiring any residues from their leaves and concentrating most trace elements in their recent growth. Unfortunately for the geochemist the rate of uptake and concentration of elements is highly dependent on the species and the season. In general sampling is conducted on one plant species and one part of the plant, usually first and second year leaves or twigs. The variation in concentration with season is so severe that sampling must be restricted in time. In general around 500 g of sample are collected and ashed prior to analysis.

The main advantage of biogeochemistry is that tree roots often tap relatively deep water tables which in glacial areas can be below the transported material and representative of bedrock. A similar effect can be obtained by sampling the near surface humic horizons

which largely reflect decayed leaves and twigs from the surrounding trees (Curtin *et al.* 1974).

8.5 **FOLLOW-UP SAMPLING**

Once an anomaly has been found during reconnaissance sampling and a possible source identified, it is necessary to define that source by more detailed sampling, by highlighting areas of elemental enrichment and eliminating background areas until the anomaly is explained and a bedrock source, hopefully a drilling target, proven. If the reconnaissance phase was stream or lake sampling, this will probably involve overburden sampling of the catchments. In the case of soil or overburden sampling it will mean increasing the density of sampling until the source of an anomaly is found, proving the source by deep overburden sampling and sampling rocks at depth or at surface when there is outcrop. A typical source can be seen in Fig. 8.15. The regional tungsten anomaly was followed up by deep overburden sampling such as that shown in Fig. 8.15b and the suboutcrop of the dyke defined. The dip of the dyke could be estimated from the occurrence of disseminated sulphides, which would also respond to induced polarization, and a drilling target outlined.

8.5.1 **Rock geochemistry**

Rock sampling is included in the techniques for follow-up because although it has been applied with some success in regional reconnaissance, it is really in detailed work, where there is good outcrop or where there is drill core, that this technique becomes most effective.

On a regional basis the most successful applications have been in the delineation of mineralized felsic plutons and of exhalative horizons. Full details are provided in reviews by Govett (1983, 1989). Plutons mineralized in copper and tungsten are usually enriched in these elements but invariably show high variability within a pluton. Tin mineralization is associated with highly evolved and altered intrusives and these are easily delineated by plotting on a K–Rb versus Rb–Sr and Mg–Li diagram. It is recommended that 30–40 samples are taken from each pluton to eliminate local effects. Studies in the modern oceans indicate haloes of the order of kilometres around black smoker fields and these have been detected around a number of terrestrial massive sulphide deposits, notably by the manganese content in the exhalative horizon.

Some care needs to be taken in the collection of rock samples. In general 1 kg samples are sufficient for base metal exploration but precise precious metal determination requires larger samples, perhaps as large as 5 kg, if the gold is present as discrete grains. Surface weathering products should be removed with a steel brush. Rock geochemistry depends on multi-element interpretation and computer based interpretation with careful subdivision of samples on the basis of lithology.

MINE SCALE USES

One of the major applications of rock geochemistry is in determining the sense of top and bottom of prospects and detecting alteration that is not obvious from visual examination. Geochemists from the former Soviet Union have been particularly active in developing an overall zonation sequence which they have used to construct multiplicative indications of younging for a variety of deposit types (Govett 1983 & 1989). Rock geochemistry has been widely applied in the search for volcanic-associated massive sulphide deposits in Canada, especially in combination with down hole e.m. methods (section 14.3.8). In particular rock geochemistry has detected the cone shaped alteration zones beneath them, which are enriched in Mg and depleted in Na, Ca and sometimes in K, and recognized distal exhalite horizons which are enriched in more mobile elements such as arsenic, antimony and in barium (for a good case history see Severin *et al.* 1989). Another major application is in the exploration of sulphide-rich epithermal gold deposits, which often show element zonation. The more volatile elements such as As, Sb and Hg are concentrated above the major gold zones whereas any copper is at depth. The successful application of these methods on a prospect demands a detailed knowledge of the subsurface geology.

8.6 SUMMARY

Exploration geologists are more likely to be directly involved in geochemical surveys than in geophysical ones and they should therefore have considerable knowledge of this technique and its application in mineral exploration. A geochemical programme naturally commences with a planning stage which involves choice of the appropriate field survey and analytical methods suitable for use for a particular area and the commodity sought. Individual areas will present very different problems, e.g. areas with deep or shallow overburdens, and therefore orientation surveys (section 8.1.1) are very desirable. If possible such a survey should be one in which a variety of sampling methods are tested over a deposit of similar geology to that of the target and in similar topographical conditions to determine which produces the best results. This done, the logistics of the major survey can be planned.

The analysis of samples collected during geochemical surveys is generally carried out by contract companies and the quality of the data they produce must be monitored and for this reason the geologist in charge must be familiar with sample collection and preparation, analytical methods and the statistical interpretion of the data obtained.

The material which will be sampled depends upon whether the overburden is residual or transported. The most widely used technique in residual areas suffering active weathering is stream sediment sampling which is a relatively simple method and allows the rapid evaluation of large areas at low cost. Panning to collect heavy mineral concentrates can be carried out at the same time thus enhancing the value of this method. In glaciated areas of difficult access lake sediment rather than stream sediment surveys may be used.

Overburden geochemistry is often a follow-up to stream sediment surveys in areas of residual and transported overburden. The method of sampling is very varied and dictated by the nature of the overburden. When the overburden is transported (glaciated terrains or sandy deserts) sampling problems are severe and solutions to them expensive.

The use of hydrogeochemistry, gases and vegetation is restricted but rock geochemistry is increasing in importance both in surface and underground work as well as by utilizing drill core samples.

8.7 FURTHER READING

Geochemistry in Mineral Exploration by Rose *et al.* (1979) remains the best starting point. A good general update on soil sampling is provided by Fletcher *et al.* (1987) with reviews of most other geochemical techniques in Garland (1989). The *Handbook of Exploration Geochemistry* series, although expensive, provides comprehensive coverage in the volumes so far published: chemical analysis (Fletcher 1981), statistics and data analysis (Howarth 1982), rock geochemistry (Govett 1983), lateritic areas (Butt & Zeegers 1992), arctic areas (Kauranne *et al.* 1992) and stream sediments (Hale & Plant 1994). Readers are strongly advised to try the practical problems in Levinson *et al.* (1987).

9 Evaluation Techniques

BARRY C. SCOTT AND MICHAEL K. G. WHATELEY

INTRODUCTION

As an exploration geologist you will be expected to be familiar with mineral deposit geology, to understand the implications of extraction on the hydrology of the mineral deposit area, to recognize the importance of collecting geotechnical data as strata control problems may have considerable impact on mine viability, to be able to propose suitable mining methods and to assess the economic viability of a deposit. To be able to cope with all these tasks you will need to be technically competent particularly in sample collection, computing and mineral resource evaluation. The basis of all geological evaluation is the sample. Poor sample collection results in unreliable evaluation.

This chapter will present some of the methods that are used in the field to obtain representative samples of the mineralized rock that will enable the geologist to undertake mineral deposit evaluation. This includes the various drilling techniques, pitting and trenching as well as face and stope sampling. Once these data have been collected they are evaluated to determine how representative they are of the whole deposit. This is achieved using statistics, a subject covered in the first part of the chapter.

The deposit's geological and assay (grade) data are prepared for presentation by the geologist. It is these data that form the basis of the evaluation. In an early phase of a project, global resource estimates, using classical methods will normally be adopted. Later, as more data become available, local estimates are calculated using geostatistical methods (often computer based). Some aspects of these evaluation methods are also presented.

Finally, to ensure that the maximum amount of information is derived from drill core (other than the obvious grade, thickness, rock type, etc.) a basic geotechnical and hydrogeological outline is given. This information may be used in assessing possible dilution of the ore upon mining, or may identify strata control or serious water problems which may occur during mining.

9.1 SAMPLING

9.1.1 Introduction to statistical concepts

POPULATION AND SAMPLE

Sampling is a scientific, selective process applied to a large mass or group (a *population*, as defined by the investigator) in order to reduce its bulk for interpretation purposes. This is achieved by identifying a component part (a *sample*) which reflects the characteristics of the parent population within acceptable limits of *accuracy, precision* and *cost effectiveness*. In the minerals industry the average grade of a tonnage of mineralized rock (the population) is estimated by taking samples which are either a few kilograms or tonnes in weight. These samples are reduced to a few grams (*the assay portion*) which are analysed for elements of interest.

Results of analysed samples plotted as a frequency curve are a pictorial representation of their distribution (Fig. 9.1). Distributions have characteristics such as mid-points and other measures which indicate the spread of the values and their symmetry. These are *parameters* if they describe a population and *statistics* if they refer to samples.

In any study the investigator wishes to know the parameters of the population (i.e. the 'true' values) but these cannot be established unless the population is taken as the sample. This is normally not possible as the population is usually several hundred thousand or millions of tonnes of mineralized rock but a *best estimate* of these parameters can be made from sampling the population, and from the statistics of this sampling. Indeed a population can be regarded as a collection of potential samples, probably several million or more, waiting to be collected.

It is fundamental in sampling that samples are representative, at all times, of the population—if they are not the results are incorrect. The failure of some mineral ventures, and losses recorded in the trading of mineral commodities, can be traced to unacceptable sampling

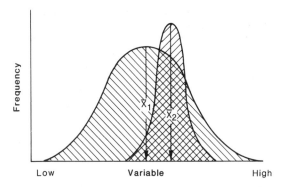

Fig. 9.1 Two normal distributions. Distribution 1 has a lower arithmetic mean ($\overline{x}_1$) than distribution 2 but higher variance (i.e. a wider spread). Distribution 2 has the reverse, a higher mean ($\overline{x}_2$) but a lower variance.

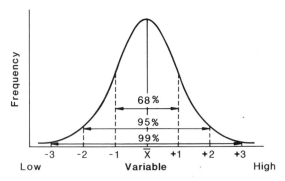

Fig. 9.2 The normal distribution: the variable has a continuous and symmetrical distribution about the mean. The curve shows areas limited and occupied by successive standard deviations.

procedures due to confusion between taking samples, representative of the object being evaluated, and specimens whose degree of representation is not known.

HOMOGENEITY AND HETEROGENEITY

Homogeneity is the property that defines a population whose constituent units are strictly identical with one another while the reverse condition is heterogeneity. Sampling of the former material can be completed by taking any group of these units as, say, the most accessible fraction. Mineralization is composed of dissimilar constituents and is heterogeneous—it is rarely (if ever) homogeneous—and correct sampling of such material has to ensure that all constituent units of the population have a uniform probability of being selected to form the sample, and the integrity of the sample is respected (see later). This is the concept of *random sampling*.

NORMAL AND ASYMMETRICAL DISTRIBUTIONS

This is a brief introduction to the subject of distributions and for details the reader is referred to standard texts such as Davis (1986) and Issaks & Srivastava (1989).

Normal distribution. In a normal distribution the distribution curve is always symmetrical and bell-shaped. By definition, the mean of a normal distribution is its mid-point and the areas under the curve on either side of this value are equal. Another characteristic of this distribution, or curve, is the spread or dispersion of values about the mean which is measured by the variance, or the square root of the variance called the standard deviation. Variance (σ^2) is the average squared deviation of all possible values from the population mean:

$$\sigma^2 = \frac{\Sigma(x_i - \overline{X})^2}{n}$$

where σ^2 = population variance, x_i = any sample value, $\overline{X}$ = population mean, n = number of samples.

Variance units are the square of the units of the original observation. A small variance indicates that observations are clustered tightly about the arithmetic mean while a large variance shows that they are scattered widely about the mean, and that their central clustering is weak. A useful property of a normal distribution is that within any specified range, areas under its curve can be exactly calculated. For example, slightly over two-thirds (68%) of all observations are within one standard deviation on either side of the arithmetic mean and 95% of all values are within ±2 (actually 1.96) standard deviations from the mean (Fig. 9.2). The mean of observations, or average, is their total sum divided by the number of observations, n.

Asymmetrical distribution. Much of the data used in geology has an asymmetrical rather than a normal distribution. Usually such distributions are skewed to the left (a positive skew) which means that the data have a preponderance of low values. Measures of such populations include the *mode* which is the value occurring with the greatest frequency (i.e. the highest probability); the *median* which is the value midway in the frequency distribution which divides the area below the distribution curve into two equal parts, and the *mean* which is the arithmetic average of all values. In asymmetrical distributions the median lies between the mode and the mean (Fig. 9.3); in normal curves these three

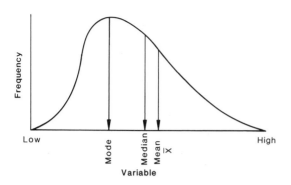

Fig. 9.3 Asymmetrical distribution, positively skewed (to the left).

Table 9.1 Critical values of 't' for different sample numbers and confidence levels

Number of samples (n)	Confidence levels 80%	90%	95%	99%
1	3.08	6.3	12.71	31.82
5	1.48	2.02	2.57	3.37
10	1.37	1.81	2.23	2.76
25	1.32	1.71	2.06	2.49
40	1.30	1.68	2.02	2.42
60	1.30	1.67	2.00	2.39
120	1.29	1.66	1.98	2.36
>120	1.28	1.65	1.96	2.33

The number of samples (n) is referred to as 'the degrees of freedom'.

measures coincide.

In asymmetrical distributions there is no comparable relationship between standard deviation (or variance) and the area under the distribution curve, as in a normal counterpart. Consequently, mathematical transformations have been used whereby skewed data are transformed to normal. Perhaps the commonest is the log normal transformation where the natural log of all values is used as the distribution and many geological data approximate to this type of distribution but this method should be used with caution. An alternative approach is to use non-parametric statistics.

Parametric and non-parametric statistics. Parametric statistics, discussed above, specify conditions regarding the nature of the population being sampled. The major concern is that the values must have a normal distribution but we know that many geological data do not have this characteristic. Non-parametric statistics, however, are independent of this requirement and thus relevant to the type of statistical testing necessary in mineral exploration. Routine non-parametric statistical tests are available as computer packaged programs.

POINT AND INTERVAL ESTIMATES OF A SAMPLE

Point estimates are the arithmetic mean ($\bar{x}$), variance (S^2) and number of samples (n). The point estimate $\bar{x}$ provides a best estimate of the population mean ($\bar{X}$) but by itself it is usually wrong and contains no information as to the size of this error as this is contained in the variance (S^2). A quantitative measure of this error, however, is obtained from a two-sided interval estimate based on the square root of the variance, the standard deviation (S).

Two-sided interval estimates. If $\bar{x}$ is the mean of n random samples taken from a normal distribution with population mean $\bar{X}$ and variance S^2, then at a 95% probability $\bar{x}$ is not more than 2 standard deviations (S) larger or smaller than $\bar{X}$. In other words, there is a 95% chance that the population mean is in this calculated interval and five chances in a hundred that it is outside. The relationship is:

$$\bar{x} + \frac{1.96\sigma}{\sqrt{n}} > \bar{X} > \bar{x} - \frac{1.96\sigma}{\sqrt{n}}.$$

This method is correct but useless because we do not know σ and the expression becomes:

$$\bar{x} + t_{2.5\%}\frac{S}{\sqrt{n}} > \bar{X} > \bar{x} - t_{2.5\%}\frac{S}{\sqrt{n}}$$

where σ is replaced by its best estimate S, and the normal distribution is replaced by a distribution which has a wider spread than the normal case. In this t distribution the shape of its distribution curve varies according to the number of samples (n) used to estimate the population parameters (Table 9.1). As n increases, however, the curve resembles that of the normal case and for all practical purposes for more than 50 samples the curves are identical. Examples of the use of this expression are given in Table 9.2 and Figure 9.4.

The width (i.e. spread) of a 2-sided interval estimate about the sample arithmetic mean depends on the term:

$$t\frac{S}{\sqrt{n}}$$

The magnitude, and width, of this term is reduced by:
1 choosing a lower confidence interval (a lower t value); an example of this approach is given in Table 9.3; the

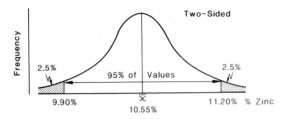

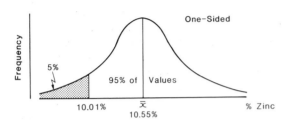

Fig. 9.4 Graphical representation of a two-sided and one-sided confidence interval at 95% confidence level. See Table 9.2 for calculations.

Table 9.2 Two-sided and one-sided confidence intervals at 95% probability

	Two-sided	One-sided
$\bar{x}$	10.55% Zn	10.55% Zn
S	3.60%	3.60%
n	121	121
$\sqrt{n}$	11	11
$\dfrac{S}{\sqrt{n}}$	0.33	0.33
$t_{2.5\%}$	1.96	NA
$t_{5\%}$	NA	1.65
$t(S/\sqrt{n})$	0.65	0.54
Upper limit	11.20%	NA
Lower limit	9.90%	10.01%

1. The total number of samples needed to halve the standard error of the mean $(S/\sqrt{n})$, and thus the range of the confidence interval, is 484 [= (2 × 11)^2].

2. With a one-sided confidence interval all the risk is placed into one side of the interval. Consequently the calculated limit is closer to the $\bar{x}$ statistic than to the corresponding limit for a two-sided interval. In the above example the upper limit could have been calculated instead of the lower.

3. See Table 9.1 for 't' values and Fig. 9.4 for a graphical representation.

Table 9.3 Effect of changing the confidence level on the range of a two-sided confidence interval

The same zinc mineralization as in Table 9.2 is used. Grade $(\bar{x})$ = 10.55% zinc. Standard Deviation (S) = 3.60%. Number of samples (n) = 121.

Confidence level	50%	80%	90%	95%
Percentile of 't'	$t_{25\%}$	$t_{10\%}$	$t_{5\%}$	$t_{2.5\%}$
Value of 't'	0.68	1.28	1.65	1.96
$t(S/\sqrt{n})$	0.22	0.42	0.54	0.65
Upper limit (U1)	10.77%	10.98%	11.10%	11.20%
Lower limit (U2)	10.33%	10.12%	10.00%	9.90%
Range (U1–U2)	0.44%	0.76%	1.10%	1.30%

1. The range becomes narrower as the confidence level decreases, and the risk of the population mean being outside this interval correspondingly decreases. At the 50% level this risk is 1 in 2, but at the 95% level it is 1 in 20.

2. See Table 9.1 for values of 't'.

disadvantage of this is that as the t value decreases the probability (i.e. chance) of the population mean being outside the calculated confidence interval increases and usually a probability of 95% is taken as a useful compromise;

2 increasing the number of samples (n), but the relationship is an inverse square root factor, thus the sample size, n, must be increased to 4n to reduce the factor 1/n by half, and to 16n to reduce the factor to one quarter, clearly the question of cost effectiveness arises here;

3 decreasing the size of the total variance, or total error $[S^2(TE)]$, of the sampling scheme. In this sense sampling error refers to variance (S^2), or standard deviation (S). A small standard deviation narrows the interval about the arithmetic mean and the smaller this interval the closer the sampling mean is to the population mean. In other words, the accuracy of the former as a best estimate of the latter is improved by reducing the variance. This is achieved by correct sampling procedures, as discussed below.

One-sided interval estimates. In this approach only one side of the estimate is calculated, either an upper or lower limit, then at 95% probability:

$$U_L = \bar{x} - t_{5\%} \frac{S}{\sqrt{n}} < \bar{X} \quad U_L = \text{Lower limit}$$

$$U_U = \bar{x} + t_{5\%} \frac{S}{\sqrt{n}} > \bar{X} \quad U_U = \text{Upper limit}$$

It is usually more important to set U_L, so that mined rock is above a certain cut-off grade. With a one-sided estimate all the risk has been placed into one side of the interval. Consequently, the calculated limit for a one-sided estimate is closer to the $\bar{X}$ statistic than it is to the corresponding limit for a two-sided interval. Examples are given in Table 9.2 and Fig. 9.4. Through interval estimates, the variability (S), sample size (n) and the mean ($\bar{X}$) are incorporated into a single quantitative statement.

9.1.2 Sampling error

GLOBAL ESTIMATION ERROR [S^2(GE)]

Essentially, sampling consists of three main steps: the actual extraction of the sample(s) from the *in situ* material comprising the population; the preparation of the assay portion which involves a mass reduction from a few kilograms (or tonnes) to a few grams for chemical analysis, and the analysis of the assay portion.

Global error includes all errors in sample extraction and preparation (total sampling error TE), and analysis (AE).

$$S^2(GE) = S^2(TE) + S^2(AE)$$

The TE includes selecting the assay portion for analysis while the analytical error proper excludes this step. Gy (1992) maintains that analysts usually complete their work in the laboratory with great accuracy and precision and that, in this strict sense, associated analytical error is negligible as compared with the variance of taking the assay portion. Consequently, for practical purposes $S^2(AE) \approx 0$ and from this $S^2(GE) \approx S^2(TE)$.

TOTAL SAMPLING ERROR [S^2(TE)]

The total sampling error (TE) includes all extraction and preparation errors. It includes errors of selection (SE) and preparation (PE): the former are inherent in both extraction and preparation while the latter are restricted to mechanical processes used in sample reduction such as crushing, etc.

$$S^2(TE) = S^2(SE) + S^2(PE)$$

Selection errors (SE) are minimized by:

1 A correct definition of the number of point increments, or samples. Sample spacing is best defined by geostatistics which is described in section 9.5.

2 A correct definition of the area sampled. Sampling of mineral deposits presents a three-dimensional problem but for practical purposes they are sampled as two-dimensional objects which equate to the surface into which samples are cut. The long axis of the sample preferably should be perpendicular to the dip of the mineralization, or at least at an angle to it, but not parallel. The cut in plan is ideally a circle, sometimes a square or rectangle, and in length should penetrate the full length of the sampled area.

3 A correct extraction and collection of the material delimited by the above cut. Usually this presents a problem with the planes of weakness in rocks, and their variation in hardness. Following from (2) and (3) an optimum sample is provided by diamond drilling which, with 100% core recovery, cuts almost perfect cylindrical lengths of rock the full length of the sampled area.

4 A correct preparation of the assay portion from the original sample, as described later.

Selection errors are of two main categories which arise from:

1 the inherent heterogeneity of the sample and

2 the selective nature of the sample extraction and its subsequent reduction in mass and size to the assay portion.

The first category comprises the *fundamental error* [S^2(FE)] and is the irreducible minimum of the total sampling error and can be estimated from the sampling model of Gy (1992), as in section 9.1.4 below. Gy (1992) separates those of the second category into seven different types which are estimated from measurements (i.e. mass and grain size) from each stage of a sample reduction system.

Preparation errors [S^2(PE)] are related to processes such as weighing, drying, crushing, grinding, etc. whose purpose is to bring successive samples into the form required by the next selection stage, and ultimate analysis. Sources of possible error are.

1 Alteration of the sample's chemical composition by overheating during drying as, for example, with coal and sulphides of mercury, arsenic, antimony and bismuth.

2 Alteration of the sample's physical condition. If the grain size of the sample is important this can be changed by drying and careless handling.

3 Once a sample is collected precautions must be taken to avoid losses during preparation of the assay portion. Losing material always introduces error (i.e. increases the variance) because the various sized fractions differ in grade and it is usually the finer grained material which is lost. Losses can be checked by weighing samples and rejects at each reduction stage.

4 Sample contamination must be avoided: sample containers and preparation circuits must be clean and free from foreign material, including the addition of dust produced by the size reduction of other samples. Accidental sources of added material include steel chips from hammers, and crushing and grinding equipment.

5 Unintentional sampling errors must be avoided such as mis-labelling and the mixing of fractions of different samples.

In essence preparation errors result from the careless treatment of samples and can be caused by the use of inadequate crushing and grinding equipment, insufficient drying capacity and inappropriate subsample splitting.

Obviously, PEs can be estimated only from the results of an operating sample reduction system and, like SEs, cannot be estimated directly for either preliminary calculations or the design of such a system. Fortunately the totality of both can be estimated from the relationship

$$S^2(TE) \le 2S^2(FE)$$

where the variance of the total sampling error does not exceed twice that of the fundamental error (Gy 1992).

SUMMARY

Sampling consists of three stages:

Extraction of the original sample from *in situ* material (i.e. the population), including its delimitation and collection.

Preparation which involves a reduction in both mass and grain size of the original sample to an assay portion for chemical analysis.

Chemical analysis of the assay portion.

Each of the above three stages is discussed in turn.

9.1.3 **Sample extraction**

Random sampling. Correct sampling technique requires a random selection of each sample from the population. At an operating mine, it is possible that sampling is non-random since it is completed from mine development which is planned on a systematic basis. Provided, however, that the mineral particles which form the population are themselves randomly distributed this objection is overcome. Many mineral deposits display trends and variations in the spatial distribution of their values and

the effect of this on the concept of their random distribution is a matter of debate.

Sample volume–variance relationship. Preferably samples should be of equal volume (i.e. length) or if volumes are unequal there must be no correlation between this and their analytical results. Although the mean of observations based on large volumes of rock should not be expected to be different from those based on small volumes (e.g. larger or smaller diameter drill core), the corresponding variance might be expected to be smaller. This is because the variability of the sample depends both on the variance and the number of samples taken (n):

$$\text{variability} \propto \frac{S^2}{n}$$

and instinctively we feel that there should be a similar relationship with the volume of the sample taken. Indeed, it has been advanced that:

$$\text{variability} \propto \frac{S^2}{\text{volume}}$$

It appears in practice, however, that there is no such simple, direct, relationship. This can be tested by splitting samples in half (e.g. mineralized parts of a drill core), analysing them separately and then combining the results as if the whole core sample had been tested. If the values from the two halves are statistically independent values from a single population, then the ratio of the variance of the values for the whole sample to the average values for the two halves should be 0.5. Results show a calculated ratio of 0.9 or greater so that little appears to be gained by analysing the whole sample rather than half of it.

If the sample volume–variance relationship were important it would be desirable to analyse all portions of a core from a drill hole rather than follow the common practice of splitting the core longitudinally to save a half as a record and use half for analysis. In sampling, however, this relationship should always be considered and, if possible, tested. This is particularly relevant if the valuable constituents have a low grade and are erratically distributed (e.g. disseminated gold and PGE values) and little is known of the variance. In such cases it is sensible either to slice the entire core longitudinally saving 10% as a record and sending 90% for analysis, or collect larger core.

This topic was discussed by Sutherland & Dale (1984) in relation to the minimum sample size for sampling placer deposits, such as alluvial diamonds, and they concluded that the major factor affecting variance was

Table 9.4 Sample density–volume relationship

Length	Pb+ Zn%	S.G.	Length × Grade	Length Length × SG	Length × Grade × SG
1.5	8.7	3.03	13.05	4.55	39.54
2.0	15.3	3.27	30.60	6.54	100.06
1.5	18.1	3.37	27.15	5.06	91.50
1.0	7.6	2.99	7.6	2.99	22.72
1.0	5.1	2.91	5.1	2.91	14.84
2.0	14.9	3.26	29.80	6.52	97.15
1.5	8.2	3.01	12.30	4.51	37.02
1.0	12.3	3.16	12.30	3.16	38.87
11.5			137.90	36.24	441.70

Average grade by volume% = 137.90 / 11.5 = 12.0%
Average grade by weight % = 441.70 / 36.24 = 12.2%

sample volume.

Sample density–volume relationship. In making re-serve calculations, grades are determined by using lin-ear %, area % and volume % relationships. Analytical data are expressed in terms of weight and unless the weight/volume ratio is unity, or varies within a narrow range, average grades calculated by volume % will be in error. This problem is illustrated in Table 9.4 which is an example of galena and sphalerite mineralization in limestone where the average grade calculated using volume percentage is lower than that calculated from weight percentage.

SAMPLING PROCEDURES

Introduction. Variance can be reduced by using well planned sampling procedures with as thorough and meticulous extraction of samples as possible. The sam-ple location is surveyed to the nearest survey point and cleaned by either chipping or washing from the rock face all oxidized and introduced material so as to expose fresh rock. A sampling team usually consists of two people, the senior of whom maintains the sampling notebook. At each location a visual description of the rock is made, noting the main rock types and minerals present, percentage of potential valuable mineral(s), rock alteration, other points of interest, and who is taking the sample. Obviously each sample has a unique number—a safe practice is to use numbered aluminium sample tags with one placed inside the plastic sample bag and the other attached by thin wire about its neck.

There are three hand sampling techniques, namely chan-nel, chip and grab sampling.

Channel sampling. A channel, usually about 10 cm wide and 5 cm deep, is cut using a hammer and chisel, or a circular saw, across the strike of the mineralization (Fig. 5.3). As the material is cut it is allowed to fall to the floor on to a plastic sheet or sample tray from which it is collected and bagged. Samples are normally 0.5–5 kg in weight, mostly 1–2 kg, and rarely taken over 2 m or so in length. There is no point in over sampling and sample spacing perhaps can be determined from the range (a) (see section 9.5.1) derived from a study of a semi-variogram.

A mineral deposit can often be resolved into distinct and separate types of mineralization. Sampling these different types as separate entities rather than as one large sample can reduce natural variation and thus the variance, and keep sample weights to a minimum. This *stratified sampling* is also of value where the separate types require different mineral processing techniques due to variation in either grain size or mineralogy. Such sample data are then more useful for mine planning and costing rather than one composite result for the whole of the mineralization.

In well jointed or well bedded rocks, etc., collection by hammering the rock face presents difficulties in that adjacent non-sample material will fall with the material being cut but this must be rejected. Additionally, it is important that a correct collection of rock masses are collected. Frequently, mineral values of interest occur in altered rocks which may well be more friable than adjacent, harder, unmineralized ground and over-collec-tion of this easily cut material will provide a biased sample.

Channel sampling provides the best technique of delimiting and extracting a sample and, consequently, provides the smallest possible contribution to the total error. The two techniques which follow are less rigor-ous, and more error prone, but less expensive.

Chip sampling. Chip samples are obtained by collect-ing rock particles chipped from a surface, either along a line or over an area, and collected as described previ-ously. In an established mine with a large database of both channel and chip sampling, statistical correlation between the two may establish a correction factor for chip sampling results and thus reduce error. Chip sam-pling is used also as an inexpensive reconnaissance tool to see if mineralization is of sufficient interest to warrant the more expensive channel sampling.

Grab sampling. In this case the samples of mineral-ized rock are not taken in place, as are channel and chip samples, but consist of already broken material. Handfuls or shovelfuls of broken rock are picked at random at some convenient location and these form the sample. It is a low cost and rapid method and best used where the mineralized rock has a low variance and mineralization and waste break into particles of about the same size. It is particularly useful as a means of quality control of mineralization at strategic sampling points such as stope outlets and in an open pit (Annels 1991).

9.1.4 Sample preparation

The ultimate purpose of sampling is to estimate the content of valuable constituents of interest in the samples (their sample mean and variance) and from this to infer their content in the population. Chemical analysis is completed on a few grams of material (the assay portion) selected from several kilograms or tonnes of sample(s). Consequently after collection sampling involves a systematic reduction of mass and grain size as an inevitable prerequisite to analysis. This reduction is of the order of times 1000 with a kilogram sample and 1 000 000 with a tonne sample.

Samples are reduced by crushing and grinding and the resultant finer grained material is split, or separated, into discrete mass components for further reduction. A sample reduction system is a sequence of stages that progressively substitutes a series of smaller and smaller samples from the original until the assay portion is obtained for chemical analysis. Such a system is essentially an alternation of size and mass reduction stages with each stage generating a new sample and a sample reject, and its own sampling error. For this reduction Gy (1992) established a relationship between the sample particle size, mass and sampling error (Box 9.1):

$$M = \frac{Cd^3}{S^2}.$$

M is the minimum sample mass in *grams,* d is the size of the coarsest top 5% of the sample in *centimetres,* C is a heterogeneity constant characteristic of the material being sampled and S^2 is the (fundamental) variance. Consequently, in reducing the mass of a sample there is a cube relationship between this reduction and its particle size if the variance is to remain constant.

In practice the observed variance of a sampling system is larger than that calculated from the above model. This is because the observed variance comprises the total sampling error (TE) while the above model calculates a particular component of this totality, the fundamental error (FE). This important point has already been re-

Box 9.1 Gy sampling reduction formula

Gy (1992) is believed to be the first to devise a relationship between the sample mass (M), its particle size (d) and the variance of the sampling error (S^2). This variance is that of the fundamental error (FE): a best estimate of the total sampling error (TE) is obtained by doubling the FE.

$$M \geq \frac{Cd^3}{S^2}$$

where: M is the minimum sample mass in *grams* and d is the particle size of the coarsest top 5% of the sample in *centimetres.* Cumulative size analyses are rarely available and from a practical viewpoint d is the size of fragments that can be visually separated out as the coarsest of the batch. S^2 is the fundamental variance and C is a heterogeneity constant characteristic of the material being sampled and = cβfg, where

    c = a mineralogical constitution factor

$$= \frac{(1-a_L)[(1-a_L)\delta A + \delta G.a_L]}{a_L}$$

  a_L is the amount of mineral of interest as a *fraction,*
  δA is the specific gravity of the mineral of interest in g cm^{-3},
  δG is the specific gravity of the gangue,
  β is a factor which represents the degree of liberation of the mineral of interest. It varies from 0.0 (no liberation) to 1.0 (perfect liberation) but practically it is seldom less than 0.1. If the liberation size (d_{lib}) is not known it is safe to use β equal to 1.0. If the liberation size is known $d \geq d_{lib}$ then $\beta = (d_{lib}/d)^{0.5}$ which is less than 1. In practice there is no unique liberation size but rather a size range.
  f is a fragment shape factor and it is assumed in the formula that the general shape is spherical in which case f equals 0.5,
  g is a size dispersion factor and cannot be disassociated from d. Practically g extends from 0.20 to 0.75 with the narrower the range of particle sizes the higher the value of g but, with the definition of d as above, g equals 0.25.

ferred to, where $S^2(TE) = 2S^2(FE)$.

DESIGN OF A SAMPLE REDUCTION SYSTEM

The choice of suitable crushing, grinding, drying and splitting equipment is critical and systems are best designed and installed by those with appropriate experience. The design of a reduction system is often neglected but it is of prime importance particularly in dealing with low grade and/or fine-grained mineralization. In particular, equipment may have to handle wet, sticky material with a high clay content which will choke normal crushers. Also the creation of dust when processing dry material is a hazard which can contaminate other samples and has to be controlled by adequate ventilation and appropriate isolation.

In any reduction system the most sensitive pieces of equipment are the crushers and grinders. Each such item works efficiently within a limited range of weight through-put and size reduction and there has to be a realistic assessment in the planning stage of the expected number and weight of samples to be processed each hour or shift. The use of inadequate comminution equipment at the best may introduce excessive preparation errors and at the worst can close the entire system.

The calculation of TE from twice the FE provides a safe estimate (see previously) from which a flexible reduction system can be designed. When it is in operation its actual total error can be calculated and compared with the initial assumption (see later).

GRAPHICAL SOLUTION OF SAMPLE REDUCTION

Introduction. Sample preparation is essentially an alternation of size and mass reduction within the parameters of the Gy formula given above which provides for a control of the variance. Such reduction is best shown graphically with log d (top particle size) contrasted with log M (sample mass), as in Fig. 9.5, where a horizontal line represents a size reduction and a vertical line a mass reduction.

These reductions have to be completed within an acceptable variance which can be taken as a relative standard deviation of less than 5%. Ideally what is required is a safety line which divides the above graph into two parts so that on one side of the line all reduction would be safe (i.e. with an acceptable variance) while on the other side unacceptable errors occur. Gy (1992) provides a mathematical expression for such a line as:

$$Mo = Kd^3.$$

Empirically Gy (1992) suggests a value of 125 000 for K but emphasizes that for low concentrations of valuable constituents (<0.01%) a higher K value should be used of up to 250 000. Consequently,

$$Mo = 125\,000\,d^3 \text{ or } \log Mo = \log K + 3 \log d.$$

This relationship on log–log graph paper (Fig. 9.5) is a straight line, with a slope of 3.0, which divides the graph into two areas.

1 To the left of this safety line Mo is > M and, consequently, from the Gy formula the fundamental sampling variance So^2 is < S^2, and is acceptable.

2 On the line, Mo = M and $So^2 = S^2$, which is acceptable. All points on the safety line correspond to a constant fundamental variance.

3 To the right of the line Mo is < M and the sampling variance is too high ($So^2 > S^2$) and unacceptable and either Mo has to be increased or d decreased to place the reduction point to the left of the safety line.

Equal variance lines. The above allocates equal variance to each sampling stage but this may not always be an optimum practice. Usually the larger the mass (M) to be sampled the larger the top particle size (d) and the higher its reduction cost. From the Gy sampling formula, however, the necessity for particle size reduction at any stage can be reduced by allocating a higher proportion of the total variance to that particular stage. With the mass (M) remaining constant this allows a coarser top particle size. Consequently it is preferable to allocate a maximum portion of the variance to the samples with the coarsest particle sizes, and less to later successive samples of finer grain size. Gy (1992) suggests that one half of the total variance is allocated to the first sampling stage, one fourth to the next, etc.

As previously, a safe estimate of the total sampling variance is calculated by

$$S^2(\text{FE}) = \frac{Cd^3}{M}$$

$$\text{Using } M = Kd^3$$

$$\text{then } S^2(\text{FE}) = \frac{Cd^3}{M} = \frac{C}{K}$$

$$\text{and } So^2(\text{TE}) = 2\frac{C}{K}$$

where C is the heterogeneity constant (Box 9.1) and K equals 125 000.

From this a series of equal variance lines can be drawn with decreasing values of K, such as K/2, K/4, etc., which correspond to increasing variances. For each stage, knowing M and the increased value of $So^2(\text{FE})$ the related coarser grain size (d) can be calculated from the Gy formula.

Variation of the safety line. The heterogeneity constant (C) contains the liberation factor β (Box 9.1). The safety lines above are calculated on the basis that this factor is at its maximum value of 1.0 which is a safe assumption but only correct when the top particle size is smaller than or equal to the liberation size (d_{lib}) of the valuable components. If d is larger than d_{lib} then β

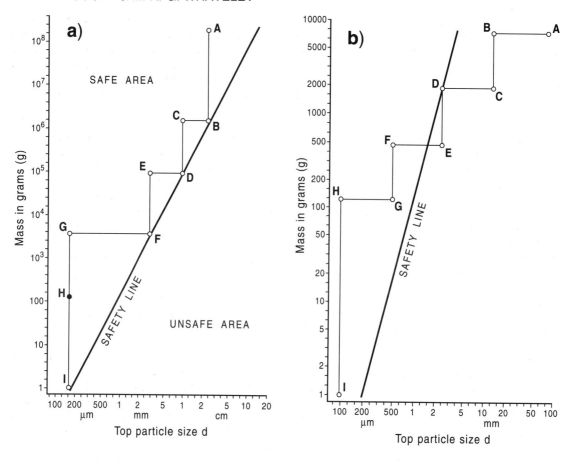

Fig. 9.5 Graphical representation of a sample reduction scheme. Fig 9.5a is an example of a safe scheme while Fig. 9.5b is unsafe. For explanation see text.

becomes $[d_{lib}/d]^{0.5}$, which is less than 1.0.

When the liberation size of the mineral of interest is not known the value of 1.0 is used as a safe procedure. This provides for the continuously straight equal variance lines as described above, with a slope of 3.0. When the liberation size is known, the above formula can be used with a β value of less than 1.0 but approaching this value as the grain size d is reduced. In this case each sampling stage has a unique safety line slope from about 2.5 increasing to 3.0 as d approaches d_{lib}, with the less steep sections in the coarser grain sizes.

SPLITTING OF SAMPLES DURING REDUCTION

In each stage after reduction in grain size the sample mass is correspondingly reduced by splitting. The sampling described in this text is concerned with small batches of <40 t and usually <1 t which are stationary in

the sense that they are not moving streams of material on, say, a conveyor belt.

Hand or mechanical shovel splitting. Splitting samples is best completed by the process of alternative shovelling which minimizes the splitting and preparation error contribution to the total error. The sample is spread either on a sheet of thick plastic or sheet iron which is on a flat, clean and smooth floor which can easily be swept. The material is piled into a cone (step 1) which is then flattened into a circular cake (step 2) and the first step is repeated three times for maximum mixing of the sample. From the final cone, shovelfuls of material are taken consecutively around its circumference and placed alternatively into two separate piles. In this way the original cone, and sample, is split into two separate but equal smaller cones and the choice as to which is the new sample is decided by the toss of a coin and not by the sampler. Further weight reduction is continued in exactly the same manner on the new sample. Alternatively, the cone may be quartered with a shovel and each quarter is further reduced in the same way. This technique is referred to as cone and quartering.

The width of the hand, or small mechanical, shovel should be at least three times the width of the diameter of the coarsest sample particle size which provides for a maximum possible size of about <100 mm with the former and <500 mm with the latter tool. Shovels should always be overflowing and working with them partially filled is not a good practice. Obviously, towards the end of a splitting operation with finer grain sizes a smaller sized tool has to be used. In order to provide an adequate number of increments shovel capacity should be between 1/30 to 1/50 of the original sample mass to provide some 30 to 50 shovelfuls, or increments, divided equally between the two splits. In a final, fine-grained 100 g sample the required asssay portion can be obtained by repeated alternative 'shovelling' with a spatula.

An example of the reduction of a bulk sample for chemical analysis is given in Figs 9.6 and 9.7. The example is from a large, low grade, disseminated gold deposit with a grade of about 1 g t^{-1}. Due to the high specific gravity of gold and the broad range of gold particle size (i.e. d_{lib} in Gy's formula) in the micrometre range, devising an adequate sub-sampling plant was difficult. The most problematical stage was the reduction of the initial bulk of 36.2 tonne samples. This was effected by splitting each such sample into two equal portions with one half (18.1 t) for sampling purposes and the other for heap leaching gold tests (Fig. 9.6). Figure 9.7 illustrates a flow sheet for reduction of 18 kg portions, which were treated in the same fashion as drilling samples, to four replicate samples each of 5 g at a grain size of 0.01 mm (Springett 1983a, b).

Instrument splitters. The Jones Riffle is the most satisfactory of stationary instrument splitters (Fig. 9.8) which divide a dry sample into two equal parts. It comprises an even number (between 12 and 20) of equally sized chutes of a slope between 45° and 60° where adjacent chutes discharge on opposite sides of the instrument. For correct implementation the material to be split is evenly spread on a rectangular scoop designed to fit the splitter width, and discharged slowly and evenly into the chutes. To prevent obstruction the chutes should be at least twice the diameter of the coarsest particle. With repeated splitting it is better practice to implement an even rather than an odd number of splitting stages, and to alternate the choice of right and left split samples. Asymmetrical splitters are now available

which split samples on a 70:30 or 80:20 basis (N. Laffoley *pers. comm.*). This speeds up sample reduction and with fewer separations reduces any errors. A sample splitter may be used at the drill site especially when percussion drilling is in progress and samples are collected at regular intervals. A 1 m section of 102 mm diameter hole produces approximately 21 kg of rock chips.

Rotary splitters capable of separating laboratory size (a few kgs) or bulk samples (a few tonnes) are also available. They use centrifugal force to split a sample into four or six containers.

The use of several riffle splitters in series, each one directly receiving the split from the preceding stage, is poor practice which can introduce sampling errors.

COMPARISON OF CALCULATED AND OBSERVED TOTAL SAMPLING ERRORS

A best estimate of the total sampling error (TE) is obtained from the fundamental error (FE)

$$S^2(TE) = 2S^2(FE)$$

This estimate is used to design a sample reduction system from which, when in operation, an observed $S^2(FE)$ can be obtained by sampling each reduction stage to establish the sample mass (M) and the maximum particle size (d), using the Gy formula. This is then compared with the previously estimated total error, on the following basis.

1 When the observed total error is smaller than or of the same order of magnitude as the estimated total error nothing should be changed in the reduction system. This provides a safety margin against the future wear of the equipment which will deliver progressively coarser material for which the sampling variance increases.

2 When the observed total error is larger than the estimated total error the former has to be reduced and particularly at the first (or coarsest) reduction stage where the variance is likely to be the major component of $S^2(FE)$. This can be achieved by retaining the same number of sampling stages but increasing the weight (i.e. the mass) treated at each separate stage, increasing the number of stages but with the same sample weight at each stage. Usually the latter choice is preferable as the crushing–grinding equipment does not have to be enlarged, as in the first option.

The total sampling error of any sample reduction system should be checked at regular intervals, say every six months depending upon throughput and use, to check that it is working satisfactorily.

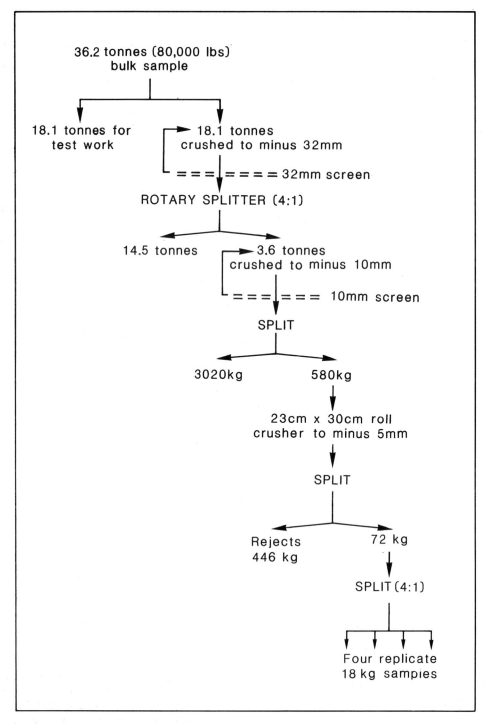

Fig. 9.6 Reduction of 36.3 tonne (80 000 lb) bulk sample by successive crushing and splitting. This involves a concomitant reduction in grain size from +15 cm to 0.5 cm. (After Springett 1983b.)

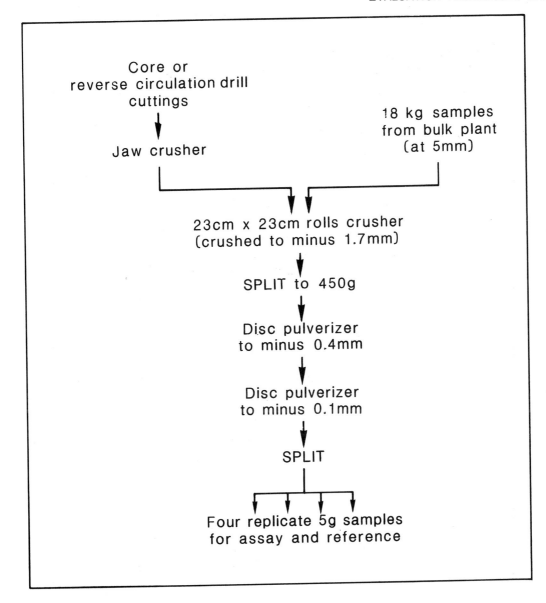

Fig. 9.7 Reduction of 18 kg (40 lb) samples at 0.5 cm size from bulk sampling plant (Fig. 9.5), and drill material, to 5 g samples at minus 0.01 cm size suitable for chemical analysis. (After Springett 1983b.)

9.1.5 Preparation of samples containing native gold and platinum

With native precious metals a problem arises from the significant difference between the density of an associated gangue (about 2.6) and that of the metals (19.0 to 21.5). Gold may be diluted with silver and copper to a density of the order of 15.0 but even at this level its density is twice that of galena, the densest of all common sulphide minerals, and six times that of quartz. This contrast means that when such valuable constituents are liberated from their gangue, flakes and nuggets will

Fig. 9.8 A Jones sample splitter. For explanation see text.

segregate rapidly under gravity and homogenization of such material is difficult (Austr. Inst. Geoscientists 1987, 1988).

CALCULATION OF THE VARIANCE OF THE FUNDAMENTAL ERROR (FE)

Under such circumstances, and with the very low concentrations of precious metals (≈ 1 ppm), Gy (1992) suggests a modification of his formula for calculating the fundamental error.

Normally, the factor $C = c\beta fg$ (see Box 9.1) but for gold and platinum Gy (1992) recommends a factor of

$$C = \frac{2.4d3}{a_L} \quad \text{and} \quad S^2 = \frac{C}{M}$$

where a_L is the concentration of the valuable constituent expressed as a fraction (i.e. 1 ppm = 1×10^{-6}).
An example of the use of this formula is given in Box 9.2.

9.1.6 **Analytical errors**

According to Gy (1992) analytical errors (AE) in the strict sense are negligible and errors which are attributed to analytical laboratories arise from the selection of the assay portion and not, on the whole, from actual analytical techniques (Section 9.1.2). Errors occur, however, and vary from laboratory to laboratory. An example is given in Table 9.5 where an artificial sample was made from gold in silica sand, at 2.74 g t^{-1} and was sent to four different laboratories for analysis (Springett 1983b). The following conclusions can be drawn.
Laboratory 1   Results are both inaccurate and imprecise.
Laboratory 2   Results are accurate but imprecise.
Laboratory 3   Results are inaccurate but precise.

Box 9.2 Variance of fundamental error [S^2(FE)] with liberated gold mineralization

Gold mineralization contains $1 g t^{-1}$ ($a_L = 10^{-6}$) and is liberated from its gangue (i.e. $\beta = 1$) with the coarsest particles about 2 mm in size (d = 0.2 cm). Under such circumstances Gy (1992) recommends a modification of his standard formula for the calculation of the fundamental variance. Normally the heterogeneity factor C equals $c\beta fg$ (Box 9.1) but for free gold he recommends $C = 2.4d^3/A_L$ and S^2(FE)$=$C/M
With a 1 tonne sample ($M = 10^6$ g), $C = 2.4(0.2)^3/10^{-6} = 19\,200$ and S^2(FE)$=1.92 \times 10^{-2}$, $S=1.4 \times 10^{-1}$, or 14%. If a normal distribution is assumed the 95% confidence level double-sided interval is ±28%, which is a low precision. If, say, a 10% precision is required at 95% confidence then a larger sample (M) is needed. At this level S(FE) equals 5×10^{-2}, and S^2(FE) equals 25×10^{-4}; then
$M = C/S^2$(FE)$= 19\,200/25 \times 10^{-4} = 768 \times 10^4$ g or 7.68 tonnes. The treatment of such large samples presents a problem but Gy (1992) recommends the use of *stratified sampling*. In this process the sample is reduced to a grain size of a few millimetres and then concentrated on a shaking table, such as the Wilfley table (Wills 1985). This produces a high density concentrate that contains all the coarse gold, an intermediate fraction, and low density tailings which is largely gangue. The intermediate fraction is recycled on the table while the tailings are sampled again on a 1% sampling ratio. In this way the 8 tonne sample is reduced to about 80 kg which is then processed as in Figs 9.6 & 9.7.

Laboratory 4   Results are accurate and precise.
Obviously only the results from Laboratory 4 are acceptable.

The scope of the control of error in chemical analysis is beyond this chapter but these errors must be considered (see section 8.2). External control is exercised by submitting duplicate samples and reference materials of similar matrices to the unknowns (section 8.2.1). It is essential that the geologist is aware of the internal control procedures of the laboratory. These usually employ the analysis of replicate samples (e.g. every, say, tenth sample is repeated) and by introducing artificially prepared reference materials (e.g. every 30th or 40th sample). Some laboratories are found to perform relatively poorly and a rigorous programme of inter laboratory checking by the use of blanks, duplicates and prepared materials is important (Reimann 1989).

9.1.7 **Sample acquisition**

BEFORE SAMPLING

Pre-sampling procedures consist of gaining as much

Table 9.5 Errors in assays on a prepared sample of 2.74 grams per tonne gold

This is an example where the population parameters are known since the material was artificially prepared. The results of the four different sets of five identical samples (A to E) each analysed separately at four different laboratories (1 to 4) are given below. After Springett (1983b).

| | LABORATORY $g\,t^{-1}$ gold | | | |
	1	2	3	4
SAMPLE				
A	5.72	2.74	4.39	2.81
B	4.18	2.40	4.01	2.78
C	5.66	2.06	3.81	2.67
D	4.42	3.43	4.01	2.67
E	3.26	2.06	3.81	2.71
Mean	4.65	2.54	4.01	2.73
S^2	0.87	0.26	0.04	0.003
S	0.93	0.51	0.21	0.06

The range of values at the 95% confidence level (which is the mean ±2S) for laboratory 1 is from 2.79 to 6.51 g t^{-1}. This range does not contain the population mean (2.74 g t^{-1}) and the variance is high. For laboratory 2 the range is 1.52 to 3.56 g t^{-1} which does contain the population mean, but the variance is high. At laboratory 3 the range is from 3.59 to 4.43 g t^{-1}. This does not contain the population mean, although the variance is acceptable. Laboratory 4 results have a range from 2.61 to 2.85 g t^{-1}, which contains the population mean and has a low variance.

Only the results from laboratory 4 are acceptable.

---

familiarity as possible with the geology and possible problems before visiting the location to be tested. Obviously if the sampling is at an operating mine these problems are already familiar and the following remarks mainly apply to work at remote prospects.

As much background information should be collected as possible from previous publications and reports. A preliminary description of the exercise serves as a source of stimulation that will help to isolate points on which further thought is needed. The task to be completed has to be briefly but adequately described.

Logistical factors are important. How much time is available for the exercise? Can the site be revisited or is this the only opportunity for sampling? What budget restraints are there? Is assistance necessary and if so of what type—surveyors, drivers, cooks, labourers, etc.? Decide on the objectives and how it is proposed to

achieve them; are they cost-effective?

What is the purpose of the sampling? Is it purely qualitative where it is necessary to prove the presence or absence of certain minerals or chemical elements, or is a statistically valid estimate of the grade of (say) a base metal prospect required? Is the study preliminary (i.e. a return visit is possible) or final (i.e. no chance of return)? Certainly before carrying out extensive systematic sampling at various locations an orientation survey should be completed to ascertain the main characteristics of the prospective target population—a preliminary mean, standard deviation, geochemical threshold and probability of error (see also section 8.3). Finally, check that the sampling procedure is sound and adequate and that the projected sample size (weight and number) is large enough for the purpose but not too large. Carefully assess the cost-effectiveness of the proposed exercise.

SAFETY

Safety aspects are paramount. Samplers should rarely, if ever, work alone and visits to mine workings (surface or underground) should be accompanied. Visits to abandoned mine workings need particular care and the workings should be investigated fully for weak rock conditions, weak roof support, water extent, noxious gases and other hazards before sampling commences. A formalized safety code will be available in all exploration offices and must be read and understood.

AFTER SAMPLING

As an example, suppose the calculated mean of a prospect from 25 samples (n = 25) is 5.72% zinc with a standard deviation of 0.35% zinc. At a 95% probability level a two-sided confidence interval is

$$5.72\% + t_{2.5\%}\frac{0.35\%}{\sqrt{25}} > \overline{X} > 5.72\% - t_{2.5\%}\frac{0.35\%}{\sqrt{25}}$$

$$= 5.72\% + 2.06\frac{0.35\%}{5} > \overline{X} > 5.72\% - 2.06\frac{0.35\%}{5}$$

$$= 5.72\% \pm 0.15\%$$ at a 95% confidence level.

That is, there are 19 chances in 20 of the population mean (i.e. the true grade) being within this range, and one chance in 20 of it being outside.

If we are dissatisfied with this result the remedy, in theory, is simple. If we wish a narrower confidence interval the site can be revisited and additional samples taken (i.e. increase n) but recall the square root relationship in the above calculation. That is, if the interval is to be halved (to become 5.72% ± 0.075%), then the total number of samples (n) required is not 50 but 100—an additional 75 samples have to be taken. Alternatively, the value of the t statistic can be reduced (see Table 9.3). At an 80% confidence level the interval is:

$$5.72\% + 1.32\frac{0.35\%}{5} > \overline{X} > 5.72\% - 1.32\frac{0.35\%}{5}$$

$$= 5.72 \pm 0.09\%.$$

In other words there are 16 chances in 20 of the population mean being within this narrower range. As it is narrower the probability of it being outside the range is higher—4 chances in 20 or quadruple the previous calculation.

Commonsense must be used in the interpretation of the results of any sampling programme and particularly with the problem of re-sampling in order to increase the n statistic. As an example, a zinc prospect 200 km from the nearest road or railway is sampled and preliminary estimates are 2.1% zinc with an estimated overall reserve tonnage of about 200 000 t. There is no point in sharpening this grade estimate by taking more samples as the low tonnage, approximate grade and location mean that this deposit is of little economic interest. There is nothing to be gained by extra work which demonstrates that a better estimate is 2.3% zinc.

9.1.8 Summary

Sample acquisition and preparation methods must be clearly defined, easy to implement and monitor, bias free, as precise and accurate as possible, and cost effective.

9.2 PITTING AND TRENCHING

In areas where soil cover is thin, the location and testing of bedrock mineralization is made relatively straightforward by the examination and sampling of outcrops. However in locations of thick cover such testing may involve a deep sampling programme by either pitting, trenching or drilling. Pitting to depths of up to 30 m is feasible and, with trenching, forms the simplest and least expensive method of deep sampling but is much more costly below the water table. For safety purposes, all pits and trenches are filled in when evaluation work is completed. Drilling penetrates to greater depth but is more expensive and requires specialized equipment and

expertise that may be supplied by a contractor. Despite their relatively shallow depth, pits and trenches have some distinct advantages over drilling in that detailed geological logging can be carried out, and large and, if necessary, undisturbed samples collected.

9.2.1 Pitting

In areas where the ground is wet, or labour is expensive, pits are best dug with a mechanical excavator. Depths of 3 m to 4 m are common and with large equipment excavation to 6 m can be achieved. In wet, soft ground any pit deeper than 1 m is dangerous and boarding must be used. Diggers excavate rapidly and pits 3–4 m deep can be dug, logged, sampled and re-filled within an hour. In tropical regions, thick lateritic soil forms ideal conditions for pitting and, provided the soil is dry, vertical pits to 30 m depth can be safely excavated. Two labourers are used and with a 1 m square pit, using simple local equipment, advances of up to 2 m per day to 10 m depth are possible, with half that rate for depths from 10 m to 20 m, and half again to 30 m depth.

9.2.2 Trenching

Trenching is usually completed at right angles to the general strike to test and sample over long lengths, as across a mineralized zone. Excavation can be either by hand, mechanical digger or by bulldozer on sloping ground. Excavated depths of up to 4 m are common.

9.3 AUGER DRILLING

Augers are hand-held or truck-mounted drills which have rods with spiral flights to bring soft material to the surface. They are used particularly to sample placer deposits. Power augers are particularly useful for deep sampling in easily penetrable material where pitting is not practicable (Barrett 1987). They vary in size from those used to dig fence post holes to large, truck-mounted rigs capable of reaching depths of up to 60 m, but depths of less than 30 m are more common. Hole diameters are from 5 to 15 cm in the larger units, although holes 1 m in diameter were drilled to evaluate the Argyle diamond deposit in Australia. In soft ground augering is rapid and sampling procedures need to be well organized to cope with the material continuously brought to the surface by the spiralling action of the auger. Considerable care is required to minimize cross-contaimantion between samples. Augers are light drills and are incapable of penetrating either hard ground or boulders. For this purpose, and holes deeper than about 60 m, heavier equipment is necessary and this is described in the next section.

9.4 OTHER DRILLING

The various methods of drilling serve different purposes at various stages of an exploration programme (Annels 1991). Early on when budgets are low, cheap drilling is required. The disadvantage of cheaper methods, such as augering, rotary or percussion drilling, is that the quality of sampling is poor with considerable mixing of different levels in the hole. Later, more expensive, but quality samples are usually collected using reverse circulation or diamond core drilling as shown below.

Purpose	Quality	Type
Reconnaissance and exploration	Low	Auger, rotary and percussion (chips)
	High	Reverse circulation (chips)
Resource or reserve evaluation	High	Reverse circulation coring in soft formations
		Diamond core drilling in hard formations

Fundamentally drilling is concerned with making a small diameter hole (usually less than 1 m and in mineral exploration only a few centimetres) in a particular geological target, which may be several hundred metres distant, to recover a representative sample. Among the available methods diamond core, rotary and percussion drilling are the most important.

ROTARY DRILLING

Rotary drilling is a non-coring method and is unequalled for drilling through soft to medium hard rocks such as limestone, chalk or mudstone. A typical rotary bit is the tricone or roller rock bit which is tipped with tungsten carbide insets. Rock chips are flushed to the surface by the drilling fluid for examination and advances of up to 100 m per hour are possible. This type of drilling is typically used in the oil industry with large diameter

holes (> 20 cm) to depths of several thousand metres with the extensive use of drilling muds. The large size of the equipment presents a mobility problem.

PERCUSSION DRILLING

Basically this is a hammer unit driven by compressed air. The hammer imparts a series of short rapid blows to the drill rods or bit and at the same time imparts a rotary motion. The drills vary in size from small hand-held units (as used in road repair work) to large truck-mounted rigs capable of drilling large diameter holes to several hundred metres depth (Fig 11.3). The units can be divided broadly into two types.

1 *Down-the-hole hammer drills.* The hammer unit is lowered into the hole attached to the lower end of the drill rods to operate a non-coring, tungsten carbide tipped, drill bit. Holes with diameters of up to 20 cm and penetration depths of up to 200 m are possible, but depths of 100 m to 150 m are more usual. Drill cuttings are flushed to the surface by compressed air and a regular and efficient source of this is vital for the successful operation of this technique. As the cuttings come to the surface they can be related to the depth of the hole. However, such direct correlation is not always reliable as holes may not be cased and material may fall from higher levels. In wet ground the lifting effect of compressed air is rapidly dissipated but special foaming agents are available to alleviate this drawback. Rigs are usually truck or track mounted and very mobile. With the latter slopes of up to 30° can be negotiated and the ability to drill on 25+° slopes partly eliminates the need to work on access roads.

2 *Top hammer drills.* As the name suggests the hammer unit, driven by compressed air, is at the top of the drill stem and the energy to the non-coring drill bit is imparted through the drill rods. These are usually lighter units than down-the-hole hammer drills and are used for holes up to 10 cm diameter and depths of up to 100 m, but more usually 20 m. Most only use light air compressors and this restricts drilling depths to at the most only a few metres below a water table, as it becomes impossible for the pressurized air to blow the heavy wet rock sludge to the surface. Usually they are mounted on either light trucks or tractors.

General remarks. Percussion drilling is a rapid and cheap method but suffers from the great disadvantage of not providing the precise location of samples, as is the case in diamond drilling. However, costs are one-third to one-half of those for diamond drilling and this technique has proved particularly useful in evaluating deposits

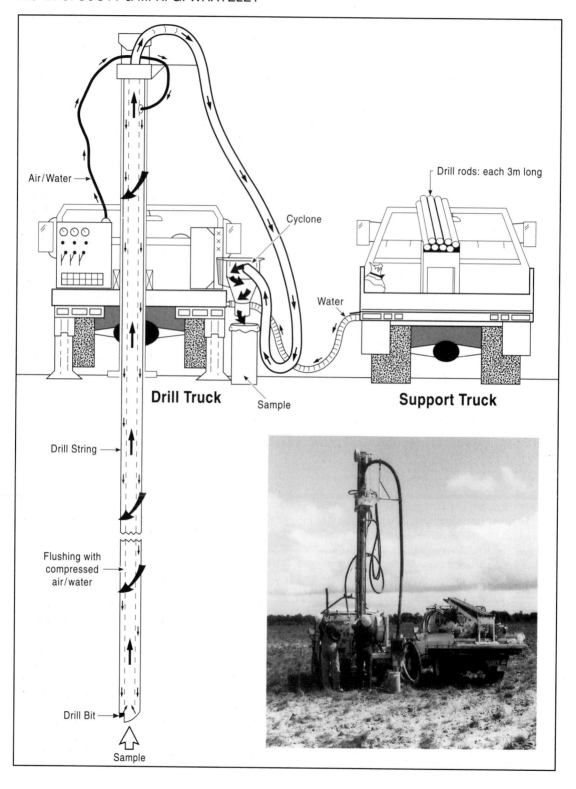

Air/Water

Cyclone

Drill rods: each 3m long

Drill Truck

Sample

Water

Support Truck

Drill String

Flushing with
compressed
air/water

Drill Bit

Sample

which present more of a sampling problem than a geological one, e.g. a porphyry copper. There is a rapid penetration rate of around a metre a minute and it is possible to drill 150-200 m in an eight hour shift. With such penetration rates and several machines, several hundred samples can be collected each day. As each 1.5 m of a 10 cm diameter hole is likely to produce about 20 to 30 kg of rock chips and dust, sample collection and examination requires a high degree of organization. Like all compressed air equipment these drills are noisy in operation.

REVERSE CIRCULATION

The reverse circulation (RC) drilling technique has been in use in exploration since the mid-seventies and can be used in unconsolidated sediments such as alluvial deposits or for drilling rock. Both air and water can be used as the drill flushing medium and both cuttings or core can be recovered. The technique employs a double walled string of drill rods (Fig. 9.9), with either a compressed air driven percussion hammer or a rotating tungsten carbide coring bit at the cutting end of the string. The medium is supplied to the cutting bit between the twin-walled drill rods and returned to the surface up the centre of the rods. In the case of percussion drilling the rock chippings are also transported to the surface up the centre of the rods and from there, via a flexible pipe, to a cyclone where they are deposited in a sample collection container (Fig. 9.9).

The advantages of using this method to collect rock chippings, rather than auger, rotary or percussion drilling, are that all the sample is collected, the method is extremely quick (up to 40 m per hour can be drilled) and there is very little contamination. The specialized rods, the need for a compressor and additional equipment makes this a more expensive drilling technique, but the additional costs are outweighed by the higher quality of sample collection. The dual nature of some RC rigs (chips and core) means that high quality core can be taken through the zone of interest without the need to mobilize a second (core) rig, thus reducing the overall drilling costs.

DIAMOND CORE DRILLING

The sample is cut from the target by a diamond-ar-

moured or impregnated bit. This produces a cylinder of rock that is recovered from the inner tube of a core barrel. The bit and core barrel are connected to the surface by a continuous length (string) of steel or aluminium alloy rods which allow the bit plus core barrel to be lowered into the hole, and pulled to the surface. They also transmit the rotary cutting motion to the diamond bit from the surface diesel power unit, and appropriate pressure to its cutting edge (Fig. 9.10 and Table 9.6).

1 *Drill bits.* Drill bits are classified as either impregnated or surface-set. The former consist of fine-grained synthetic or industrial grade diamonds within a metallic cement while the latter have individual diamonds sized by their number per carat. In general, impregnated bits are suitable for tough compact rocks such as chert, while surface-set varieties with large individual diamonds are used for softer rocks such as limestone (Fig. 9.11).

Diamond bits will penetrate any rock in time but, because of their high cost and the need to maximize core advance and core recovery with minimum bit wear, the choice of bit requires considerable experience and judgement. Second-hand (i.e. used) surface set bits also have a diamond salvage value and consequently are not used to destruction. Drill bit diameters are classified with either a letter code (American practice) or in millimetres (European practice) (Table 9.7).

2 *Core barrels.* As the cylinder of rock (the core) is cut by the circular motion of the drill bit it is forced up into the core barrel by the advancing drill rods. Core barrels are classified by the length of core contained and are usually 1.5 m to 3.0 m in length but can be as long as 6 m. They are normally double-tubed in the sense that in order to improve core recovery an inner core barrel is independent of the motion of the drill rods and does not rotate. Triple-tubed barrels can be used in poor ground.

Previously, to recover core the barrel had to be removed from the hole by pulling the entire length of drill rods to the surface, a time consuming process. Wire-line drilling (Q series core) is now standard practice; in this method the barrel is pulled to the surface inside the connecting drill rods using a thin steel cable. This has the advantage of saving time but often necessitates having a smaller diameter core (Table 9.7).

3 *Circulating medium.* Usually water is circulated down the inside of the drill rods, washing over the cutting surface of the drill bit and returning to surface through the narrow space between the outside of the rods and the wall of the drill hole. The purpose of this action is to lubricate the bit, cool it and remove crushed and ground rock fragments from the bit surface. In friable and soft rock this action may flush away part of the sample and

---

Fig. 9.9 *(Opposite)* Schematic diagram and photograph showing a reverse circulation drill. The drill was exploring for mineral sands in Western Australia (White 1989).

Fig. 9.10 A diamond drill. From left to right the components are the power, control, drill and water pump units. Reproduced by permission of Diamond Boart Craelius Ltd.

Table 9.6 Diamond drilling rigs

In mineral exploration most drill holes are less than 450 m in length. A typical drill rig for this purpose is shown in Fig. 9.10 but many models are in use. Some of the specifications for the drill in Fig. 9.10 are:

Drill unit weight	890 kg	Length of holes	AQ (48 mm) 425 m
Power unit weight	790 kg		BQ (60 mm) 350 m
Motor rating	40 kW at 2200 rpm^{-1}		NQ (76 mm) 250 m
Flush pump flow	76 litres min^{-1}		
Flush pump pressure	5 Mpa		

The drill is a fully hydraulic rig. It is suitable for both underground and surface units and is powered by either an air, diesel or electric motor. To a certain extent penetration rates can be increased by an increase in pressure on the drill bit. This is regulated by hydraulic rams in the drill unit at surface and transmitted to the bit by the drill rods. A limit is reached when all the diamond chips in the bit's surface have been pressed as far as possible into the rock. Higher penetration rates can then only be achieved by increasing the bit rotation speed. Excessively high bit pressure, with low rotation speed, results in abnormally high wear of the bit. Conversely, low pressure with high rotation results in low penetration and polishing (i.e. blunting) of the diamonds.

In the above unit, drill water would flush through a 60 mm diameter hole at about 50 cm sec^{-1}. In a 350 m length hole return water would take about 12 minutes to pass from the drill bit to the surface. The ascending velocity of the fluid should be greater than the settling velocity of the largest rock particles generated by the cutting and grinding action of the bit.

Fig. 9.11 On the left-hand side are several types of surface-set diamond bits, at the right are bits with tungsten carbide insets and at the back are impregnated bits. Reproduced by permission of Diamond Boart Craelius Ltd.

this can sometimes be alleviated with the addition of specialized additives. Water may be used in combination with various clays and other additives which, in addition to the above functions, have added uses such as sealing the rock face of the drill hole. The science and use of such drilling fluids is a subject in itself and is much developed in the oil industry which drills significantly larger diameter and deeper holes than are required in mineral exploration (Chilingarian & Vorabutr 1981).

4 *Casing.* Cylindrical casing is used to seal the rock face of the hole. It provides a steel tube in which the drill string can operate in safety and prevents accidents such as rock collapse and either loss or influx of water. Casing, and drill bits, are sized so that the next lower size (i.e. smaller diameter) will pass through the larger size in which the hole had been previously drilled.

5 *Speed and cost of drilling.* Drilling machines used in mineral exploration usually have a capacity of up to 2000 m and may be inclined from horizontal to vertical. The rate of advance depends upon the type of drill rig, the bit, the hole diameter, usually the larger the hole

Table 9.7 A selection of core, drill hole and drill bit sizes

Size	Hole diameter (mm)	Core diameter (mm)	Core area (% of hole area)
METRIC SIZES			
Conventional drilling			
36T	36.3	21.7	36
56T2	56.3	41.7	55
66T2	66.3	51.7	61
Wire-line drilling			
56ST	56.3	35.3	39
AMERICAN SIZES			
Conventional drilling			
AX	48	30.1	39
BX	59.9	42.0	49
NX	75.7	54.7	52
Wire-line drilling			
AQ	48	27	31
BQ	60	36.5	36
NQ	75.8	47.6	39
HQ	96.1	63.5	44

The hole diameter is the diameter of rock cut by the drill bit. Part of this is removed by the cutting and grinding action of the bit and water flushed to the surface, while the central part is retained as core in the core barrel.

diameter the slower the advance; depth of hole, the deeper the hole the slower the advance; rock type being drilled, hard or soft, friable, well jointed, etc.; and the skill of the driller.

The driller must decide on the best volume of drilling fluid to be flushed over the drill bit; the right pressure to apply at its cutting surface, this can be adjusted using hydraulic rams at the surface; the right number of revolutions per minute; and the correct choice of drill bit. Until recently there was no direct method of knowing what operating conditions existed at the drill bit which are the critical factors if drilling efficiency is to be optimized. Recent developments in 'measurements while drilling (MWD)' have overcome this and allow a variety of direct measurements such as temperature, pressure at the bit face, rate of water flow, etc. to be made (Dickinson *et al.* 1986, Prain 1989). Another technique being developed is a retractable bit that can be raised, replaced and lowered through the string of drill rods. At present when a worn bit requires replacement the complete drill string has to be withdrawn, which is a time consuming process

(Jenner 1986, Morris 1986). However, modern rigs with automated rod handling can pull the string and unscrew the rods at very high speed.

Drilling advance rates of up to 10 m an hour are possible but depend a great deal upon the skill of the driller and rock conditions. Costs vary from US$50 to $80 a metre for holes up to 300 m long and from $70 to $150 a metre for lengths up to 1000 m. Standard references for diamond drilling are Cummings & Wickland (1985), the Australian Drillers Guide (1985), The Management of Drilling Projects (1981) and the Longyear Handbook.

9.4.1 Logging of drill-hole samples

Information from drill holes comes from the following main sources: rock, core, or chips; down-the-hole geophysical equipment; instruments inside the hole (see MWD above); and performance of the drilling machinery. In this section we are only concerned with geological logging, but the geologist on site at a drill location must be familiar with all sources of information. The collection of geotechnical data from core is discussed in section 9.6.

GEOLOGICAL LOGGING

Effective *core recovery* is essential—that is, the length or volume (weight) of sample recovered divided by the length or volume (weight) drilled expressed as a percentage. If recovery is less than 85–90% the value of the core is doubtful as mineralized and altered rock zones are frequently most friable and the first to be ground away and lost during drilling. The core is not then representative of the rock drilled, it is not a true sample and it is probably misleading.

1 *Core drilling.* Core logging begins at the drill site where decisions have to be made as to whether the hole is to be either continued or abandoned. Wetted core is more easily examined, using either a hand lens or a binocular microscope. Most organizations have a standard procedure for core logging and a standard terminology to describe geological features.

The first logging at a drill site is usually completed rapidly with a more detailed examination of the core at a later date. Nevertheless, the main structural features should be recorded (fracture spacing and orientation) and a lithological description (colour, texture, mineralogy, rock alteration and rock name) with other details such as core recovery and the location of excessive core loss (when say > 5%). The description should be system-

Fig. 9.12 Diamond drill core being examined and stored in a wooden core box. Note the excellent core recovery. Reproduced by permission of Diamond Boart Craelius Ltd.

atic and as quantitative as possible, qualitative descriptions should be avoided.

Core is stored in slotted wooden, plastic or metal boxes short enough to allow two persons to lift and stack them easily (Figs 9.12 & 12.10). Core is collected for a variety of purposes other than geological description, e.g. metallurgical testing and assaying. For these latter purposes the core is measured into appropriate lengths (remembering the principle of stratified sampling) and divided or split into two equal halves either by a diamond saw or a mechanical splitter. Half the core is sent for assay or other investigations whilst the other half is returned to the core box for record purposes. Obviously structural features have to be recorded before splitting and a good practice is to photograph wet core, box by box, before logging it, to produce a permanent photographic record (see Fig. 12.10).

Rock chips and dust ('sludge') can be collected during core drilling; they represent the rock cut away by the diamond drill bit. Drilling with air circulation in relatively shallow holes (as in most percussion drilling) delivers cuttings to the surface within a minute or so.

However, with core drilling, water circulation, and longer holes, there is an appreciable time lag before cuttings reach the surface. It is estimated that from a depth of 1000 m cuttings can take 20 to 30 minutes to reach the surface, with the inherent danger of differential settlement in the column of rising water due to differences in mineral and rock specific gravities and shape. Consequently rock sludge is rarely examined during core drilling.

Obtaining core is expensive so it is sensible to retain it for future examination. However, adequate and long term storage involves time, space and expense but the value of the contained information is important particularly as during mining some drill locations may be permanently lost.

2 *Non-core drilling.* In non-core drilling the chips and dust are usually collected at 1–2 m intervals, dried and separately bagged at the drill site. After washing they are relatively easy to examine with the use of a hand lens and binocular microscope. Samples can be panned so as to recover a heavy mineral concentrate. It is a good practice to sprinkle, and glue, a sample of rock chips and panned concentrates from each sample interval on to a board so that a continuous visual representation of the hole can be made. Again, descriptions must be systematic and quantitative.

9.4.2 Drilling contracts

Drilling can be carried out with either in-house (company) equipment or it can be contracted out to specialist drilling companies. In the latter case the conditions of drilling, amount of work required and the cost will be specified in a written contract. The purpose of drilling is to obtain a representative sample of the target mineralization in a cost-effective manner. Consequently, the choice of drilling equipment is crucial and much depends upon the experience of the project manager. Unless the drilling conditions are well known test work should, if possible, be carried out to compare different drilling methods before any large scale programme begins (Box 9.3).

The main items of cost in a contract are as follows.

1 Mobilization and transport of equipment to the drilling site. This can vary from movement along a major road by truck to transport by helicopter.

2 Setting-up at each site and movement between successive borehole locations. Again costs can vary greatly depending on distance and terrain.

3 A basic cost per metre of hole drilled.

4 Optional items costed individually, e.g. cementing

Box 9.3 Comparative sampling results from two different drilling methods.

If drilling conditions are not known a programme of test work to compare and contrast appropriate drilling procedures may be desirable before any large scale sampling programme is commenced, as in the following example.

Silver mineralization 12.2 m thick consisting of fine-grained dolomite and quartz with minor chalcopyrite, galena and sphalerite occurs in limestone. The silver is principally acanthite but with some native silver. It has been tested by diamond and percussion drilling, and also by mining from a test shaft.

Diamond drilling: BQ core, diameter 36.5 mm with samples every 1 m over the total length of 12.2 m; core recovery was 93%. The core was split for assay and the total weight of the overall sample was 10 kg. The weighted assay average was 62 g t^{-1} silver. The main concern in diamond drilling was the possibility of high grade, friable silver mineralization being ground by the diamond bit and washed away by the circulating water.

Percussion drilling: With a 1.6 cm diameter hole samples were collected every 1 m; material recovery was 87%. All the recovered material was taken for the analysis and totalled 255 kg. The weighted assay average was 86 g t^{-1}. Whilst this method was cheaper than core drilling, there was the possibility that the high specific gravity silver minerals might sink in the column of water in the bore hole and not be fully recovered.

Bulk sample: From a 1.1 m square test shaft all the material was taken for assay, totalling 51.8 t. The weighted assay average was 78 g t^{-1}.

Accepting the bulk sample result as the best estimate of silver grade the cheaper percussion drilling method can be used to sample the deposit rather than the more expensive core drilling. (After Springett 1983b.)

holes, casing holes, surveying, etc.

5 Demobilization and return of equipment to driller's depot.

All items of cost should be detailed in the contract. There may be a difference of interest between the driller and the company representative on site, normally a geologist, who is there to see that drilling proceeds according to the company's plan and the contract. The driller may be paid by distance drilled during each shift whilst the geologist in charge is more concerned with adequate core recovery and that the hole is proceeding to its desired target. The company representative usually signs documents at the completion of each shift where progress and problems are described and it is on the basis of these documents that final cash payment is made. Consequently, it is important for this representative to be thoroughly familiar with the contract and the problems which can arise at a drilling site, particularly if in a remote location.

The client (i.e. the company) is at liberty to fix specifications such as, say, a plus 90% core recovery, a vertical hole with less than 5° deviation, etc. It is then at the discretion of the drilling company as to whether or not these requirements are accepted, but if they are and the conditions are not met the failure is remedied at the expense of the contractor.

9.4.3 **Borehole surveying**

Drills are usually unpredictable wanderers. In section holes drilled at a low angle to rock structures tend to follow that structure while holes at a higher angle tend to become perpendicular to the feature. A similar behaviour can be expected from alternations of hard and soft rock (e.g. layers of chert in chalk). Steep holes are likely to flatten and all holes tend to spiral counter clockwise with the rotation of the drill rods. However, wandering also depends upon the pressure on the drilling bit and its rate of revolution, so that two drill holes in similar geological situations may take different paths (see Fig. 16.10d). Holes with a length of 1000 m have been found to be several 100 m off course—while individual lengths of drill rod are essentially inflexible a drill string several hundred metres in length will bend and curve. Since drill holes are used for sampling at depth, if the orientation of the hole is not known the location of the sample is similarly unknown, and geological projections based on these samples will be in error.

Borehole surveys are conducted as a matter of routine in all holes and there are a variety of instruments for this purpose. They most commonly indicate the direction (azimuth) and inclination of the hole at selected intervals, commonly every 100 m, using a small magnetic or gyroscopic compass. The calculation of corrected positions at successive depths is a straightforward mathematical procedure, if one knows the location of the top of the hole (X, Y & Z co-ordinates) and the initial hole inclination.

Deliberate deflection can be achieved either by wedging holes (say, 1° per wedge) or using a hydraulically driven downhole motor which acts as a directional control device (Cooper & Sternberg, 1988). This relatively new directional-drilling tool was originally designed for the oil industry. The downhole motor is powered by the circulating fluids passed down the centre of the rod string which drives the bit without drill-rod rotation. A spring deflection shoe located just above the bit acts as a directional control device and delivers a constant side pressure to the drill hole wall forcing the bit to move (and deflect) in the opposite direction. The amount of spring loading in the shoe is pre-set before the unit is lowered into the hole and, obviously, progress is monitored by closely spaced surveys. The deflection unit is a special tool and is not used in normal drilling.

9.4.4 **Drill hole patterns**

The pattern of drilling in an exploration programme depends primarily upon the intended use of the data. In reconnaissance work where the geology is poorly known the first holes may be isolated from each other and drilled for geological orientation purposes. In exploration for sedimentary uranium deposits, coal and borates reconnaissance holes may be drilled up to 10 km apart to locate sedimentary formations of interest and obtain structural data. Later holes may be a few kilometres apart whilst drilling a specific target calls for 'fences' or lines of holes, at a close spacing of 100 m to 200 m. However, the location and spacing of holes in the last eventuality are guided by detailed geological mapping, geochemical, geophysical and geostatistical results (see section 5.2).

Generally speaking a systematic grid of drill holes (and samples) taken normal to the expected mineralized zone is a preferred pattern. This provides a good statistical coverage and geological cross sections can be made with a minimum of projection. Such fences may well be 200 to 400 m apart with individual holes spaced at 100 to 200 m intervals, which allows room for systematic infilling if the results justify extra work. The inclination of individual holes is obviously important and generally they should be drilled at right angles to the expected average dip of the mineralization.

Before drilling a hole it is recommended that a section be drawn along its projected length, allowing for hole deviation if this is possible. As the hole progresses the section is modified. Drilling is an expensive and a time-consuming part of mineral exploration. The objective is to drill a precise number of holes and provide the exact number of intersections needed to demonstrate the grade and tonnage (dimensions) of the mineralization at an appropriate level of accuracy and precision. Unfortunately such optimization is rarely, if ever, possible. At second best an iterative procedure is followed whereby a series of successively smaller drilling programmes are completed with each followed by a re-assessment of all existing data—a process of successive approximations. The main problem is that in the calculation of mineral resources and reserves the zone of influence of each sample is not known until a minimum amount of work is completed. If two adjacent samples (taking a drill hole as a sample) cannot be correlated at an acceptable confidence level, then neither has an acceptable zone of influence in the intervening space and further sampling is necessary. Conversely where adjacent samples show appropriate correlation further sampling is not required.

There are several methods of assigning zones of influence to either successive samples or individual boreholes: the mean-square (dn) successive difference test, the use of correlation coefficients or geostatistics using the range (a) of a semi-variogram. However, a relatively large nugget effect (section 9.5.1) in the semi-variogram indicates an element of uncertainty which would call for a reduction in spacing. Such an effect is apparent in the evaluation of low grade disseminated gold deposits where drilling may be required on a 50 m grid basis or less.

9.5 MINERAL RESOURCE AND ORE RESERVE ESTIMATION, GRADE CALCULATIONS

During exploration and initial evaluation of a base metal, industrial mineral or coal deposit, the principal emphasis is placed upon its geology and the estimation of the quality and quantity of the resources present. Data are collected from several sampling programmes as described above, including trenching, pitting, percussion drilling, reverse circulation drilling and diamond drilling as well as during underground development. The object of this sampling is to provide a mineral inventory. Once the mining engineers and financial analysts have established that the mineral can be mined at a profit, or in the case of industrial minerals can be marketed at a profit, it can then be referred to as an ore reserve.

The above brief definition of ore and ore reserves is controversial. Good accounts of the various definitions are summarized by Riddler (1994), Armitage & Potts (1994) and Jakubiak & Smakorkski (1994). Mining engineers tend to refer to reserves that are extractable at a profit and therefore there are losses associated with dilution and loss in the processing. Geologists are interested in the amount and quality of *in situ* mineralization within a geologically defined envelope. *In situ* geological resources tend to be slightly greater than the mineable reserves. Discussions of the subject are given by

Taylor (1989), Annels (1991) and Whateley & Harvey (1994). Commonly used definitions of reserves and resources for mineral and coal deposits are given by the United States Geological Survey (USGS 1976), and take into account the increasing degree of confidence in the resources and of the financial feasibility of mining them.

Before geological resource calculations can proceed, a study of the mineralization envelope is required. This envelope can often be defined by readily identifiable geological boundaries. In some cases inferences and projections must be made and borne in mind when assessing confidence in the reserves. Some deposits can only be delineated by selecting an assay cut-off to define geographically and quantitatively the potential mineralized limits (see Chapter 15, where an assay cut-off is used to define the outline of the Trinity Silver Mine and where waste blocks are identified within the deposit for separate disposal). Initially this is highly subjective and in later stages when confidence is higher, the ore limits will be carefully calculated from conceptual mining, metallurgical, cost and marketing data.

Geologists must provide the basis for investment decisions in mining using the data provided from the above sampling. It is their responsibility to ensure that the data base that they are using is valid. Once the geological and assay (grade) cut-offs have been established, usually in discussions with the mining engineers, it is possible to start the evaluation procedure.

9.5.1 Tonnage calculation methods

CLASSICAL STATISTICAL METHODS

The use of classical statistical methods in ore reserve calculations is generally restricted to a global estimation of volume or grade within the mineralization envelope. The simplest statistical estimate involves calculating the mean of a series of values, which gives an average value within the geologically defined area. Calculation of the variance will give a measure of the error of estimation of the mean (Davis JC 1986). When using these non-spatial statistical methods it is important that the samples are independent of one another, i.e. random. If sample locations were chosen because of some assumption or geological knowledge of the deposit then the results may contain bias. Separate calculations must also be made of different geological areas to ensure that the results are meaningful.

The data are usually plotted on frequency distribution graphs (histograms) (Figs 8.5, 9.1 & 15.10) and scatter diagrams (correlation graphs) (Fig. 8.7). The distributions are usually found to approximate to a Gaussian or to a log–normal distribution. Cumulative frequency distributions can also be plotted on normal probability paper and a straight line on such a plot represents a normal distribution. Significant departures from a straight line may indicate the presence of more than one grade zone and thus more than one population within the data. Each population should be treated separately (cf section 8.3.1).

CONVENTIONAL METHODS

In calculating the resource or reserve potential of a deposit, one formula, or a variation of it, is used throughout, namely

$$T = A \times Th \times BD \text{ where:}$$

T	=	tonnage (in tonnes),
A	=	area of influence on a plan or section in km^2,
Th	=	thickness of the deposit within the area of influence in m,

BD = bulk density. This includes the volume of the pore spaces. It is obtained by laboratory measurement of field samples.

The area of influence is derived from a plan or section of the geologically defined deposit. The conventional methods commonly used for obtaining these areas are: thickness contours (constructed manually), polygons, triangles, cross-sections or a random stratified grid (Fig. 9.13). Popoff (1966) outlined the principles and conventional methods of reserve calculation in some detail. The choice of method depends upon the shape, dimensions and complexity of the mineral deposit, and the type, dimensions and pattern of spacing of the sampling information. These methods have various drawbacks which relate to the assumptions on which they are based, especially the area of influence of the sampling data, and generally do not take into account any correlation of mineralization between sample points nor quantify any error of estimation.

Large errors in estimation of thickness (or grade) can therefore be made when assuming that the thickness or

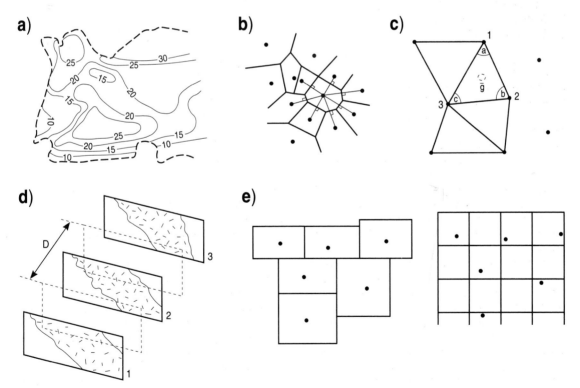

Fig. 9.13 Conventional methods of estimating the area of influence of a sample using (a) isopachs, (b) polygons, (c) triangles, (d) cross sections or (e) a random stratified grid. (From Whateley 1992.)

grade of a block is equal to the thickness or grade of a single sample point about which the block has been drawn. Care should be taken when using the cross-sectional method that the appropriate formula is used for computing volume between sections, as significant errors in tonnage estimation can be introduced where the area and shapes on adjacent sections vary considerably. Manual contouring methods are less prone to this estimation error by virtue of the fact that contours are constructed by linear interpolation, thereby smoothing the data irregularities.

The inverse power of the distance $(1/D^n)$ method can also be used to calculate reserves. The deposit is divided into a series of regular blocks within the geologically defined boundary, and the available data are used to calculate the thickness value for the centre of each block. Near sample points are given greater weighting than points further away . The weighted average value for each block is calculated using the following general formula:

$$Th = \frac{\sum (V \times f)}{\sum f}$$

where:

Th = the thickness at any block centre,
V = known thickness value at a sample point,
f = $1/D^n$ weighting function,
n = the power to which the distance (D) is raised.

The geologist has to decide how many of the available data are to be used. This is normally decided by choosing a distance factor for the search area (usually a distance, a, derived from a semi-variogram), the power factor which should be employed (normally D^2), and how many sample points should be used to calculate the centre point for each block. The time needed to calculate many hundred block values on a complex deposit requires the aid of a computer. This method begins to take the spatial distribution of data points into account in the calculations.

Most of these methods have found application in the mining industry, some with considerable success. The results from one method can be cross-checked using one of the other methods.

GEOSTATISTICAL METHODS

Semi-variograms. The variables in a deposit (grade, thickness, etc.) are a function of the geological environment. Changing geological and structural conditions result in variations in grade or quality and thickness between deposits and even within one deposit. However, samples which are taken close together tend to reflect the same geological conditions, and have similar thickness and quality. As the sample distance increases, the similarity, or degree of correlation, decreases, until at some distance there will be no correlation. Usually the thickness variable correlates over a greater distance than

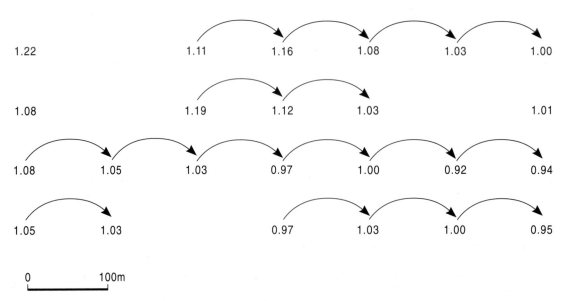

Fig. 9.14 A portion of a borehole grid with arrows used to identify all the pairs 100 m apart in an E–W direction. (From Whateley 1992.)

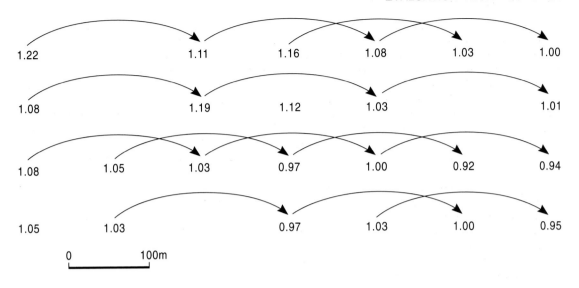

Fig. 9.15 A portion of a borehole grid identifying all the pairs 200 m apart in an E–W direction. (From Whateley 1992.)

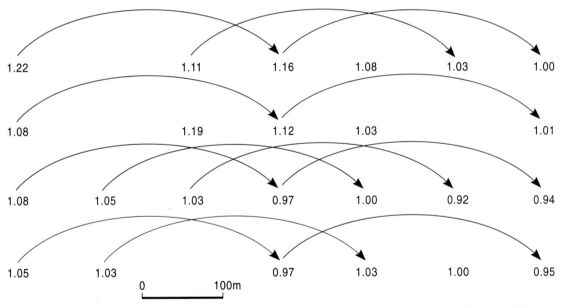

Fig. 9.16 A portion of a borehole grid illustrating some of the pairs 300 m apart in an E–W direction. (From Whateley 1992.)

coal quality variables (Whateley 1991, 1992) or grade.

Geostatistical methods quantify this concept of spatial variability within a deposit and display it in the form of a semi-variogram. It is assumed that the variability between two samples depends upon the distance between them and their relative orientation. By definition this variability (semi-variance or $\gamma(h)$) is represented as

half the average squared difference between samples that are a given separation or lag distance (h) apart (Formula [9.1]).

In the example (Fig. 9.14), the sample grid shows thickness values of a stratiform mineral deposit or coal seam at 100 m centres. Suppose we call this distance between the samples (in the first instance 100 m) and their orientation, h. We can calculate the semi-variance from the difference between samples at different distances (lags) and in different orientations e.g. at 200 m (Fig. 9.15) and 300 m (Fig. 9.16), etc. These examples are all shown with an E–W orientation, but the principle equally applies with N–S, NE–SW, etc. directions. We calculate the semi-variance of the distances for as many different values of h as possible using the following formula:

$$\gamma(h) = \frac{\displaystyle\sum_{i=1}^{n(h)} (z_{(i)} - z_{(i+h)})^2}{2n(h)} \qquad [9.1]$$

where: $\gamma(h)$ = semi-variance,
n(h) = number of pairs used in the calculation,
z = grade, thickness, or whatever,
(i) = the position of one sample in the pair,
(i+h) = the position of the other sample in the pair, h metres away from z(i).

Using the above formula [9.1], the $\gamma(h)$ values calculated for lag 1 (h=100 m) are as follows:

$(1.11 - 1.16)^2 + (1.16 - 1.08)^2 + (1.08 - 1.03)^2 +$
$(1.03 - 1.00)^2 + (1.19 - 1.12)^2 + (1.12 - 1.03)^2 +$
$(1.08 - 1.05)^2 + (1.05 - 1.03)^2 + (1.03 - 0.97)^2 +$
$(0.97 - 1.00)^2 + (1.00 - 0.92)^2 + (0.92 - 0.94)^2 +$
$(1.05 - 1.03)^2 + (0.97 - 1.03)^2 + (1.03 - 1.00)^2 +$
$(1.00 - 0.95)^2 = 0.0481$
$(100) = 0.0481/[2 \times 16] = 0.0015 \ (m)^2$

Using the same formula the $\gamma(h)$ values for lag 1 to lag 6 (h=600 m) are as follows:

(100) = 0.0015, (200) = 0.0037, (300) = 0.0036,
(400) = 0.0055, (500) = 0.0108, (600) = 0.0104.

The values thus calculated are presented in a graphical form as a semi-variogram. The horizontal axis shows the distance between the pairs, while the vertical axis displays the values of $\gamma(h)$ (Fig. 9.17).

Once a semi-variogram has been calculated, it must be interpreted by fitting a 'model' to it. This model will help to identify the characteristics of the deposit. An example of a model semi-variogram is shown in Fig. 9.18. The model fitted to the experimental data has the mathematical form shown in equations [9.2] and [9.3]. It is known as a spherical scheme model and is the most common type used, although other types do exist (Journel & Huijbregts 1978, David 1988, Isaaks & Srivastava 1989, Annels 1991). C_0 is added where a nugget effect exists.

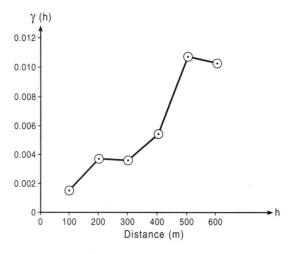

Fig. 9.17 A plot of the experimental semi-variogram constructed from the data derived from the portion of the borehole grid shown in Figs 9.14 to 9.16. The calculations are shown in the text.

The semi-variogram starts at 0 on both axes (Fig. 9.17). At zero separation (h=0) there is no variance, but

$$\gamma(h) = C_0 + C(\frac{3h}{2a} - \frac{1(h)^3}{2(a)^3}) \qquad \text{(for h < a)} [9.2]$$

$$\gamma(h) = C_0 + C \qquad \text{(for h > a)} [9.3]$$

it is not usual to sample the same place twice! Even at relatively close spacings there are small differences. Variability increases with separation distance (h). This is seen on the semi-variogram where a rapid rate of change in variability is marked by a steep gradient until a point where the rate of change decreases and the gradient is 0. Beyond this point sample values are independent and have variability equal to the theoretical variance of sample values (Fluor Mining and Metals Inc. 1978, Journel & Huijbregts 1978, Isaaks & Srivastava 1989). This variability is termed the sill ($C + C_0$ on Fig. 9.18) of the semi-variogram, and the point at which this sill value is reached is termed the range (a) of the semi-variogram. There is often a discontinuity near the origin, and this is called the nugget variance or nugget effect (C_0). This is generally attributable to differences in sample values over very small distances, e.g. two halves of core and can include inaccuracies in sampling and assaying, and the associated random errors.

The range a on Figure 9.18 is 500 m. One can see that if the drill spacing is greater than 500 m then the data from each sample point would not show any correlation, they would in effect be random sample points. If, how-

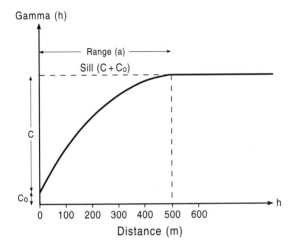

Fig. 9.18 The idealized shape of a semi-variogram (spherical model). (From Whateley 1992.)

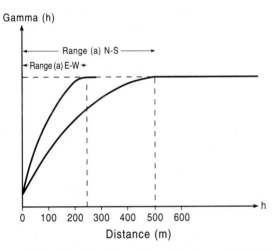

Fig. 9.19 The idealized shape of a semi-variogram (spherical model) illustrating the concept of anisotropy within a deposit. (From Whateley 1992.)

ever, one were to drill at a spacing less than 500 m, then the semi-variogram shows us that the data will, to a greater or lesser extent (depending on their distance apart), show correlation. Holes 100 m apart will show less variance than holes 400 m apart. A spacing therefore at or just less than 500 m, i.e. the range of the semi-variogram, will include such correlation.

When h is $\geq$ a then the data show no correlation and this gives a random semi-variogram. This can be interpreted to mean that sample spacing is too great.

It is important to note that the rate of increase in variability with distance may depend on the direction of a vector separating the samples. In a coal deposit through which a channel passes, the variability in the coal quality and thickness in a direction normal to the channel (i.e. across any 'zoning') will increase faster than the variability parallel to the channel (Fig. 9.19). The different semi-variograms in the two directions will have different ranges of influence and as such represent anisotropy or 'hidden' structure within the deposit.

The squared differences between samples are prone to fluctuate widely, but theoretical reasoning and practical experience have shown that a minimum of 30 evenly distributed sample points are necessary to obtain a reliable semi-variogram (Whitchurch *et al.* 1987). This means that in the early stages of an exploration programme a semi-variogram may be difficult to calculate. In this case typical semi-variogram models from similar deposits can be used as a guide to the sample spacing in the early phases of an exploration programme, until sufficient samples have been collected from the deposit under investigation. This is analogous to using the genetic model to guide exploration.

Estimation variance. No matter which method is used to estimate the properties of a deposit from borehole data, whether it is manual contouring or computer methods, some errors will always be introduced. The advantage of geostatistical methods over conventional contouring methods is their ability to quantify the potential size of such errors. The size of the error depends on factors such as the size of the block being estimated (Whitchurch *et al.* 1987), the continuity of the data and the distance of the point (block) being estimated from the borehole. The value to be estimated (Z*) generally differs from its estimator (Z) because there is an implicit error of estimation in Z–Z* (Journel & Huijbregts 1978). The expected squared difference between Z and Z* is known as the estimation variance (σ_e^2)

$$\sigma_e^2 = E([Z–Z^*]^2) \qquad [9.4]$$

with E = expected value.

The estimation variance can be derived from the semi-variogram

$$2\gamma(h) = E(z_{(i)} - z_{(i+h)})^2 \qquad [9.5]$$

By rewriting equation [9.5] in terms of the semi-variogram, as described by Journel & Huijbregts (1978), the estimation variance can then be calculated.

Knudsen (1988) explained that this calculation takes

into account the main factors affecting the reliability of an estimate. The calculation measures the closeness of the samples to the grid node being estimated. As the samples get farther from the grid node, this term increases, and the magnitude of the estimation variance will increase. It also takes into account the size of the block being estimated. As the size of the block increases the estimation variance will decrease. In addition it measures the spatial relationship of the samples to each other. If the samples are clustered then the reliability of the estimate will decrease. The variability of the data also affects the reliability of the estimate, but since the continuity (C) is measured by the semi-variogram this factor is already taken into account. The estimation variance, σ_e^2, is a linear function of the weights established (see Kriging) by kriging (Journel & Huijbregts 1978). Kriging determines the optimal set of weights which minimize the variance. These weighting factors are also used to estimate the value of a point or block. As it is rare to know the actual value (Z) for a point or block it is not possible to calculate Z–Z*(cf cross validation below).

Kriging variance. Once the kriging weights for each block have been calculated (see Kriging) the variances can be calculated. This calculated estimation variance is also known as the kriging variance. Once the kriging variance is determined it is possible to calculate the precision with which we know the various properties of the deposit we are investigating, by obtaining confidence limits (σ_e) for critical parameters. Knudsen & Kim (1978) have shown that the errors have a normal distribution. This allows the 95% confidence limit, $\pm 2(\sigma_e)$, to be calculated (sections 8.3.1 & 9.1)

$$\sqrt{\sigma_e^2} = \sigma_e$$

95% confidence limit $= Z^* \pm 2(\sigma_e)$.

Estimation variance and confidence limit are extremely useful for estimating the reliability of a series of block estimates or a set of contours (Whateley 1991). An acceptable confidence limit is set and areas which fall outside this parameter are considered unreliable and should have additional borehole data collected. Thus the optimal location of additional holes is determined. The confidence limits for critical parameters such as Au grade or ash content of coal can be obtained. Areas which have unacceptable variations may require additional drilling, or in a mining situation it may lead to a change in the mine plan or a change in the storage and blending strategy.

Kriging. Kriging is a geostatistical method used to estimate the value of a point or a block as a linear combination of the available samples in or near that block. The principles of kriging are discussed by Clark (1979), Journel & Huijbregts (1978), David (1988), Isaaks & Srivastava (1989) and Annels (1991) among others. The kriging weights are calculated such that they result in the minimum estimation variance by making extensive use of the semi-variogram. The kriging estimator can be considered as unbiased under the constraint that the sum of the weights is one. These weights are then used to estimated the values of thickness, elevation, grade, etc. for a series of blocks.

Kriging gives more weight to closer samples than more distant ones, continuity and anisotropy are considered, the geometry of the data points and the character of the variable being estimated are taken into account, and the size of the blocks being estimated is also assessed.

Having estimated a reliable set of regular data values, these values are contoured as are the corresponding estimation variances. Areas with relatively high estimation variances can be studied to see whether there are any data errors, or whether additional drilling is required to reduce the estimation variance.

Cross validation. Cross validation is a geostatistical technique which is used to verify the kriging procedure, to ensure that the weights being used do minimize the estimation variance. Each of the known points is estimated by kriging using between four and twelve of the closest data points. In this case both the known value (Z) and the estimated value (Z*) are known, and the experimental error, Z–Z* can be calculated, as well as the theoretical estimation (or kriging) variance (σ_e^2). By plotting Z against Z*, it is possible to test the semi-variogram model used to undertake the cross validation procedure. If the values show a high correlation coefficient (near 1), we can assume that the variables selected from the semi-variogram provide the best parameters for the solution of the system of linear equations used in the kriging matrices to provide the weighting factors for the block estimate and estimation variance calculations (Journel & Huijbregts 1978).

9.5.2 Grade calculation methods

CONVENTIONAL METHODS

Often borehole chips or cores, or channels, etc. are sampled on a significantly smaller interval than a bench height in an open pit or a stope width in an underground mine. It is the geologists' responsiblity to calculate a composite grade or quality value over a predetermined interval from the samples taken from the cores or channels. In an open pit the predetermined interval is often a bench height. In some instances where selective mining is to take place, such as in a coal deposit, the sample composites are calculated for coal and waste separately (see Chapter 12 for a more detailed description of sampling and weighted average quality calculations). In an underground vein mine, the composite usually represents the minimum mining width (e.g. section 16.7). Sometimes a thickness × grade accumulation is used to calculate a minimum mining width (e.g. section 13.6)

The general formula given above for calculating the weighted average thickness value for the $1/D^n$ block model, also applies when calculating weighted average grade or quality values. The weighting function changes though, and three variables, length, SG and area are used. For example when compositing several core samples for a particular bench level in a proposed open pit, length of sample is used. This is particularly important when the geologist sampling the core, has done so using geological criteria to choose sample lengths and samples are of different lengths (e.g. Fig.16.11). Thus:

$$\overline{G} = \frac{\sum L \times G}{\sum L}$$

where:

$\overline{G}$ = weighted average grade of each borehole on that bench,

G = grade of each core sample,

L = length of sample.

When samples are to be composited from, say core, and the rocks are of significantly different densities, then the weighting factor should have bulk density (BD) included. Thus:

$$\overline{G} = \frac{\sum BD \times L \times G}{\sum BD \times L} .$$

Area of influence can also be used as part of the weighting function. For example, to calculate the weighted average grade of a stratiform deposit which is to be assessed by polygons of different sizes, the area of each polygon is used as the weighting function. Thus:

$$\overline{G} = \frac{\sum A \times G}{\sum A}$$

where A = area of influence of each polygon.

COMPUTATIONAL METHODS

In large ore bodies, or deposits which have a large amount of data, such as Trinity Silver Mine (see Chapter 15), computers must be used to manipulate the data. Commercial software packages are available (SURPAC, DATAMINE) which can perform all the conventional methods described above. Computers offer speed and repeatability, as well as the chance to vary parameters to undertake sensitivity analyses. In addition computers can use exactly the same procedures as described for geostatistical mineral resource estimation methods. As well as using thickness to calculate volume and (with SG) tonnage, grade or quality can be estimated along with the estimation error.

9.6 GEOTECHNICAL DATA

During the sampling and evaluation of a mineral deposit, the exploration geologist is in an ideal position to initiate the geotechnical investigation of that deposit. Later on, when the geotechnical and mining engineers are invited on to the investigating team, they will need to know some of the properties of the rocks, such as strength, degree of weathering and the nature of the discontinuities within the rock, in order to begin designing the rock-related components of the engineering scheme (haulages, underground chambers, pit slopes, etc.). It is important to investigate some of the geotechnical properties of the overburden and of the rocks in which the mineralization is found, during the exploration phase of a project, because a decision has to be made as to whether it is feasible to mine the potential ore before financial decisions are taken in the feasibility study which may follow.

Geotechnical investigations essentially cover three aspects: soil, rock and subsurface water. A good field guide for the description of these is given by West (1991). Soil refers to the uncemented material at or near the surface, and can include weathered or broken rock. Soil is mainly investigated for civil engineering purposes such as foundations of buildings. Rock is the solid material which is excavated during mining to reach and extract the ore. Subsurface water is part of the soil and the rock, and is found in the pore spaces and in joints, fractures and fissures, etc.

There are two forms in which the data can be collected; (1) mainly subjective data obtained by logging core and the sides of excavations (Table 9.8), and (2) quantitative data obtained by measuring rock and joint strengths and also virgin rock stress, etc., either in the

Table 9.8 A summary of the description and classification of rock masses for engineering purposes. (Adapted from Engineering Group Working Party 1977, Brown 1981.)

Descriptive indices of rock material
Consideration should be given to rock type, colour, grain size, texture and fabric, weathering (condition), alteration and strength

Rock condition (weathering)

Fresh	No visible signs of rock material weathering
Slightly weathered	Discolouration indicates weathering of rock material and discontinuity surfaces
Weathered	Half the rock material is decomposed or disaggregated. All rock material is discoloured or stained
Highly weathered	More than half the rock material is decomposed or disaggregated. Discoloured rock material is present as a discontinuous framework. Shows severe loss of strength and can be excavated with a hammer
Completely weathered	Most rock material is decomposed or disaggregated to a soil with only fragments of rock remaining

Rock material strength

Extremely strong	Rock can only be chipped with geological hammer. Approximate unconfined compressive strength >250 MPa
Very strong	Very hard rock which breaks with difficulty. Rock rings under hammer. Approximate unconfined compressive strenth 100–250 MPa
Strong	Hard rock. Hand specimen breaks with firm blows of a hammer, 50–100 MPa
Medium	Specimen cannot be scraped or peeled with a pocket strong knife, but can be fractured with a single firm blow of a hammer, 25–50 MPa
Weak	Soft rock which can be peeled by a pocket knife with difficulty, and indentations made up to 5 mm with sharp end of hammer, 5–25 MPa
Very weak	Very soft rock. Rock is friable, can be peeled with a knife and can be broken by finger pressure, 1–5.0 Mpa

laboratory or *in situ*. Subjective data helps with the decision making fairly early in a project, but these data will always need quantification. It is therefore necessary to study and record the direction and properties of joints, cleavage, cleats, bedding fissures and faults, etc., and to study and record the mechanical properties (mechanics of deformation and fracture under load), petrology and fabric of the rock between the discontinuities.

It is clear from the above that structural information is extremely important in mine planning. In the initial phases of a programme most of the structural data come from core. The geologist can log fracture spacing, attitude and fracture infill at the rig site. A subjective description of the quality of the core can be given using the Rock Quality Designation (RQD) (Table 9.8). Pieces of core greater than 10 cm are measured and their length

summed. This is divided by the total length of the core run and expressed as a percentage. A low percentage means a poor rock while a high percentage means a good quality rock.

Bieniawski (1976) developed a geomechanical classification scheme using five criteria; strength of the intact rock, RQD, joint spacing, conditions of the joints and ground water conditions. He assigned a rating value to each of these variables, and by summing the values of the ratings determined for the individual properties he obtained an overall rock mass rating (RMR). Determination of the RMR of an unsupported excavation for example, can be used to determine the stand up time of that excavation.

In order to provide information that will assist the geotechnical and mining engineers it is necessary to

Table 9.8 *(Continued)*

Descriptive indices of discontinuities
Consideration should be given to type, number of discontinuities, location and orientation, frequency of spacing between discontinuities, separation or aperture of discontinuity surfaces, persistence and extent, infilling and the nature of the surfaces

Rock quality designation

Sound RQD >100	Rock material has no joints or cracks. size range >100 cm
Fissured RQD 90–100	Rock material has random joints. Cores break along these joints. Lightly broken. Size range 30–100 cm
Jointed RQD 50–90	Rock material consists of intact rock fragments separated from each other by joints. Broken. Size range 10–30 cm
Fractured RQD 1–50	Rock fragments separated by very close joints. Core lengths < twice NX core diameter. Very broken. Size range 2.5–10 cm
Shattered RQD <1	Rock material is of gravel size or smaller. Extremely broken. Size range <2.5 cm

RQD in per cent = $\dfrac{\text{Length of core in pieces 10 cm and longer}}{\text{Length of core run}} \times 100$

RQD diagnostic description

> 90%	Excellent
75–90	Good
50–75	Fair
25–50	Poor
< 25%	Very poor

understand the stress–strain characteristics of the rocks. This is especially important in relation to the mining method that is proposed for the rocks, the way in which the rocks will respond in the long term to support (in underground mining), or in deep open pits how deformation characteristics affect the stability of a slope. As these rock properties become better understood, it is possible for the geotechnical and mining engineers to improve the design of underground mines, the way in which excavations are supported, the efficiency of mining machinery and the safety of the mine environment.

9.6.1 **Quantitative assessment of rock**

There are numerous good textbooks which cover the properties of rocks from an engineering point of view (Krynine & Judd 1957, Goodman 1976, 1989, Farmer 1983, Brady & Brown 1985), the latter reference being particularly relevant to underground mining. The main properties which are used in the engineering classification of rocks for their quantitative assessment are poros-

ity, permeability, specific gravity (bulk density), durability and slakability, sonic velocity and rock strength. Porosity and permeability are important in the assessment of the subsurface water. Specific gravity is required to determine the mass of the rock. Durability and slakability tests reveal what effect alternate wetting and drying will have on surface or near surface exposures. Under conditions of seasonal humidity some rocks have been known to disaggregate completely (Obert & Duval 1967). Sonic velocity tests give an indication of the fracture intensity of the rock. Probably the most important property of rock is its strength.

ROCK STRENGTH

The fundamental parameters which define rock strength are stress and strain, and the relationship between stress and strain. Stress (σ) is a force acting on a unit area. It may be hydrostatic when the force is equal in all directions, compressional when the force is directed towards a plane, tensional when directed away from a plane,

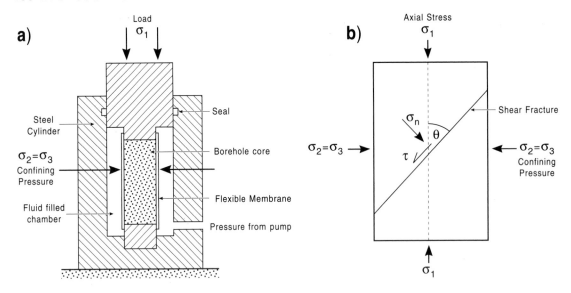

Fig. 9.20 (a) Schematic diagram of the elements of triaxial test equipment. (b) Diagram illustrating shear failure on a plane, an example of the results of the triaxial test on a piece of core. σ_1 is the major principal compressive stress, $\sigma_2 = \sigma_3$ = intermediate and minor principal stresses respectively. σ_n is the normal stress perpendicular to the fracture, and τ is the shear stress parallel to the fracture.

torsional (twisting), or shear stress (τ) when the forces are directed towards each other but not necessarily in line. Strain (ε) is the response of a material to stress by producing a deformation, i.e. a change in shape, length or volume. The linear relationship between stress and strain is know as Hooke's Law and the constant (E) connecting them is Young's modulus. It is also called the modulus of elasticity, and gives a measure of stiffness.

The strength of a rock sample can be tested (usually in the laboratory) in several ways. Unconfined (uniaxial) compressive strength is measured when a cylinder (usually core) of rock is loaded to a point of failure. The load immediately prior to failure indicates the rock's strength. Tensile strength is measured by 'pulling' a cylinder of rock apart. The tensile strength of rock is generally less than one tenth of its compressive strength. Shear strength is measured as a component of failure in a triaxial compressive shear test. A cylinder of rock is placed inside a jacketed cylinder in a fluid filled chamber. Lateral as well as vertical pressure is applied and the rock is tested to destruction (Fig. 9.20). The strength is that measured immediately prior to failure.

FAILURE

Observations on rocks in the field, in mines and in open pits indicate that rocks which on some occasions appear strong and brittle may, under other circumstances, display plastic flow. Rocks which become highly loaded may yield to the point of fracture and collapse. A rock is said to be permanently strained when it has been deformed beyond its elastic limit.

With increasing load in a uniaxial (unconfined) compressive test, the slope of a stress–strain curve displays a linear elastic response (Fig. 9.21). Microfracturing within the rock occurs before the rock breaks, at which point (the yield point) the stress–strain curve starts to show inelastic behaviour followed by a rapid loss in load with increasing strain. If the load is released before the yield point is reached, the rock will return to its original form. Brittle rocks will shear and ductile rocks will flow, both types of behaviour resulting in irreversible changes.

The results of the uniaxial (unconfined) compressive tests described above give a measure of rock strength and are convenient to make. The conditions that the rock was in prior to being removed by drilling (or mining) were somewhat different, because the rock had been surrounded (confined) by the rest of the rock mass. When a rock is confined its compressive strength increases. The triaxial compressive strength test gives a better measure of the effects of confining pressure at depth (Fig. 9.20). Core that has been subjected to triaxial testing, fractures in a characteristic way (Fig.

9.20b). Several failure criteria have been developed for rocks, the simplest being represented by Coulomb's shear strength criterion that can be developed on a fracture plane such as the one shown in Fig. 9.20b and is represented by

$$\tau = \sigma_n \tan \theta + c$$

where:

τ	=	total shearing resistance,
σ_n	=	normal stress acting at right angles to the failure plane,
c	=	cohesive strength,
$\tan \theta$	=	coefficient of internal friction where
θ	=	angle between stress fracture induced and σ_1.

The theory behind this equation is explained by Farmer (1983) and Brady & Brown (1985).

Triaxial test results are used to construct a Mohr diagram (Fig. 9.22). The symbol σ_3 represents lateral confining pressures in successive tests, while σ_1 represents the axial load required to break the rock. Circles are drawn using the distances between the different values of σ_3 and σ_1 as the diameter. The common tangent to these circles is known as the Coulomb strength envelope. The angle θ can be determined from the diagram, rather than having to measure the fracture angle on the tested rock specimen. A value for c can also be determined from the diagram.

The expected confining pressures and axial loads that might be encountered in a mine can be calculated. Any sample whose test results plot as a Mohr diagram that falls below the tangent should not fail. A sample which failed in a test with lower confining stresses but similar axial stresses, would give rise to a circle which would cut the tangent suggesting that the critical stresses have been exceeded and failure might occur (Peters 1987). This does not take into account failure that might take place along discontinuities, nor does it consider the effects of anisotropic stress systems in the rock mass or changes that will occur with time.

The effect of discontinuities is probably more important than being able to measure the rock strength. The discontinuities may be weak and could result in failure at strengths below that of the rock itself.

IN SITU STRESS DETERMINATION

Most *in situ* stress determinations are made before mining takes place usually from a borehole drilled from an adit. Two common procedures are employed in the

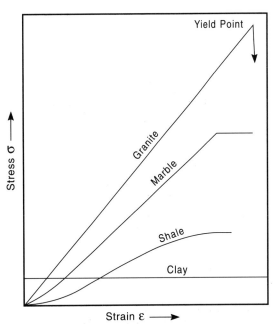

Fig. 9.21 A simplified stress–strain diagram illustrating the idealized deformational behaviour of some rock types. That of granite represents an idealized perfectly elastic material. Clay is a typical material that behaves as a plastic material which will not deform until a certain stress is achieved. Marble is an example of an elastoplastic material and shale represents a ductile material where stress is not proportional to strain. (Modified after Peters 1987.)

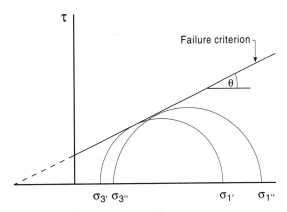

Fig. 9.22 An example of a Mohr diagram.

determination of *in situ* stress. The first is based on the measurement of deformation on the borehole wall induced by overcoring (Fig. 9.23) and the second by measuring the component of pressure in a borehole or slot needed to balance the *in situ* stress. In the overcoring system, a strain gauge is fixed to the borehole wall using

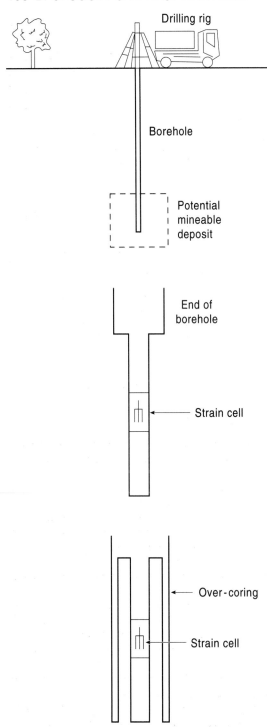

Fig.9.23 A diagrammatic representation of the principle of overcoring.

a suitable epoxy resin. Overcoring with a core barrel of larger size causes stress relief in the vicinity of the strain cell which induces changes on the strains registered in the gauges (Brady & Brown 1985). These measurements are used to help in mine design.

Additional tests are carried out during mining using various types of strain gauges. Some mines, e.g. Kidd Creek, have their own rock mechanics, backfill and mine research departments whose rock mechanics instruments are regarded as an indispensible tool for detecting and predicting ground instability (Thiann 1983). Excellent examples of the use of rock mechanics in both underground and surface mining are given by Hoek & Brown (1980) and Hoek & Bray (1977) respectively.

9.6.2 Hydrogeology

During an exploration programme it is important to note where the water table is encountered in a borehole because if one were to mine this deposit later, then it is essential to know where the water is, how much water there is and the water pressure throughout the proposed mine. Useful tests can be performed on boreholes include measuring the draw down of the water level by pumping, and upon ceasing pumping measuring the rate of recovery. Parker tests in boreholes are used to obtain a quantitative estimate of the contribution a particular bed or joint may make to the water inflow at a site (Price 1985).A good description of the basic fieldwork necessary to understand the hydrogeology of an area is given by Brassington (1988). Two properties of the rocks that can be measured are the porosity and the permeability. Porosity describes how much water a rock can hold in the voids between grains and in joints, etc. Permeability is a measure of how easily that water can flow through and out of the rock. The flow of water is described empirically by Darcy's law, and the relative ease of the flow is called the hydraulic conductivity. Darcy's law is a simple formula

$$v = Ki$$

where:

v	=	the specific discharge or the velocity of laminar flow of water through a porous medium (the critical Reynold's number at which turbulence development is seldom reached),
K	=	hydraulic conductivity,
i	=	hydraulic gradient (the rate at which the head of water changes laterally with dis -tance).

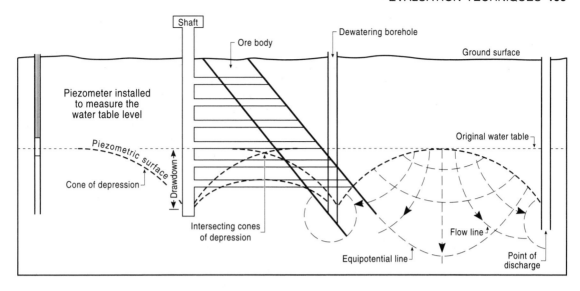

Fig. 9.24 Diagram to illustrate dewatering cones and a piezometric surface. The shaft and dewatering borehole are sited close enough to each other for their cones of depression to intersect and produce a lowering of the water table which effectively results in the dewatering of part of the mine. The flow net illustrates the flow of underground water through rocks of uniform permeability. Flow takes place towards the points of discharge, which in this case are pumps at the bottom of the shaft and dewatering boreholes. (Modified after McLean & Gribble 1985.)

The hydraulic conductivity can be measured in boreholes by undertaking pumping tests. The head of water at a point (e.g. a borehole) within a rock mass is the level to which the water rises when the borehole is drilled. The tube inserted in a borehole to measure this water level is called a piezometer. Several boreholes may intersect points on a piezometric surface (Fig. 9.24), and the maximum tilt (i) of this surface from the horizontal defines the hydraulic gradient (McLean & Gribble 1985).

Groundwater moves slowly under the influence of gravity from areas of high water table to low areas (Fig. 9.24). The pumps lower the piezometric surface in the proximity of the borehole creating a difference in pressure, with the pressure much less at the point of discharge (the pumps) than in the areas between the pumps. This results in the water flowing in curved paths towards the pumps.

Rocks which are permeable and allow water to flow freely through the joints, fractures and pores are known as aquifers. Rocks which have a low permeability and are impervious and do not allow water to flow through them are known as aquicludes.

Having found an orebody, finding water for the mill is often the next concern. A mill requires a considerable volume of water and it can happen that the amount of

water available may make or break a project.

9.7 SUMMARY

Mineralization is composed of dissimilar constituents, is heterogeneous and much mineralization data may have an asymmetrical rather than a normal distribution. Simple statistical sampling methods used for investigating homogeneous populations may therefore give a highly misleading result if used to interpret mineralization data where in many cases a sample can represent only a minuscule proportion of the whole. The pitfalls and problems that this may cause have been discussed and emphasis placed on the need to pay careful attention to sampling error.

Sample extraction must be random and preferably of equal volume but as the samples may amount to several kilograms or even tonnes in weight there must be a systematic reduction of mass and grain size before analysis can take place. Sample reducing systems require thoughtful design and must be carefully cleaned between each operation. Gold and platinum-bearing samples present special problems because of their density which may lead to their partial segregation from lighter gangue material during sample preparation.

Samples may be acquired by pitting and trenching and

by various types of drilling. If coring methods are used then effective core recovery is essential and the core logging must be most methodically performed. Drilling contracts should be carefully thought out and based on sound knowledge of the drilling conditions in the area concerned. Drill holes deviate from a straight line and must be surveyed. Drilling patterns must be carefully planned and the geologist in charge ready to modify them as results come to hand.

The analytical data derived from the above investigations are used in mineral resource and ore reserve estimations, and for grade calculations. Once again classical statistical methods have only a limited application and geostatistical methods must be used. The sample collection work provides a great opportunity to initiate geotechnical investigations and this should be capitalized on. The collection of geotechnical data will provide mining engineers with essential information for the design of open pit and underground workings. It will also provide a starting point from which the hydrogeology of the target area can be elucidated

9.8 FURTHER READING

A textbook which covers statistical analysis is the eminently readable Davis (1981) while Issaks & Srivasta (1989) give a very readable account of geostatistics. The mathematical detail of geostatistical estimations are eleborated by Journel & Huijbregts (1978). Annels (1991) covers this subject as well as drilling, sampling and resource estimation. Discussions of the problems of resource definition are covered by Annels (1991) and Whateley & Harvey (1994).

An introductory text for geotechnical investigations is written by West (1991), while Hoek & Brown (1980) and Hoek & Bray (1977) are still the most sought after books that describe the application of rock mechanics to underground and surface mines respectively. Brassington (1988) gives a good description of the basic field work necessary to understand the hydrogeology of an area.

9.9 APPENDIX: USE OF GY'S SAMPLING FORMULA

1 Introduction
These are worked examples of the calculation of the total variance (TE) of sphalerite and chalcopyrite mineralization and the two-sided confidence of that mineralization. These are followed by a worked example of how to calculate the mass of a sample to be taken knowing the confidence limits.

Using the short form of the Gy formula for the fundamental variance (section 9.1.4)

$$So^2(FE) = \frac{C}{K} \text{ . Taking K as 125 000}$$

$$So^2(FE) = \frac{C}{125\ 000}$$

and the variance of the total error (TE) is estimated as twice that of the fundamental error (FE). It is assumed that the liberation factor β is constant and equals unity, which is a safe assumption.

2 Calculation of total variance (TE) and double-sided confidence interval: sphalerite.
A prospect contains sphalerite mineralization of mean grade 7.0% zinc in a gangue of density 2.7 kg m^{-3}. Sphalerite contains 67% zinc with a specific gravity of 4.0 kg m^{-3}.
(a) Fundamental variance of each sampling stage (see also Box 9.1). Critical content (a_L) = 7% zinc = 10.5% sphalerite as a fraction = 0.105.

Constitution factor (c)

$$= \frac{(1-0.105)}{0.105}[(1-0.105)4.0 + (0.105)2.7]$$

$$= 8.52(3.58 + 0.28) = 32.89, \text{ say } 33.$$

Heterogeneity constant (C) = cβfg

$$= 33 \times 1 \times 0.5 \times 0.5 = 4.13.$$

Fundamental variance

$$= \frac{C}{K} = \frac{4.13}{125000} = 33 \times 10^{-6}.$$

(b) Total variance (TE) of the sampling scheme.
At each reduction stage the total variance is equal to double the fundamental variance. The complete sampling scheme consists of four reduction stages (Fig. 9.5a) each ending on the safety line (see text). Consequently, the total variance equals eight times (2 × 4) the fundamental variance as calculated above.

Total variance (TE) = 8 × (33 × 10^{-6}) = 264 × 10^{-6}.

Relative standard deviation = $\sqrt{(264 \times 10^{-6})}$ = 16.2 × 10^{-6}

Absolute standard deviation as % of mineral = 10.5% sphalerite × 1.6% = 0.17% sphalerite.

Absolute standard deviation as % of metal = 0.17% sphalerite × 67% = 0.11% zinc.

The confidence level at 95% probability is ±2 SD = ±2(0.11%) = 0.22% zinc.

The two-sided confidence level is then **7% ± 0.22 zinc** at 95% probability, which is an acceptable result. Generally, a relative standard deviation of <5% is acceptable.

3 Calculation of total variance(TE) and double-sided confidence interval: chalcopyrite mineralization.
A prospect contains chalcopyrite mineralization of grade 0.7% copper ($\bar{x}$) in gangue of density 2.7 kg m^{-3}. Chalcopyrite contains 33% copper with a density of 4.2 kg m^{-3}.
(a) Fundamental variance of each sampling stage.
Critical content (a_L) = 0.7% = 2.1% chalcopyrite as a fraction = 0.021.

Constitution factor (c)

$$= \frac{1-0.021}{0.021}\left[1-0.021\ 4.2 + 0.021\ 2.7\right]$$

$$= 46.62(4.11 + 0.06) = 194.4, \text{ say } 195.$$

Heterogeneity constant (C) = cβfg = 195 × 1 × 0.5 × 0.5 = 24.38.

Fundamental variance

$$= \frac{C}{K} = \frac{24.38}{125000} = 195 \times 10^{-3}.$$

(b) Total variance (TE) of the sampling scheme.
As above but, suppose, the system consists of five reduction stages each ending on the safety line. The total error variance now equals ten times (2 × 5) the fundamental variance.

Total variance(TE) = 10 × (195 × 10^{-6}) = 1950 × 10^{-6}.

Relative standard deviation = (1959 × 10^{-6})$^{-2}$ = 44 × 10^{-3} = 4.4%.

Absolute standard deviation as % of mineral = 2.1 % chalcopyrite × 4.4% = 0.09% chalcopyrite.

Absolute standard deviation as % of metal = 0.09% × 33% = 0.03% copper.

The confidence at 95% probability is ±2 SD = ±2(0.03 %) = 0.06% copper.

The two-sided confidence level is then **0.7% ± 0.06%** copper at 95% probability, which is an acceptable result. Generally a relative standard deviation of < 5% is acceptable.

4 Calculation of the weight of a sample (M) to be taken knowing the confidence interval.
(a) Introduction (see section 9.1.4)

$$M = \frac{Cd^3}{S^2(FE)}$$

(b) Sphalerite mineralization.
Sphalerite mineralization assays 7% zinc and the confidence level required is ± 0.2% zinc at 95% probability. The sphalerite is liberated from its gangue at a particle size of 150 μm: the gangue density is 2.7 kg m^{-3}. Sampling is undertaken during crushing when the top particle size (d) is 25mm.

Critical content (a_t) = 7% zinc = 10.5% sphalerite = 0.105 as a fraction.

Constitution factor(c) as in previous example = 33.

Liberation factor (β) = (0.015/2.5)0.5 = (0.006)0.5 = 0.077.

Heterogeneity constant(C) = cβfg = 33 × 0.077 × 0.5 × 0.25 = 0.32.

Top size (d) = (2.5)^3 =15.62.
Cd^3 = 0.32 × 15.62 = 5.0.

Fundamental variance (FE).
Relative standard deviation 2S=$\frac{0.2\%}{7}$ = 0.029. Therefore S =0.014.

Mass of sample=$\frac{Cd^3}{S^2(FE)} = \frac{5.0}{(0.014)^2} = 25.5$ kg.

This is the sample weight to constrain the fundamental variance. For the total variance a sample weight of twice this quantity is required [i.e. S^2(TE) ≤ 2S^2(FE)], or **51 kg.**
To illustrate the benefits of sampling material when it is in its most finely divided state, assume that the same limit (± 0.2% zinc) is required when sampling after

grinding to its liberation size when d is 0.015 cm and β equals 1.0:

$C = 33 \times 1 \times 0.5 \times 0.25 = 4.13$,

$d^3 = (0.015)^3 = 3.3 \times 10^{-6}$,

$Cd^3 = 13.63 \times 10^{-6}$ and

M = 6.9 grams, say 7g.

10 Feasibility Studies

BARRY C. SCOTT

10.1 INTRODUCTION

It is said that mineralization is found, orebodies are defined and mines are made. Previous chapters have outlined how mineralized rock is located by a correct application of geology, including geophysics and geochemistry. Having located such mineralization its definition follows and this leads to a calculation of mineral reserves where a grade and tonnage is demonstrated at an appropriate level of accuracy and precision. We have discussed mineralization and mineral reserves (in Chapters 3 & 9)—the question now becomes how does a volume of mineralized rock become designated as ore and by what process is this ore selected to become a producing mine? A mine will come into existence if it produces and sells something of value. What value does a certain tonnage of mineralized rock or ore have? What is meant by value in this context? How is this value calculated and evaluated? These questions are considered in this chapter.

10.2 VALUE OF MINERALIZATION

The value of mineralization is a function of several factors.

10.2.1 Mine life and production rate

MINE LIFE

Other things being equal the greater the tonnage of mineable mineralization the longer the mine life and the greater its value. Obviously

$$\text{Mine life in years} = \frac{\text{Total mineable reserve}}{\text{Average annual production}}$$

Factors relating to mine life include the following.
Legal limitations. Exploitation of the mineralization may be under the constraints of a lease granted by its former owner such as the government of the state concerned. If the lease has an expiry date then the probability of obtaining an extension has to be seriously considered.

Market forecast. What will be the future demand and value for the minerals produced? This important factor is considered separately below.
Political forecast. If a lease agreement has been negotiated with the local government how long is this government likely to remain in power? Will a new government wish to either cancel or modify this agreement? (See also section 1.4.4).
Financial return. The development of a mine is a form of capital investment. The main concern of the investor is for this money to be returned with an acceptable level of profit and the optimization of this financial return will probably decide the number of years of production. As a very general rule the minimum life of a mine should be seven to ten years.

PRODUCTION RATE

A commonly used guide for the Average Annual Production Rate (AAPR) is

$$\text{AAPR} = \frac{(\text{Mineral reserve})^{0.75}}{6.5}$$

This is a broad generalization but for mineable reserves of 20 Mt the rate is 1.5 Mt a^{-1}, and for 10 Mt a rate of 0.9 Mt a^{-1}. Factors relevant to the annual production rate include the following.
Type of mining. The two basic types of mining are either underground (Fig. 10.1) or from the surface (open pit), (Fig. 10.2). As the former can be more than three times the cost of the latter, underground extraction may require at least three times the content of valuable components in the mineralization to meet this higher cost. This in turn implies a lower tonnage, but a higher grade of mineable reserve and a lower annual production rate.

A basic characteristic of open pit mining is the surface removal of large tonnages of rock (overburden) in order to expose the ore for extraction (Fig. 10.2). This removal of waste rock and ore usually proceeds simultaneously and the resultant stripping ratio of tonnage of waste removed to tonnage of ore is a fundamental economic specification. With a ratio of 4:1, 4 t of waste has to be removed for every 1 t of ore recovered; if the mining costs were US$2 t^{-1} then the extraction cost per tonne of

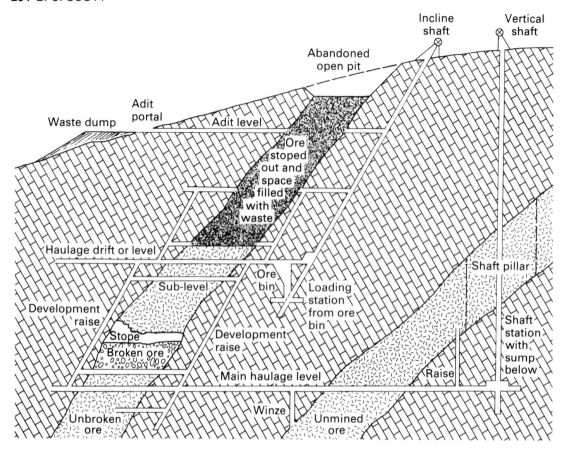

Fig. 10.1 Mining terminology. Ore was first extracted at the mineralized outcrop using an open pit. Later an adit was driven into the hillside to intersect the ore at a lower level. An inclined shaft was sunk to mine from still deeper levels and, eventually, a vertical shaft was developed. Ore is developed underground by driving main haulage and access drifts at various levels and connecting them by raises from which sub-levels are developed. It is mined from the lower sub-level by extracting the mineralization upwards to form a stope. Broken ore can be left in a stope to form a working platform for the miners and to support the walls against collapse (shrinkage stoping). Alternatively, it can be withdrawn and the stope left as a void (open stoping) or the resultant space can be filled with fine-grained waste rock material brought from surface (cut-and-fill stoping). Ore between the main haulage and sub-level is left as a remnant to support the former access until the level is abandoned. (From Barnes, 1988, *Ores and Minerals,* Open University Press, with permission.)

ore is US$10 t^{-1}. Conversely, if, say, the stripping ratio were 8:1 at the same mining cost then the extraction cost per tonne of ore would be US$18 t^{-1}.

At the time of writing open pit mining costs vary from about US$1 to US$3 t^{-1} of rock extracted. An economy of scale is apparent in that the higher charge is for production levels of lower than 10 000 t p.d., while the lower cost refers to outputs exceeding 40 000 t p.d., ore plus waste rock.

Underground mining is generally on a smaller scale than open pit mining —an output of 10 000 t p.d. would be modest for an open pit but large for an underground mine. This generally lower level of production is in part related to the more selective nature of underground extraction in that large tonnages of waste rock do not have to be removed in order to expose and extract ore.

Underground mining costs vary greatly dependent upon the mining method employed but in large orebodies where production is mechanized, costs per tonne of ore extracted can be comparable with those of an open pit. At the other end of the scale, in narrow veins using labour intensive methods, costs can be as high as US$30 to 40 t^{-1}.

Generally underground mining costs are higher than

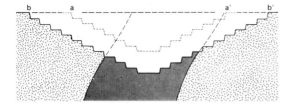

Fig. 10.2 Development of an open pit mine. During the early stages of development (a–a') more ore is removed than waste rock and the waste to ore ratio is 0.7:1. That is, for every tonne of ore removed 0.7 tonnes of waste has also to be taken. The waste rock is extracted in order to maintain a stable slope on the sides of the open pit, normally between 30° and 45°. Obviously the greater the slope the less the waste rock to be taken, and the lower the waste to ore ratio. As the pit becomes deeper the ratio will increase; at stage b–b' it is 1.6:1. (From Barnes, 1988, *Ores and Minerals*, Open University Press, with permission).

those at the surface. However as the depth of an open pit increases, the ratio of waste to ore mined becomes greater and more waste rock per tonne of ore is extracted (Fig. 10.2). This has the effect of increasing the overall extraction cost until, eventually, it can equal underground mining costs.

Scale of operations. The economies of scale mean that usually the larger the annual output the lower the production cost per tonne of mineralization mined. Obviously there is an upper limit to this general principle otherwise most mines would be worked out in their first year of production!

Accessibility of the mineralization. How accessible is the mineralization? There are limitations on open pit and shaft depths, and shaft size, which restrict output. What type of mining extraction is to be followed? Underground methods vary considerably in their nature and scope dependent upon the geology of the mineralization and the host rocks. Some methods are suitable for large scale production in wide zones of mineralization whilst others are designed for small workings in narrow veins.

Capital expenditure limitations. There may well be restrictions on the availability of capital for developing and equipping the mine. Consequently it may not be possible to achieve the maximum desirable rate of annual production despite the presence of adequate and accessible mineable reserves.

10.2.2 Quality of mineralization

Grade is but one facet of quality, albeit an important one,

and other factors such as mineralogy and grain size enter into the definition (section 2.2). Untreated ore is rarely a saleable product and usually some method of concentration has to be employed to separate waste rock and gangue from the valuable constituents (Wills 1988). Concentration plants vary from low cost (separating sand from gravel) to high cost, complex plants as in the separation of lead and zinc sulphides (Fig. 10.3).

10.2.3 Location of the mineralization

The valuable products have to be taken to market for sale, and supplies for the mining operation follow the same route in reverse. Local availability of power, water supply and skilled labour must be considered, and local housing, educational and recreational facilities for the workforce. Clearly mineralization adjacent to existing facilities has greater value than that in a remote, inhospitable location.

10.2.4 Market conditions

Increasingly the sale of minerals extracted from a mine is its *only* income. From this revenue the mining company has to pay back the capital borrowed to develop the mine and pay the related interest, production and processing costs, transport of the product to market, product marketing, dividends to shareholders, local and national taxes, and many other items (Fig. 10.4).

Can the product be sold and at what price? It cannot be assumed that demand is growing sufficiently fast to absorb all feasible new production and there has to be a reasonable assessment of future demand and price over the life of the mine, or at least its first ten years, whichever is the shorter. The product price (section 1.2.3) is the most important single variable in a feasibility study but the most difficult to predict.

Many mineral products are sold on commodity exchanges, in London and New York, where a price is set each working day and considerable fluctuations occur (Figs 1.5 & 1.6). Most mine owners have little or no control over these prices which are vital for the well-being of producing mines. Market principles suggest that during a period of low prices production costs should be minimized and revenue maximized. One way of achieving this would be to extract the maximum possible tonnage of easily accessible mineralization at the highest grade. Alternatively during periods of higher price tonnages of lower grade material could be extracted. Mining systems, however, have a great deal of inertia and production rates cannot be varied so easily.

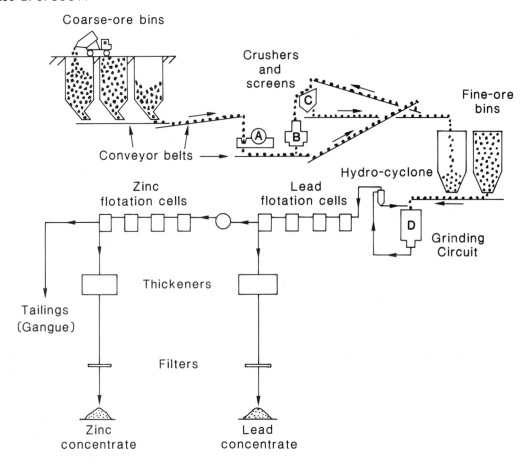

Fig. 10.3 Separation of lead and zinc sulphides. The ore has an average grade of 9.3% zinc and 2.0% lead, as sphalerite and galena respectively. At the main site the broken ore at 15 cm size is crushed via a primary (A) and a secondary crusher (B) to less than 40 mm size. The former crusher is in open circuit in that no material is returned for re-crushing. The secondary, finer, crusher is in a closed circuit and the crushed rock passes over screens (C) on which only material less than 40 mm size passes through to the fine-ore bins. Oversize is returned to the crusher B. The crushed ore is then ground, in water, to less than 0.075 mm, which is the effective liberation size of the contained sulphides, in the grinding circuit D. This is in a closed circuit with hydrocyclones, which separate oversize material (>0.075 mm) and return it for further grinding and size reduction, in D. The undersize is now a finely ground powder suspended in water and forms a slurry (about 40% solids) which can be pumped for treatment in the flotation units, or cells. In flotation the surface properties of the liberated sulphides are used whereby air bubbles become attached to individual particles of sphalerite which float to the surface of the cells and are skimmed off as a froth, whilst all other minerals and rock particles sink. Each sulphide has different surface properties depending upon the pH of the slurry and added chemicals. In this way sulphides are separated from their gangue, and from each other. A sphalerite concentrate is produced with, say 54% zinc, and galena concentrate with 65% lead. The separation process is reasonably efficient and recovers 91% of the contained zinc values but only 77% of the lead. The waste rock, or tailings in the form of a fine slurry, contains about 0.2% zinc and 0.9% lead. (From Evans 1993 and Wills 1988.)

Present day mineral and metal prices are available in publications such as the *Mining Journal, Engineering and Mining Journal* and *Industrial Minerals* which summarize average weekly and monthly commodity prices, and provide annual reviews.

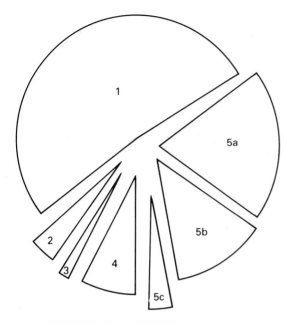

DISTRIBUTION OF INCOME

1 Wages, salaries and benefits to employees
2 Taxes and imposts to governments
3 Interest to providers of loan capital
4 Dividends to shareholders
5 Reinvested in R.G.C. Group
   a) depreciation
   b) exploration and evaluation
   c) retained earnings

Fig. 10.4 Distribution of revenue in 1984 by Renison Goldfields Consolidated Ltd. (After *Consolidated Gold Fields PLC Annual Report* 1985.)

10.2.5 **Economic climate**

What is the general business sentiment? Are there years of prosperity ahead (rising commodity prices) or recession (fall of demand and prices)? What are predictions for inflation over the life of the mine? Apart from rising equipment and labour costs, inflation may have a disasterous effect if it gives rise to higher interest rates on capital borrowed to develop the mine.

As far as mining companies are concerned economic cycles are doubly important because in times of recession, and inflation, they suffer falling mineral prices, rising interest rates on loans, but operating costs continue to rise (Kernet 1991). In the late 1970s a number of large copper deposits had been discovered and were brought into production just as the world economy entered a period of recession. This led to a dramatic fall in the copper price to a low of US$0.54 lb^{-1} in June 1982 and many producers incurred heavy financial losses on their operations. The reverse, however, happened in the late 1980s when booming economies took the copper prices from US$0.54 lb^{-1} in 1986 to US$1.68 lb^{-1} in 1988. Conversely, in 1989 a new recession commenced which has continued into the early 1990s with depressed mineral prices and low profitability—in 1992 copper averaged $1.04 lb^{-1} and fell below $1.00 lb^{-1} in 1993.

10.2.6 **Political stability of host country**

Fundamentally a company developing a mineral property wishes to have its invested capital returned plus a healthy profit otherwise the investment will be taken elsewhere. There are many other necessary and desirable objectives which can be pursued in establishing a new mine such as safety and health of employees, efficient mineral extraction, care for the environment, payment of local and national taxes, etc., but, rarely, can these be achieved unless the mine is an attractive investment. The government of the host country may have a different order of priorities, e.g. continuity of employment, social objectives, and taxes to central government for national social policies while dividends to shareholders and repayment of loans, which were used to develop the mine, may be given a low order of priority. In addition an inflow of foreign investment is sometimes disliked with the thought of foreign shareholders controlling a national resource of the country. Thus it is essential to operate within a framework of acceptable law and, if necessary, a mutually agreed contract between developer and host. This may require delicate negotiation and it has to produce a true partnership otherwise the agreement will not last, particularly if there is a change of government or leadership. There are known deposits of proved mineralization which have no value because of the impossibility of achieving such a satisfactory agreement (see also section 1.4.4).

10.2.7 **Environmental factors**

Approaches for a mining lease in most developed countries may be greeted by objections concerning possible blasting vibrations, dust, noise, atmospheric and water pollution, and increased traffic density. With modern equipment and design most, if not all, of these can be overcome but at a cost. This cost is, in effect, an added operating charge which reduces the profitability of a mine, and hence its value (see also section 1.4.2).

10.2.8 **Government controls**

Any potential mine will have to operate within a national mining law and a fiscal policy enforced by an inspectorate. Fiscal policy is usually of prime importance for it decides items such as taxation of profits, import duties on equipment and transfer of dividends and capital abroad if foreign loans and investors are involved. If this transfer cannot be guaranteed then the value of the proposed mine, and the mineralization, will be considerably less (see also section 1.4.3).

10.2.9 **Trade union policy**

It is a foolish investor who does not consult the local trade unions before starting a new mining project. Union disputes can stop a mine as effectively as machinery breakdowns and continual strikes and disputes over working practices can make a mine unprofitable. In other words industrial disputes can change mineable ore to mineralized rock of no value. In several countries trade unions are, in effect, a branch of the government and if a contract is agreed between host and developer this will include a trade union agreement.

10.2.10 **Value of mineralization— a summary**

Mineralization is only of value if a saleable product can be produced from it. 'Value' is a financial concept and is related to the several factors discussed above rather than to just a single consideration of grade and tonnage. The main factors which enhance the value of mineralization are an increase in tonnage, grade, mineral recovery, product sales price and political stability, together with decreases in mining cost, ore beneficiation cost, transport cost to market, capital cost of development and taxes. Of course not all these factors are under the direct control of a mining company wishing to develop mineralization into mineable ore.

10.3 **VALUATION OF MINERALIZATION**

The factors in section 10.2 can be expressed in terms of revenue and expenditure (income and cost) of a proposed mining operation. All relevant items are brought together in an annual summary which is referred to as a *cash flow*. In any particular mineral project:

Cash Flow = Cash into the project (revenue) minus cash leaving the project (expenditure)

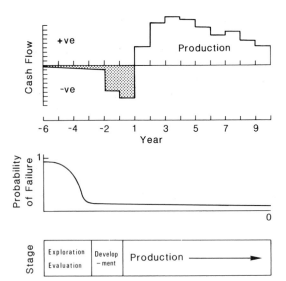

Fig. 10.5 The cash flow of a mineral project over fifteen years. During the first four years the mineralization is evaluated and over the next two years the mine is built at a total capital cost of US$250M. Over these six years the cash flow is negative in that there is no revenue generated. In the first year of production (Year 1) revenue commences and continues during the life of the mine. From this revenue is paid all the mine operating costs and repayment of the capital loans, plus interest, which were used to develop the mining operation. The lenders of capital require their money back within the first three to five years of production. Also shown is the probability of failure of the project, and its main stages. Obviously the size of the database increases with time and the probability of failure decreases correspondingly.

= (Revenue)–(Mining Cost + Ore Beneficiation Costs +Transport Cost + Sales Cost + Capital Costs + Interest Payments + Taxes).

Cash flow (Fig. 10.5 and Example 10.1) is calculated on a yearly basis over either a ten-year period or the expected life of the mine, whichever is the shorter. It is a financial model of all relevant factors considered in section 10.2 and is the process by which the value of a tonnage of mineralization is calculated to determine whether this tonnage is mineable ore. Such mineralized rock only has value by virtue of its ability to produce a series of annual positive cash flows (revenue exceeding expenditure) over a term of years. Another way of expressing this is to say that the value of a tonnage of

Example 10.1 Cash flow calculation.

From the results of an order of magnitude feasibility study on an occurrence of sphalerite and galena mineralization, the first three years of a ten year cash flow are given below. The plan is to mine the mineralization underground and produce galena and sphalerite concentrates at the main site. These would be sold to a lead and zinc smelter for metal production.

Calculations are for year end; interest payments on borrowed capital to build the project are excluded. Units are US$1000 and 1000 tonnes.

(a) Total Initial Capital (CAPEX) to establish the mine is $140 600. Construction will take two years: Years -1 and -2.

(b) Ore production is 2 400 t p.a. but only 1 600 t p.a. in the first year of production (Year 1). Ore grade is 9.3% zinc and 2.0% lead.

(c) Ore treatment is as in Fig. 10.3.

(d) Revenue: the selling price of zinc is taken as $0.43 lb^{-1} and for lead as $0.21 lb^{-1}. Translating this value to ore gives a revenue of $46.6 per tonne.

(e) Operating costs per tonne:

Mining	$7.25
Mineral processing	8.10
Overheads	4.35
	$19.70 per tonne ore

The cost includes transporting the sphalerite and galena concentrates to a local port, and loading charges. The concentrates are sold to the smelter at the port and it is then the smelter company's responsibility to ship the concentrates to their plant.

(f) There is a royalty payment to the previous mineral owners of 4.5% of gross revenue less production cost.

(g) Calculation f is made before the payment of tax.

	Year-2	Year-1	Year 1	Year 2	Year 3
1 Capital expenditure					
(a) Initial CAPEX	(50800)	(89800)			
(b) Replacement CAPEX			795	1590	3180
(c) Cumulative CAPEX	(50800)	(140600)			
2 Revenue					
(d) Production (Tonnes)			1600	2400	2400
(e) Gross revenue (US$)			74560	111840	111840
3 Operating costs					
(f) Production cost			31520	47280	47280
(g) Royalty [4.5% (e-f)]			1940	2905	2905
(h) Total cost [b+f+g]			34255	51775	53365
4 Operating margin [e-h]			40305	60065	58475
5 Cash flow before tax	(50800)	(89800)	40305	60065	58475

Comment. The initial CAPEX of $140 600 is recovered towards the end of the third year of production. Consequently, the project has passed the first financial hurdle (CAPEX has to be recovered in the first three to five years of production). The mineralization has a value which would be defined on a 10 year cash flow by the methods outlined in section 10.4.

mineralized rock is the value today of all future annual cash flows arising from mine production. How these values are determined is discussed in section 10.4.

The basic data for the financial model, or cash flow, is collected in a feasibility study which is considered next. Some geologists may reject this involvement with finance. However it has to be remembered that while all acknowledge the expertise of mine-making (i.e. trans-

posing mineralized rock into mineable ore) which lies with geologists and other engineers, the ultimate success or failure of a project is measured in money earned. Decisions are made with whatever geological information is available and it is the geologist's responsibility to see that as much critical information as possible is taken into account. Many geological assumptions are included in financial decisions regarding the development of a mineral property and geologists should participate fully in these financial deliberations.

10.3.1 Feasibility study or project evaluation

A feasibility study is a report on the geological, mining, metallurgical, marketing and construction aspects of the project (O'Neil 1982, Potts 1985, Gentry 1988, Goode *et al.* 1991, Smith 1991). From these inputs capital and operating costs are derived to prepare estimated cash flows from which the value of the mineralization in question is determined. It provides a base on which economic and social decisions can be made and it includes four main aspects; strategic or political, technical, financial and social. There is every variation in the detail of feasibility studies from a single sheet compiled in a few minutes to a formal study involving a team of geologists and mining engineers for several months or years.

TYPES OF FEASIBILITY STUDY

Order of magnitude and exploration review. This is an assessment to see if there is a possibility of the mineralization on a property being developed into an operating mine. No detailed engineering design is made and the probable method of mineral extraction and beneficiation is assessed from the known geometry of the mineralization. Operating and capital costs are calculated from empirical factors and comparison with similar operations. Much information is also taken from sources such as the publications of state and national bureaux of mines and from reports of projects in the mining literature. Such examples and much useful data is included in O'Hara (1980), Mular (1982), Bureau of Mines (USA) Cost Estimating System Handbook (1987), Camm (1991), Goode *et al.* (1991), Smith (1991), Craig Smith (1992) and references such as the *Annual Review Reference Manual and Buyers Guide* of the *Canadian Mining Journal*. The main purpose of such a feasibility review is to decide whether it is worthwhile proceeding with the expense of a further stage of exploration. Such reviews also emphasize topics on which there is a lack of information and their accuracy is entirely dependent on data availability, but ± 30% would be acceptable. Several of these studies are completed during the life of an exploration project, each improving in accuracy as the database grows.

Preliminary or project definition. This more detailed

approach, with an accuracy of ±15% to ±25%, is the principal project evaluation process. Complete engineering plans are not prepared but the mine design must be specified in sufficient detail to permit this level of costing. Among other features alternative methods of mining and processing the mineralization, the effects of various levels of output, the mine infrastructure requirements, labour requirements and environmental impact are examined. Experimental mining and a pilot beneficiation plant using bulk samples may be required. Such a study can cost several hundred thousand dollars and is usually completed, in the larger mineral companies, by an in-house evaluation group.

By the time a company commissions a study of this type it may have already spent significant sums of money on the project depending upon its location and geological complexity. Indeed on large projects several million dollars may have been spent, but geologists must be realistic about terminating a project. It is recognized in the mineral industry that the chances of the successful establishment of a new mine are low—that is, the conversion of mineralized rock into mineable ore is a remote chance. A large and successful Canadian mining company investigated some one thousand mineral projects over a forty year period and of these only seventy eight warranted major work, only eighteen were brought into production but eleven did not return the company's capital investment, so only seven were viable from the thousand projects considered.

Definitive or project design. Detailed engineering and planning (± 10% accuracy) are undertaken only when it is almost certain that the project will proceed. This is the final stage before major capital expenditure is incurred and involves producing detailed plans, work schedules, etc., so that orders can be placed for equipment and the mine development can begin. These investigations are thorough, penetrating and usually completed by outside consultant companies who specialize in this type of work. Such a study usually takes several months or years, depending upon the initial database, and can cost several million dollars for a large project.

This final feasibility study is the document given to merchant banks (the 'bankable document') who are being asked to finance the project (see later). The banks, in turn, appoint specialists who evaluate the project and prepare their own, internal, assessment.

Detailed studies. These are carried out on operating mines when an increase in output is proposed following a discovery of additional ore, an increase in commodity price, a change in the mining method or the introduction of different machinery. Current and historical statistics and costs are immediately relevant and the accuracy of such estimates can be within ±5%.

Consideration has to be given to the readership of feasibility studies as there are different levels of understanding regarding the details of a project and differences of emphasis are always possible. Broadly is it intended for internal or external (outside the company) use? If the former, is it for an operational or an administrative group? External readers could be a government department, prospective investors or banks who may loan money for the project.

A feasibility study is a technique for evaluating a new mineral project. If a mine is in production another approach could be to value the issued shares of the controlling company but this is beyond the present scope (see Kernet 1991).

10.4 MINERAL PROJECT VALUATION AND SELECTION CRITERIA

Previous sections have considered the value of mineralization in absolute terms (i.e. a sum of money expressed as a cash flow) and this is the basis of mineral project evaluation. Another aspect of value is its use in a relative sense, that is the ranking of the favourability of a group of similar projects. This relative value is based upon ranking each project's absolute value and is particularly useful when a company has more projects than it can manage. Any company has finite managerial and financial resources and a choice has to be made between two or more alternatives based on some tangible measurement of economic value or return. There are four main techniques, discussed below, all of which are based on the annual cash flow for the project in question.

10.4.1 **Payback**

This is a simple method which ranks mineral projects in order of their value by the number of years of production required to recover the initial capital investment from the project cash flow. In Example 10.2 the projects can be ranked in order of payback B, D, C and A. The method serves as a preliminary screening process but it is inadequate as a selection criterion for it fails to consider earnings after payback and does not take into account the time value of money (section 10.4.3). It is useful in areas of political instability where the recovery of the initial investment within a short period of time is particularly

important and here project B would take precedence.

10.4.2 **Operating margin to initial investment ratio**

There are several variations on this theme but it produces a dimensionless number which indicates the amount of cash flow generated per dollar invested. It is an indication as to how financially safe the investment is and the larger the number the higher the value of the mineralization. It is easy to calculate and considers the whole life of the project compared with payback which considers only the first few years. All factors in the cash flow (such as revenue, operating cost, taxes, etc.) are taken into consideration and all affect the ratio number. However, as in payback, the time value of money is not taken into consideration. As a simple example consider spending $100 today to receive $300 in three years, or spending $100 today to receive $400 in ten years. The increase in the ratio favours the latter yet most would prefer the former. Using this approach the projects in Example 10.2 can be ranked in order of value as C, D, B and A.

10.4.3 **Techniques using the time value of money**

These techniques are commonly used in the financial evaluation of mineralization and mineral projects. The opening theme is that money has a time value (Wanless 1982). Disregarding inflation, money is worth more today than it will be at some future date because it can be put to work over that period. If a dollar were invested today at an interest rate of 15% compounded annually, it would amount to $1.00 \times (1 + 0.15)$ or $1.15 after one year and $1.00 \times (1.15)^5 = \$2.01$ after five years. This is the compound interest formula:

$$S = P(1 + i)^n$$

where S is the sum of money after n periods of interest payment, i is the interest rate and P the initial investment value. In this example we can say that the five-year future value of the $1 invested today at 15% is $2.01. Reversing this viewpoint from the calculation of future values to one of value today, what is the value today of $2.01(S) if this is a single cash flow occurring in five years' time at the accepted rate of interest of 15%? The answer is obviously $1.00 and the present value (P) is a variation of the compound interest formula above:

$$P = \frac{S}{(1+n)^n} = \frac{\$2.01}{(1+0.15)^5} = 2.01(0.497) = \$1.00$$

Example 10.2 Mineral project evaluation and selection criteria.

Mineralization at four similar prospects has been investigated and an order of magnitude feasibility study has been completed on each. At each location it is thought that a mine can be constructed in one year. Production commences in Year 1. Which, if any, are worth retaining? Cash flow values are in arbitrary units.

Year		A	B	C	D
Initial investment (CAPEX)	-1	(3100)	(2225)	(2350)	(2100)
Operating margin	1	500	2000	150	1100
	2	500	1000	450	900
	3	1000	500	1000	1150
	4	2000	500	3400	950
Payback (years)		3.6	1.2	3.2	2.1
Operating margin / CAPEX		1.3	1.8	2.1	1.9
NPV at 15% discount		-560	+835	+656	+780
DCF ROR		11%	40%	21%	30%

The value of the mineralization at Prospect A does not reach the minimum DCF return of 15% and is discarded. The remaining three meet this requirement and can be ranked in value:

Criteria	Most favourable	Middle	Least
Payback	B	D	C
Op. margin / CAPEX	C	D	B
NPV	B	D	C
DCF ROR	B	D	C

The mineralization at prospect B has the greatest value although it is not the lowest cost to develop, but it has the great virtue of generating a large cash flow at the beginning of the project. Prospect A, the least favourable, has the largest CAPEX and a low cash flow in Years 1 and 2. At a discount rate of 15% the mineralization at Prospect B is worth 835 units, C is worth 656 units and D 780 units. On the basis of this study Prospect B is the one to be retained.

This expression is the present value discount factor, more usually termed the discount factor, and tables of calculated factors are readily available (Table 10.1). From the formula, and the table, it is evident that this factor decreases with increasing interest rates and number of years which causes two important characteristics in valuing mineralization and mineral projects. Firstly, as discount factors are highest in the early years, the discounted value of any project is enhanced by generating high cash flows at the beginning of the project. That is, the ability to extract higher grade and more accessible (lower mining cost) ore at the beginning of a project can significantly enhance its value. Not all mineralization is amenable to this arrangement but if this can be achieved

it will provide a higher value than a comparable location in which this is not possible. Secondly, discount factors decrease with time and by convention and convenience cash flows are not calculated beyond usually a ten-year interval as their contribution to the value becomes minimal. This has the added advantage of limiting future predictions. In cash flow calculations it is difficult to predict with some degree of accuracy what is to happen next year—several years ahead is in the realm of speculation although best estimates have to be made.

Thus the value of money today is not the same as money received at some future date. This concept is concerned only with the interest that money can earn over this future intervening period and is not concerned

Table 10.1 A Selection of present value discount factors

	Years	1	2	3	4	5	10	15
Discount rate	5%	0.952	0.907	0.864	0.823	0.784	0.614	0.481
	10%	0.909	0.826	0.751	0.683	0.621	0.386	0.240
	15%	0.870	0.756	0.658	0.572	0.498	0.247	0.123
	20%	0.833	0.644	0.579	0.482	0.402	0.162	0.065

For explanation see text.
The table demonstrates the decrease in the magnitude of the discount factors with increasing years and increasing discount rate, governed by the basic formula:

$$\text{Discount factor} \ = \ \frac{1}{(1-i)^n}$$

where i= interest or discount rate (as a fraction) and n= number of years.

with inflation. In this calculation money received at some time in the future is assumed to have the same purchasing power as money received today; it has constant purchasing power and is referred to as *constant money*. The effect of inflation on project value, however, is important and is considered in section 10.6.

NET PRESENT VALUE (NPV)

The second formula above is used to determine the present value of the expected cash flow from a project at an agreed rate of interest. The cash flow from each year is discounted (i.e. multiplied by the factor from Table 10.1 appropriate to the year and interest rate). A summation of all these yearly present values over the review period provides the Present Value (PV) of the evaluation model. From this PV the initial capital investment is deducted to give the Net Present Value (NPV). Some cases are given in Example 10.2.

The NPV method includes three indicators regarding the value of a mineral project, provided that the net present value summation is a positive sum:
(a) the initial capital investment is returned,
(b) the financial return on this investment is the specified interest rate,
(c) the net present value summation provides a bonus payment which is sometimes called the acquisition value of the mineralization concerned. Projects can be ranked according to the size of this bonus provided that the same interest rate is used throughout (Example

10.2). The reason why project B is superior to projects D and C is that the cash flows in project B are larger in the earlier years and the initial investment is returned sooner.

The selection of an appropriate interest rate is critical in the application of NPV as a valuation and ranking technique, as this discounts the cash flow and determines the net present value. A minimum rate is equal to the cost of initial capital used to develop the project. Additional factors such as market conditions, tax environment, political stability, pay-back period, etc. have to be considered over the proposed life of the project. This is a matter of company policy and usually the interest rate for discounting varies from 5% to 15% *above* the interest rate of the required initial capital investment. In times of high interest rates this discount rate is particularly onerous.

DISCOUNTED CASH FLOW RATE OF RETURN (DCF ROR)

This technique is a special case of NPV where the interest rate chosen is that which will exactly discount the future cash flows of a project to a present value equal to the initial capital investment (i.e. the NPV is zero). This DCF return is employed for screening and ranking alternative projects and is commonly used in industry. Since this discount rate is not known at the beginning of a calculation an iterative process has to be used which is ideally suited for computer processing. A very approximate first estimate can be obtained by dividing the total initial capital expenditure by the average annual cash flow, and dividing the result into 0.7 but it is dependent upon the shape of the overall cash flow (Example 10.2). Also over a narrow range of discount rates the DCF ROR value correlates as a straight line (Example 10.3). Consequently by calculating a series (3 to 4) of positive and negative NPVs at selected discount rates the DCF rate can be seen graphically at the point where NPV equals zero.

Generally speaking mining companies finance mineralization for development which has a value equal to or exceeding a DCF rate of return of 15% after tax or 20% before tax has been deducted.

COMMENT

These quantitative economic modelling techniques have contributed much to an improvement in the process of investment decision making. During the 1980s, however, there was an increasing concern with the totality of mineral projects and a greater emphasis on aspects such as business risks associated with rapidly rising capital requirements due to inflation, the unpredictibility of future economic conditions, sophisticated financing arrangements, and host government attitudes.

Some mining organizations have used comparative production cost ranking to reduce investment risk. They require that the production cost to market for a new project is in the lower quartile of all major primary producers of that commodity. This is based on the premise that if the commodity prices fall, and there is a corresponding reduction of project revenue, then their operation will be protected by a cushion of other and higher cost producers who, it is assumed, will be forced to reduce production at an early date thus stabilizing the commodity price. It is worth noting, however, that a project with such a low production cost would have a favourable investment rate of return (either NPV or DCF) and would be ranked highly by these techniques.

The NPV method of valuation requires management to specify their desired rate of return and indicates the excess or deficiency in the cash flow above or below this rate while DCF provides the project's rate of return and does not consider the desired rate. As a comparison between the two it is said that NPV provides a conservative ranking of projects as compared with DCF.

Long versus short run considerations have to be assessed. As an example compare two occurrences of mineralization, one with a fifteen-year life and a 15% DCF and another with a five-year life and a 20% DCF.

If the short life, high DCF project is selected, management must consider the business opportunity at the end of the project. What is the probability of finding and developing a series of such short life, high return mineral deposits? The long-term future of the company may be better served by choosing the mineral project with the lower DCF. Lastly, there is the condition of selecting a series of small projects with high DCF. Small projects often require the same management time and attention as larger ones. With a finite amount of time its effective use will tend towards large projects with, perhaps, lower DCF rather than a series of smaller schemes.

In summary, the objective of valuation is to summarize and convey to management in a single determinant a quantitative summary of the value of a mineral deposit. Clearly there is no single perfect method of assessing this and good management uses every available relevant method and is aware of their inherent weaknesses and strengths. In the final analysis in the evaluation of mineral projects there is no substitute for sound managerial judgement.

10.5 RISK

Risk pervades our entire life and the way we act and it can be described in two ways—either with qualitative expressions (it's a sure thing, we have a fair chance in the Upper Palaeozoic, etc.) or in a quantitative sense using probability. A probability of 1.0 means that the event *will* occur while 0.0 means that it will *never* happen; negative probabilities do not exist. A probability of 0.31 means that there are 31 chances in 100 that the event will occur and 69 in 100 that it will not. What is an acceptable risk? Risk is very much a personal assessment. For instance many people would not work in an underground mine because it is seen as being dangerous yet every day accept higher risks such as dying from fire in the home or in a road accident (Rothschild 1978).

In the previous section four methods of analysing cash flows in the valuation of mineralization were presented. The projected cash flows were assumed to occur (i.e. probabililty = 1.0) and they did not include quantitative statements of risk: they were treated as decisions under certainty. However the valuation of mineralization involves the introduction of many factors (grade, tonnage, mining cost, tranport cost, etc.) which are not constants (i.e. there is not a single, unique value) but variables. Consequently mineral valuation comprises a series of uncertain decisions and the risk of these decisions being wrong can be assessed.

10.5.1 **Risk analysis**

QUALITATIVE ASSESSMENT

Adjustment of discount rates for NPV & DCF. A commonly used overall method of allowing for risk is to use an abnormally high discount rate in the valuation:

Safe Long Term Rate + Risk Premium Rate = Risk Rate.

For instance if the accepted minimum valuation rate is 15% DCF and there is a perceived risk an added rate is included to bring this minimum to, say 23%. This is the simplest method of allowing for risk but also the least satisfactory as the added rate is a matter of personal judgement; its determination with any degree of accuracy is impossible.

Adjustment of costs. A danger here is that of over adjusting. Consider a mineral valuation where as a safety factor it is decided to increase the best estimate of operating costs, to reduce ore grade and selling prices and apply a high discount rate as an additional risk allowance. It will have to be a remarkable project to survive such treatment.

A better approach is to calculate a *base case* from the most accurate information available. Risk factors are then added in a *sensitivity analysis* (Example 10.3) by considering the variable components one at a time while all others are kept constant. In any evaluation certain components have a greater effect upon the size of the cash flow (and hence value) than others and the purpose of a sensitivity analysis is to identify them so that further investigations can improve their reliability (i.e. make them less risky). Commonly grade is varied incrementally, and other components such as extraction cost, mineral recovery, etc.

The most significant component is revenue which is dependent upon the future sales price of the product and this is the most difficult forecast. Capital and operating costs only move upwards but are not likely to increase at a rate exceeding 15% a year. However selling prices, and particularly for those commodities sold on exchanges, may decrease or increase and in extreme instances can either double or halve in a year (Figs 1.5 & 1.6). Forecasting commodity price trends is a specialist operation and there are groups who provide this service on a consultancy basis, such as Metals and Minerals Research Services, London. If no reliable price trend predictions are available a break-even price can be calculated from a cash flow at a required rate of return— this is the price at which operating and financing costs can be met and the mine can continue in production.

Once this minimum price is known, the possibility of its being maintained over a standard ten-year period can be assessed.

These sensitivity results are best appreciated graphically with variation in the single component concerned plotted against the related DCF ROR (Example 10.3)

QUANTITATIVE ASSESSMENT

Discussion on components used in the calculation of a cash flow have not included any quantitative statement of risk; each component has been thought of as if it were a constant with a probability of occurring of 1.0. Most of these components are variables, a concept readily apparent from considering the grade of a deposit which is an average value within its own distribution and probability of occurrence. The same principle applies to estimates of mining cost, recovery, selling price, etc.

By using these component distributions in the calculation of a cash flow with their probability of occurrence, instead of fixed values with no risk element, the probability distribution of DCF or NPV can be calculated and also the probability of a certain rate of return (Wanless 1982, Goode *et al.* 1991).

10.6 **INFLATION**

Mining is among the most capital intensive of industries, and ranks near the top of the industrial sectors in this respect. This is related to the complex technology of modern production systems; high investment in major infrastructures such as town sites, railways and ports; and burgeoning expenditure on environmental protection. To offset this trend projects became larger (i.e. higher production) to benefit from an economy of scale which would either control or reduce increasing capital (and production) costs per unit of production. Generally speaking, however, the capital cost of mineral projects rose significantly during the 1970s and 1980s. This inherent increase was exacerbated by inflation during the same period. As an example, an inflation rate of 7% a year for five years will increase estimates by 40% in Year 5. The impact of high inflation rates on capital (and operating costs) for projects with a long pre-production period of several years can be considerable.

Related to inflation, monetary exchange rates are an important item to consider in mine evaluation. When a company operates a mine under one currency but sells its products in another changes in exchange rates between them can have serious consequences for cash flows.

When inflation was in low single figures its effect on

Example 10.3 Sensitivity analysis.

This example is taken from a feasibility study completed on a coal mine in western Canada. In the study it was apparent that two of the critical factors were variation in production cost and in tonnage of coal sold as these could both have a serious effect on the cash flow. These factors were varied and expressed as DCF ROR.

1 Variation in mine production cost

Base case	+ 5 %	+ 10 %	- 5 %	- 10 %
26.6%	22.3%	18.5%	31.5%	36.9%

2 Variation in tonnage of coal produced sold

Base case	+ 5 %	+ 10 %	- 5 %	- 10 %
26.6%	33.9%	42.9%	20.2%	14.5%

Obviously the lower the tonnage produced the lower the revenue and the lower the DCF ROR etc. None of the variations bring the DCF ROR to less than the hurdle of 15%. Results are best presented graphically.

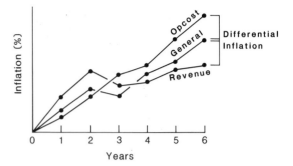

Fig. 10.6 Differential inflation. Revenue is increasing at a slower rate than general inflation, while the reverse is true with mine operating costs (opcost). There is a positive differential inflation with the former and negative with the latter.

the valuation of mineral projects was negligible and could be ignored but with higher rates its effect on calculated rates of return may be considerable. The value of money, measured by what it will buy, becomes progressively less as time passes. The overall rate of increase in price (or fall in the value of money) is measured by a gross national product deflator and is the weighted average of the various price increases within a particular national economy. There are other measures of inflation (Camm 1991) which may be more appropriate in particular circumstances such as retail and wholesale price indices, the indices of capital goods in various categories, and wage indices for a country, an industry or a particular skill.

The differential inflation (or escalation) which exists between the various components in the cash flow is an important factor and while inflation may be assumed always to be a positive factor, escalation can be either positive or negative as the particular component moves at a rate either faster or slower than the measure of general inflation (Fig. 10.6).

10.6.1 **Constant and current money**

Cash flows may be calculated in either *constant* or *current* money terms. Constant money is assumed to have the same purchasing power throughout the valua-tion period and thus one year's money may be compared directly with that of any other. This is the basis on which DCF and NPV are calculated, as above and in Examples 10.1, 10.2 and 10.3.

In current money the figures used in calculations are adjusted for the anticipated rate of inflation each year and are, in essence, the figures which are expected to be entered into the books of account for that year and represent the net available profit or loss. Money from different years is not directly comparable and DCF and NPV derived from this type of cash flow are misleading.

If the rate of inflation is the same for capital and operating costs, and revenue (i.e. there is no escalation) and there are no tax or other financial complications (which is seldom, if ever, the case) the DCF yield may be correctly estimated by completing the calculation in constant money terms, without regard to inflation. This is because, if the components which form the cash flow are inflated at the inflation rate and the resultant flow is then deflated for DCF and NPV calculations at the same rate, this will produce a flow identical with that produced by ignoring inflation altogether.

Opinions differ on the treatment of inflation in mineral project evaluation but a safe method is to use it as a type of sensitivity analysis on the Base Case. Separate specific cost and revenue items which correlate well with published cost indices can be used over a limited period of, say four years ahead, and thus account can be taken of differential inflation over this period. Beyond this separate predictions are unrealistic and it is advisable to use only a single, uniform rate for the complete cash flow for the remainder of the project duration. The Base Case cash flow prediction is in constant money whilst the inflation model will be in current money.

In most projects the greater part of the finance to build a mine is obtained from banks. Banks test the ability of a project cash flow to repay these loans, and related interest payments, by applying their own estimates of future inflation rates to the company Base Case. Many mineral companies then prefer to prepare cash flows for their Base Case, and sensitivity analysis, in constant money terms and leave the preparations of inflated flows in current money to the loan-making banks (Gentry 1988).

10.7 MINERAL PROJECT FINANCE

The development of mineralization into a producing mine requires a skilful combination of financial and technical expertise. For the purpose of this section assume that a definitive feasibility study (section 10.3.1) has been completed and the company concerned intends to proceed with the development of the defined mineralization.

10.7.1 Financing of mineral projects

TRADITIONAL FINANCING

Generally finance to develop mineral projects (Institution of Mining & Metallurgy 1987) is obtained from three sources (Potts 1985).

Equity. This is finance provided by the owners (i.e. shareholders) of the company developing the project through their purchase of shares in the company when it is floated on a stock exchange. Throughout the life of the mine these shares may be bought and sold and if the mine is successful they may be sold at a higher price than their original value. Another return to shareholders is dividends on each share which are paid out of profits after other financial commitments have been met. A successful mine may pay dividends from the start of production to the end of its life but for many mines fluctuating mineral prices mean that dividend amounts vary dramatically from year to year. Usually a reduction in the dividend leads to a drop in the share price (Kernet 1991).

Debt. In this case money can be supplied by sources outside the company, usually a group of banks. Debt finance places the company in a fundamentally different position than that of equity financing. Lenders may have the power to force it to cease trading (i.e. close down) if either interest charges or loans are not paid in a previously agreed manner. Thus, ultimately, control of the

company is in the hands of the suppliers of finance and not the mining company itself.

Retained profits. A successful company may retain some of its profit and not distribute all of it in the form of dividends to shareholders. In this way a source of finance can be accumulated within a company that is preparing to develop a mineral property.

The traditional means of financing mineral development in the first half of this century was a combination of the issue of equity, debt finance and the use of retained profit. This method was adequate while the capital cost of development was millions or tens of millions of dollars. During the last three to four decades the capital cost and size of major mineral projects has increased rapidly and large projects now cost several hundreds of millions of dollars, possibly a billion dollars. With these levels of expenditure very few, if any, mineral companies are able to finance new ventures using traditional methods. Thus companies have sought other methods of financing which made an optimum use of their financial strength and technical expertise but preserved their borrowing capability as much as possible. One method of achieving these ends is project finance.

PROJECT FINANCE

Project financing differs fundamentally from the traditional type. The organization providing the finance for the project looks either wholly or substantially to the cash flow of the project as the source from which the loans (and interest) are repaid and to its assets as security for the loan. In this way mineral projects are financed on their own merit rather than from the cash flow of the mining company that is promoting the scheme (the sponsor).

Lenders to the project like to see security attached to the revenue of the cash flow in the form of firm sales contracts for the mineral products, and it is common for them to form a consortium to spread any lending risk as widely as practicable. Project finance is then a type of non-recourse borrowing, which is not dependent upon the sponsor's credit. However, for this the sponsoring company isolates the new project from its other operations (Fig. 10.7) and usually has to provide written guarantees that it will be brought to a specified level of production and managed effectively.

Banks prefer to have a safety margin as protection against a deterioration in the project cash flow. They will therefore agree to finance only a proportion (say 60% to 80%) of the cost of a project with the sponsor

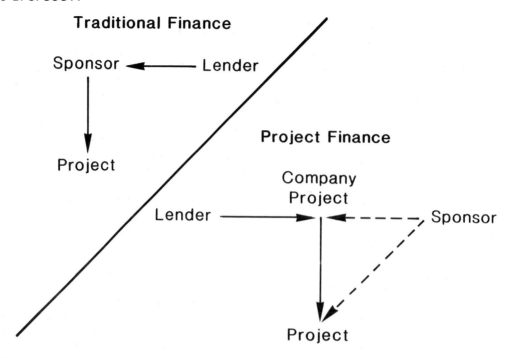

Fig. 10.7 Traditional and project financing of capital requirements for mine development. For explanation see text.

providing the remainder: full debt financing is rare. Two important considerations which decide this proportion of finance are the length of the payback period (see earlier) and the quality of the management who are to bring the new mine into production. If the loan is to be repaid over a relatively short period of time (say three to five years) a bank will be more likely to lend a larger proportion of the total capital cost. Management is perhaps the paramount factor in the evaluation of a project—bad management can destroy a project which good management could turn it into the next RTZ Corporation plc!

As a final comment it should be remembered that the loaning banks and the sponsor have a difference of emphasis in project development. A mining company will wish to achieve a positive return on its investment (i.e. the equity of the project company) to satisfy the demands of its shareholders. A bank, however, will wish to ensure that its money is returned on time with interest and loan repayments taking priority over dividend payments to shareholders. The payment of dividends, however, can have considerable effect on the price of the project company shares traded on stock exchanges.

Consequently a compromise is usually reached where, whilst the major proportion of the cash flow is reserved for project loan repayment, a smaller part (say 20%) is placed at the disposal of the project company.

It is a matter of opinion as to whether cash flows used in DCF analysis should include drawdown and repayment of debt (i.e. project finance), and interest payments. If not then projects are evaluated as if they are financed by 100% equity although in practice this will not happen (i.e. financing will be a combination of loan and equity). Such a procedure is said to separate investment decisions from financing decisions (i.e. repayment of debt, etc.). It is a fundamental principle that a basically poor project cannot be transformed into an attractive one by the manner in which it is financed. If an all-equity evaluation demonstrates a strong cash flow and meets standard parameters (see earlier) then after this it can be determined how it is best financed.

Banks are not inherently concerned with the rate of return of the proposed project. Their return is more or less fixed: it is the interest rate at which they agree to lend money. Thus banks are more concerned with protection of their loans and they tend to examine cash flow predicted in the feasibility study from this perspective. A criterion is that the net present value at an agreed rate of interest must be at least twice the total initial loan plus interest payments. Thus there is the obvious implication that a bank will lend money for a mineral project provided that its loan and interest payments are secure, even if actual project cash flows fall to half their estimated level.

10.7.2 **Risk in project finance**

Project loans are provided on an agreed time schedule of draw-down and repayment, based on the technical data and cash flows in the final feasibility report. The interest rate on loans is usually a premium rate (say 3%) above an agreed national base rate. As this base rate fluctuates the loan interest rate will also change and in times of inflation with high rates (as in 1990 and 1991) repayments can be onerous for a project company

Even the best of feasibility studies may be incorrect in actual practice and events which both sponsors and lenders particularly look to during the construction and operation of a new mineral project are as follows.

DURING CONSTRUCTION

It is probable that the greatest risk in project financing occurs during the construction phase. This phase may legitimately take several years and during this time the loan is being used and no revenue is being generated. Mine construction is usually contracted out to major civil engineering companies and some form of completion guarantee from the managing contractor is crucial. This is usually to complete and commission the project in a specified time period and to make penalty payments to provide funds if construction is not completed on time. Complications which can delay construction include an increase in capital costs due to inflation and/or technological difficulties, delays in the delivery of plant and equipment, strikes, delays caused by abnormal weather, financial failure of a major contractor and poor management.

A delay in completion and subsequent operation postpones the generation of the cash flow from which loans and interest were to be repaid. It also creates a requirement for additional money over and above that originally negotiated in order to finance this delay period. Obviously the key factors are the quality of the final project feasibility report and the reliability of the managing contractor.

DURING PRODUCTION

Adverse factors during production include errors in the quantity and quality of ore reserve estimation, technical failure of equipment, fall in the price of the mineral product, foreign exchange fluctuations such as a currency devaluation, poor labour relations which provoke industrial strife, environmental difficulties, government interference in the running of the project of which the final condition is expropriation. Any of these events may cause delay in the repayment of the loan. In cases of grave default the lenders have the ultimate right to take over the project and replace its management.

GENERAL

Banks, in reviewing a feasibility study, pay particular attention to the degree of confidence which is attached to estimates of mineable grade and tonnage, annual production rate, capital and operating costs. While banks may be prepared to accept the risk of commodity prices falling below those projected they are not prepared to accept technical risks. Of increasing importance to banks and project sponsors is the potential liability in the event of inadvertent damage to the environment and the manner in which a feasibility study examines environmental issues is reviewed with great care.

10.7.3 **Ok Tedi Mine**

The Ok Tedi Mine, Papua New Guinea (PNG) provides an example of the valuation of mineralization and associated project financing. This porphyry copper deposit was discovered in 1968 by Kennecott Copper Corporation (USA). Negotiations between this company and the host government on the development of the project broke down in 1975 and Kennecott left PNG. Further exploration and development work was continued by the PNG Government. In 1976 Broken Hill Proprietary (Australia) (BHP) agreed to evaluate the deposit and established that more than 250 Mt of mineralization at 0.7% copper (in the form of chalcopyrite) occurred in the porphyry and 115 Mt of material at 1.2 g t^{-1} gold in the capping gossan. A definitive feasibility study was completed in 1979. A consortium was formed which included BHP, Amoco Minerals (a subsidiary of the Standard Oil Company of Indiana, USA) and a German group of companies that largely represented copper smelting activities. This sponsoring consortium, managed by BHP, presented proposals for the development of the deposit to the PNG Government in November 1979 which were accepted in early 1980 (Pintz 1984). The proposals, based on the feasibility study, were that development would take place in four stages for the production of gold metal, and copper concentrates for export to Germany and other locations. A copper smelter and refinery was not envisaged in PNG.

Construction:	1981 to 1984.

Production Stage 1: 1984 to 1986. — To produce 22 500 t d^{-1} gossan at 2.9 g t^{-1} gold to produce about 22 000 kg of gold a year, depending on the grade and recovery.

Production Stage 2: 1986 to 1989. — Gold production to continue as previously, but also to mine copper ore (average grade 0.7% copper) at a commencement rate of 7 000 t d^{-1} and expanding to 60 000 t d^{-1} by the end of 1989. Stripping ratio for the overall open pit is 2.5:1.

Production Stage 3: Post 1990. — Gold mining is phased out and mine copper production continued at 60 000 t d^{-1} ore. At an average grade of 0.7% copper, 80% recovery and 300 production days a year this is equivalent to 100 800 t a^{-1} copper metal in shipped chalcopyrite concentrates containing 25% to 27% copper.

In the early years low cost material with high value (the gold ore) was produced in order to repay rapidly the project finance. In later years after this gold ore had become exhausted copper ore would be mined from a large open pit. Copper sulphide concentrates only were to be produced on site and these concentrates sold to a German copper smelter group.

A separate project company was formed, Ok Tedi Mining Co. Ltd., with the following shareholders and sponsors: BHP 30% and managers, German Group 20%, AMOCO 30%, PNG Government 20%.

In 1981 the estimated total cost of Stage 1 was US$855 000 000 with a project debt to equity ratio of 70:30. Consequently the amount of equity required from the sponsors was US$256 000 000 and the remaining $599 000 000 was raised as project finance from a consortium of banks in several countries.

Construction was completed on time and Production Stage 1 was successfully concluded. However, the com-

mencement of Stage 2 coincided with a decrease in the price of copper and the ensuing delays in project development caused renegotiation of several aspects of the initial agreement between the project company and the host government. The project is continuing.

10.8 SUMMARY

In the valuation and ranking of mineralization, mineral properties and projects a large number of organizations proceed to the stage of developing cash flows in constant money which are then tested by sensitivity analysis. A minimum acceptance rate of return (the "hurdle" rate) is stated below which projects are not considered, such as 20% DCF ROR before tax and 15% DCF ROR after tax. This usually means that the initial capital investment is returned in the first four years or so of mineral production. Few companies consider inflation and even fewer consider a quantitative assessment of risk. See also section 10.2.10.

10.9 FURTHER READING

A number of text books and articles are given in the reference list. The O' Hara (1980) article and the Mular (1982) textbook provide much useful basic data on the evaluation of orebodies, and the estimation of mining costs, but are now rather dated. A more relevant source is the *US Bureau of Mines Cost Estimating System Handbook* (1987), which provides a rigorous basis for cost estimating. A good system for order of magnitude estimates is in Camm (1991) and Craig Smith (1992). Two other articles of note are the description of the Palabora copper open pit in South Africa (Crosson 1984), and the Neves Corvo copper–zinc–lead–tin project in Portugal (Bailey & Hodson 1991).

In the author's opinion there is a scarcity of suitable textbooks and articles on the financial aspects of mineral evaluation, and Wanless (1982) remains a best choice. Aspects of the financing of mineral projects were presented in nine papers at a meeting 'Finance for the mining industry' under the auspices of the Institution of Mining and Metallurgy in 1987. Case studies of mineral evaluation and related mine development are comparatively rare but probably the best published is that of the Ok Tedi copper-gold open pit mine (Pintz 1984) in Papua New Guinea.

Lord Rothschild's (1978) short paper on 'Risk' is a readable and excellent introduction to this important topic.

Part II—Case studies

11 Cliffe Hill Quarry, Leicestershire— Development of Aggregate Reserves

MICHAEL K. G. WHATELEY AND WILLIAM L. BARRETT

11.1 INTRODUCTION

The annual production and use of construction aggregates in the UK, including both sand and gravel and hard rock, amounts to nearly 300 Mt. The materials are used for road building, construction, civil engineering, concrete, house building, chemicals and other specialized applications. Within such a wide market rocks with particular properties are better suited to certain applications than others. The physical properties of the rocks are defined in terms of their response to different tests, most of which are covered by the relevant British Standards. Some international, American and local (but not British Standards) tests are also used.

Tarmac Quarry Products Ltd. (Tarmac) is one of the leading aggregate, ready mixed concrete and waste disposal companies in the UK and it also operates in Europe and Scandinavia. The group as a whole has Construction, Building Materials, Housing, Industrial and American Divisions. The group turnover in 1989 was £3500M. Tarmac operates about 100 active quarries and sandpits in the UK, spread from Ullapool in northwest Scotland to the south coast of England.

The biggest market for aggregates in the UK is the south-east of England where ironically there is no hard rock and demands have historically been met from sand and gravel sources (both land based and marine dredged) and by supplies railed in from elsewhere in England. Notwithstanding the expected efforts to expand local sand and gravel output from both land and marine sources, an aggregates supply shortfall of between 15 and 30 Mt p.a. is predicted for the next decade for south-eastern England. The deficiency in indigenous aggregates will have to be satisfied by increasing imports into the region from elsewhere in the UK and also from overseas. Leicestershire is well placed as a supplier and has suitable quality rock resources to benefit from this demand.

In the mid 1970s it became apparent that the profitable microdiorite (markfieldite) quarry of Cliffe Hill in Leicestershire (Fig. 11.1) belonging to Tarmac Roadstone Limited, East Midlands was running short of recoverable reserves. Significant reserves remained at the site, but these were sterilized beneath the processing plant (Fig. 11.2).

The markfieldites of the Charnwood Forest area occur as a number of relatively small igneous bodies intruded 550 Ma ago into late Proterozoic pyroclastics and metasediments. Markfieldite differs from other microdiorites in having a granophyric groundmass and it has a general uniformity and strength which puts it amongst the best and most consistent general purpose construction aggregate materials in the UK. An impressive durability means that the markfieldite can also be used for all classes of railway track ballast, as well as road stone.

Charnwood Forest contains major faults, e.g. the Thringstone Fault, which brings Carboniferous rocks against the Proterozoic sequence to the west of Stud Farm (Fig. 11.1). Knowing the location of faults is important in any quarry. Associated minor faulting is also important to quarrying operations. In Cliffe Hill Quarry one apparently minor shear zone was mapped. As a minor feature it could easily have been missed during core logging, however it offset the Precambrian basement by almost 60 m (Bell & Hopkins, 1988).

In order to maintain its production from the English Midlands, and to maximize the exploitation of a valuable national resource, Tarmac initiated a search for alternative sources of similar quality markfieldite in the neighbouring areas. This search was conducted in a number of phases with each subsequent and more costly exercise only being undertaken if clearly justified by the results of the preceding one. All costs quoted in connection with this case study relate to values current at the time of the expenditure.

This case study gives a clear example of the procedure adopted by one company in the exploration for and development of a hard rock resource. Readers wishing to expand their reading into sand and gravel and limestone resources are referred to British Geological Survey publications on the procedures for the assessment of conglomerate resources (Piper & Rogers 1980) and limestone resources (Cox et al. 1977). Additional background reading can be found in the book by Collins & Fox (1985).

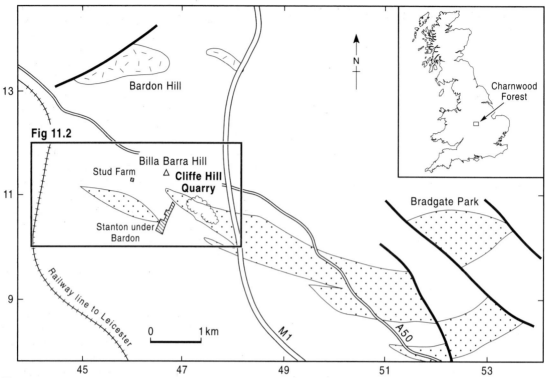

Fig. 11.1 Distribution of markfieldite (a microdiorite) in the southern part of Charnwood Forest, Leicestershire, UK. Stippled areas are known and projected areas of markfieldite as given in Evans (1968). Hatched areas are other intrusive rocks. The grid coordinates are in km.

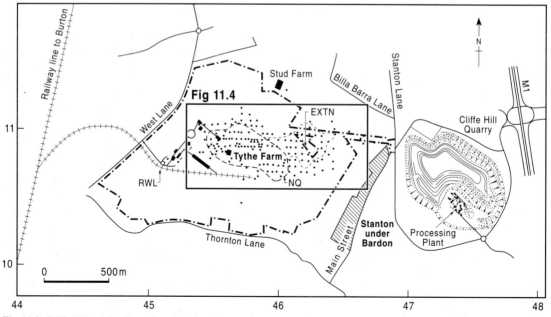

Fig. 11.2 Cliffe Hill and Stud Farm Quarries, showing the location of some of the borehole sites (dots) and the proposed new railway line (RWL). The pecked and dotted line represents the outline of the area included in the planning permission. The pecked line represents the outline of the proposed new quarry (NQ) with the possible extension shown to the east (EXTN).

11.2 **PHASE 1—PRELIMINARY DESK STUDY**

Cliffe Hill Quarry is adjacent to the M1 Motorway and a railway giving it excellent access to major markets in the south-east of England. The initial desk study in 1977 concentrated on extensions to Cliffe Hill Quarry and adjacent land and mineral right holdings on Stud Farm. The study included a detailed literature search which indicated that both Billa Barra Hill to the north-west, and Stud Farm around 1 km to the west of the existing quarry might be of potential interest (Fig. 11.1). The 1:63 360 and 1:10 560 Geological Survey maps and the account by Evans (1968) of the Precambrian rocks of Charnwood Forest pointed to additional areas of diorite near Stud Farm. A well sunk on Tythe Farm (Fig. 11.2) at the end of the last century intersected markfieldite at a relatively shallow depth.

Subsequent ground surveys eliminated the former site as being composed of fine-grained tuff, volcanic breccia and metasediments, but revealed scattered float boulders of markfieldite in the soils at the latter site. In view of the total absence of exposures at Stud Farm, a few pits were dug using a mechanical digger, and although these failed to reach bedrock they confirmed the existence of further pieces of markfieldite in the soils. The term bedrock in this chapter refers to the unconformity between the Triassic Mercia Mudstones and the markfieldite. The mudstones are also referred to as overburden which generally includes glacial deposits as well.

The above exercises could be regarded as separate phases or sub-phases, but since they took place in swift succession they have been regarded collectively as Phase 1. A cost (excluding overheads and geologist's time) of around £1500 was incurred for this work.

11.3 **PHASE 2—INITIAL FOLLOW-UP WORK**

The preliminary desk study had indicated areas of interest, but the Stud Farm holdings were insufficient to support a viable quarry (Bell & Hopkins 1988). Negotiations were entered into with the owner of adjacent land. An agreement was reached which allowed Tarmac access to the land for exploratory drilling. A small exploration drilling budget was approved and in mid-1978 nineteen, 150 mm continuous flight auger holes were drilled with a Dando 250 top drive, multi-purpose drill rig. These holes, drilled alongside established farm tracks over the higher parts of the farm, proved that rock, harder than could be penetrated with the augers, existed at exploitable depths over an area large enough to contain around 20 Mt of material. The results from this drilling enabled the geologists to establish overburden distribution, overburden types, water-bearing zones, the depth of weathering and the gradient of the bedrock surface (Bell & Hopkins 1988). At this stage there was no proof of the nature of the harder material since no samples were recovered. Nevertheless, it was assumed that the impenetrable rock was markfieldite and, although the drill pattern was irregular, it showed that there was only thin overburden. Geologists calculated the reserves and overburden ratios for a theoretical quarry, and within the margins of error in such calculations, there appeared to be sufficient reserves for a viable quarry. The auger drillng in Phase 2 is estimated to have cost between £2000 and £2500.

11.4 **PHASE 3—DETAILED FOLLOW-UP WORK**

Phase 3 of the Cliffe Hill Quarry project included all the exploration drilling, a limited amount of geophysics, the testing of the core samples, the design of the landscaping works, the detailed specification of the processing plant and the submission of the original planning application, and a number of subsequent amendments to it, to the relevant local government offices.

11.4.1 **Drilling**

The Estates and Environment Department of Tarmac acts as a contract drilling company to the operating divisions and the Stud Farm work, although welcome, caused a number of problems. With the continuing land acquisition programme rig-time was becoming scarce and although it had been hoped to undertake any additional work at Stud Farm in-house, the rapid exploration success meant that the follow-up work could not be fitted into the programme of the existing, company owned rigs. The company therefore took the decision to purchase a brand new six cylinder, 100 hp top drive, multipurpose, drilling rig (Fig. 11.3) to evaluate fully the resource potential at Stud Farm. A rig of the type used, together with a back-up vehicle and the relevant in-hole equipment, would have cost around £120 000, but this figure is included as depreciation in the total drilling costs given below.

Drilling, using open hole methods in the overburden and coring at various diameters in the bedrock, commenced in August 1979 and continued with some breaks

Fig. 11.3 A top drive, multipurpose, flight auger drilling rig used to evaluate the resources on the Stud Farm property adjacent to Cliffe Hill Quarry.

until December 1989. During this period about 240 boreholes totalling around 17 000 m were drilled, mainly on a 50 × 50 m grid (Fig. 11.2). A planning application for Stud Farm was prepared in parallel with the exploration work.

The experience gained while drilling the original holes during phase 2 indicated that certain aspects of the drilling would have to be improved if sufficient data were to be collected. The indications were that there was between 10 m and 50 m of overburden. Although the flight augers could probably cope with this depth, removing the clay from the auger and clearing the hole proved slow and difficult. The clay was also expansive and, when wetted by the water flush from a diamond bit, the hole tended to close and trap the core barrel. These related problems were ultimately solved by using a drag bit and water flush. A drag bit is a bladed bit which is used when the sticky material such as clay and marl would clog up air flush bits. The drag bit drilling technique produced an open hole to bedrock almost as fast as the augers could when cleaning time was taken into account. In addition the clay had to some extent expanded due to the water flush. By drilling with

sufficient annulus size (the gap between the drill rods and the sidewall up which the drilling medium [air, water or foam] carries the drill cuttings to the surface) and casing placement to bedrock, a good core recovery was usually possible (Bell & Hopkins 1988).

During the early drilling programme the driller remarked on the apparent low penetration rate of the core barrel. At the time this had been assumed to be the result of operating in an uncased hole. Once the main programme began it became apparent that, although the high rotation speed improved penetration, the productivity was lower than expected. A series of checks and experiments narrowed the problem down to polishing of the impregnated core bits. After a series of trials, conducted in conjunction with one of the leading British bit manufacturers, a bit matrix which gave the optimum balance between bit wear and penetration was developed. Once this problem had been solved the drilling proceeded at a rapid rate (Bell & Hopkins 1988). During the summer of 1981 a series of boreholes was drilled along a line joining Stud Farm with Cliffe Hill Quarry (Fig. 11.2). The aim was to investigate the route of a proposed tunnel to link the two quarries. The purpose of the boreholes was twofold; (i) to delineate the bedrock-overburden interface, in order to ensure that the tunnel which was planned to link Cliffe Hill Quarry to Stud Farm (section 11.4.5) remained in solid rock throughout its length, and (ii) to identify the engineering properties of the bedrock with particular reference to jointing and faulting. Some heavy faulting was identified in two boreholes.

Between September 1981 and August 1982 the drilling was concentrated on two areas of the site. (i) Investigation of the proposed plant site. Some ten boreholes were drilled where access permitted. (ii) Drilling of a newly acquired area to the north-east of Stud Farm was carried out. The results confirmed the already familiar picture of highly variable overburden thicknesses, with the bedrock often 'dipping away' beneath rapidly thickening Triassic Mercia Mudstone overburden.

Between April 1983 and March 1985 drilling was carried out on a 50 m grid across the area of the proposed quarry (Fig. 11.2) with the aim of:
(i) confirming the geophysical interpretation,
(ii) identifying the volume of weathered material, and
(iii) building up detailed knowledge of the faults, rock quality, etc., to enable detailed quarry plans to be drawn up.

The drilling results were used to produce the overburden isopach map (depth to bedrock) (Fig. 11.4). This map was initially drawn independently of the geophysi-

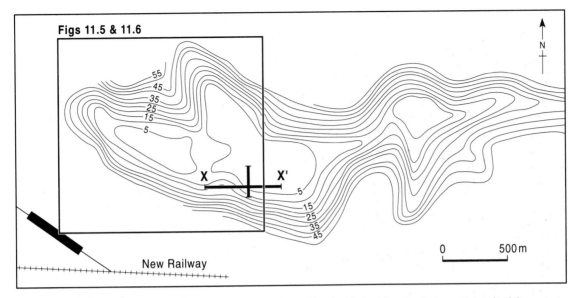

Fig. 11.4 Overburden isopach map on the Stud Farm property drawn from data derived from borehole results, conductivity measurements and resistivity soundings.

cal data and clearly shows the success of the previous electromagnetic and resistivity surveys in revealing the distribution of the overburden (Figs 11.5 & 11.6). Although some inaccuracy is evident in the deeper areas, with the resistivity method underestimating thickness of overburden, the two shallow ridges and steep sides were accurately delineated. Drilling indicated that these features can be attributed to faulted blocks of markfieldite alternating with late Proterozoic metasediments, as seen in Cliffe Hill Quarry.

11.4.2 Geophysics

As indicated previously, the planning application was being prepared in parallel with the exploration work and the work was co-ordinated by a team of geologists, engineers, estates surveyors and landscape architects. The continual acquisition of additional borehole data during the planning of the project meant that occasionally parts of the scheme had to be redesigned. However, the boreholes revealed that the areal extent of the markfieldite was somewhat larger than originally anticipated. It became apparent that a geophysical investigation of the extent and form of the intrusion was essential. It was hoped that this knowledge would reduce the number of major interpretational changes that might be made in the future and also enable the drilling to be more effectively planned.

REFRACTION SEISMICS

Following a trial seismic refraction survey in December 1980, a full survey was carried out across the whole of the site in February 1981. A standard 12 channel refraction seismic technique was employed, with a Nimbus seismograph recording on 12 geophones the first arrivals of the ground waves produced from the explosive source. The aim was to produce an overall picture of the bedrock topography.

The results of the survey indicated that the overburden was thin near the topographically high centre of the property, and that overburden increased in all directions away from this centre. Unfortunately the major geophysical surface did not correlate well with the surface of the markfieldite as identified by borehole information. It was assumed that this was because the seismic method was delineating the base of a weathered or fractured layer. In this particular deposit the depth of weathering is up to 20 m on the higher ground, but significantly less in the bedrock on the flanks of the hill where the older rocks were more deeply buried below the marls of the Triassic Marcia Mudstone. Much of the weathered rock has to be quarried conventionally and can be used for lower specification purposes. The seismic refraction method in this application had the effect of making the bedrock surface appear generally deeper than it is and also less variable at depth.

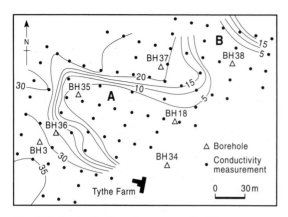

Fig. 11.5 Overburden isopach map on the Stud Farm property drawn from data derived from conductivity measurements.

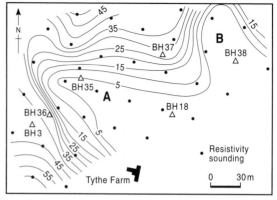

Fig. 11.6 Overburden isopach map on the Stud Farm property drawn from data derived from resistivity soundings.

ELECTRICAL RESISTIVITY AND ELECTROMAGNETICS

Planning permission was granted in August 1983 by the Leicestershire County Planning Department. Immediate priority was given to the detailed investigation of the site for development. An accurate overburden volume figure was required and insufficient boreholes had been drilled. Further geophysical surveys were commissioned to obtain additional detail of the overburden distribution.

Overburden thicknesses were required on a 50 m grid across the area investigated in Phase 1, and it was decided to use a combination of electrical resistivity and electromagnetic techniques to produce the detail required.

The area was traversed using the Geonics EM34 Ground Conductivity Meter (20 m and 40 m coil separations). The Mercia Mudstone, which composes most of the overburden, is considerably more conductive than the underlying markfieldite. Thus an increased thickness of the overburden produced an increased conductivity reading. The readings, reflecting the degree of conductivity of the different materials, were then used to produce a conductivity contour map. Therefore, by traversing the site it was possible to produce a qualitative picture of overburden variation. Although the assumption of a two-layer model of overburden and bedrock oversimplified the situation somewhat, e.g. an additional boulder clay layer, etc. could be present, the marked contrast in conductivity meant that the method proved to be very successful in delineating the shallower (<20 m deep) bedrock features (Fig. 11.5). Although borehole information allowed some correlation of con-

ductivity with depth, the results remained essentially qualitative and a resistivity survey was carried out to quantify the data (Fig. 11.6).

The quantification of the conductivity contours was attempted by taking a series of resistivity soundings (electrical depth probes) along sectors of equal conductivity. This technique produces a more accurate depth reading at a point, by avoiding significant lateral variations in overburden thickness. The resistivity sounding methods used employed the BGS multicore offset sounding cable (a Werner configuration) with an Aberm Terrameter. An electrical current is passed through two electrodes and the potential difference is recorded across a further pass of electrodes to produce a resistance value in the standard array. The curves produced are computer processed to give depths to bedrock at the measurement sites.

With the highly variable nature of the overburden thickness, the electromagnetic results proved to be very important for accurately positioning the resistivity lines so that they did not cross any sharp lateral variations in bedrock topography. This greatly increased the quality of the resistivity data.

Combining the electromagnetic and resistivity results with borehole data produced an overburden isopach map (Fig. 11.4). The most important features identified from this plan are the two shallow 'ridges' (marked A and B on Figs. 11.5 & 11.6) away from which the overburden thickness increases rapidly. Neither of these features had been identified by the earliest drilling and, because of their narrow nature, might have remained undetected for some time. Their presence resulted in a significant reduction in calculated overburden volume.

100m

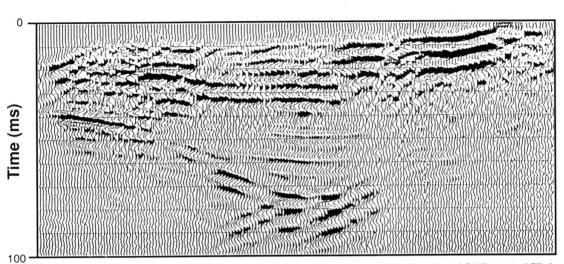

Fig. 11.7 A migrated section of part of line X–X' (Fig. 11.4), showing an image of an ancient wadi cut in markfieldite (m) and filled with Triassic mudstone (T). (After Ali & Hill 1991.)

REFLECTION SEISMICS

Recent modification to seismic source, detector, recording equipment and field techniques has improved the seismic reflection method of exploration to enable it to be used sucessfully for exploration at shallow depths. The system has been used recently (Ali & Hill 1991) to generate, extract and record high-resolution seismic data which eliminated noise, in particular low frequency, high amplitude, source-generated ground roll. High-frequency energy was needed to identify and distinguish them from other interfering waves.

Shallow reflection seismic methods were used on the Stud Farm property to locate and image the unconformity between the Triassic Mercia Mudstones and the markfieldite. It was assumed that the homogeneity of the overlying mudstone would result in no reflections from within their sequence, and therefore make the detection of reflections from the unconformity easier (Hill 1990). However, some reflections were obtained from within the mudstone and these are thought to represent sandy and calcareous horizons (Fig. 11.7).

Data were recorded along two lines (Fig. 11.4), usually with six-fold coverage and 1 m geophone spacing. Despite poor weather and poorly consolidated near surface materials, some reflected waves were recognized. A probable function was chosen based on experience gained in a nearby quarry, with similar geology (Ali & Hill 1991) and stacking of the data carried out. Careful

processing and correction of the data produced a section on which the unconformity is clearly imaged as a steep sided, ancient wadi (Fig. 11.7).

11.4.3 Sample testing

In any deposit, and particularly in a new one, it is essential to determine the precise physical characteristics of the materials, both in order to establish the potential markets and to determine the type of processing plant that will be required. In some deposits a detailed chemical profile of the material may also be necessary.

The intrusive rocks of this part of Leicestershire are known to be hard, durable and mainly consistent, but it is nevertheless essential to carry out a large number of British Standard and other special tests to quantify any variations and to establish the precise characteristics of the materials. The core samples obtained were of two sizes (35 mm and 60 mm in diameter) and all were subjected, where sample quantity permitted, to the following tests as stipulated in British Standard No 812 (1975): Relative Density, Water Absorption, Aggregate Impact Value (AIV) and Aggregate Crushing Value (ACV). A few samples were also tested under the same standard for Aggregate Abrasion Value (AAV) and Polished Stone Value (PSV).

The Leicestershire quarries have historically supplied stone to the profitable British Rail High Speed

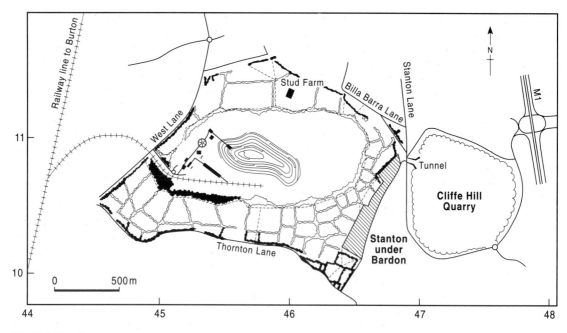

Fig. 11.8 Plan showing the proposed development of the new quarry (on the left), the current workings in Cliffe Hill Quarry and the proposed field pattern and associated landscaping of the new quarry.

Track ballast market. In order to qualify for this market the stone must achieve a very low and consistent value in a non-British Standard wet attrition test. A larger size material is required for this test and consequently only the 60 mm cores could be used. As testing progressed, it proved possible to establish a usable correlation between the water absorption values determined on all samples and the wet attrition values.

Over 120 samples of core were tested for some, or all, of the properties mentioned above. The results provided the data necessary to determine which markets could be penetrated and what type of basic plant would be required to produce the materials to satisfy those outlets in terms of both anticipated quantity and product specification.

11.4.4 Reserve estimation

The confidence in the resources grew as the amount of data increased and several refinements of the estimation method evolved as follows.

1 After the initial phase of flight auger drilling an inferred resource estimate (USGS 1976) of some 20 Mt $\pm$ 100% was quoted.

2 By the time the first planning application was submitted, a generalized bedrock and deposit shape had been defined by widely spaced boreholes and some typical cross-sections had been drawn. A resource estimate based on the area of the cross-section, extended to the mid point between sections likely to be recovered using an average quarry configuration, gave an enhanced figure with a confidence limit of $\pm$ 20%.

3 By 1989 with much of the relevant drilling completed, two sets of cross sections (N–S & E–W) at 50 m intervals were drawn. With a provisional quarry development scheme superimposed, a more precise reserve estimate was produced. The extractable area on each cross section was multiplied by the distance to the midpoint between the adjacent sections on each side. The N–S sections were considered best but the figures were also verified using the E–W sections. In all cases the volumes were converted to tonnages by multiplying the volume by the average relative density of the rock in a saturated, surface dried condition. In this case the value 2.7 was used for the fresh rock. The final calculations were considered more precise and were accorded a confidence limit of $\pm$ 10%.

11.4.5 Quarry planning

As the drilling and sample testing continued with a progressive clarification of the combined overburden and weathered rock thicknesses, and of the overall structure, shape and quality of the deposit, three possible operating schemes were devised for developing the new quarry in conjunction with recovering the sizeable reserves which would become available following the dismantling of the plant at the old quarry.

Bell & Hopkins (1988) described these three options as follows. A straightforward option to work out Cliffe Hill Quarry completely, while towards the end of its life, developing Stud Farm as its natural successor. This scheme had several drawbacks. The dimensions of both quarries would impose a ceiling on potential output. The extension of Cliffe Hill would require a road diversion and relocation of the existing plant and unfortunately the only feasible site for the latter would position it closer to an adjacent village. This potential plant site was also the only area suitable for overburden disposal. Even if these problems could be resolved for Cliffe Hill it would be necessary to install a second new plant when Stud Farm opened. This scheme was costed out on the basis of capital investment and the return it would yield. The main stumbling block proved to be the cost of establishing two plants, each with a relatively short life. These would have been economically viable above certain quarry outputs but neither quarry could supply the required outputs on a long term basis.

The second option was the creation of a quarry at Stud Farm at an early date with a new processing plant and quarry. Once this was commissioned the existing Cliffe Hill plant would be demolished but crushing and screening would continue using a mobile plant. An analysis of this proposed development indicated that a mobile operation of economic size at Cliffe Hill could only be located in one place and this would result in a substantial sterilization of reserves. There was also some doubt as to whether the mobile plant was capable of producing the different product types required. The financial analysis indicated that production costs using a mobile plant, which could possibly require replacement several times during the Cliffe Hill reserve extraction, combined with the additional cost of a new plant at Stud Farm, made the financial return marginal.

The third option was the most radical. This envisaged a new high capacity processing plant at Stud Farm which would be fed from both the Stud Farm and Cliffe Hill Quarries (Figs 11.2 & 11.8). The latter when extended would be linked directly to the new plant by an underground tunnel (to ensure the least disturbance to the inhabitants of Stanton under Bardon) and act as a satellite producer of primary crushed rock. Working both quarries simultaneously could supply considerably more output than a single quarry. The necessary output levels also meant that each quarry would be operating at an optimum level from the point of view of production costs. This proposed scheme was originally received by the management with some scepticism, but when costed, it proved capable of meeting the company's financial investment and return criteria and was agreed to by all the relevant company experts before being submitted to the local planning authority in October 1980. Comprehensive but positive and wide ranging discussions between the company and the Leicestershire county planners followed before consent was granted in August 1983.

During the discussions it became clear that the third option favoured by the company would not be supported by the planners on the basis that simultaneous working on both sides of Stanton under Bardon (Fig. 11.2) would be too disruptive to the villagers. The planners did however, insist that all winnable reserves at both sites should be recovered at some stage. In the light of these factors, the company opted for developing a full scale operation to work out the Stud Farm site completely, following the completion of which, the remaining reserves at the old quarry would be recovered via some suitable connection, for processing at Stud Farm.

Once planning permission had been granted, Tarmac undertook a full financial feasibility study, after which further applications were submitted in 1986, both to permit an acceleration of the development of the new quarry, and for the construction of a direct rail link from the site to the Leicester to Burton railway line (Fig. 11.2). Both proposals received planning consent in the same year. Numerous controls and legally binding agreements are associated with the planning consents, and were formulated in discussions between the planning authority and the company prior to the consents being granted.

An approximate estimate of some of the major costs for Phase 3 (as defined above), under a number of headings are tabulated in Table 11.1.

11.5 PHASE 4—PROJECT IMPLEMENTATION

The Phase 3 drilling, testing and other evaluation exercises having proved, to an acceptable risk level, the existence of an economically viable quality and quantity of material, the major cost operations (collectively designated as Phase 4) commenced in 1986.

In anticipation of the commencement of major operations, extensive site investigation surveys were carried

Table 11.1 Some major phase 3 costs

Cost centre	Order of magnitude of expenditure in 1987–1988 £ values
1 Exploration drilling to end of 1986	280 000
2 Sample testing to the same date	25 000
3 Geological and geophysical surveys	80 000
4 Landscape design work	75 000
5 Engineering design work and research	100 000
6 Financial and administrative considerations	75 000
7 Compilation and submission of planning application	20 000
Approximate total of major costs	655 000

out to provide the data for the compilation of the statutory Mines & Quarries (Tips) Regulation 9 reports, without which bulk excavations and depositions on the vast scales envisaged are prohibited in the UK. Such reports must be with the Inspectorate at least thirty days before the commencement of the operations to which they relate. The quarry development as a whole will necessitate the excavation of several million cubic metres of overburden (down to a depth of 40 m) and the disposal of this material into natural looking landforms on the surrounding lands.

The major operations constituting Phase 4 of the development were: further drilling and testing, overburden removal, landform creation, detailed site investigation for foundation design for the plant site (Barrett 1992), plant site excavation, quarry development, blasting, road and rail access construction, the cultivation and restoration of the completed overburden disposal areas, and the erection of the plants and ancillary structures. A brief summary of the Phase 4 progress on an annual basis is as follows.

In 1986 earthworks commenced at the new site, with the clearing of hedges, the lifting and storage of topsoil from approximately half the site and the excavation of around 1.6×10^6 m^3 of overburden, mainly from the proposed plant site at the western extremity. The exca-

vated material was used to create a new landform on the south-western perimeter of the site (Fig. 11.8). Two electricity power lines were also re-routed during this period.

In 1987 earthworks continued with the stripping of a further 1.6×10^6 m^3 of overburden. This operation exposed bedrock over a workable area and completed the excavation of the plant area. The overburden was disposed of along the south-western perimeter to complete this new landform. Some 8 ha of new landform were cultivated and seeded, and rock removal to form the primary crusher slot and platform commenced. Highway improvements were undertaken, and erection of the plant and associated infrastructure commenced towards the year's end.

During 1988 the erection of the processing plant, the asphalt and ready mixed concrete plants, the electrical sub-stations, weighbridges and offices was started and continued throughout the year. The development of quarry benches and further overburden stripping were also carried out in 1988. The overburden was deposited to create a new landform along the northern and western perimeter of the site. All completed landform sections were cultivated and seeded and tree planting commenced around the main entrance to the quarry.

In 1989 the erection of the plants and ancillary structures was essentially completed and plant commissioning commenced towards the end of the year. Further quarry development was carried out to create the approximately 200 m of 15 m high faces that will be needed to sustain the anticipated production. The rail link was also completed and officially opened in October (Figs 11.2 & 11.8).

Cultivation and planting works continued on the completed landforms and the Central Electricity Generating Board commenced the erection of new pylons to enable the 132 kV power line to be re-routed in 1990.

1990 saw most of the planned works completed and several million tonnes of rock produced. In the future further scheduled overburden removal and bank building phases will be required, and are incorporated with appropriate safeguards in the planning consents.

Phase 4 costs are shown in Table 11.2.

11.6 SUMMARY

The ultimate total expenditure to develop a site from a greenfield into a productive and profitable quarry is high. The importance, therefore, of carrying out sufficient exploration and testing to guarantee as nearly as possible the predicted quality and quantity of material is

Table 11.2 Some major phase 4 costs

Cost centre	Order of magnitude of expenditure in 1988–1989 £ values
1 Overburden excavation and disposal	15 000 000
2 Post phase 3 drilling and testing	192 000
3 Landscaping design, cultivation, supervision	375 000
4 Plant purchase, erection, commissioning	30 000 000
5 Road improvements and rail link	3 000 000
6 Quarry development and sundry other costs	4 000 000
Approximate total of major costs	52 567 000

Table 11.3 Summary of costs by phases

Phase	Cost in 1987-1989 £ values
Phase 1—Preliminary desk study	1 500
Phase 2—Initial follow-up (auger drilling)	2 500
Phase 3—Detailed follow-up (coring, testing, etc.)	650 000
Phase 4—Project implementation	52 567 000
Total	53 221 000

essential. The rapidly increasing cost of each phase (Table 11.3) emphasizes the necessity for a logical approach, with each subsequent phase only being undertaken if clearly warranted by the results from the preceding one. In this particular case exploration costs accounted for approximately 1.25% of the total costs of the development of the quarry operation.

12 Soma Lignite Basin, Turkey

MICHAEL K.G. WHATELEY

12.1 INTRODUCTION

The variables in a coal deposit (quality, thickness, etc.) are a function of the geological environment. Excellent descriptions of all aspects of coal deposits and the assessment of these deposits are given by Ward (1984), Scott (1987) and Whateley & Spears (1995). Coal or lignite is a heterogenous, organoclastic sedimentary rock mainly composed of lithified plant debris (Ward 1984), which was deposited in layers and may have vertical and lateral facies changes. These changes reflect variations in vegetation type, climate, clastic input, plant decomposition rates, structural setting, water table, etc. Coal originated as a wet spongy peat, which after burial underwent compaction and diagenesis (coalification). The coalification process proceeds at different rates in different structural settings. Coal which has been subjected to low pressure and temperature is referred to as a low rank coal, such as lignite. Coal which has been subjected to high pressure and temperature is referred to as a high rank coal, such as anthracite. Bituminous coal is a medium rank coal. Thus the properties of coal are almost entirely a reflection of the original depositional environment and diagenetic history.

Some of the properties of coal depend on the nature of the original plant material or macerals (Stach 1982, Cohen *et al.* 1987). These properties are measured using microscopic techniques. Large scale, subsurface changes are determined by studying the lithology of the coal and coal-bearing strata in borehole cores and down-the-hole geophysical logs. These data form the basis for any investigation into the resource potential of a coal deposit.

The quality of a coal deposit must be assessed. This is undertaken by sampling outcrops, borehole cores (Whateley 1992), mine faces, etc. (section 9.1.3), and sending the coal to the laboratory. The coal is subjected to proximate analyses to determine the moisture, ash, volatile and fixed carbon contents (Ward 1984). Additional tests which are often requested are sulphur content, calorific value (CV) and specific gravity (SG). The quality of coal will determine the end use to which it is put, e.g. steam coal to be burnt in a power station to generate electricity or metallurgical coal for steel making, although other uses are possible.

Since the energy crisis caused by the OPEC oil price rises which started in 1973, there has been a worldwide increase in the search for alternative fossil fuels. Lignite, bituminous coal and anthracite are important sources of energy. The exploration for and exploitation of these fuels is important for developing countries which wish to reduce their reliance on imported oil by building coal fired power stations at or near their own coal mines.

In 1982 Golder Associates (UK) Ltd (GA) undertook a World Bank funded feasibility study of the Soma Isiklar lignite deposit in western Turkey (Golder Associates 1983) for the state owned coal company Turkiye Komur Isletmeleri Kurumu (TKI). The lignite from the Soma Mine was used as feedstock for the ageing 44 MW Soma A Power Station. It was the intention to increase the mine production from 1.0 Mt to 2.5 Mt a^{-1}, 2.0 Mt of which were to come from the expanded surface mine. The increased tonnage was needed to feed the newly constucted 660 MW Soma B Power Station. Excess lignite would be used for domestic and industrial purposes.

The GA report deals with the geological interpretation of the deposit, assessment of the reserves in terms of tonnage and quality, assessment of the geotechnical aspects of the mine area, and production of mine plans for the proposed surface and underground mines. Most of this chapter is derived from that report, although the geostatistical reserve estimations have been subsequently completed at Leicester University as student dissertations (Lebrun 1987, Zarraq 1987). The existing surface mine had suffered a major footwall failure in 1980. The steeply dipping, shaly, footwall sediments underwent non-circular failure on weak horizons over several hundred metres along strike and buried the working face. One of the intentions of the study was to establish the reasons for this failure and to incorporate safety factors into the new mine design that would reduce the risk of a repeat of this type of failure.

It is the intention in this chapter to outline the way in which the geologists, who were involved on this study,

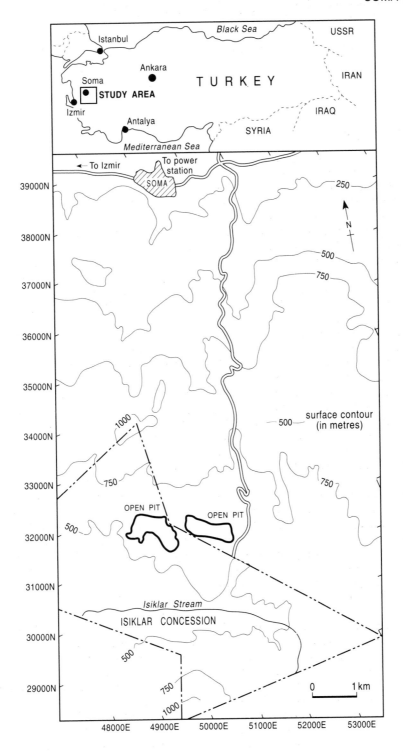

Fig. 12.1 Location diagram of the Soma study area.

collected the data, presented the data for the mining and geotechnical engineers, and calculated the reserve and quality parameters. The geotechnical and mining aspects are also described. To simplify this chapter only the thick, lowermost seam in the area designated the open pit, is described. GA also reported on the underground mining potential of the Soma Basin, which they reported might be economically mineable in the future.

12.1.1 Location

The Soma lignite deposit is in the Manisa Province of western Turkey, 10 km south of the town of Soma (Fig. 12.1). A high (1100 m) ridge separates the proposed mine site from the town. The site is on the south facing slope of this ridge, with elevations ranging from 750 m in the north to 310 m in the south.

12.1.2 Turkish mining rights

Mining and operating rights vary in different parts of the world. Before any foreign or local company can start operations it is necessary to understand the local legal system. In Turkey all mineral and coal rights are deemed to be owned by the State and are not considered to be part of the land where they are found. The Ministry of Energy and Natural Resources administers and implements the mining laws and regulations, and grants exploration permits, exploitation permits and leases.

12.2 EXPLORATION PROGRAMMES

12.2.1 Previous work

The lignite in the Soma area has been mined since 1913, first for local domestic and industrial consumption and later as feed for the Soma A power station. TKI took over lignite production on 1979.

The database for this deposit was derived from four drilling programmes (Table 12.1), field mapping and a feasibility study conducted by TKI in 1981–82. The earliest study was conducted by Nebert (1978), which included lithological logs of thirty-four cored boreholes (Table 12.1) and two geological maps of the Soma area with cross sections. Cored lignite was analysed for moisture and ash contents, calorific value, and in some cases volatile matter and sulphur content. The holes were drilled at 500 m centres to establish the resource potential of the basin.

The second exploration programme was supervised by Otto Gold GmbH. Open hole drilling was used with

Table 12.1 Summary of drilling exploration programmes

Drilling programmes (#)	Number of holes	Total meterage	Drilling date
200	34	8 790	1960
100	50	10825	1976
300	91	547	1981
400	29	8631	1982
Total	122	29 793	

predetermined intervals cored to include lignite seams. Lignite cores were analysed for ash and moisture content, and calorific value. The sulphur and volatile content were determined on a few samples. The programme was designed to identify lateral and down dip lignite limits, as well as to undertake infill drilling at 250 m centres within the basin.

TKI initiated the third drilling programme as part of their preliminary mine feasibility study. They also used rotary drilling techniques with spot coring. Full proximate analyses and calorific value determinations were done on these cores, and occasionally sulphur analyses. These holes were drilled to obtain additional information in areas of structural complexity or areas with a paucity of data. TKI produced isopach, structure contour, isoquality and polygonal reserve maps as well as reserve tables, surface and underground mine plans, manpower schedules, etc.

A fourth drilling programme was carried out during the feasibility study. Five of the holes were fully cored for geotechnical studies, while the remainder were rotary drilled and spot cored. Lignite cores were analysed for proximates, calorific value and specific gravity. Hardgrove grindability tests (section 12.6) and size analyses were also carried out on samples from the existing open pit. Although 122 boreholes were drilled, only 83 of them intersected the lowest lignite seam, or the horizon at which the equivalent of the lignite seam occurred. This was due to common drilling problems such as loss or sticking of rods in the hole, or burning the bit in at zones of serious and sudden water loss. This resulted in an overall density of 11 holes km^{-2}, equivalent to a rectangular grid roughly 330 m × 250 m although some holes are closer and some farther apart than this. This is considered to have given sufficient

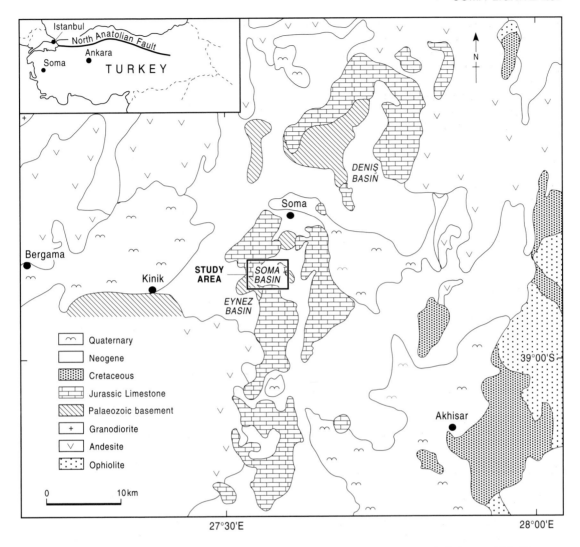

Fig. 12.2 Generalized geology of the Soma area (taken from the 1:250 000 Izmir sheet regional geological map published by MTA). The inset shows the regional structural setting for the Soma Basin in relation to the major structural feature of northern Turkey, the North Anatolian Fault (NAF).

density to classify the lignite in terms of measured (= proved) reserves (USGS 1976).

12.2.2 Core recovery

One of the major problems of assessing lignite or coal deposits is that of core recovery. If core is lost during drilling there is no way of determining the quality of the lost core and often the better quality, more brittle, bright sections are lost on these occasions. It is usual to have a drilling contract that requires the drillers to recover at least 95% of the core in the seam. Recoveries less than this (within reason) usually require redrilling. Techniques that help to improve core recoveries include the use of large diameter wireline (section 9.4.5) or air flush core barrels (e.g. HQ series with a nominal hole diameter of 96.1 mm) and the use of triple tube core barrels (Cummings & Wickland 1985, Berkman 1989).

12.2.3 **Geophysical logging**

Down-the-hole (DTH) geophysical logging (section 7.13) was first used in the oil industry but soon found its way into coal exploration (Ellis 1988). It is now used routinely in the evaluation of coal deposits because geophysical logs can help reduce drilling costs by enabling the use of cheaper, rotary, open hole drilling of, say 80% of the holes. Good comparison of open hole and cored holes is achieved with the use of DTH logs. They can also help in identifying the top and bottom of the seams, partings within the seams, lithological changes, in checking on core recovery and the depth of each hole and ensuring that drillers do not claim for more metres than they actually drilled. Seams often have characteristic geophysical signatures which, in structurally complex areas, can help with seam correlation.

Typical DTH geophysical logs used on coal and lignite exploration programmes are natural gamma, density, neutron, caliper and resistivity, as well as sonic and slim line dip meter logs (Ellis 1988). Shale, mudstone and marl usually have a high natural gamma response while coal has a low response, with sharp contacts often being observed. Coarsening-upward or fining-upward sequences in the clastic sections of the logs can also be inferred from the gamma logs. The density log, as its name implies, reflects the change in density of the rocks, with coal and lignite having low densities and shale and sandstone having higher densities. The neutron tool is used in estimating porosity, and the resistivity tool may be used to indicate bed boundaries. The caliper tool defines the size of the hole. The sonic log is also used in estimating the rock quality (fracture frequency), and the dip meter can be used to interpret sedimentary structures (Selley 1989).

A geophysical log showing the typical responses that the rocks in the Soma basin give, is described in section 12.3.2.

12.2.4 **Sampling**

It is essential to establish a procedure for sampling core and working faces so that continuity is maintained throughout an exploration project and into the mining phase. For example, at Soma the whole lignite sequence was sampled, including all parting material. The core was split and one half was retained in the field. All lithological layers in the seam >30 cm thick were sampled. Thinner layers were included with adjacent layers until a minimum thickness of 30 cm was obtained. The 30 cm limit was used as it was considered by the

study team that this was the minimum thickness of parting that could be mined as waste in the open pit (see also sections 12.6.1 to 12.6.3). This ensured that weighted average estimates for run-of-mine (ROM) lignite could be made.

12.2.5 **Grouting**

Where deep coal is likely to be mined by underground methods, all holes drilled prior to mining should be sealed using pressure grouting. This will reduce the potential water inrush hazard in underground mining.

12.3 **GEOLOGY**

12.3.1 **Geological setting**

The basement in western Turkey consists of Precambrian, Palaeozoic and Mesozoic sedimentary and igneous rock (Campbell 1971, Brinkmann 1976) which have been subjected to various structural and metamorphic episodes. Turkey has a great variety of structures, the largest of which is the North Anatolian Fault (NAF), a major strike-slip fault system (Fig. 12.2). The NAF was formed in the late Serravallian (Sengor *et al.* 1985). At this time westerly strike-slip movement of the west Anatolian Extensional Province took place. At the same time a series of NE trending grabens, such as the Soma Graben, began to form in western Turkey. These grabens began to fill with Serravallian (Samartian) sediments, which contain thick lignite deposits.

12.3.2 **Geology of the Soma Basin**

The Soma Basin contains thick deposits of Miocene and Pliocene sediments (Fig. 12.3), which range in age from Serravallian (Samartian) to Pontian (Gökçen 1982), but does not contain volcanic rocks. The Tertiary sediments rest unconformably on Mesozoic basement rocks. The stratigraphy of the basin is summarized in Fig. 12.4. It contains two thick Miocene lignite seams designated the KM2 and KM3 but this case history deals only with the KM2 seam.

The Mesozoic basement forms a rugged topography around the basin of limestone hills rising to 350 m above the Tertiary sediments. Boreholes which reached the basement confirmed the presence of limestone below the sediments in the basin. At Soma, the top of the basement appears to consist of debris flow deposits.

The Miocene deposits have been divided into three formations (Fig. 12.4). The basal Turgut Formation is

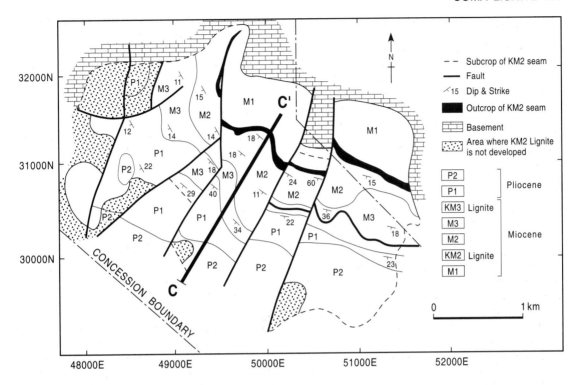

Fig. 12.3 Simplified geological map of the Soma Basin. The detail of the limit of the lignite in the south-south-west is unclear as no information was available due to the great depth of the lignite. Block numbers refer to the proposed mining blocks used in reserve and quality estimation procedures (see text).

predominantly immature sandstone and conglomerate, but there is a gradational upward fining of the sediments to the lignite horizon. Increasing amounts of sandy clay, clayey silt and carbonaceous clay appear towards the top of the formation until these grade into the KM2 seam. The basal contact of the seam is gradational and is usually placed where the first recognizable lignite appears with <55% ash content. The lignite is hard, black and bright, has numerous cleats (typical close spaced jointing of coal and lignite) and breaks with a concoidal fracture. Using the ASTM classification, the calorific value of this material would more properly result in it being termed a sub-bituminous B coal, but it is referred to as a lignite in the local terminology.

There is an extremely sharp contact between the KM2 and the overlying Sekköy Formation. This formation is a massive, hard marlstone and geophysical logs identify this contact clearly. The KM2 has a distinctly low natural gamma signature in contrast to the very high signature of the marl. The sharp change from peat (lignite) formation to the marlstone is believed to be the result of a change to an arid climate (Sengor *et al.* 1985).

In cores and in the open pit, the contact with the overlying freshwater limestone of the Yatagan Formation is unconformable and marked by a colour change. The gamma log shows a distinct change from the very high signature of the marl to a medium response in the limestone. The Yatagan Formation contains the KM3 Lignite which is interbedded with a series of thick calcareous clastic units, making the economic assessment of this unit more difficult. The fresh water limestone is conformably overlain by massive marlstone the basal half of which contains thin, laterally impersistent, lignite seams. Gökçen (1982) considered these marlstones to belong to the Upper Miocene Yatagan Formation, although Golder Associates (1983) placed the Miocene-Pliocene contact at the top of the freshwater limestone.

12.3.3 Structure of the Soma Basin

The Soma Basin is a fault-controlled graben, which has been subjected to regional tilting to the south-west. The basin formed in response to stike-slip faulting along the

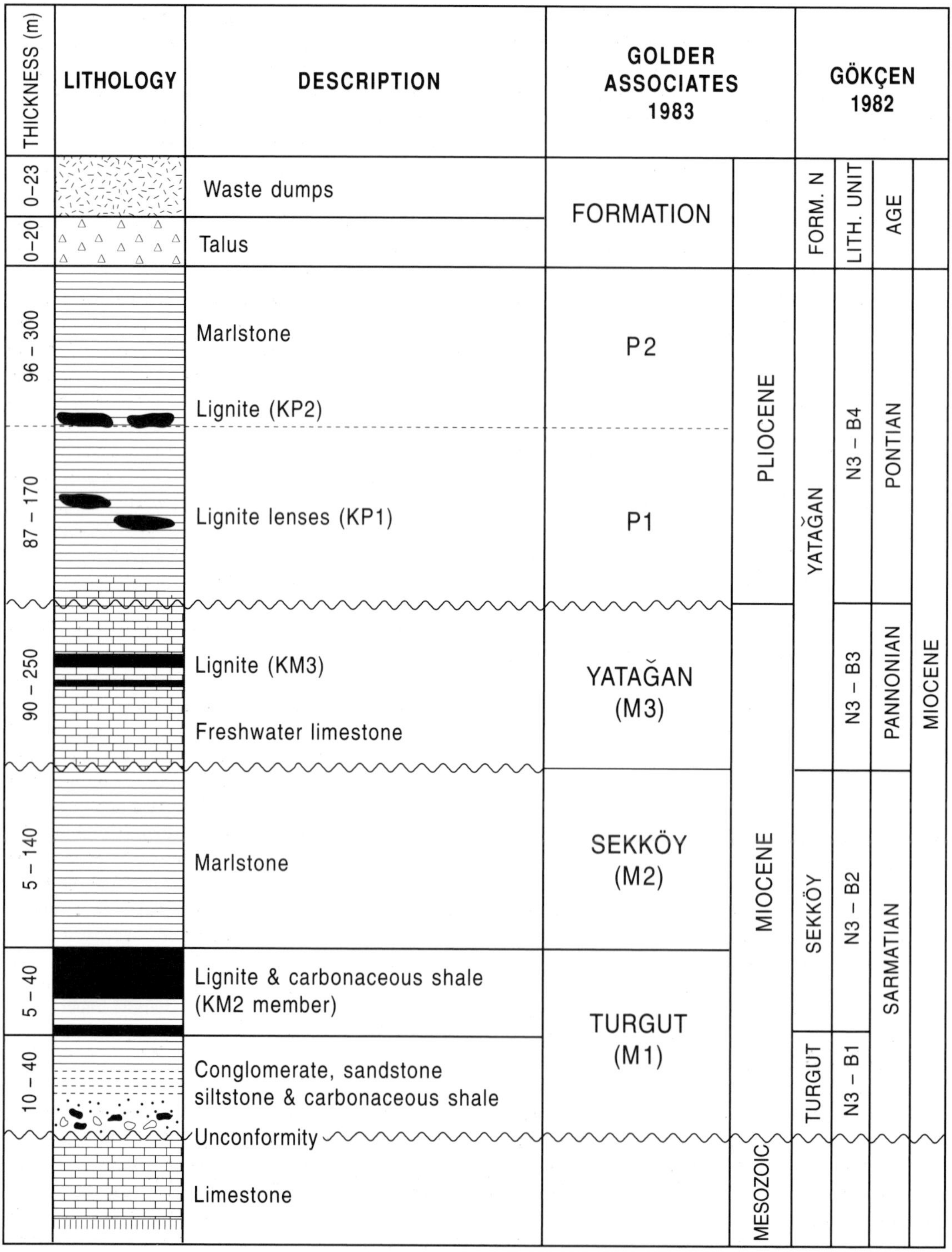

Fig. 12.4 A simplified stratigraphical column for the Soma Basin. The sediments have not been dated with any certainty. Comparison of traditional and recent microfossil dating (Gökçen pers. comm.) shows that more detailed work is still required.

North Anatolian Fault Zone during the mid Miocene (Sengor *et al.* 1985). The faulting in the Soma Basin trends NE–SW (Fig. 12.3). The major faults were defined by elevation differences in the structure contour map of the top of the KM2 seam. It is generally accepted that a disruption to the trend of a set of structural contours, that are expected to behave in a uniform way, may indicate the presence of a fault (Annels 1991, Gribble 1993, 1994). Fault positions were interpolated as lying between certain boreholes and, as mining proceeds and more information becomes available, then the position of these faults will be defined more accurately. These faults have throws of up to 150 m and displacements of up to 300 m.

Faults of this size will effect the design of both surface and underground mines. The faults at Soma were used to define mining blocks, Blocks A, B, C, D and E (Fig. 12.8). What is more difficult to determine is the amount of secondary faulting and fracturing formed in association with the major faulting. These minor, subparallel faults may have throws of only a few metres. It is difficult to plot these faults from drilling results, but they may have a significant effect on underground mining operations.

The sediments were deposited around the arcuate northern basin rim, and they dip to the SW at an average of 20°. These dips vary within each block, e.g. near the northern rim of the basin the dips are significantly steeper. A more accurate assessment of the dips was made by interpreting the structure contour map of the top of the KM2 seam (Section 12.4).

12.3.4 Depositional model for the Soma Basin

A depositional model of the Soma Basin was proposed after the isopach and ash content maps of the KM2 seam, detailed borehole logs, and stratigraphical sections were examined and interpreted in the light of ancient and modern analogues, in the manner described by McCabe (1984, 1987, 1991). The basin originated as a NE–SW fault-controlled graben in the basement rocks, which gradually infilled as the basin subsided at varying rates. Steep peripheral gradients around the north of the basin at the time of initial deposition resulted in rapid transportation of sand and gravel to the centre of the basin by high energy run-off. These debris flow deposits alternate with poorly sorted, coarse-grained sandstone. Gradual lessening of the gradient, with subsequent lowering of the energy regime resulted in sediment load fractionation and deposition of finer sediments in the

basin (sandstone and siltstone). The overall result is a fining-upward sequence from gravelly sandstone at the base to siltstone and mudstone at the top (Fig. 12.4). Temporal variations in depositional conditions are indicated by the repetition of these fining-upward sequences and rapid lateral variation. In the partially filled basin, conditions were conducive to plant growth which resulted in the formation of the thick, laterally variable KM2 lignite seam. The lignite is generally shaly at the base but improves in quality (i.e. has a lower ash content) towards the top.

Two distinct sub-basins have been recognized, the eastern Demir Basin and the western Seri Basin, separated by the Elmcik High (Fig. 12.5). The plant forming ecosystem or mire (Moore 1987) formed around the northern rim of the basin, but carbonaceous material was transported to the more distal portions of the basin to the south and south-west. Fine-grained sediment was also transported into the basin, and deposited mainly along the northern rim of the Seri Basin, where the lignite is strongly banded having many non-lignitic partings. The overall result is a gradual facies change from low ash lignite to carbonaceous shale basinward. There is a pronounced thinning of the lignite over the Elmacik High, suggesting that this area was an active high during peat formation.

A climatic change from warm humid conditions favouring peat formation to an arid climate resulted in a sudden and basin wide change of sedimentation to marl deposition. The fine grain size and the calcareous nature of the material suggests deposition by low energy input into a low energy water body. The resulting marlstone (Sekköy Formation) is unconformably overlain by the freshwater limestone of the Yatagan Formation. The KM3 lignite occurs at varying levels within the limestone, but is areally restricted. The KM3 has a significantly higher ash content than the KM2 lignite. Alternating marlstone and limestone continues upwards, with occasional thin, laterally impersistent seams developing.

12.4 DATA ASSESSMENT

In order to establish the reserve and quality criteria that were required for the mine design and financial analysis of the project, several maps had to be constructed, such as structure contour maps, isopach and isoquality maps as well as stripping ratio maps and cross-sections. All the borehole logs were examined, the seams were correlated and the top and bottom of the KM2 seam were established. Lignitic material was frequently rejected

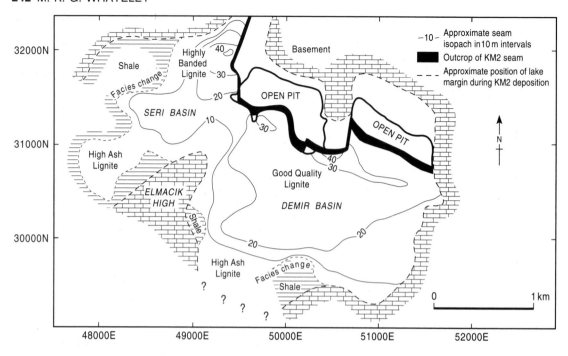

Fig. 12.5 A palaeogeographical interpretation of the Soma Basin at the time that the KM2 seam was being formed.

from the base of the seam because of poor quality. The top of the seam was readily identified at the sharp contact of the lignite with the overlying marlstone, but the basal contact was often drawn only after the quality data had been assessed. Once the seam thickness was established it was possible to construct the maps.

Since this study was completed several integrated computer packages have become available that calculate and plot contours quickly and accurately (Whateley 1991), e.g. Datamine, Surpac, Borsurv, PCExplore, etc.). During this study a planimeter was used to measure areas (e.g. between contours for tonnage calculations [sections 12.4.2 & 12.7]). Modern software packages have various user definable methods of contour construction. Two contoured surfaces (or digital terrain models [DTMs]) can be superimposed and volumes calculated, e.g. DTMs of the top and bottom of a seam, or DTMs of the surface and the top of a seam for overburden volume. The assessment of all the data up to and including the mine design and the financial analysis can now be undertaken using these computer packages. Manual methods (described in section 9.5) are still widely used by many exploration and mining companies.

12.4.1 **Structure contour maps**

Three structure contour maps were drawn for this study: (1) on the top of the KM2 seam, (2) on the bottom of the KM2 seam and (3) on the top of the Mesozoic Basement. The KM2 seam has the greatest areal distribution, a very well defined upper contact and the most detailed and reliable data base. The elevation of the top of the KM2 seam was plotted manually at each data point and contours were constructed by linear interpolation between points. The contours were drawn at 20 m intervals (Fig. 12.6). The strike of the lignite seam at depth was assumed to be similar to that of the overlying sediments at the surface. Where the structure contours differed significantly from this, a fault was inferred. By checking against the surface geological map and cross-sections it was possible to establish the fault pattern in the basin. The structure contours were then redrawn between the faults to give the pattern shown (Fig. 12.6). It was important to determine the structure of the basin in order to assist the mining engineers with their design of the optimum open pit, by avoiding areas of unstable ground near faults and loss of lignite near these fault zones.

Once the structural pattern was established, it could be transferred to the structure maps of the base of the

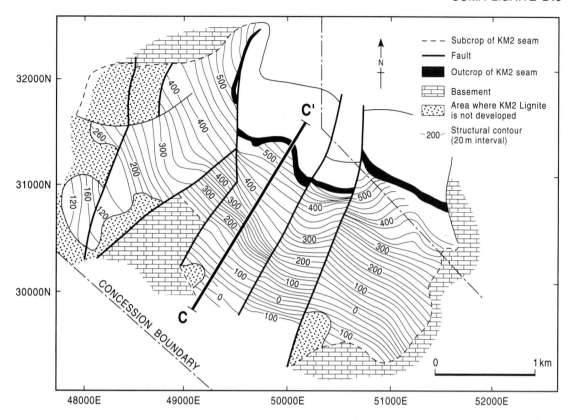

Fig. 12.6 A structure contour map of the top of the KM2 seam, Soma Basin.

KM2 and the top of the basement. The structure contour map of the base of the KM2 was constructed by placing the isopach map of the KM2 on top of the structure contour map of the top of the KM2. Where the two sets of contours intersected, the seam thickness was subtracted from the top of seam elevation to give the new elevations. These contours were drawn at 20 m intervals, and were only drawn in the area designed to be the open pit, for use in mine design and to identify areas of potential slope failure.

The structure contour map of the top of the basement was also produced by linear interpolation. Only a small number of boreholes penetrated the basement, so the contours were constructed at 50 m intervals. The structure contour map of the top of the KM2 was then placed on the basement map and then the intersections of contours of the same value outlined the limits of the KM2 seam (Fig. 12.3).

12.4.2 Isopach maps

For this study, three isopach maps were drawn manu-ally, namely: (1) for the KM2 seam (Fig. 12.7), (2) for the overburden material, and (3) for the sediments between the KM2 seam and the basement (the Turgut Formation). The thickness of the KM2 was determined in each hole between the sharp upper contact with the overlying marlstone and the base of the seam. The isopachs show the thickness of KM2 that could be extracted by surface and underground mining methods, and include waste partings within the lignite which could be separated as waste during mining (section 12.4.4). The isopach map was constructed at 5 m inter-vals by linear interpolation. The seam ranges in thick-ness from 2 to 57 m, with an average in the open pit area of some 17 m. The extreme thickness of 57 m is the result of a borehole intersecting a steeply dipping part of the seam. The seam isopach map formed the basis on which the reserve estimates were based, as well as being used to calculate the structure contours at the base of the KM2 seam.

The isopach map of the waste material above the KM2 seam (the overburden) was constructed by placing the map of the surface topography over the structure contour

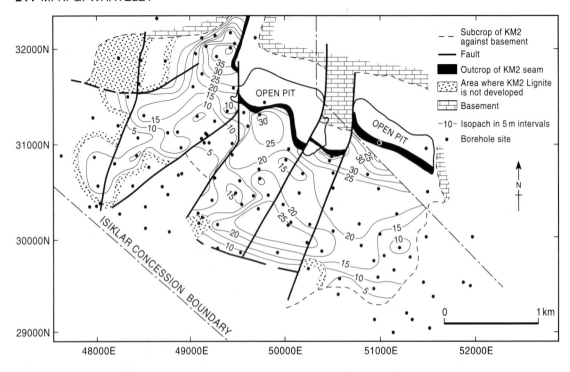

Fig. 12.7 An isopach map of the KM2 seam. Location of all the boreholes drilled in the Soma Basin are also shown.

map of the top of the KM2 seam. Where the respective contours intersected, the elevation of the top of the seam was subtracted from the elevation of the topographical contour and the resulting thickness value plotted. This method of constructing the overburden map provided more data points than would have been available had the borehole values alone been used, resulting in a more reliable map. The overburden above the KM2 seam varies from a few metres near the existing workings, to approximately 150 m at the deepest part of the proposed surface mine. The map was used to construct the stripping ratio map needed in the mine planning exercise.

A detailed contour map was constructed during the mine design phase of the project for use by the mining engineers to calculate the volume of overburden material which had to be removed at various stages of mining from the various mining blocks (see section 12.3.3). Contours were constructed and a planimeter was used to measure the area between adjacent overburden thickness contours which were then multiplied by the applicable vertical overburden thickness. Products were summed to obtain the total volume for each block (Table 12.2). Pit slope volumes were calculated in a similar manner assuming a final pit slope angle of 45°.

An isopach map of the Turgut Formation was constructed by subtracting the values of the structure contours at the base of the KM2 seam from the structure contours at the top of the basement. This isopach map was only drawn for areas of potential open pit development to assist the geotechnical engineers in predicting the areas of potential footwall failure in future operations (see section 12.5.1).

12.4.3 Stripping ratio maps

The stripping ratio map, or overburden ratio map shows the overburden material in m^3 as a ratio per tonne of lignite, prior to applying selective mining criteria. This map (Fig. 12.8) was constructed by placing the overburden isopach map on the isopach map of the lignite. Where the contours intersected, the thickness of the lignite was multiplied by a specific gravity of 1.73 to obtain tonnes. The specific gravity was obtained by averaging the values derived from laboratory analyses of lignite from the open pit area of the mine. The product was divided into the overburden thickness to calculate the stripping ratio. Contours were then constructed at unit intervals between 1 and 10.

Table 12.2 Estimated overburden volumes at the proposed Soma Isiklar open pit mine. For explanation of M bank m^3 see section 12.8.2

Block	Pit area			Slopes			Totals
	Area (km^2)	Thickness (m)	Volume (M bank m^3)	Area (km^2)	Thickness (m)	Volume (M bank m^3)	Volume (M bank m^3)
A	0.146	78.9	11.52	0.148	62.7	9.28	20.80
B	0.412	90.1	37.12	0.189	54.6	10.32	47.44
D	0.072	87.6	6.31	0.096	60.3	5.79	12.10
E	0.882	84.8	74.79	0.357	70.1	25.24	100.03
Totals (mean)	1.512	(85.8)	129.74	0.790	(64.1)	50.63	180.37

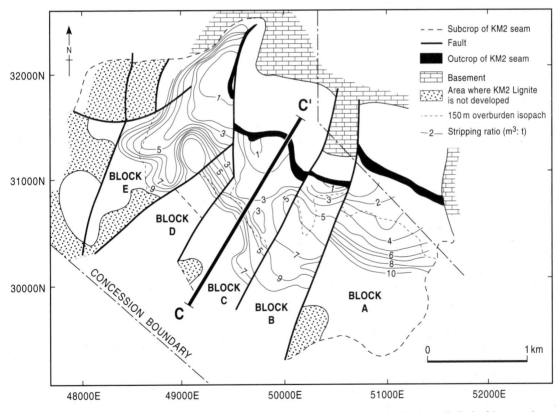

Fig. 12.8 A stripping ratio map showing the 150 m overburden isopach line used to determine the down-dip limit of the open pit mine.

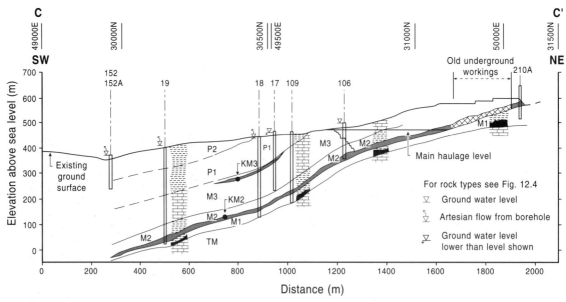

Fig. 12.9 An example of a down-dip section used to help in correlating the stratigraphy and to illustrate the hydrogeological information derived from boreholes.

The stripping ratio map provides a guide to areas where mining would be preferentially started, i.e. where the ratios are lowest. The financial study suggested that open pit mining could be carried out economically where the stripping ratio is less than 7:1. As a general guide, the 7:1 stripping ratio in coal or lignite mining provides a limit to surface mining, although different companies operate different criteria and this figure is variable.

12.4.4 Cross–sections

East–west, north–south and down-dip sections (Fig. 12.9) were drawn at regular intervals across the basin. They were constructed by plotting the data from the structure contour and isopach maps. Boreholes which fell on or close to the section lines were also plotted. The E–W and N–S sections were cross checked by comparing and correlating the intersection points. The sections were used to verify the structural information and the seam correlations. They were not used for reserve estimation (Reedman 1979), although this is commonly done by some coal mining companies.

12.5 GEOTECHNICAL INVESTIGATION

12.5.1 Investigation programme

It is always advisable to initiate geotechnical investiga-

tions as soon as possible in an exploration programme in order to avoid expensive duplicate drilling of boreholes. In this study the following programme was carried out.

1 *Geotechnical logging* of five fully cored boreholes. In addition to the geological descriptions, the logs included the rock quality designation (RQD) (section 9.6) and evaluation of the point load strength (Brown 1981), which can be used to to estimate the uniaxial compressive strength. Where a point load strength test was not available, the field description was used as a basis for estimating the strength class (Table 9.8). All the cores were fully labelled and photographed (Fig. 12.10). This is particularly important where cores are prone to deterioration.

2 *Geotechnical testing* of borehole cores and samples collected from the existing open pit.

3 *Structural mapping* of overburden exposures in the existing open pit.

4 *Measurement of water levels* in exploration boreholes and measurement of flows from artesian boreholes.

5 *Sensitivity analyses* of known failures altering the variables such as position of the piezometric surface or material strength to establish typical shear strength values for design from back analysis of existing footwall failures in the open pit.

These data were used in geotechnical analyses to provide preliminary design guidelines for the proposed open pit.

Fig. 12.10 A photograph of a box of core drilled during this study. All core recovered during this study was photographed before the core had been tested and sampled by the geotechnical engineers and before the lignite had been sampled. This ensured that a full record of the *in situ* core was available. All core box lids had a detailed descriptive label attached which gave the borehole number, box number, depths, date drilled and a B&W and colour chart to ensure that each film was processed to the same standard.

12.5.2 Geotechnical conditions

OVERBURDEN

The overburden consists predominantly of limestone and marlstone with lesser amounts of mudstone and siltstone. The limestone and marlstone are moderately strong, with mean uniaxial compressive strengths of the order of 80 MPa. The siltstone and mudstone are weaker rocks with compressive strengths generally less than 30 MPa. The overburden rocks are bedded and jointed and the bedding is generally parallel to that in the lignite seam and dips range between 15° and 45° to the southwest. Jointing is mainly vertical.

LIGNITE

Point load strengths of selected lumps of lignite are about 1 MPa. However, the lignite is highly cleated and friable and the mass strength would be considerably lower.

FOOTWALL

The footwall of the KM2 seam consists of variable thicknesses of mudstone, siltstone and immature sandstone overlying basement limestone. The immediate footwall has a high clay content and is generally weak, with a high slaking potential. Typical residual shear strengths for this material are c = 0 to 30 kPa, and Φ = 14 to 17°. A series of back analyses was carried out on exposed footwall material in the existing open pit. These analyses confirmed that the above shear strength values were realistic for design purposes, and that footwall stability is highly dependent on groundwater conditions and the local dip of the footwall.

12.5.3 Hydrogeological conditions

Standing water levels in exploration boreholes were measured using piezometers. The levels were generally close to the surface and, in the topographically lower areas, flowing artesian boreholes occurred. In some cases, artesian flows were high ($7 \, l \, s^{-1}$) and flows could be maintained for long periods. One of the main aquifers in the strata appeared to lie in the vicinity of the KM2 seam.

12.5.4 Implications for open pit mining

The overburden rocks are moderately strong, requiring drilling and blasting prior to excavation. Controlled blasting procedures would be required on all final walls for good slope stability. Practical mining considerations would limit the maximum wall slope to about 55° between haulroads and an overall slope of 45° for the final highwall of the pit, including provision for 20 m wide internal haul roads. The overall slope angle of the advancing face is determined by the equipment clearance and access requirements but should not exceed 40° for stability and safety. This could be steepened to 45° at the final limit when controlled blasting procedures are used. The footwall dip means that footwall failures could occur in the new pit. In order to minimize the risk, a strike advance mining method was recommended (section 12.8.1).

Dewatering the new open pit by means of boreholes would probably be required, especially prior to the initial box cut excavation. A continuous programme of borehole dewatering will probably be required, to reduce water pressure in the basement thus avoiding the possibility of footwall heave. An excellent description of the processes involved in the dewatering of a large open pit mine is given by Cameron & Middlemis (1994).

12.6 LIGNITE QUALITY

Coal seams consist of multiple layers of carbonaceous material and non-combustible rocks (waste), such as clay, shale, marlstone, sandstone, etc. These interbedded sediments are not continuous and at Soma can be regarded as lenses within the range of the borehole spacing. This inevitably means that difficulties arise when seam quality and reserve quantities are being considered. Two alternatives are available when considering the approach to be adopted when estimating lignite quality (and reserves). The first is to assume that the zone within the lignite seam containing most of the lignite layers is mined in total without any attempt at selectively mining waste partings. This gives the *in situ* lignite quality data. The second is to consider mining the lignite layers selectively, aiming to produce a run of mine (ROM) product which is of acceptable quality.

This gives the mineable lignite quality data.

The non-selective mining approach will result in a ROM product which would not generally meet the power station specification and would be highly variable in composition. This could be homogenized in a blending stockpile and upgraded in a washing plant, which is costly and results in losses. Selective mining in the open pit will be more expensive than bulk mining since time will be lost in moving machinery and there will also be unavoidable losses and dilution of lignite associated with this method. At Soma it was necessary to adopt the selective mining approach in the lignite quality and reserve estimations, because the lignite quality is low and further quality losses caused by bulk mining would be unacceptable.

Lignite quality was assessed by evaluating the data provided with the borehole logs obtained during the first three drilling programmes (Table 12.1). Additional quality data came from samples of core submitted to the MTA laboratory in Ankara during the feasibility study. The core from the earlier programmes was analysed for ash, moisture (on an as received basis) and calorific value. In some cases volatile matter and sulphur content were also determined. In the last drilling programme the lignite core was analysed for proximate analyses on an as received basis, specific gravity and calorific value. As the power station stockpiles the lignite in the open, the as received analyses approximate more closely to the quality of the lignite that is actually burnt. Lignite samples which do not have the surface moisture removed by air drying are said to have been analysed on an as received basis. Samples which are air dried until the mass of the samples remains constant (all surface moisture is assumed to have evaporated) before being analysed are reported on an air dried basis (section 12.6.4).

These data were placed on a computer data base to facilitate the assessment of the lignite quality, using down hole weighted averaging techniques. The samples were weighted by sample length and specific gravity in order to calculate the weighted average quality for each

Table 12.3 Assumed ash and specific gravity values for the rock types in the Soma Isiklar Basin

Rock type	Assumed ash content (%)	Assumed SG
Lignite		1.40
Clayey lignite		1.70
Lignitic limestone	75	2.00
Marlstone	75	2.30
Clay	75	2.40

composite. To ensure that a true weighted average was obtained, quality values had to be assigned to the partings within the seam. Minor partings are often not sampled during the exploration phase of a programme, but during mining these partings will often be incorporated with the lignite, thus reducing the quality of the product. Partings were sampled in the last drilling programme and these values were assumed to apply to similar rock types that had not been sampled earlier. Similarly, where the specific gravity of the lignite had not been determined, a value was assumed (Table 12.3). Holes where the core recovery was less than 75% were omitted as not being representative.

The possible expansion of the lignite mine was investigated for the purpose of increasing the supply of lignite to the adjacent thermal power stations. The lignite is crushed and fed directly into the furnace as a powdered fuel. The solid fuel specifications for steam-raising plants usually includes the Hardgrove Grindability Index. This test gives an indication of the ease with which a material will be crushed (Ward 1984). A low value indicates a hard rock while a high number indicates a relatively soft rock. One would expect a high number (>63.2) for a lignite, but the values obtained in tests ranged between 33 and 57. This probably reflects the amount of parting material that is included with the lignite. Sieve analyses were also carried out, and these showed that over 60% of the ROM production in the open pit is +30 mm. The domestic and industrial markets require a lump product, which the mine can easily supply.

12.6.1 Borehole sampling

Cores from the seam were split and one half was retained in the field. The entire thickness of the lignite seam was sampled including all parting material. All layers within the seam which were >30 cm were sampled as individual samples. Thinner layers were included with adjacent layers until a minimum thickness of 30 cm was obtained. A maximum of 2 m of core was sampled at one time.

12.6.2 Selection criteria—surface mine

In appraising the viability of using surface mining methods the following steps were taken. The weighted average quality of the mineable lignite seam was obtained by classifying any sample with <55% ash content as lignite and samples with >55% ash as waste. The 'rules' that were applied to each sample to obtain the composite mineable thickness are shown in Fig. 12.11. The quality

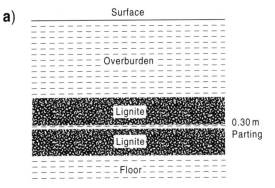

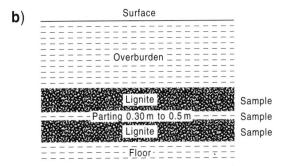

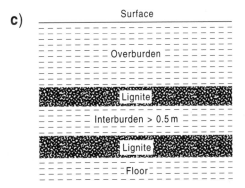

Fig. 12.11 An illustration of the sampling 'rules' that were applied to all the lignite core recovered during the study in the Soma Basin. (Modified after Jagger 1977.)

of the mineable reserves was calculated on all material within the seam except (1) waste partings >50 cm, and (2) lignite piles of <30 cm with partings on either side so that the total thickness of lignite plus partings above and below is >50 cm. In addition, dilution and loss of lignite during mining was taken into consideration. At each interface between the lignite and a parting, a

Table 12.4 Examples of selective mining evaluation in the proposed open pit at Soma Isklar, Turkey

Borehole Number	210	320
Lignite, vertical thickness (m)	18.40	11.80
Waste rejected from seam	0.00	3.00
Mineable vertical thickness (m)	18.40	8.80
Number of waste partings	0	3
Number of interface	2	8
Dilution (m)	0.2	0.8
In situ CV (kcal kg^{-1})	3524	2071
Undiluted CV (kcal kg^{-1})	3524	2679
Mineable CV (kcal kg^{-1})	3485	2435
In situ ash content (%)	27.6	39.9
Undiluted ash content (%)	27.6	29.7
Mineable ash content (%)	28.1	33.8
In situ SG	1.55	1.90
Undiluted SG	1.55	1.70
Mineable SG	1.56	1.78
Recovery by thickness (%)	100	75

dilution by volume of 10 cm of parting was included replacing a lignite loss of 10 cm. These interfaces were taken into account at the top and bottom of the seam as well. The quality of the undiluted mineable reserves was calculated in a similar way to that of the mineable reserves, but with no dilution or loss of lignite taken into consideration. Table 12.4 gives some examples of the changes in quality which arose by applying these selection criteria.

Once the areas to be mined on an annual basis had been outlined, the quality in each area was estimated using the boreholes within a 450 m radius. The mineable qualities, examples of which are seen in Table 12.4, were used to determine this quality release schedule for the open pit.

Inverse distance weighting (section 9.5.1) was reviewed and compared to a simple weighted averaging technique. The two methods indicated little difference in the qualities. The 450 m radius of influence was chosen following a study of semi-variograms of thickness, calorific value and cumulations of thickness multiplied by calorific value. The range indicated on these semi-variograms was approximately 560 m. The distance of 450 m was selected because it is just greater than the normal 2/3 to 3/4 of this range which is normally selected (Annels 1991), and generally three or more boreholes fell within the radius.

12.6.3 Selection criteria—underground mine

This section has been included to show that different mining methods require different selection criteria. During the study, alternative underground mining methods suitable for the extraction of a medium dipping, thick seam, were examined. In-seam mining (a) and cross-seam (b) mining were considered (Fig. 12.12). As

Table 12.5 Examples of selective mining evaluation in the proposed underground mine at Soma

Borehole number	218	302
Lignite vertical thickness (m)	14.70	12.10
Lignite in waste partings (m)	0.00	0.75
Waste rejected from seam (m)	0.00	2.35
Mineable vertical thickness (m)	14.70	9.00
In situ CV (kcal kg^{-1})	3662	2400
Mineable CV (kcal kg^{-1})	3595	2500
In situ ash content (%)	24.5	39.5
Mineable ash content (%)	25.2	37.6
In situ SG	1.49	1.73
Mineable SG	1.51	1.70
Recovery by thickness (%)	100	74

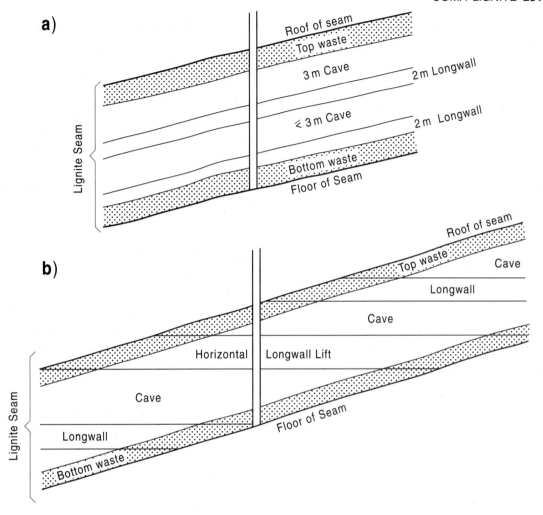

Fig. 12.12 A sketch of the two underground mining methods proposed for the deep lignite in the Soma Basin. (a) In-seam mining, in which the lignite is mined down-dip and (b) cross-seam mining, in which lignite is mined horizontally.

a computer data base was being used, it was possible to try sensitivity analyses on the various selection criteria. These were used to assist in the selection of the best mining methods and in the calculation of the ROM tonnages and qualities. The weighted average quality of the *in situ* and mineable lignite was calculated using tabulated data examples of which are shown on Table 12.5.

IN-SEAM MINING

The mineable lignite was determined once waste part-

ings (material with >55% ash content and/or <1800 kcal kg^{-1}) >1.5m thick were rejected. Occasionally thin lignite beds within the thick waste partings were also rejected. The top and bottom of the *in situ* and mineable lignite are the same. It was assumed that there would be a 100% recovery of the lignite in the longwall slice, and 60% recovery of lignite and 40% dilution by waste in the caved zones above the longwall zones. The mineable quality takes into account the lignite losses and waste dilution which would occur during caving. It was considered that selective mining could be implemented above and below a waste parting >1.5 m thick.

Table 12.6 Average quality of the lignite in the Soma Basin

	SG	CV (kcal kg^{-1})	Ash (%)	Moisture (%)
OPEN PIT				
In Situ	1.70	3069	32.7	13.5
Mineable	1.73	3103	33.0	13.5
Block A	1.73	4143	20.9	13.8
Block C	1.73	3104	32.5	14.4
Block D	1.73	2516	39.1	14.0
Block E	1.73	2816	36.6	12.9
UNDERGROUND				
In Situ	1.66	3337	30.0	16.2
Mineable	1.66	3467	27.5	16.2

CROSS-SEAM MINING

The mineable lignite was determined once the top and bottom waste material were excluded. As this method is less selective, dilution of the lignite is inevitable. This is accounted for by expecting a 60% recovery of lignite and a 40% dilution by waste.

12.6.4 Summary of lignite quality—surface mine

Once the limits of the potential open pit had been established using the outcrop line and the stripping ratio limit down dip, it was possible to calculate the average quality of the *in situ* and the mineable lignite. The weighted average quality data are shown in Table 12.6. The *in situ* lignite quality includes small areas of lignite which may be included in the lignite resource of the basin but were considered to be inaccessible to open pit mining. All the analyses were reported on an as received basis. The air dried moisture content was calculated for several of the lignite samples from the final series of holes. The results showed that the inherent moisture (air dried) was approximately 8% and the surface moisture was approximately 7%, totalling 15% on an as received basis. The moisture content does not precisely represent that of the mined lignite because of the presence of drilling fluid in the core samples and also the additional moisture that would be picked up during mining.

The sulphur content of the KM2 seam was analysed in a few cases. The results indicate that the KM2 is a low sulphur lignite. The weighted average is as follows:

combustible sulphur 0.53% and sulphur in the ash 0.42%, giving a total sulphur at 0.95%.

12.6.5 Summary of lignite quality—underground mine

The areas down dip of the 7:1 stripping ratio line defined that part of the basin that could be potentially mined by underground methods. Within this area it was possible to calculate the average *in situ* lignite quality and the mineable lignite quality values (Table 12.6).

12.7 LIGNITE RESERVE ESTIMATES

The lignite reserves were calculated, by assessing the data provided from borehole logs (Whateley 1992) obtained by drilling (Table 12.1), within the limits of the open pit. These limits were established between the outcrop line and the 7:1 stripping ratio line. On the NW and the E sides of the pit, the limit is determined at the depositional edge of the seam. This limit was determined by using the basement and basal KM2 structure contour maps (section 12.4.1). The pit outline was placed on the isopach map of the total vertical thickness of the KM2 seam, and the area between each isopach was measured with the aid of a planimeter (cf. section 12.4). The areas were multiplied by the average vertical thickness between the isopach lines (the thickness value at the midpoint) to obtain the *in situ* volume. The volume was multiplied by the average specific gravity to obtain the *in situ* tonnage.

The mineable lignite tonnage was calculated from the *in situ* tonnage by applying a recovery factor, examples of which are given in the final row of Table 12.4. The weighted average recovery for the open pit is 91%, but this varies from as low as 45% in borehole 208 to 100% in many of the remaining holes. The recoveries of mineable lignite from the total lignite will change during mining depending on the local geology. Development drilling immediately in advance of production will determine these recoveries. The reserve figures were calculated as follows: *in situ* reserves 49.4 Mt, mineable reserves 40.6 Mt and kriged estimate of *in situ* reserves 46.3 Mt (LeBrun, 1987).

12.7.1 Comparison of estimation methods

Zarraq (1987) undertook a comparison of reserve estimation methods using the Soma data. He calculated the *in situ* reserves for the whole basin using polygons, manual contours and kriging. His reserves estimates are

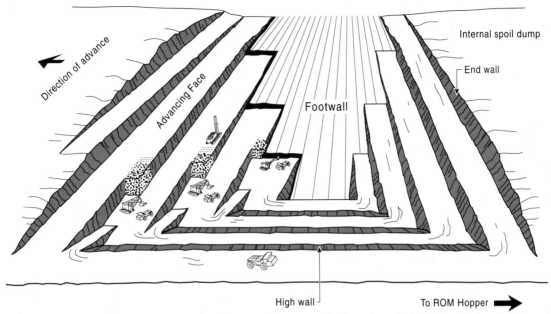

Fig. 12.13 A sketch of a truck and shovel, terrace mining operation proposed for the surface mine in the Soma Basin.

as follows:

Method	Tonnage
Polygons	117.8 Mt
Manual contouring	102.5 Mt
Kriging (150 × 150 m blocks)	109.4 Mt
Statistical mean (area × average thickness)	117.5 Mt

The polygonal method has probably overestimated the reserve slightly because there are some large polygons which have thick lignite associated with them, e.g. the polygon centred on hole 208 has an area of 187 630 m^2 with 17.95 m of lignite, while the average area of all the polygons is only 70 000 m^2. Polygons also create artificial boundaries which do not exist in nature. The only way to reduce the estimation variance is to increase the sample density, which costs time and money. However, this is a quick method of providing a global estimation.

Manual contouring smooths the data by virtue of the linear interpolation used to construct the map. It appears to weight the large area of low values excessively, which may explain the lower reserve figure. Although the kriged estimate of reserves does take the spatial rela-tionships into account, the estimation variances were generally high. Kriging gives the best estimate for each block with the lowest estimation variance, so the advantage of kriging is that one knows how reliable each block estimate is (section 9.5.1).

12.7.2 **Confidence in the reserves**

The confidence in the mineable reserve estimate for the open pit area was calculated using the mean and standard deviation for the thickness and specific gravity derived from the boreholes drilled in the area designated to be the surface mine.

The formula $\frac{tS}{n}$ was applied where

S = standard deviation, n = number of boreholes and t = t value for n–1 degrees of freedom at the 90% confidence level.

The results of this calculation were expressed as a percentage of the mean. The percentages of the mean results were then used in the following formula:

$$G = A^2 + B^2$$

where A = percentage of the mean for thickness, B = percentage of the mean for S.G. and G = global confidence in the average expressed as a percentage.

The confidences as a percentage of the mean for the

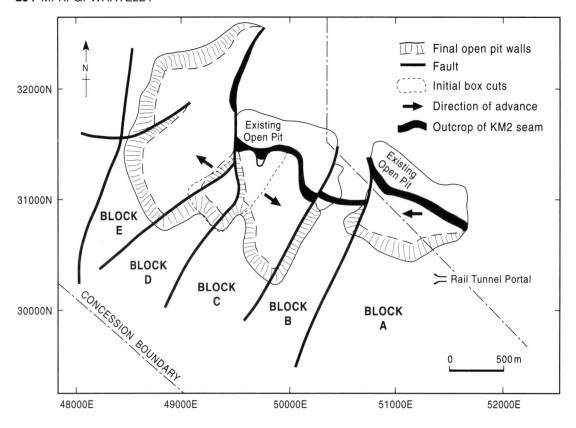

Fig. 12.14 The outline of the final open pit design determined by the 150 m overburden isopach.

open pit area are as follows: Thickness (A) 15.9%, S.G. (B) 5.9% and global confidence (G) 17.0%. These results indicate that the reserves calculated from thickness and S.G. are sufficiently well known that they can be classified as proved within the confidence limits.

12.8 SURFACE MINE EVALUATION

This evaluation project called for a production rate of 2.13 Mt ROM lignite each year. After rejection of some waste dilution by rotary breakers, the delivered output is expected to be 2 Mt p.a. This rate of production was considered appropriate to exploit the mineable reserves over a mine life of 21 years. Reserves are contained in four fault bounded blocks (A,C,D, and E), but it was not considered advisable to include block B, because the lignite has been partly extracted using now abandoned underground workings. The presence of the faults and the varying dip angle of the footwall dictated the mining method and box-cut locations.

12.8.1 Selection of mining method

The hard massive marlstone which forms the overburden will require blasting before removal can take place and this rules out the use of a dragline or a bucket wheel excavator. Rear-dump trucks and face shovels will have to be used. Three alternative mining configurations may be considered using this equipment: (1) advance down the dip, (2) advance up the dip or (3) advance along strike. The third method, also known as terrace mining, was considered the most suitable with the particular geological and geotechnical conditions which exist at Soma (section 12.5.4).

Following this method a box-cut would be excavated down the full dip of the deposit from the outcrop until an economic stripping limit, or practical mining depth limit, is reached. This study concluded that a maximum

---

Fig. 12.15 *(Opposite)* Details of (a) the proposed open pit bench design, (b) haul road and (c) lignite mining operation.

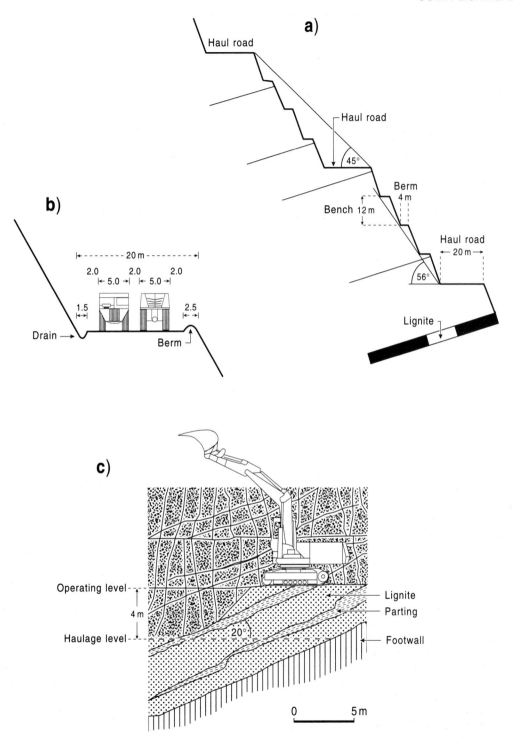

mining depth of 150 m was feasible, although constraints were generally related to the geological configuration and consideration of maintaining footwall stability, rather than to stripping ratio economics. Advance could then be in one or both directions along the strike line, depending on the box-cut location within the proposed mining area. Waste disposal within the excavation is practical (and usually desirable) as the box-cut excavation is enlarged. Horizontal benches would be formed in the overburden along the advancing face and along the highwall formed on the deep side of the excavation. As much spoil as possible from the advancing face would be removed along the highwall benches for disposal to form an internal spoil dump within the excavation and behind the advancing face; the remainder having to be dumped outside the pit. The front of the spoil dump and the advancing face then advance in unison as mining continues along the strike of the deposit (Fig. 12.13). The final highwall is expected to have a maximum slope angle of 45°.

The application of this mining method will bring several advantages. (1) A minimum area of footwall clay will be uncovered at any time, thus reducing the risk of footwall failure, (2) internal dumping of waste reduces transport costs, helps stabilize the footwall and begins reclamation at an early stage of mining, and (3) the stripping ratio is more or less constant over the life of the mine, and hence the mining costs are stabilized.

12.8.2 Mine design

BOX-CUT LOCATIONS

Two box-cuts were proposed to take account of the adverse geological factors (Fig. 12.14). The first was sited in blocks C and D, close to the centre of the open pit reserve. This location was dictated by consideration of local stripping ratios, seam dip, footwall clay thickness and the presence of faulting. The central site means that mining will eventually advance both to the E and to the W. It was proposed that the second box-cut should be excavated at the western end of block A. Advance would be towards the W.

SLOPES AND ACCESS

A bench height of 12 m was proposed with excavation from horizontal benches in the overburden and lignite, with connecting ramps at up to 8% grade located in the advancing faces (Fig. 12.13). The highwall should have an overall slope angle of 45°, formed by the 12 m bench,

and by berms and haulroads inside the pit for internal waste dumping (Fig. 12.15). Controlled blasting will be required for stable slopes.

Advancing faces will require an overall slope angle of 10° to 20°, depending on the current mining activity. A working width of 65 m was proposed for each bench for independent drilling and blasting, loading and haulage activities. Lignite excavation would be from sub-benches 4 m high, subdividing the 12 m overburden benches for ease of excavation and loading (Fig. 12.15).

HAUL ROADS

A haul road width of 20 m will be needed to provide room for trucks to pass with adequate side and centre clearances (Fig. 12.15). Spoil disposed of within the pit will need be transported on haul roads at 48 m vertical intervals formed in the highwall. Ramps between haul roads, and to the rim of the pit would be formed in the advancing faces. Spoil disposal outside the pit will be via three ramps to main haulage routes on the pit rim. Lignite will be transported to the ROM lignite plant by a main haul road from the pit to the south portal of the main rail tunnel (Fig. 12.14) through the mountain and then by overland conveyer belt to the power stations near Soma (Fig. 12.1) to the north.

SPOIL DISPOSAL

One-third of the 180 M bank m^3 of overburden (Table 12.2) could be backfilled into the mining excavation, with the remainder being disposed of in two dumps outside the pit. Bank m^3 refers to the volume of rock *in situ*. Once mined the rock 'swells' to occupy a greater volume although the increase in the void spaces results in a mass reduction m^{-3}.

12.8.3 Mining equipment

Large items of open pit equipment were not manufactured in Turkey at the time of this study. Calculations of equipment utilization, mechanical availability and productivity were made to decide what equipment would have to be imported. The overburden equipment requirements were assessed to be three rotary blasthole rigs (250 mm diameter holes), and electrically powered rope shovels for overburden stripping, assisted by front end loaders. Hydraulic face shovels were recommended for lignite loading. Rear dump trucks were selected for both overburden and lignite transport. Supporting equipment included tracked and wheeled bulldozers and a conventional fleet of pit and haul road maintenance vehicles.

12.9 **SUMMARY**

Available data on the reserve tonnage and quality of the Soma deposit were adequate for a feasibility evaluation, which suggested that the reserves would support a projected life of 21 years at the proposed rate of production and specified quality. Additional data were required on the ground water regime before forecasts could be made of the effects of ground water pressure and flows on mining. This highlights the need for careful planning of any exploration drilling programme to maximize the data that can be gained from any drilling.

12.10 **FURTHER READING**

Ward (1984) covers a wide range of topics and his book is intended for senior students and professional geologists. He discusses coal exploitation, mining, processing and utilization in a clear manner. Other papers on these subjects can be found in the *Bulletin de la Société géologique de France, 162 (1991)*. The Special Publications of the Geological Society edited by Annels (1991) and Whateley & Spears (1995) give some excellent case histories of coal deposit evaluation. *Stach's Textbook of Coal Petrology* (1982) is well referenced and gives detailed information on the microscopic properties of coal. The sedimentology of coal and coal-bearing strata is covered in a series of papers in a Geological Society of London Special Publication edited by Scott (1987). Further papers on the sedimentology of coal-bearing strata are to be found in the IAS Special Publication edited by Rahmani & Flores (1984), and in the *International Journal of Coal Geology*.

13 Witwatersrand Conglomerate Gold— West Rand

CHARLES J. MOON AND MICHAEL K. G. WHATELEY

13.1 INTRODUCTION

Alluvial placer deposits are renowned as gold producers, particularly from rich deposits in the Yukon, the Mother Lode area of the forty-niners in northern California and the modern day rush in the Sierra Pelada area of the Amazon. Even today much Russian production comes from the Lena and Magadan alluvial deposits in northeast Siberia. However modern alluvial gold output is considerably less that from deposits in quartz pebble conglomerates that appear to be fossilized placers. These deposits are restricted in geological time to the late Archaean and early Proterozoic and dominated by one basin, the Witwatersrand Basin in South Africa. Other gold producing quartz pebble conglomerates occur in the Tarkwa area of Ghana, Jacobina and Moeda in the São Francisco Craton of Brazil while the Elliot Lake area in Ontario has been a significant producer of uranium from similar rocks and carries minor gold.

This case history concentrates on the Witwatersrand (Afrikaans for ridge of white waters and often abbreviated to Rand or Wits) Basin that has been responsible for the production of 37% of all gold produced in modern times as well as significant uranium. This basin was the source of 611 t in 1992, about 30% of annual world gold production. Mining in the Witwatersrand Basin is from large tabular orebodies that form the basis of some of the largest metalliferous underground mines in the world and which reach more than 3500 m below surface. This combination of depth and scale has been a great challenge to mining engineers and it has produced a long and interesting exploration history involving large expenditures on deep drilling and geophysics.

13.2 GEOLOGY

Mineralized conglomerates are found in a variety of settings within the Kaapvaal Craton (Fig. 13.1) but the vast majority of production has come from the upper part of a 7000 m sedimentary sequence termed the Witwatersrand Supergroup. These sediments were deposited within a 350 by 200 km basin that formed relatively soon after the consolidation of the Kaapvaal Craton from a series of greenstone belts and granitic intrusives. The Witwatersrand sedimentation was part of a much longer depositional history which began with sedimentation in a probable rift setting of bimodal volcanics and limited sediments, termed the Dominion Group (Stanistreet & McCarthy 1991). Recent age determinations date volcanics within the succession at 3074 ± 6 and underlying granites at 3120 ± 6 Ma (Armstrong et al. in De Beer & Eglinton 1991). The basal sediments are conglomeratic and supported minor gold production and substantial uranium operations. These Dominion Group sediments form structural remnants overlain by Witwatersrand Supergroup sediments.

13.2.1 Stratigraphy of the Witwatersrand Supergroup

The Witwatersrand Supergroup is divided into two, the lower West Rand Group and the upper Central Rand Group, which contains most of the gold deposits.

The West Rand Group consists of about equal proportions of shale and sandstone and 250 m of volcanics, the Crown lavas (Tankard et al. 1982). A number of iron-rich shales are present in the lower part of the group and these give a magnetic response used in exploration. The depositional environment was littoral or sub-tidal on a stable shelf as indicated by the general upward coarsening nature of the macrocycles. However there are a number of upward fining cycles with conglomerate at their base, which represent more fluvial conditions and contain mineralization. Despite their alluring names, the Bonanza, Coronation and Promise Reefs (reef is used as a synonym for mineralized placer throughout the Witwatersrand) have produced only 35 t of gold, although their mineralogy is similar to placers in the Central Rand Group (Meyer et al. 1990). Age dates on detrital zircons give a lower age limit for deposition of 2990 ± 20 Ma and an upper limit of 2914 ± 8 Ma on the Crown Lava (De Beer & Eglington 1991).

The later stages of the deposition of the Witwatersrand Supergroup reflect major changes in the style of sedimentation together with changes in the shape and size of the basin of deposition (Stanistreet & McCarthy 1991).

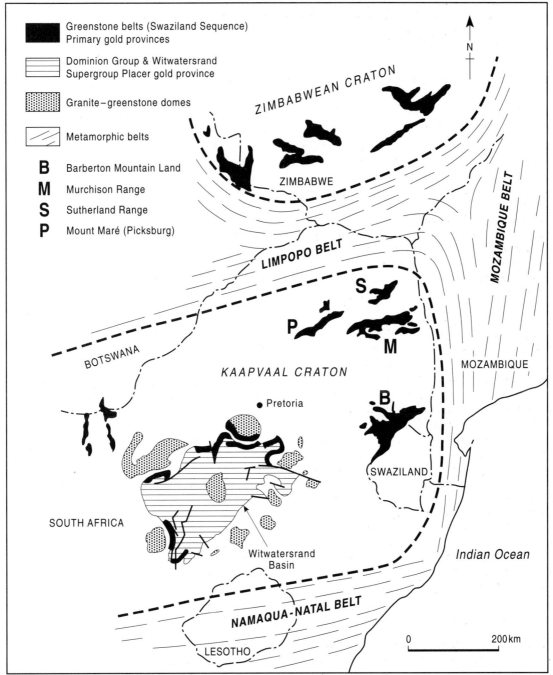

Fig. 13.1 Regional tectonic setting of the Witwatersrand Basin. Mesozoic and younger cover has been removed. (After Saager 1981.)

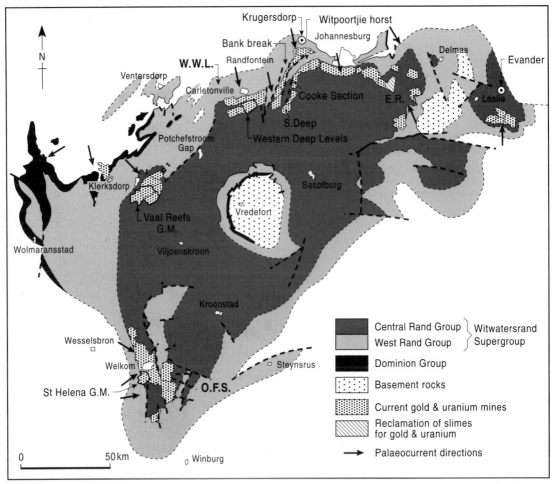

Fig. 13.2 Witwatersrand Basin with later cover removed, E.R. indicates location of East Rand Basin (Fig. 3.7). (Modified from Camisani-Calzolari *et al.* 1985.)

The Central Rand Group consists predominantly of coarse-grained subgreywacke with less than 10% conglomerate, silt and minor lava (Tankard *et al.* 1982). The overall depositional setting is of fan delta complexes prograding into a closed basin, perhaps containing a lake or inland sea. The change in depositional style reflects the increasing importance of tectonism, certainly of folding, probably of faulting. Burke *et al.* (1986) suggested that the overall tectonic setting is that of a foreland basin related to collisional tectonics in the area of the Limpopo river to the north. Dating of detrital zircons, shows that the Eldorado Formation, at the top of the group, is younger than 2910 ± 5 Ma and older than overlying Ventersdorp lavas dated at 2714 ± 8 Ma (Robb *et al.* 1990). As there is no metamorphic event of this age

known from the Limpopo collisional belt, the foreland basin theory is in doubt.

The fan delta complexes have distinct entry points into the basin as indicated by palaeocurrent data, pebble size and composition and isopach maps; these correspond with the location of the major goldfields. At present, eight major goldfields are recognized, from south to north-east, the Orange Free State (OFS or Welkom), Klerksdorp, West Wits Line (or Far West Rand), West Rand, Central Rand, East Rand, South Rand and Evander (Fig. 13.2). The areas between the goldfields are known as gaps, notably the Bothaville and Potchefstroom Gaps. Recent exploration has shown that there is significant mineralization in these gaps, although at great depth. However the south side of the basin seems to be essentially barren.

Regional correlation between the different reefs is difficult as they are developed on unconformities and some reefs are developed in only one goldfield, as might be anticipated from the discrete sediment entry points. However broad correlations are possible as shown in Fig. 13.3. The reefs can be divided into two types, sheet placers and channelized placers (Mullins 1986). Sheet placers have good continuity and are gravel lags overlain by mature sand, whereas the channelized placers have well defined channels filled with lenticular bodies of sand and gravel. Examples of channelized placers are the Kimberley Reef and the Ventersdorp Contact Reef (VCR) whereas the Vaal Reef, Basal Reef and the Carbon Leader are sheet-like. Where both sheet-like and channelized placers are developed on the same unconformity, as in the case of the Steyn Placer in the OFS Goldfield, the channelized placers are more proximal to the source area and change into sheet-like placers over a distance of 20 km. The stratigraphical location of the major reefs is shown in Fig. 13.3.

13.2.2 Overlying lithologies

Little of the known extent of the Witwatersrand Basin sediments is exposed and most is covered by later rocks, both Precambrian and Phanerozoic: lavas and sediments of the Ventersdorp Supergroup, and sediments of the Transvaal and Karoo Sequences.

The uppermost major placer in the Witwatersrand Basin, the Ventersdorp Contact Reef, is preserved by a cover of Ventersdorp lavas and appears to have formed at a time of volcanism and block faulting. It is economically important in the West Wits Line and Klerksdorp areas, where it is heavily channelized. Although other placers have developed higher in the Ventersdorp succession, and under similar conditions, they are of little economic importance. The remainder of the Ventersdorp sequence consists largely of tholeiitic and high magnesia basalts with lesser alluvial fan and playa lake sediments.

The Transvaal Sequence was deposited from around 2400 to 2100 Ma over a wide area of the Kaapvaal Craton. The important parts of the sequence as far as exploration and mining are concerned are the basal conglomeratic unit, known as the Black Reef, and the overlying dolomites of the Malmani Subgroup. The Black Reef has been mined in a number of areas, notably on the East Rand and in the Klerksdorp areas. Mineable areas correlate with areas where the Black Reef cuts Central Rand Group mineralization and it seems certain

that the Black Reef mineralization is the result of reworking or remobilization of gold from below (Papenfus 1964). The dolomites overlie most of the mining area on the West Rand and Klerksdorp areas and cause considerable problems as they contain large amounts of water and are prone to the formation of sinkholes. One sinkhole swallowed the primary crusher at West Driefontein in 1962.

The youngest major sedimentary sequence found on the Witwatersrand consists of sediments of the Karoo Supergroup deposited from Carboniferous to Triassic times in glacial, marginal marine to fluvial conditions. These are important in two respects. The Ecca Group sediments are coal-bearing and besides providing a cheap source of energy have probably generated the methane that is found in a number of mines and which poses a threat of underground explosions. The Karoo sediments also have a high thermal gradient and give rise to a blanketing effect which raises the overall thermal gradient and reduces the ultimate working depths in both the O.F.S. and Evander Goldfields.

Witwatersrand sediments have been intruded by a variety of dykes, which cause considerable disruption to mining. The oldest intrusives are dykes of Ventersdorp age and the youngest lamprophyres of post-Karoo age but the most important are 'Pilanesberg' age dykes, up to 30 m wide, on the West Rand which divide the dolomites into separate compartments for pumping purposes (Tucker & Viljoen 1986).

13.2.3 Structure and metamorphism

Although the Witwatersrand sequence has clearly been deformed and metamorphosed, there has been little emphasis on making a synthesis until the last few years. An exception is faulting of the reefs which has been examined in detail as this is a major control on the day to day working of the mines. The type of deformation varies through the basin but the most important for mining is block faulting which forms the boundaries of mining blocks. The main activation of these faults was in mid Ventersdorp times when there was also thrust faulting at the margin of the OFS Goldfield (Minter *et al.* 1986). The major faults were, however, reactivated in post-Transvaal but pre-Karoo times, although the exact timing is unclear. In the east of the basin the sediments are affected by the intrusion of the Bushveld Complex which has tilted them.

The occurrence of a pyrophyllite–chloritoid–chlorite–muscovite–quartz–pyrite assemblage within the pelitic sediments indicates a regional greenschist metamor-

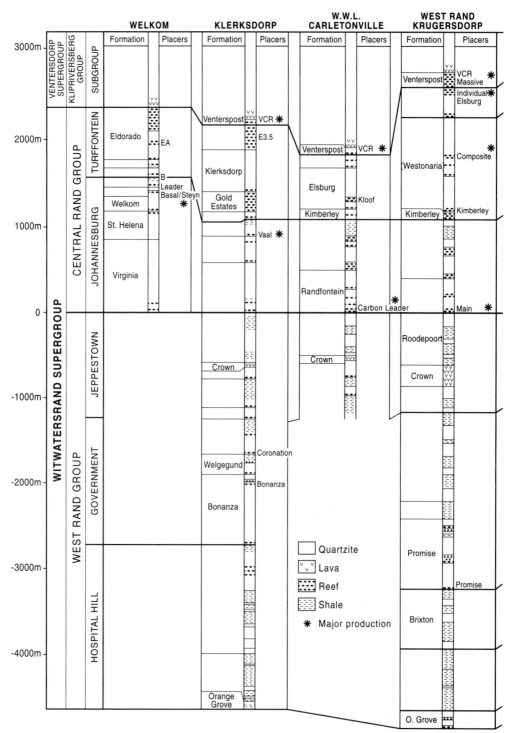

Fig. 13.3 Stratigraphical columns of the Witwatersrand Supergroup in the main gold producing areas. Note the general correlations between goldfields and major producing conglomerates. (After Tankard *et al.* 1982.)

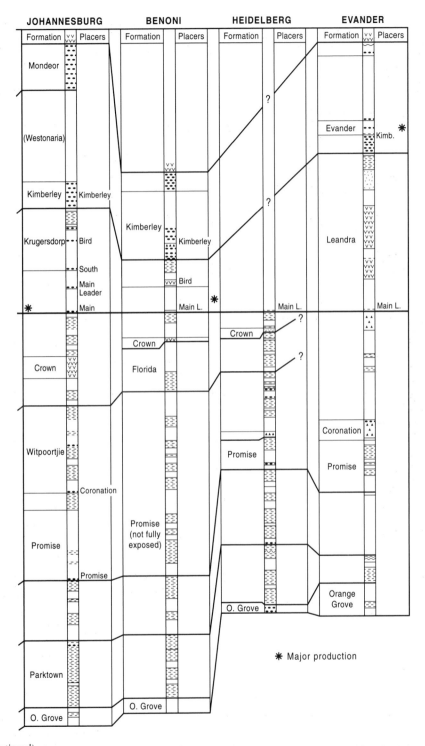

Fig. 13.3 *(Continued)*

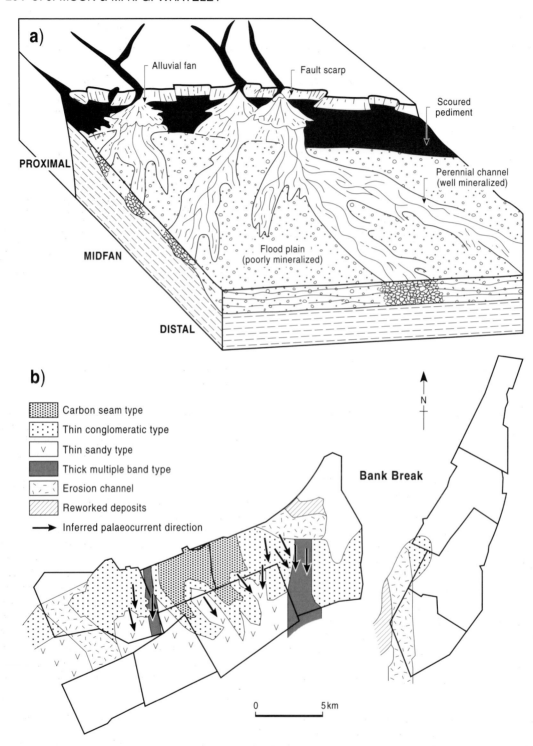

a) Alluvial fan, Fault scarp, Scoured pediment, PROXIMAL, Perennial channel (well mineralized), MIDFAN, Flood plain (poorly mineralized), DISTAL

b)

Carbon seam type
Thin conglomeratic type
Thin sandy type
Thick multiple band type
Erosion channel
Reworked deposits
→ Inferred palaeocurrent direction

Bank Break

N

0   5km

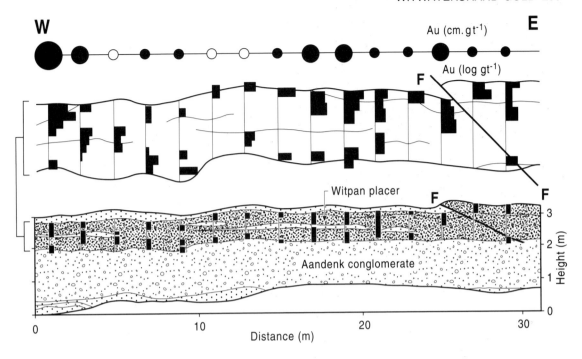

Fig. 13.5 Relation between grade and location in the Witpan placer, OFS Goldfield. The Witpan placer is stratigraphically above the B Reef of Fig. 13.3. The exact grades were omitted for commercial reasons. (After Jordaan & Austin 1986.)

phism at temperatures of 350 ± 50 °C (Phillips *et al.* 1987). Such metamorphism could account for the local remobilization of gold.

13.2.4 Distribution of mineralization and relation to sedimentology

Gold and uranium grades are controlled by sedimentological factors on both a small and a large scale. The diagram in Fig. 13.4 shows a typical relationship between grade and sedimentology for a channelized reef, with higher grade areas restricted to the main channels. The delineation of these payshoots may be the only way to mine certain conglomerates economically (Viljoen 1990). The distribution of facies can also be mapped in the sheet-like placers as shown for the Carbon Leader in Fig. 13.4b. High grade gold and uranium deposits are associated with the carbon seams and thin conglomerates, particularly in the West

Driefontein area. Mapping of these sedimentary structures is also important as they form homogeneous geostatistical domains for the purpose of grade calculations. In general the oligomictic placers have higher grades than the polymictic conglomerates, which contain shale and chert as well as vein quartz clasts, presumably as a result of greater reworking.

On a smaller scale, variations in sedimentary structures control grade distribution. This is exemplified by Fig. 13.5 which shows the distribution of gold through the channelized A reef (Witpan Placer) in the OFS Goldfield. In this case gold is concentrated at the top and base of the conglomerate unit. Gold is also associated with heavy mineral bands within the unit. In the sheet conglomerates gold is closely associated with carbon, if present, and the basal few centimetres may contain most of the gold in the reef. The remainder of the reef, although mostly waste, must be mined to expose the working face using a minimum stoping height of 80 cm.

Fig. 13.4 *(Opposite)* (a) Sketch showing a reconstruction of depositional conditions in a channelized reef and relation to grade. Generalized from the Composite Reef. (From Tucker & Viljoen 1986.)
(b) Facies of the Carbon Leader on the West Wits Line. Driefontein is the mine immediately west of the Bank Break. (From Viljoen 1990.)

Fig. 13.6 Underground phograph of a typical conglomerate. The UE1a Upper Elsburg at Cooke Section, Randfontein Estates Gold Mine. Courtesy of the Sedimentological Section of the Geological Society of South Africa.

13.2.5 Mineralogy

The conglomerates mined (Fig. 13.6) are predominantly oligomictic with clasts of vein quartz. The matrix consists largely of secondary quartz and phyllosilicates, including sericite, chlorite, pyrophyllite, muscovite and chloritoid. The opaque minerals have been the object of intense study and more than 70 are known; a review is given by Feather & Koen (1975). Pyrite is the predominant opaque mineral with important, but lesser amounts, of gold, uraninite, carbon, brannerite, arsenopyrite, cobaltite, galena, pyrrhotite, gersdorffite, chromite and zircon.

Pyrite forms 3% of a typical reef and is present largely as rounded grains. The larger of these are known as buckshot pyrite, being similar to the shot used in cartridges. Pyrite occurs as grains with recognizable partings and as porous grains, often filled with chalcopyrite or pyrrhotite. Some concretionary grains are also present. At least some of the pyrite was formed by pyritization of other minerals and rocks, such as ilmenite or ferruginous pebbles. Pyrrhotite probably formed largely as a result of the metamorphism of pyrite and is most frequent near dykes. Uraninite is present mainly as rounded grains of apparently detrital origin. Gold occurs largely in the fine-grained matrix between pebbles as plate-like, euhedral or spongy grains. The majority of gold is in the range 0.075–0.15 mm and is not often visible. Very fine-grained gold is found in the carbon seams. The majority of gold shows signs of remobilization and replacement relations with the fine-grained matrix. It also is commonly associated with secondary sulphides and sulpharsenides. Typically gold grains contain about 10% silver.

Uraninite is the dominant uranium mineral although the titanium-rich uranium mineral, brannerite, predominates in the Vaal Reef, presumably forming as a result of a reaction between ilmenite and uranium in solution. Carbon, or more correctly kerogen, is common in a number of the sheet conglomerates and is closely associated with high uranium and gold grades. Hallbauer (1975) suggests that it formed from lichen and fungi.

Chromite is widely distributed in the reefs and shows a sympathetic relationship with zircon. Both appear to be detrital. Two rare detrital minerals are diamond and osmium-iridium alloys. Authigenic sulphides are patchily distributed and include cobaltite, gersdorffite, sphalerite and chalcopyrite. This assemblage is associated with remobilized gold.

13.2.6 Theories of genesis

The origin of the mineralization has been a matter of bitter debate between those who believe it to be essentially sedimentary, developed either by normal placer forming mechanisms, the *palaeoplacer theory*, or by precipitation of gold in suitable environments, the *synsedimentary theory*; and the hydrothermalists who believe that the conglomerates were merely the porous sediments within which gold and sulphides were deposited from hydrothermal solutions derived either from a magmatic source or from dehydration of the West Rand Group shales. The full background to the debate can be found in Pretorius (1991).

The key arguments in favour of a placer origin are the strong spatial correlation between gold, uraninite and unequivocally detrital zircon and the intimate relationship between the heavy minerals and the sedimentary structures and environment. Placer advocates also point to the equal intensity of mineralization in porous conglomerates and less porous pyritic quartzites. The problems for the placerists are mainly mineralogical: the very small particle size of the gold, its hackly shape and low fineness (i.e. high % Ag), which are more typical of hydrothermal deposits, the presence of pyrite, uraninite and the absence of black sands typical of modern placers. The placerists invoke metamorphism to explain the shapes of the gold grains and explain the lack of magnetite as the result of intense reworking (Reimer & Mossman 1990) or transformation into pyrite after burial. The problem of detrital pyrite and uraninite is somewhat more difficult to explain as these minerals are unstable in most modern environments, although they are known from glacially fed rivers, such as the Indus in Pakistan (for discussion see Maynard *et al.* 1991). However it seems very probable that the Archaean atmosphere contained less oxygen than at present and that both uraninite and pyrite would be stable under these conditions.

Another major problem is the source of the large quantity of gold and uranium in the basin, far more than is likely to have been derived from any Archaean greenstone belt, certainly more than from the most productive South African greenstone belt, Barberton. While it is certain that greenstone belts were part of the source area, as evidenced by the occurrence of platinum group minerals in the Evander area, it seems likely that the gold and uranium could also have been derived from altered granites. Studies by Klemd & Hallbauer (1986) and Robb & Meyer (1990) have demonstrated the presence of hydrothermally altered granites with gold and uranium mineralization in the immediate hinterland of the Witwatersrand Basin. A neat twist to the placer theory is the suggestion by Hutchinson & Viljoen (1988) that the gold could have been derived from reworking of low grade exhalative gold deposits in the West Rand Group. The syngeneticists such as Reimer (1984) suggest that the gold was carried in solution into the basin and precipitated in carbon-rich conglomerate beds.

The principal arguments for a hydrothermal origin are that the gold is ragged, sometimes replaces pebbles, may be contained within and replace pyrite that has replaced pebbles, and is associated with a suite of hydrothermal ore minerals and alteration products (sericite and chlorite) typical of hydrothermal orebodies the world over. Previous important advocates of the hydrothermal origin were Graton (1930), Bateman (1950) and Davidson (1965). The hydrothermal theory has been recently given a new breath of life by Phillips *et al.* (1987) who pointed out the lack of metamorphic studies and demonstrated that the shales show evidence of metamorphism to greenschist facies similar in many ways to the conditions of formation of greenstone belt gold deposits. They also suggested that in some areas faulting may have controlled gold distribution.

The modified placer theory, which emphasizes the control on the occurrence of ore minerals by placer forming mechanisms but accepts some modification by metamorphism, has been widely used in exploration and is favoured by the majority of geologists working in gold mining and exploration. If you are lucky enough to make an underground visit it is unlikely that you will offered the option of a hydrothermal tour!

13.3 ECONOMIC BACKGROUND

13.3.1 Profitability

The economics of mining in the Witwatersrand have been largely governed by the price of gold. Unlike most other commodities the gold price is not controlled by supply and demand for industrial use but by the perception of its worth. As gold is transportable and scarce the metal has been used as a way of moving and storing wealth. Its price was regulated for long periods of time by inter-governmental agreement, enforceable by the large reserves held by reserve banks such as the US hoard at Fort Knox. These large stocks were initially held when currencies were directly convertible into gold. This convertibility, the gold standard, was abandoned by the UK in 1931 and by the USA in 1933. From 1935 to 1968 the gold price was fixed at $35 oz^{-1} but in 1968 President Johnson allowed the gold price to float for private buyers and in 1971 President Nixon removed the fixed link between the dollar and gold and left market demand to determine the daily price. The 1970s saw extreme fluctuations in the value of gold as it was used as a hedge against inflation with sharp increases in price (Fig. 1.8) following the increases in the oil price in 1973 and 1980. The gold price has remained in the $300–400 range in the rest of the 1980s, steadily depreciating against an inflating dollar as production has increased. The sharp increase in the gold price in 1931 set off the major exploration programmes that led to the discovery of two new goldfields in South Africa. That of the 1980s led to the prospecting boom of 1985–1989.

Long term plots (Fig. 13.7) of the overall profitability of the Witwatersrand mines reflect these changes in the gold price, changes in the South African rand–US dollar exchange rate and increases in working costs. The overall picture is relatively healthy with spectacular profits in the early 1980s. Since then working costs have escalated sharply as the overall recovered grade has declined, largely as a result of the exhaustion of higher grade deposits (Fig. 13.7), and South Africa has suffered sustained high inflation rates. In particular, working costs have increased due to higher wages and the increasing depth and complexity of mining. As the political situation changes and legalized discrimination has been eliminated, gold mines have not been willing or able to maintain wage differentials based on skin colour or migrant labour and overall labour costs have increased sharply. Reductions in unit costs will only come

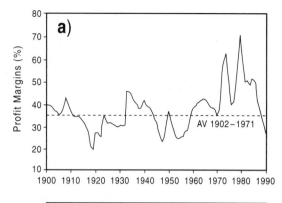

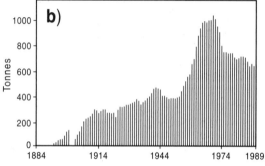

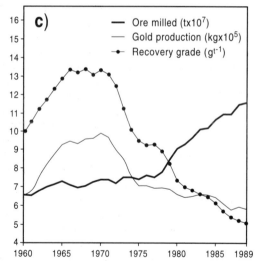

Fig. 13.7 Plots of (a) profit margins, (b) gold production and (c) gold production, tonnes milled and grade for the Witwatersrand Basin. (From Willis 1992.)

through increases in productivity, in particular through mechanization of mining.

The taxation arrangements under which the mines operate are complex but favour the treatment of large tonnages of low grade ore. A typical taxation scheme consists of two elements, lease taxation and normal taxation. In the case of Western Deep Levels the lease formula is y = 15–90/x where y is percentage of lease taxation after a 5% capital allowance and x is the ratio of taxable profit to mining revenue as a percentage. The normal taxation formula is y = 71–409/x where y is the percentage of tax payable and x the ratio as a percentage of mining profits to revenue. In 1989 the mine made R354M profit on revenue of R1238M and paid tax of R9M largely because of capital expenditure of R258M. In contrast it paid R207M in tax in 1986 when revenues were R1011M and working profit R611M.

Under the old mining laws an existing mine was ring fenced and it was not possible to offset capital expenditure on a new mine against profits on the existing property. This condition was relaxed under the 1991 budget arrangements and should encourage development.

13.3.2 Organization of mining

Mining on the Witwatersrand requires large amounts of capital and has been dominated by large companies since its inception. Six companies or mining houses control all the major mines, although a considerable proportion of the equity of individual mines is held by individual and institutional holders. Share prices for individual mines are quoted on the Johannesburg and London Stock Exchanges. Three of the mining houses, Johannesburg Consolidated Investments (JCI), Goldfields of South Africa (GFSA) and Rand Mines, were founded on the early success on the Rand and backed by capital from the diamond discoveries in Kimberley. Present day mining is dominated by the Anglo American Corporation which besides controlling output of 252 t p. a. of gold from its own mines, has considerable stakes in all the other mining houses and in several of their mines (Anglo American Corp. 1991). The other major companies are Anglovaal and Genmin, the mining subsidiary of Gencor.

13.3.3 Mineral rights

Mineral rights on the Witwatersrand are largely privately held, either by gold mining companies or in unmined areas by farmers or descendants of the original landowners. The value of these assets is usually appreciable and purchase values may be as high as R10 000 ha⁻¹; options to purchase are negotiated, usually lasting for three or five years (Scott-Russell *et al.* 1990). The

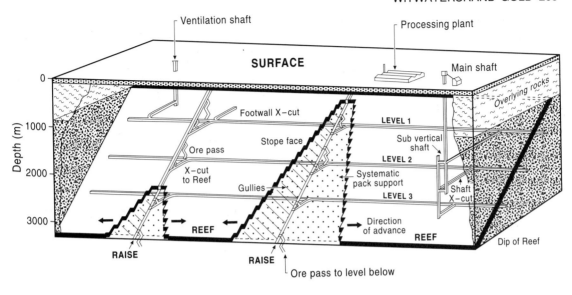

Fig. 13.8 Schematic diagram showing a typical Witwatersrand mine. (From International Gold Mining Newletter April 1990.)

competitive nature of exploration means that options are often not easy to obtain and option costs may rocket as soon as news leaks out of a company's interest. In many cases the mineral rights have been divided into various shares and agreement is required by all parties. However further subdivision is not possible since 1991, a very positive development of the 1991 mining act.

13.3.4 Mining methods

The generally narrow mining widths, availability of cheap labour, moderate dip of the reefs and great depths of most mines have led to the development of a very specific mining method, known as breast stoping. In this method (Fig. 13.8), development is kept in the footwall of the reef and a raise is cut between levels. From this raise gullies are cut at a 45° angle to the raise and the stope developed in a Christmas Tree manner. The stopes advance by jackleg drilling and blasting of a 2 m length of stope on each shift. Ore is scraped out from the advancing face along the gully using mechanical scrapers and the stope is cleaned carefully, often using high pressure jets, as much gold is contained in dust in the cracks in the floor of the stope. The scraper transfers the ore to footwall boxholes whence it is trammed to the main ore pass and the primary underground crusher. If the stope takes in more than one level then it is known as a longwall. The hanging wall of the advancing stope is supported by hydraulic props, later reinforced by timber packs.

Mechanized mining of narrow stopes is not easy and mechanization has been most successful in wide stopes, such as those in the Elsburg reefs on the West Rand, which are often 2 m or thicker. This was first introduced in 1984, and now produces about 65% of JCI's tonnage. Trackless mining is seen as a way of reducing the working costs, using specialized machinery instead of the labour intensive conventional methods. Trackless mining involves the substitution of tracks, locomotives and railcars by free-steering, rubber-tyred (or caterpillar tracked) equipment for drilling, loading and tramming of the ore, waste, men and materials. The substitutions involved are shown in Table 13.1.

The potential benefits of such conversions are lower operating costs, higher productivity, greater flexibility and improved grade control and safety (particularly in stopes, on haulages and declines and at grizzlies). There is a concomitant change in the work force, with the large semi- or unskilled work force replaced by a much smaller, highly skilled work force.

There are a number of variations on the way trackless mining is carried out, but the general use of room (bord) and pillar mining using Tamrock and Atlas Copco drilling rigs and LHD vehicles is common to most. There are two stages of production, namely primary development and stoping followed by secondary extraction of pillars.

Table 13.1 Equipment substitution in trackless mining (LHD = Load, haul, dump vehicles)

Operation	Conventional mining	Trackless mining
Drilling	Pneumatic and hydraulic hand-held drills	Hydraulic drill rigs
Stope cleaning	Scraper winches	LHDs
Gully cleaning	Gully winches	Trucks or LHDs
Tramming	Locomotives and railcars	Trucks or LHDs
Light haulage	Locomotives and railcars	Utility vehicles

During primary development and stoping, access ramps are driven on reef between two levels approximately 150 m apart. Stoping commences with the development of rooms approximately 17 m apart off the access ramp. Pillars 10 m × 7 m are left between rooms. Drilling of blast holes is by means of boom drill rigs. The reef is blasted out into the heading making the broken ore accessible to the LHD vehicles. Seventy-one per cent of the ore is extracted during this stage of mining, the remaining 29% is left in the pillars. Secondary partial extraction of the pillars takes place once primary stoping is complete. The pillars are reduced to a final size of 5 m × 5 m. The area is supported by roof bolts, and the broken ore is again removed by LHD vehicle.

A major consideration at great depths is rock stability. The quartzites have considerable strength and deform in a brittle manner, causing rockbursts (explosive disintegration of rock). These events cause disruption to production and are a major cause of fatal accidents. Rockburst frequency can be reduced by back filling of stopes and careful monitoring of stress build ups.

13.4 EXPLORATION MODELS

Exploration has been carried out on a variety of scales ranging from delineating extensions of known pay shoots within a mine to the search for a new goldfield. As most of the shallow areas have been explored one of the key decisions to make before optioning land is to decide the depth limit of economic mining and exploration. The deepest mining planned in detail at present is about 4000 m but options have been taken over targets at up to 6000 m and holes drilled to 5500 m. According to Viljoen (1990) good intersections have been made at 5000 m and although there is of course a stratigraphical limit, mineralization is present at great depths. The major constraints are the stability of mine openings, rock temperature (at 6000 m it is likely to be >85° C) and the costs of mining at these depths. At present, temperature problems are overcome by circulating refrigerated air and chilled water and amounts of these have to be increased substantially at 5000–6000 m. Currently the deepest single lift shaft is less than 3000 m and mining to 5000–6000 m would require two further shafts underground, termed secondary or sub-vertical shafts and tertiary shafts. Therefore every tonne of ore and materials would have to be transhipped at least twice. At 3000 m depth it can take a miner one and a half hours to reach the mining face and consequently production time is reduced. Mining at 6000 m may require a revision of shift patterns.

Most exploration has taken place within the known extent of the Witwatersrand Basin although there have been a number of searches for similar rocks outside it. The latter search has particularly concentrated on areas where lithologies similar to those in the basin are known either in outcrop or at depth. The possibility of West Rand Group correlatives is usually based on magnetic anomalies and the exploration model is that of a basin within mineable depth and preferably closer to postulated source areas. Examples of Witwatersrand lithologies at depth are known from deep drill holes and from surface outcrops in the Eastern Transvaal and Northern Natal.

Within the basin, exploration is aimed at detecting the continuation of known reefs or new reefs within known goldfields or of detecting new goldfields in the intervening (gap) areas. The 1980s saw a great deal of exploration in the gap areas based on the premise that previous deep drilling has missed potential goldfields due to lack of understanding of the structure or controls on sedimentation. The fundamental information required is an understanding of the subsurface stratigraphy and the potential depths of the target. Upthrown blocks of the Central Rand Group are also a target.

The detection of extensions to known goldfields, either down dip or laterally, is a less difficult task as considerable information is available from the other mines. In this case it is possible to collate the grade distribution within an individual reef and project payshoots across structures as shown in Fig. 13.9. In this example sedimentation is controlled by synsedimentary

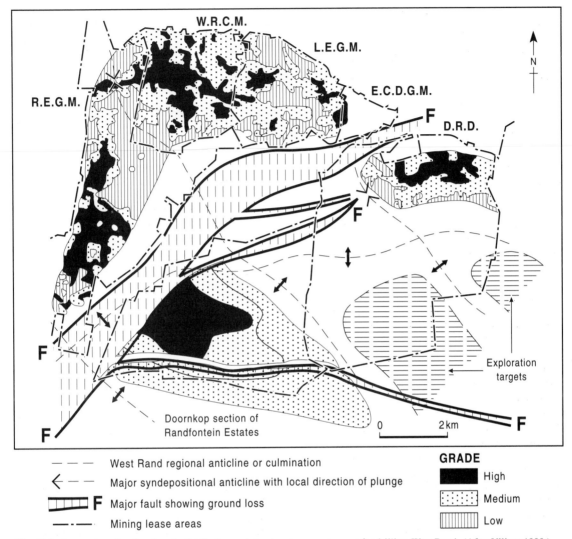

Fig. 13.9 Integration of regional grade distribution and structure to generate areas for drilling, West Rand. (After Viljoen 1990.)

structures with the highs forming areas of non-deposition or of low grade. The target areas in the structural lows can be clearly established by structural contours and isocons of grade distribution.

In 1990, JCI were spending 5% of their exploration budget on the search for new basins, 7% on exploration for new goldfields, 12% on basin edge deposits and 26% on down dip extensions. The remaining 50% was spent in the search for extensions to known mines (Scott-Russell *et al.* 1990).

Environmental constraints on exploration have, in the

past, been minimal as the Witwatersrand Basin mainly underlies flat farmland, largely used for maize and cattle production. In spite of the obvious problems generated by the extensive use of cyanide, there has been little concern about pollution. Forstner & Wittman (1979) however demonstrate that there is major contamination of surface waters associated with oxidation of pyrite and that high Zn, Pb, Co, Ni and Mn are present as a result of this oxidation and the cyanidation process. Cyanide itself is not a problem as it breaks down under the influence of ultraviolet light and levels are within those

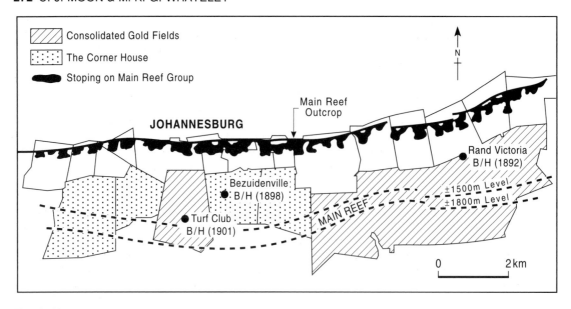

Fig. 13.10 Exploration and mining on the Main Reef, about 1900. (After Viljoen 1990.)

specified in the South African Mines and Works Act. Also of immediate concern has been the dust generated by the erosion of the slimes dams. This can be countered by revegetation (Friend 1990). The rehabilitation of surface workings has now been given legal force in the Water Act and the Minerals Act of 1991.

13.5 EXPLORATION METHODS

The original discovery of gold-bearing conglomerate was made in 1886 on the farm Langlaagte by two itinerant prospectors, George Walker and George Harrison, who were resting on their way to the Barberton greenstone goldfield (Werdmuller 1986). While engaged in building a house for the owner they went for a walk and Harrison, who had been a miner in Australia, recognized the potential of the iron-rich conglomerate. He crushed and panned a sample that gave a long tail of gold. They were no doubt encouraged to prospect by the occurrence of vein gold deposits in the surrounding area, notably in Transvaal Supergroup sediments.

13.5.1 Surface mapping and down-dip projection

After the original discovery the surface exposures were quickly evaluated by surface prospecting and the outcrop of the conglomerates mapped. In contrast to some

expectations the conglomerates were not a surface enrichment but continued down dip. However this presented a major problem as the surface iron oxides turned to pyrite at around 35 m and, as all prospectors knew, this meant poor recoveries of gold by the recovery process then used, amalgamation of the gold with mercury. Fortunately the cyanide process of extraction had been developed by three Scots chemists in 1887 and was immediately demonstrated to be effective. In contrast to ores from greenstone belts, Witwatersrand ore gives good recoveries due to the presence of gold as small free grains with very little or no gold contained in arsenopyrite, pyrite or tellurides. The first major trial of cyanidation on tailings from the amalgamation process from the Robinson Mine yielded 6000 oz (190 kg) of gold from 10 000 tons. The success of cyanidation prompted new interest in the possible depth continuation of the reefs. This depth continuation was tested by drilling and the first drill hole in April 1890 intersected the South and Main reefs at 517 ft (157 m) and 581 ft (177 m) with grades of 9 oz 12 pennyweights (dwt) t^{-1} (328 g t^{-1}) and 11 dwt t^{-1} (18.8 g t^{-1}) per ton. Further drill holes, notably the Rand Victoria borehole (Fig. 13.10) suggested in 1895 that the reef decreased in dip from the 60–80° at outcrop to 15° at 2350 ft (715 m) and that grades persisted at depth. The large potential of the area down dip was recognized by the mining houses with capital to finance mining at depth and much of the area down dip was pegged by Goldfields and Corner House companies (Fig. 13.10). This down dip area provided the basis for a number of major operations of which the largest, Crown Mines, eventually produced 163 Mt of ore and 1412 t of gold, an average grade of 8.7 g t⁻¹ Au.

For the forty years after the initial discovery exploration and mining concentrated on the areas down dip from the outcrop of conglomerates on the Central, West and East Rand. Exploration involved drilling and development down dip from outcrop; by 1930 the Village Main Mine had reached 710 m below collar. In addition, although the average reef was payable, it was noticed that there were considerable variations in grade. Pioneering studies by Pirow (1920) and Reinecke (1927) demonstrated that these higher grade areas formed payshoots and could be correlated with larger pebbles and thicker reefs. Reinecke proposed that the control on the payshoots was sedimentary and paralleled the direction of cross-bedding.

13.5.2 Magnetics and the West Wits Line

The next stage in the exploration history was the search for blind deposits under later Transvaal and Karoo cover to the west of the West Rand Fault which forms the western limit of the West Rand Goldfield. Although a number of geologists were convinced of the continuation of the Main Reef to the west of the known outcrop and it had been tested by drilling in the period 1899–1904, results were unconvincing. A shaft sunk in 1910 to test these reefs had to be stopped at 30 m due to an uncontrollable inflow of water from the Transvaal dolomites.

Wildcat drilling was prohibitively expensive and the search for the subcrop of the Main Reef was an unsolved puzzle for the period 1910–1930. The solution to this problem was found by a German geophysicist, Rudolph Krahmann. One day in 1930 at a motor rally he noticed that the rocks of the West Rand Group contained magnetite and he thought that it would be possible to trace these shales under cover using newly developed magnetometers. After checking the theory over exposed West Rand Group sediments five major and four minor anomalies were found within the sequence (for fuller details see Englebrecht 1986 or the very readable account of Cartwright 1967). The depth penetration of the method was soon tested with backing from Goldfields of South Africa who had also acquired the rights to the cementation process that allowed shaft sinking through water-bearing formations. Finance also became more easily available with the increase in the gold price from £4.20 to £6 per ounce. The initial eleven drill holes (Fig.

13.11) down dip from the Middlevlei inlier not only intersected what was believed to be the equivalent of the Main Reef but a further conglomerate at the base of the Ventersdorp lavas, termed the Ventersdorp Contact Reef. These holes gave rise to the Venterspost and Libanon Mines, but the real excitement came with results further to the west between the major Bank Fault, subparallel and west of the West Rand Fault, and Mooi River (Fig. 13.11). Drilling on the farm Driefontein began with three holes designed to cut the Main Reef. The first two holes were to the north of the subcrop of the Main Reef against the dolomites while the third cut poor values within the Main Reef but intersected a carbon seam with visible gold some 50–60 m below the Main Reef. The significance of this was not appreciated at first as core recovery in the very fragile carbon seams was incomplete due to grinding of core. This carbon seam later became known as the Carbon Leader Reef. In the original calculations of viability a grade of 13.6 g t⁻¹ over 1.1 m was assumed, based on previous working on the Main Reef. When the first shaft was sunk by Rand Mines on the farm, Blyvooruitzicht, a small piece of ground that Goldfields did not own, the nature of the discovery soon became clear. Underground sampling around the shaft area in 1941 returned a grade of 69 g t⁻¹ Au over 70 cm compared with a drillhole grade of 23.1 g t⁻¹ over 55 cm. The mine management at first refused to believe the results and had the shaft resampled only to yield the same results. Similar grades were obtained when the GFSA mine at West Driefontein opened in 1952. Exploration to the west of the Bank break, along an area known as the West Wits Line, largely by GFSA, led to the development of ten mines. Although magnetics led to the establishment of the subcrop of the Main Reef, the structural complexity of the area was not appreciated until the 1960s when the angular unconformity between the Main Reef and the Ventersdorp lavas was described. In the north and east of the area, the upper parts of the Central Rand Group have been removed by pre-Ventersdorp erosion. In addition parts of the Carbon Leader have been removed by erosion and distinct barren areas can be recognized (Englebrecht *et al.* 1986).

When the importance of the finds on the immediate subcrop of the West Wits Line became apparent companies were quick to appreciate the down dip potential. In particular Anglo American Corp. made a deliberate decision in 1943 to explore for deposits at 3000–4000 m and obtained the mineral rights to the area now known as Western Deep Levels (Fig. 13.2). This is currently the deepest mine in the world and was based on twelve drill

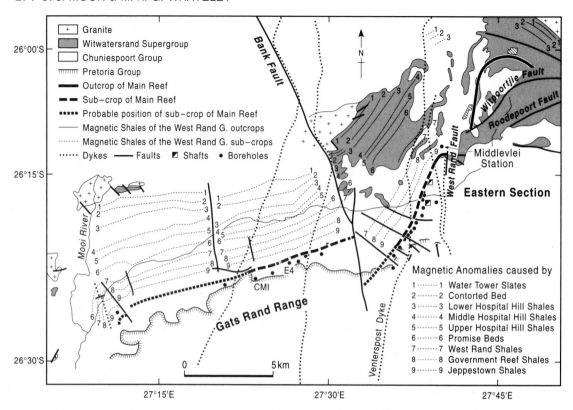

Fig. 13.11 Exploration on the West Witwatersrand line 1936 (after Englebrecht 1986). Drill hole E4 is the Carbon Leader discovery hole on the farm Driefontein, now (West) Driefontein Mine and CM1 the discovery hole on the farm Blyvooruitzicht.

holes with an average intersection in the Carbon Leader of 3780 cm-g t[-1] and additional mineralization in the VCR. Production in 1990 was 38.5 t of gold.

Although the magnetic method was outstandingly successful on the West Wits Line, detailed studies in other areas of the basin have encountered problems in interpretation. In particular the lateral and vertical facies variations of the beds are not understood. Magnetic anomalies can also be due to beds, other than Witwatersrand sediments, such as magnetite rich layers in basement (Corner and Wilsher 1989).

13.5.3 **Gravity and the discovery of the Orange Free State Goldfield**

The increase in the gold price in the early 1930s encouraged companies to explore areas even more remote than

the West Wits Line. In particular Anglo American made a discovery (now the giant Vaal Reefs Mine complex) in the Klerksdorp area, immediately north of the Vaal River, which forms the border between the Transvaal and Orange Free State provinces (Fig. 13.2). This discovery encouraged other companies, particularly Union Corporation to explore in this area and also to the south of the Vaal River, where it was theorized that the Central Rand Group would extend under cover. In the Klerksdorp area Union Corporation had demonstrated for the first time that gravity methods could be used to predict the depth of the Central Rand Group as their density was significantly lower at 2.65 kg m[-3] than the Ventersdorp lavas at 2.85 kg m[-3]. This technique was selected to improve interpretation south of the Vaal River.

The area to the south of the Vaal River, particularly that around the village of Odendaalrust (north of

Fig. 13.12 *(Opposite)* Gravity and magnetic anomalies over the St. Helena Gold Mine, OFS Goldfield. Note how the discovery hole (SH7) is on the flank of the main anomaly. (From Roux 1969.)

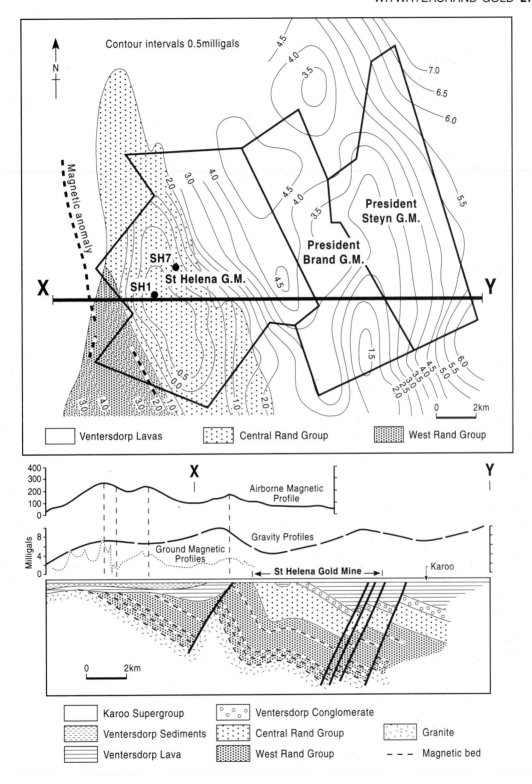

Contour intervals 0.5milligals

Magnetic anomaly

7.0
6.5
6.0
5.5
4.5
4.0
3.5

President
Steyn G.M.

President
Brand G.M.

SH7
St Helena G.M.
SH1

X — Y

0   2km

☐ Ventersdorp Lavas   ⠿ Central Rand Group   ▨ West Rand Group

400
300
200
100
0

X — Y

Airborne Magnetic Profile

Gravity Profiles

8

Milligals

4

Ground Magnetic Profiles

St Helena Gold Mine

Karoo

0   2km

☐ Karoo Supergroup   ⌾ Ventersdorp Conglomerate

▦ Ventersdorp Sediments   ⠿ Central Rand Group   ⬚ Granite

☰ Ventersdorp Lava   ▨ West Rand Group   – – – Magnetic bed

Wesselbron, Fig. 13.2), had attracted the attention of a number of prospectors as it had been demonstrated as early as 1910 that gold could be panned from conglomerates that are stratigraphically part of the Ventersdorp Supergroup. A small venture took up the submittal from the original prospector and drilled a hole on the farm Aandenk to test the conglomerates. Although it was obvious from the beginning that the conglomerates were stratigraphically part of the Ventersdorp Supergroup, the venture persisted and intersected the top of the Central Rand Group quartzites in 1934 at a depth of 2726 ft (831 m). The hole eventually bottomed at 4046 ft (1234 m) with a best intersection of 120 in-dwt t^{-1} (521 cm-g t^{-1}) when they ran out of funds.

On the basis of the results from the Aandenk borehole and the success of the gravity method a company, Western Holdings, was floated by South African Townships to prospect in the Southern Free State in collaboration with Union Corporation. Intensive geophysical surveys showed a significant gravity anomaly over the farm St Helena flanked to the west by magnetic anomalies (Fig. 13.12). The first hole SH1, drilled in 1938, indeed cut Central Rand group quartzites at 302 m and was drilled for another 1850 m cutting three reefs at 200–300 in-dwt t^{-1} (860–1300 cm-g t^{-1}). Further drilling to the west confirmed the location of the magnetic shales. By this stage a number of other companies were drilling in the area but the stratigraphical control of the mineralization was unknown. After routine assaying of a conglomerate one of these companies obtained an assay of 6.9 g t^{-1} over 1.05 m in what is now known to be the major gold bearing conglomerate. Within a few months Western Holdings intersected payable values of the same reef in hole SH7 at 59 g t^{-1} over 1.5 m at a depth of 348 m. These were the first intersections of the Basal Reef (Fig. 13.3), which has been the main producing placer in the Orange Free State Goldfield. Proving the exact positions of the reefs was however difficult as the goldfield is cut by a number of major faults and interpretation of gravity data was not as simple as first thought. Gravity anomalies could reflect the depth of the Central Rand Group quartzites but they also could be due to faulting, folding and changes in the lithology of the Ventersdorp Volcanics. However experience showed that the Central Rand Group anomalies could be recognized by their sharp gradients whereas changes due to the Ventersdorp were much more gradual. Further exploration and exploitation of the goldfield was delayed by World War Two but resumption of exploration after the war confirmed the potential of the area including a phenomenal intersection on the farm Geduld of 3377.5

dwts t^{-1} (5775 g t^{-1}) over 6 inches (15 cm). Although not the active partner in the original exploration venture, Anglo American Corp. became the dominant operator in the OFS goldfield by taking over Western Holdings and several other companies immediately after 1945.

Similar techniques were successful in the discovery by Union Corporation of the Evander Sub-basin, about 100 km ESE of Johannesburg (Fig. 13.2). Although the area had been known to contain concealed Central Rand Group rocks it was only the advent of airborne magnetic and ground gravity surveys that unravelled the structure. Persistence with a drill programme eventually proved that the Kimberley Reef was the only payable reef and led to the establishment of four mines (Tweedie 1986).

13.5.4 Reflection seismics, gaps and upthrown blocks

One of the major advances of the 1980s was the use of seismic reflection surveys in detailing the structure and location of the Central Rand Group rocks at depth. Although use of the method was mooted by Roux in 1969, it was the impact of improved technology and lack of work in the oil industry that prompted a renewed trial by oilfield service companies in collaboration with the major mining houses. Initial logging of boreholes showed that there was a contrast in seismic velocities as confirmed by initial surface surveys (Pretorius et al. 1989). Although it is not possible in reconnaissance surveys to define the individual reefs, the surveys are able to define major contacts, particularly the base and top of the Central Rand Group and the contact between the Transvaal dolomites and the Pretoria shales overlying them. By 1989 Anglo American had covered the basin based on more than 10 000 line-km of 2-D survey and further coverage concentrated on the detailed follow up of target areas. Other mining houses use seismic surveys extensively and JCI (Campbell & Crotty 1990) have used 3-D surveys in planning mining layouts and the location of individual longwall faces at their South Deep prospect (section 13.7.2). The cost of 3-D surveys limits their use; Campbell & Crotty report that the survey at the South Deep prospect cost R2.2M, the cost of 1½ undeflected drill holes.

Pretorius et al.(1989) defined three different types of seismic survey used by Anglo American during the 1980s.

1 Reconnaissance 2-D surveys of stratigraphy and structure on public roads concentrating on the gap areas where the target beds are relatively shallow.

2 Reconnaissance 2-D of deeper areas within the basin

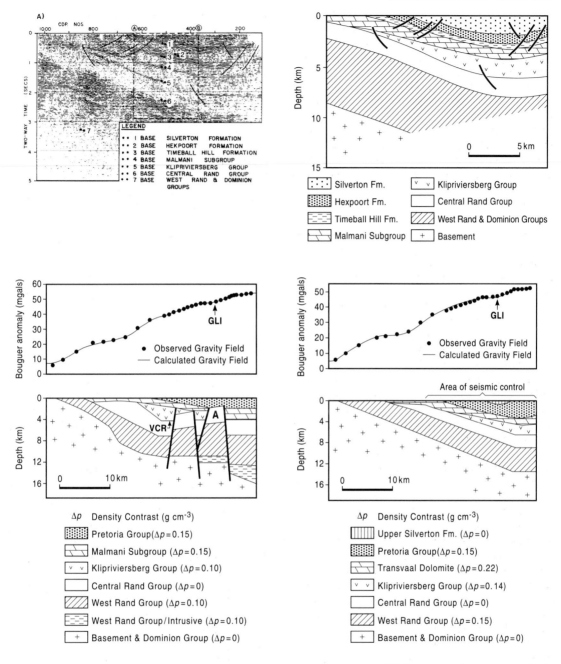

Fig. 13.13(a) Use of reflection seismics in discriminating between two possible gravity models for targets on the western margin of the Witwatersrand Basin. The left-hand model used a block of upthrown Central Rand Group (A) to explain the gravity low while the right-hand model, confirmed by seismics, invoked near surface Transvaal sediments. (From Pretorius *et al.* 1989 with permission.)

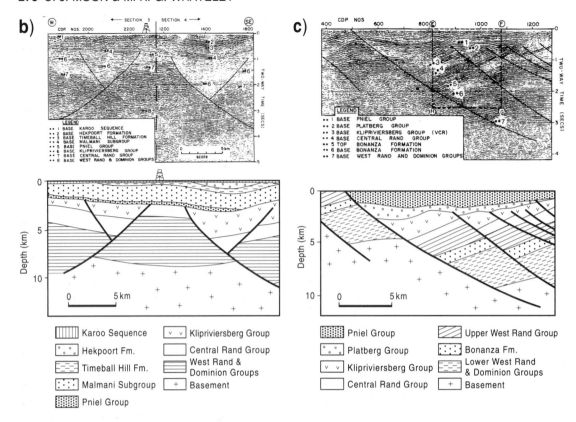

Fig. 13.13(b) and (c) Typical uses of seismic reflection data. (b) detection of upthrown blocks and (c) detection of unsuspected upthrown blocks adjacent to a lease area. (From Pretorius *et al.* 1989.)

where targets might be upthrown to mineable depths by faulting.

3 Detailed 2-D surveys of extensions to reserves on existing mines.

Few examples of seismic sections have yet been published but the paper by Pretorius *et al.* (1989) provides examples of the use of the technique in reconnaissance exploration. The area shown in Figure 13.13a is on the western edge of the basin and the seismic survey clearly delineates the upper and lower contacts of the Central Rand Group and faulting within the Transvaal sediments. The projected depth of the VCR at the upper contact is over 5000 m and drill targets can be assigned on this basis. The contrast between the seismic section and the previous interpretation of the gravity data is striking (Fig. 13.13a) In the original gravity interpretation the 3 mgal anomaly superimposed on the strong regional gradient was attributed to a horst of Central Rand Group sediments at a depth of around 2 km whereas the seismic section showed that there was no

major faulting, that the Central Rand target was at 5 km and the gravity low was the result of changes in lithology of the Pretoria Group sediments. Thus seismic and gravity data can be successfully integrated.

Figure 13.13b shows the result of a reconnaissance seismic reflection survey in the middle of the basin that detected an upthrown block of Central Rand Group sediments that are at mineable depth (<3000 m); this was confirmed by later drilling.

The third type of survey is exemplified by the seismic survey which defined the structure at the margins of an established mine (Fig. 13.13c). From this section an additional area of Central Rand Group sediments at mineable depths was established.

13.5.5 Drilling

One of the major financial constraints on exploration is drilling. A 6000 m deep hole drilled from surface may take 18 months to complete and cost >R4M. A typical 15

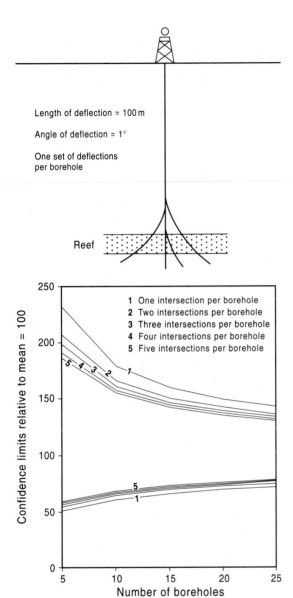

Length of deflection ≃ 100 m

Angle of deflection ≃ 1°

One set of deflections per borehole

Reef

1 One intersection per borehole
2 Two intersections per borehole
3 Three intersections per borehole
4 Four intersections per borehole
5 Five intersections per borehole

Fig. 13.14 Schematic diagram of deflections and plot of confidence limits for increasing number of drill holes and deflections per hole. There is considerable variation between placers, the case shown is the Vaal Reef. (From Magri 1987.)

hole programme on a lease of 7000 ha at 3–4000 m target depth would cost around R18M and take twenty years to complete if the holes were drilled sequentially. This poses considerable problems, not only of capital, but of decision making as mineral rights options are only three or five years in length and a decision to exercise the option must be taken at the end of the period. Clearly programmes must have several rigs drilling simultaneously or drilling must be accelerated. Each drill hole will be deflected to provide a number of reef intersections and improve confidence in grade estimation (Fig. 13.14, Table 13.2)

Perhaps the key decision to be made is approximately how many intersections need to be made before a confident estimate can be made of a lease's value or lack of prospectivity. A quantitative approach to the problem has been given by Magri (1987), whose results are summarized in Fig. 13.14. Using an approach similar to Krige's t estimator to predict the mean of log-normally distributed data, he has produced confidence limits of estimates of the mean grade for reefs of immediate interest to his company (Anglovaal) based on knowledge of the grade distribution in mined blocks. These clearly show that making further drill intersections after a certain cut-off, which varies for each reef, provides little extra information. In the example shown this is twelve or thirteen. In the same way it is possible to optimize the number of deflections from each drill hole, usually four are made. Magri suggested that the optimum approach to drilling an area is to decide on a minimum number of holes necessary to give a picture of an area, say 10 for a 7000 ha lease and review the data when drilling is complete. If the grade of the mean less confidence limit for the prospect is greater than the pay cut-off, then the programme should be continued. If not, it should be abandoned.

The application of drilling rigs used for oil and gas exploration has helped to speed exploration and to increase the accuracy of deflections from the mother hole.

13.5.6 Underground exploration

Considerable underground exploration is necessary before a mining layout can be designed that will minimize losses of production from faulting and dykes. Fig. 13.15 shows the layout of a typical underground drilling programme in a faulted area.

13.5.7 Re-treating old mine dumps

One of the most significant innovations on the Witwatersrand has been the major projects to re-treat old slimes dams and sand tips. The reasons for the occurrence of the gold in waste are many; gold may be enclosed in pyrite or cyanide-resistant minerals that were not crushed or the operation of the plant was not

Table 13.2 Costs of drillholes and deflections, in 000 Rand 1987 value (1R=US$0.3) (Magri 1987)

Depth	Number of deflections						
	1	2	3	4	5	7	9
1000	93	118	131	144	158	188	220
1500	150	176	190	205	220	253	288
2000	220	249	267	286	306	348	393
2500	303	337	359	382	407	457	511
3000	408	446	473	501	530	592	658
3500	545	590	624	660	698	777	862
4000	714	768	812	858	905	1004	1111
4500	907	962	1007	1053	1102	1204	1315

optimal for the mineralogy treated. The average gold content in the waste is probably about 0.3 ppm but waste from the older mines on the East and Central Rand contains approximately 0.7 ppm. Advances in metallurgical techniques and increases in the gold price have allowed profitable reworking of this waste and by 1986 about 11 t of gold per year were produced from this source.

13.6 RESERVE ESTIMATION

Reserve estimation on a gold mine occurs in two phases, namely those resources calculated during exploration and those reserves estimated during exploitation. Ore reserves calculated during exploitation refer to those blocks of ore that have been developed sufficiently to be made available for mining. Usually, a raise would have been driven and sampled. The samples selected for use in the reserve calculation are those which are above the cut-off assigned according to the mine's valuation method (Janisch 1986, Lane 1988). The valuation of a developed reserve block has been traditionally carried out using the average grade of all the samples taken along the raises and the drives. This is steadily giving way to kriging methods (Krige 1978, Janisch 1986).

In exploration, resource estimation is not so precise. It is essential to identify the blocks which could be mined. These blocks would most probably be bounded by faults. Hence the importance of establishing the structure of an area from the borehole information, using structural contour maps and a series of cross sections. More recently this information has been supplemented with information from 2-D and 3-D seismic surveys (Campbell & Crotty 1990).

The average thickness and grade for each block must be calculated. Each borehole has an average of four deflections which intersect the reef, often in random directions (Fig. 13.14) and this gives an unbiased sampling of the reef within a radius of about 5 m of the original hole. When an inclined hole intersects a dipping reef, only an apparent dip of the reef can be seen in the core, so corrections must be made, using the following formula, to calculate the true thickness, to be used in the tonnage calculations.

$$W = \cos b \ \{C(\operatorname{Sin} d - [\cos (90-e) \tan b]\}$$

where W= true reef thickness, b= reef dip, C= apparent reef thickness, d= borehole inclination, and e= azimuth of reef dip.

A Sichel's t average grade value (Sichel 1966) is calculated for each cluster of acceptable sample values. Each value thus calculated is either assigned to the block into which the borehole falls (modified polygonal method; Chapter 9), or the averages of all the holes, also calculated using the Sichel's t estimator, is used as the average grade of the reef for the entire area being assessed.

Usually a mine has a minimum mineable reef thickness, often of 1 metre. Where reef intersections are less than 1 m the grade is recalculated to reflect the additional waste rock that would have to mined to achieve the minimum mineable stoping height and this height (in this example, 1 m) is used in tonnage calculations. Often, to establish whether the minimum mining height

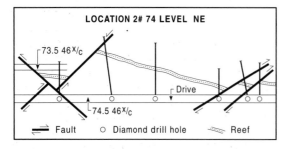

Fig. 13.15 Section showing typical underground drilling to determine reef location and fault throws, Vaal Reefs Mine. The length of the section is approximately 70 m . (After Thompson 1982.)

is going to be economic to mine, an accumulation or grade thickness product is calculated, such that $a_x = W_x \times g_x$. This accumulation is the product of the true thickness W in the sample at x and the average grade, g_x. In the gold mines, thickness is quoted in centimetres, the grade in g t^{-1}, and the product in cm g t^{-1}. Where a sample is shown to have 35 g t^{-1} over 10 cm, the accumulation is 350 cm g t^{-1}. To recalculate the grade over the minimum mining width of 100 cm (1 m) the accumulation is divided by the mining width and in this case the grade over 100 cm is 3.5 g t^{-1}. If the cut-off grade for this particular mine happened to be 3 g t^{-1}, then this block would be considered payable and the block included in the reserve calculations using the 100 cm mining width. The tonnage is calculated for a block which has a borehole at the centre using the following formula:

$$M = W \times A \times r$$

where M = mass in tonnes, W = average true thickness of the reef (with a minimum of 1 m), A = area of the block and r = specific gravity.

The tonnage for a particular area is then calculated by summing the tonnage for each block.

13.7 EXPLORATION AND DEVELOPMENT OF THE SOUTHERN PART OF THE WEST RAND GOLDFIELD

As discussed in section 13.4, the easily discovered, near surface deposits have been known, and often exploited, for many years. The object of more recent exploration is to find the deeper, more stratigraphically, sedimentologically and structurally complex deposits. In areas with active mining, careful analysis and reinterpretation of geological data, including data supplied in company annual reports, can lead to new discoveries being made in areas previously thought to be barren. The area discussed in this case history is the southern extension of the West Rand Goldfield mined immediately down dip from the outcrop of the Central Rand Group. The original mines are Randfontein Estates, West Rand Consolidated, Luipardsvlei and East Champ d'Or (Fig. 13.16). Four deposits have been discovered to the south of these mines and to the east of a major structural break known as the Witpoortjie Horst. They are the Cooke and Doornfontein sections of Randfontein Estates, Western Areas and the South Deep prospect (SDPA). The exploration of Cooke Section will be discussed in detail as it is one of the few detailed, published, exploration case histories for the Witwatersrand (Stewart 1981) and that of South Deep as it shows the application of methods

developed in the 1980s.

Development of Cooke Section only started in the mid 1970s after almost a decade of exploratory drilling, but nearly ninety years after the discovery of gold in the region. Explanations for this late discovery are that these gold-bearing rocks are concealed beneath younger sediments, the structure is more complex than in other areas, the gold-bearing conglomerates are highly variable and these conglomerates are in a part of the sequence that had not before yielded economic quantities of gold. This latter point, in conjunction with the fact that the conglomerates which normally produced gold in other parts of the basin were not here considered to be economic, meant that other areas were explored in preference to the Cooke Section.

13.7.1 Cooke Section—previous exploration

The area had been geologically mapped by Mellor (1917). This map formed the basis of exploration until 1962 when Cousins extended this mapping southward. Mapping (Fig. 13.16) showed that Witwatersrand Supergroup rocks crop out with an approximate E–W strike from the East Rand through Johannesburg to Krugersdorp in the west (Fig. 13.2). In the west the strike of the Witwatersrand rocks starts to swing to the south-west, but the outcrop is covered by younger sediments of the Transvaal Supergroup and the Witwatersrand rocks are upthrown on to a horst, the Witpoortjie Horst, by major faulting (Fig. 13.17).

Attempts were made to trace extensions of the West Rand by drilling from as early as the beginning of this century and resulted in the recognition of the West Wits Line (section 13.5.2). Further drilling was carried out in the area to the south of Randfontein Estates between 1915 and 1950 (Fig. 13.18). This work showed that the Witpoortjie Horst had a nearer N–S strike than previously thought. This drilling also provided more detailed information about the variations in the generalized stratigraphical column (Fig. 13.19). For example, the Main Reef zone was expected to be highly productive, but nothing significant was intersected in the Cooke Section. Unfortunately, the Middle Elsburg Reefs, which were subsequently shown to be the main gold-bearing reefs of this area, were only poorly developed in these holes. One hole was drilled in what is now known to be a high grade part of Cooke Section but did not penetrate the reef as it was cut out by a dyke.

Ten holes were drilled in the early 1950s (Fig. 13.18) in Cooke Section. Some were intended to obtain infor-

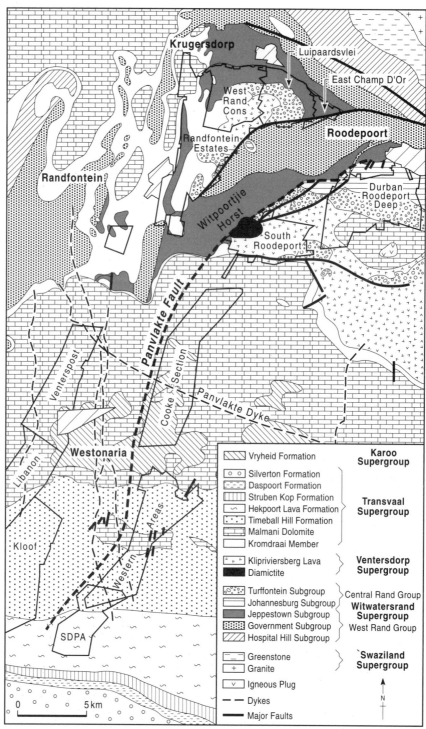

Fig. 13. 16 Surface geology of the West Rand. (From Stewart 1981.)

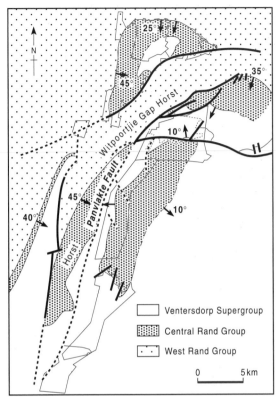

Fig. 13.17 Sub-Transvaal geology of the West Rand. Note the absence of Central Rand Group sediments on the Witpoortjie Horst (From Lednor 1986.)

mation about the potential of the Main Reef zone on the horst while two of the holes were drilled to test the potential of the reefs on the eastern, downthrown side of the Witpoortjie Horst. Holes were also drilled farther south, and as they immediately intersected good gold values in the Elsburg Reefs, on what is now Western Areas Gold Mine, further drilling on the Cooke Section was deferred until 1961.

Before the 1950s, borehole cores were only assayed for gold content. With the advent of interest in uranium in the 1950s, the possibility of extracting the uranium in addition to the gold was considered. Sampling and assaying of conglomerates for uranium showed that the increase in uranium price had made the uranium an attractive proposition which could make a deposit with marginal gold values, payable. The existing data were reviewed and the results of one of the holes (GB2) drilled in the downthrown block, showed gold and uranium values which were considered worth following up (Fig. 13.18).

GEOPHYSICS

Gravity and airborne and ground magnetic surveys were carried out across the Cooke Section to obtain data that would assist in the interpretation of the structure. Despite the success of these methods elsewhere in the Witwatersrand Basin (section 13.6) neither method proved directly successful in this area and drilling was used to further evaluate the Cooke Section. The density difference between the Witwatersrand quartzites and the overlying Ventersdorp lavas was masked by the thick sequence of younger Transvaal Supergroup dolomites, which made the gravity method unsuccessful. Magnetics were used to try to establish the presence of the magnetic shale in the West Rand Group. Unfortunately the combined thickness of the overlying Central Rand Group and Transvaal Supergroup rocks was sufficient to effectively mask the magnetic signal from these shales.

DRILLING

One hole drilled near the hole GB2 had revealed promising gold grades. This new hole intersected 1.32 m of Elsburg Reef at 968 m which assayed 20 g t^{-1} Au and 690 ppm U_3O_8 . This success led to the planning of a full drilling programme in which 88 holes were completed between 1960 and 1970 (Fig. 13.18). The holes, drilled on a grid, were spaced approximately 800 m in a N–S (strike) direction and as close as 300 m in an E–W (dip) direction.

At that time it was standard practice to drill these holes using narrow diameter (NX or BX), diamond bits. The unconsolidated overburden was drilled using larger diameter, percussion rigs (also known as drilling a pilot hole). Pilot holes are now drilled to significant depths, with much time and cost saving (Scott-Russell *et al.* 1990). The pilot hole would have steel casing inserted to prevent collapse of the unconsolidated material into the hole. Once coring commenced in the hard rock, this continued until the final depth was reached and a complete section of core would be available for the geologist to log. To save time and expense, four deflections were usually drilled from the original hole to obtain additional reef intersections. On average each hole took 4–5 months to drill.

All the core was taken to a central location were it was logged. After logging, most of the Transvaal Supergroup dolomites and the Ventersdorp Supergroup lavas were discarded, but the Witwatersrand Supergroup rocks were stored, in case they were needed at some future date for further study. The entire succession of

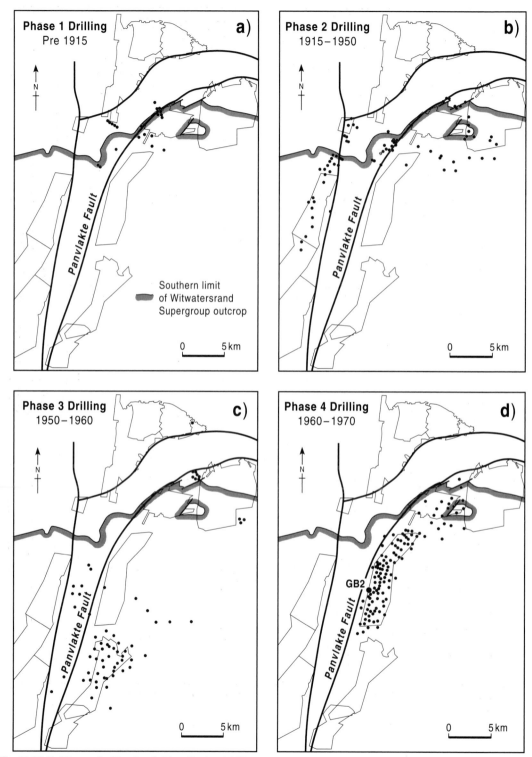

Fig. 13.18 Drilling on the West Rand. (From Tucker & Viljoen 1986.)

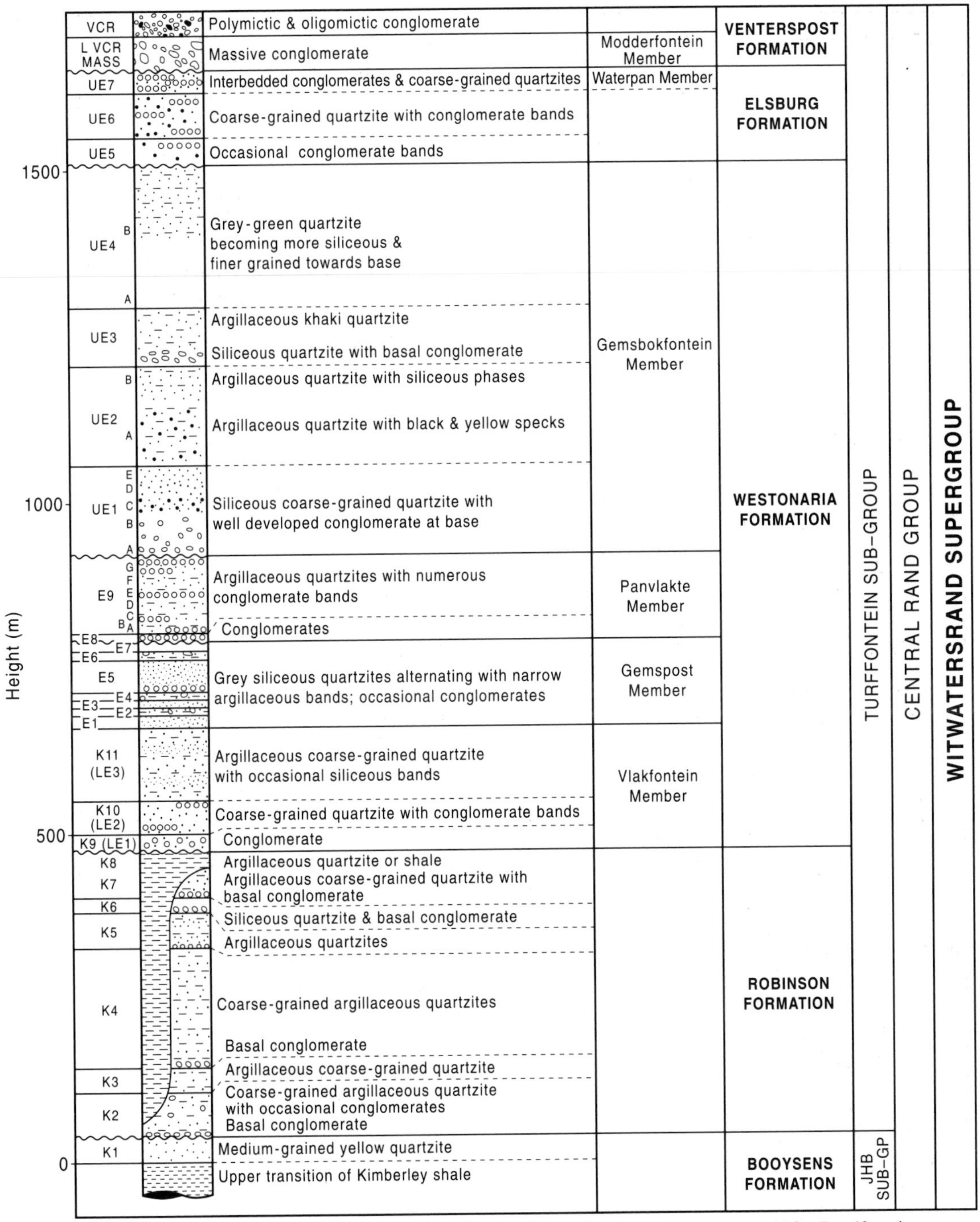

VCR	Polymictic & oligomictic conglomerate		**VENTERSPOST**	
L VCR MASS	Massive conglomerate	Modderfontein Member	**FORMATION**	
UE7	Interbedded conglomerates & coarse-grained quartzites	Waterpan Member		
UE6	Coarse-grained quartzite with conglomerate bands		**ELSBURG FORMATION**	
UE5	Occasional conglomerate bands			
UE4 B / A	Grey-green quartzite becoming more siliceous & finer grained towards base			
UE3	Argillaceous khaki quartzite / Siliceous quartzite with basal conglomerate	Gemsbokfontein Member		
UE2 B / A	Argillaceous quartzite with siliceous phases / Argillaceous quartzite with black & yellow specks			
UE1 E D C B A	Siliceous coarse-grained quartzite with well developed conglomerate at base		**WESTONARIA FORMATION**	
E9 G F E D C B A	Argillaceous quartzites with numerous conglomerate bands / Conglomerates	Panvlakte Member		
E8 E7 E6				
E5 E4 E3 E2 E1	Grey siliceous quartzites alternating with narrow argillaceous bands; occasional conglomerates	Gemspost Member		
K11 (LE3)	Argillaceous coarse-grained quartzite with occasional siliceous bands	Vlakfontein Member		
K10 (LE2)	Coarse-grained quartzite with conglomerate bands			
K9 (LE1)	Conglomerate			
K8	Argillaceous quartzite or shale			
K7	Argillaceous coarse-grained quartzite with basal conglomerate			
K6	Siliceous quartzite & basal conglomerate			
K5	Argillaceous quartzites			
K4	Coarse-grained argillaceous quartzites / Basal conglomerate		**ROBINSON FORMATION**	
K3	Argillaceous coarse-grained quartzite			
K2	Coarse-grained argillaceous quartzite with occasional conglomerates / Basal conglomerate			
K1	Medium-grained yellow quartzite / Upper transition of Kimberley shale		**BOOYSENS FORMATION**	

Height (m): 1500, 1000, 500, 0

TURFFONTEIN SUB-GROUP · CENTRAL RAND GROUP · WITWATERSRAND SUPERGROUP · JHB SUB-GP

Fig. 13.19 Detailed stratigraphical column for the Central Rand Group in the Randfontein Estates area. (After Randfontein Estates Gold Mine *pers. comm.*)

Witwatersrand Supergroup rocks was then sampled, at 30 cm intervals where gold and uranium grades were expected, and at 60 cm intervals where no assay values were expected. Fire assaying was used to determine the gold content and radiometric methods were used to determine the U_3O_8 content, with suitable check assays undertaken both within one laboratory and between laboratories.

INTERPRETATION OF RESULTS

The unusual sequences intersected by the first few boreholes made it difficult to interpret the stratigraphy and the structure. Work commenced on compiling a detailed stratigraphical column relevant to this area. It was found that the initial logging had not been done in sufficient detail to compile a specific stratigraphical column. Cores were retrieved from the store on their trays and two holes were laid out side by side for re-examination. In this way detailed comparisons could be made and this was repeated many times. Marker bands, mainly fine-grained quartzites, were identified which could be used for correlation, and, after many months and many attempts, a stratigraphical column was produced which was tested against new holes. The difficulties can be gauged from Fig. 13.20 which shows the correlation between three holes (note the numerous unconformities). The process was further complicated by the presence of faults and dykes. Eventually a reliable stratigraphical column was generated and each hole could then be divided into reef zones. At this stage it was possible to study the grade and geometry of individual reef zones, using grade distribution and isopach maps, and to determine the structure of the area, using structure contour maps. The sedimentological controls on the gold and uranium distribution mean that additional data can be gained by comparing a reef isopach map with the overlying inter-reef interval (Fig. 13.21) to identify possible depositional trends, or by investigating pebble size, and conglomerate to quartzite or gold to uranium ratios.

The drilling showed that the cover of Transvaal Supergroup dolomites varied in thickness from 100 m at the northern boundary of Cooke Section to approximately 600 m in the south. Below the dolomites older rocks were faulted. The dominant feature is the N–S Panvlakte Fault, which forms the eastern boundary of the Witpoortjie Horst. This horst was active during the later part of Witwatersrand sedimentation. Most of the Witwatersrand rocks had been eroded from the horst during the pre-Transvaal peneplanation. To the east of

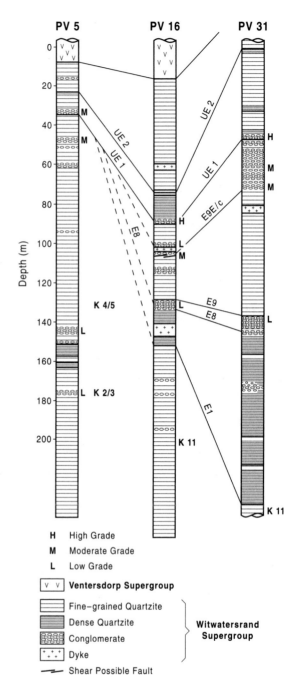

Fig. 13.20 Logs of three drill holes indicating the difficulties of correlating between holes. For the location of holes see Fig. 13.21. (After Stewart 1981.)

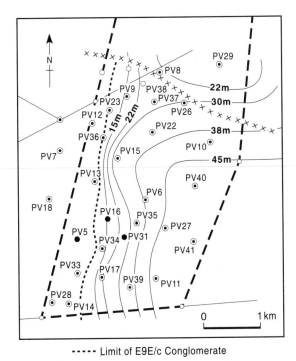

Fig. 13.21 Typical isopach map used to interpret drilling, in this case the E9c to UE1a unit. (After Stewart 1981.)

this fault, an almost entire succession of Witwatersrand rocks, and some Ventersdorp Supergroup lavas, were preserved on Cooke Section (Fig. 13.22). A gentle, N–S trending anticline exists, which causes the sediments to dip from just a few degrees up to 45° to the east.

Of all the recognized reefs on Cooke Section (Figs 13.20 & 13.22) only three composite units, or packages, are currently considered economically important, the UE1a–E9Gd, the E8 and the K7–K9 packages. The former is the most important reef package that is currently being mined with only minor stoping on the E8 and E9 reefs.

DEVELOPMENT OF COOKE SECTION

In early 1973, shaft sinking was completed on the Cooke Section No. 1 Main Shaft at a final depth of 867 m. Towards the end of the year stoping began in the area adjacent to the shaft. By the end of 1973 some 15 544 t of ore had been mined and hoisted to the surface. This material had an average grade of 17.7 g t^{-1} Au. Cooke Section No. 2 Main Shaft was commissioned in 1977, and in 1980 the Cooke Section No. 3 Shaft was started. It was finished, at a final depth of 1373 m, in 1983.

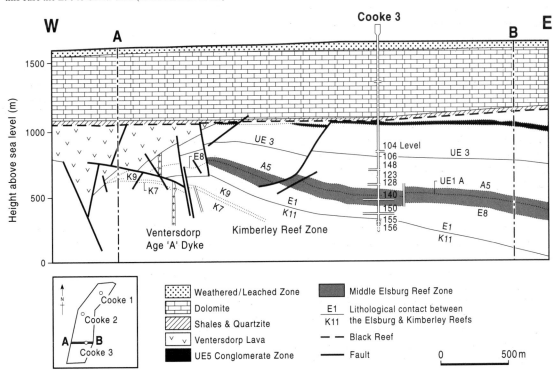

Fig. 13.22 Section through Cooke Section Mine. (After Randfontein Estates Gold Mine *pers. comm.* 1991.)

Cooke Section has evolved into a large relatively low grade producer. By 1992 around 5 Mt of ore were being mined at an average grade of 8 g t^{-1} of gold. The income from uranium has become of very minor importance with the decline in the price of that metal. Trackless mining now accounts for 70% of production and the mine is one of the most efficient on the Witwatersrand. Cooke Section has been a major success due to the nature of the reefs and its inception at a time of the sharp increase in the price of gold.

13.7.2 South Deep

This prospect is one of the major undeveloped deposits in the Witwatersrand Basin with resources of more than 1000 t of contained gold. *In situ* reserves, using cut-offs based on a $400 oz^{-1} gold price, are 115.9 Mt at 9.0 g t^{-1} Au (South Deep 1990). The potential of the area was first recognized during the drilling of Western Areas in the mid-1960s. However the depth (>2500 m) and the exploration of other prospects in the area delayed systematic exploration until the late 1970s. It was quickly recognized that the deposit was unusually thick and provided opportunities for mechanized wide orebody mining as well as conventional stoping. The reef packages of interest are the Massive and Individual Elsburg Reefs as well as the Ventersdorp Contact Reef (Figs 13.3 & 13.19). The thicker Massive and Individual Elsburg Reefs lie to the east of a NNE trending feature known as the upper Elsburg shoreline. The Individual Elsburg Reefs, that are up to nine in number, merge within 750 m of the shoreline to form a composite reef which can be up to 70 m thick. To the west of the shoreline only the Ventersdorp Contact Reef is developed. The initial exploration was by conventional drilling from the surface (Fig. 13.23) and about 200 intersections were used in the feasibility study.

In order to make evaluation more accurate a 3-D seismic survey was shot between August 1987 and September 1988. A nominal grid of 25 × 50 m reflectors was used to generate a 1:10 000 Ventersdorp Contact Reef (VCR) structure contour map. All faults over 250 m in strike length and 25 m throw are shown; these features can be extrapolated from the VCR to the Elsburgs and form the basis of mine layouts and ore reserve determinations that are more accurate than could be estimated from drilling (Campbell & Crotty 1990).

WIDE OREBODY MINING

In the South Deep Project Area, to the west of the shoreline, near the palaeoshoreline, the reefs are narrow and will be exploited by narrow orebody stoping techniques, as described above. To the east of the shoreline, the reefs increase in thickness from 2 m to over 100 m (Fig. 13.24), and it is in these areas that the mechanized, wide orebody mining systems will be used. Room and pillar mining can be used to mine reefs up to 8 m thick, but for greater thicknesses, vertical crater mining will be employed (Tregoning & Barton 1990). Vertical crater mining can be developed initially in the same way as room and pillar mining. Two stopes will be established at the top and bottom of the proposed mining block (Fig. 13.25). Each reef block is planned to be 150 m wide in a strike direction and 250 m wide in a dip direction. The upper excavation is called the drilling drive and the lower, the cleaning drive. Retreat mining by means of slots up to 40 m high (between the two levels), 50 m long and up to 7 m wide will be practised. The stopes are to be cleared using remote-controlled LHDs, and a vacuum or hydraulic cleaning system used for sweeping before backfilling. The cemented, crushed waste rock, backfill will be placed in the mined out slots as a permanent support.

FUTURE DEVELOPMENT

South Deep has an advantage over other new prospects in that the target reefs are a continuation of those mined on the south section of Western Areas and test mining of the area is possible without the cost of sinking a new shaft. A twin haulage way has been driven from Western Areas' south shaft to the proposed site of South Deep's main shaft, a distance of about 2000 m at a cost of around R70 M. Underground drilling from the haulage (Fig. 13.26) and development has shown that grades obtained from surface drilling err on the low side but has confirmed the structure deduced from the seismic survey. Relatively limited mining is underway to extract the reef in the area around the proposed main shaft. Besides producing revenue from around 30 000 t per month the development will destress the shaft pillar. Development is also underway to enable underground testing of the wide orebody by drilling three holes at 40 m intervals. The initial development has been financed by floating a company (South Deep Exploration Limited) on the Johannesburg Stock Exchange. However in spite of the reserves, no decision has yet (1993) been taken to go ahead with the development of full scale production, initially estimated to have a capital cost of R2170 M.

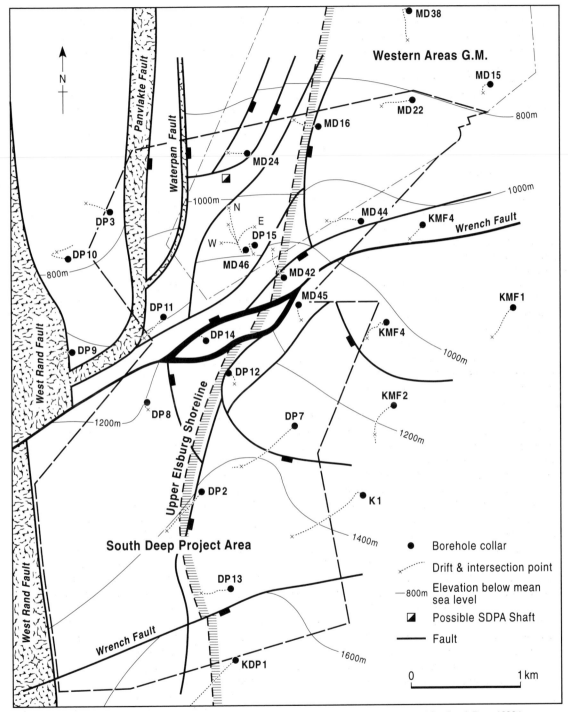

Fig. 13.23 Drill plan and simplified structure contours on VCR of the South Deep Prospect Area. (After South Deep 1990.)

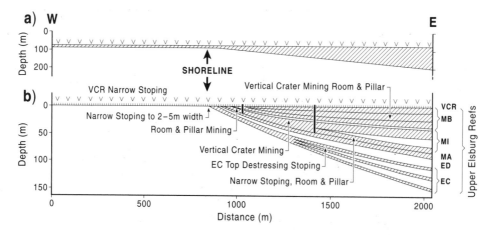

Fig. 13.24 Generalized section showing the relation of proposed mining methods to reef thickness. (After Tregonning & Barton 1990.)

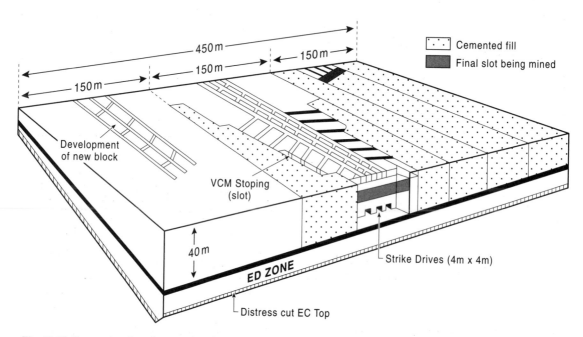

Fig. 13.25 Perspective view of proposed vertical crater mining. (After Tregonning & Barton 1990.)

13.8 CONCLUSIONS

The Witwatersrand Basin with its large tabular orebodies is a unique gold occurrence. Exploration has evolved from the testing of down dip extensions from outcrop through the recognition of blind districts, such as the West Wits Line, to the definition of deep targets that can

support a large underground mine. Geophysical methods have proved very successful in delineating the stratigraphical setting of the blind deposits. Magnetic and gravity methods made a major contribution in the 1930s to 1960s as have reflection seismics in the 1980s. Definition of mineable areas depends on an understanding of the sedimentological and structural controls on ore

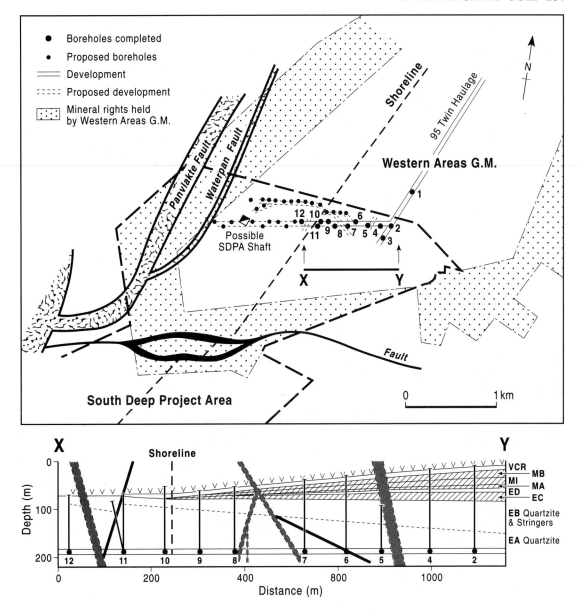

Fig. 13.26 Plan and section showing initial development at South Deep from Western Areas gold mine. Note the drilling to confirm grade and reef location. (After South Deep 1992.)

distribution as revealed by the very limited information available from deep drilling. Evaluation of this limited information led to the development of geostatistical methods that provide reliable estimates of grade and tonnage before a decision was made to commit large

expenditures to shaft sinking. The discovery of the South Deep deposit shows that large deposits remain to be discovered, although they are likely to be very few in number.

14 A Volcanic-associated Massive Sulphide Deposit—Kidd Creek, Ontario

ANTHONY M. EVANS AND CHARLES J. MOON

14.1 INTRODUCTION

Before reviewing the geology of Kidd Creek it is necessary to describe and discuss the general features and different types of massive sulphide deposits as a good knowledge of these is essential in designing an exploration programme to find more deposits of this type. The term volcanic-associated massive sulphide deposits is something of a mouthful and in this chapter will often be abbreviated VMS deposits.

14.2 VOLCANIC-ASSOCIATED MASSIVE SULPHIDE DEPOSITS

14.2.1 Morphology

These deposits are generally stratiform, i.e. lenticular to sheetlike bodies (Fig. 3.8) developed at the interfaces between volcanic units or at volcanic–sedimentary interfaces. Though often originally roughly circular or oval in plan deformation may change them to blade-like, rod-shaped or amoeba-like orebodies, which may even be wrapped up and stood on end, and then be mistaken for replacement pipes, e.g. the Horne Orebody of Noranda, Quebec. The massive ore is usually underlain by a stockwork that may itself be ore grade and which appears to have been the feeder channel up which mineralizing fluids penetrated to form the overlying massive sulphide deposit.

14.2.2 Size, grade, mineralogy and textures

Data from five metallogenic provinces are given in Table 14.1 (from Large *et al.* 1987 and Lydon 1989). The majority of world deposits are small and about 80% of all known deposits fall in the size range 0.1–10 Mt. Of these about a half contain less than 1 Mt (Sangster 1980). Average figures of this type tend to hide the fact that this is a deposit type that can be very big or rich or both and then very profitable to exploit. Examples of such deposits are given in Table 14.2.

The mineralogy of these deposits is fairly simple and often consists of over 90% iron sulphide, usually as pyrite, although pyrrhotite is well developed in some. Chalcopyrite, sphalerite and galena may be major constituents, depending on the deposit class, bornite and chalcocite are occasionally important, arsenopyrite, magnetite and tetrahedrite–tennantite may be present in minor amounts. With increasing magnetite content these ores grade to massive oxide ores (Solomon 1976). The gangue is principally quartz, but occasionally carbonate is developed and chlorite and sericite may be locally important. Their mineralogy results in these deposits having a high density and some, e.g. Aljustrel and Neves-Corvo in Portugal, give marked gravity anomalies, a point of great exploration significance (section 7.4).

The vast majority of massive sulphide deposits are zoned. Galena and sphalerite are more abundant in the upper half of the orebodies whereas chalcopyrite increases towards the footwall and grades downward into chalcopyrite stockwork ore (Figs 3.8 & 14.7). This zoning pattern is only well developed in the polymetallic deposits. As the number of mineral phases decreases, so the zonation tends to become obscure and may not be in evidence at all in deposits dominated by pyrite or pyrite-chalcopyrite. From a study of nineteen deposits Huston & Large (1989) showed that gold tends to have two distinct occurrences: (1) an Au–Zn–Pb–Ag association found in the upper part of zinc-rich massive sulphide lenses and (2) an Au–Cu association commonly present in the stockwork and lower portions of copper-rich deposits. In each of the nineteen deposits one of these occurrences dominates, almost to the exclusion of the other. From their study of gold distribution in some orebodies of the Iberian Pyrite Belt Strauss and Beck (1990) arrived at very similar conclusions.

Textures vary with the degree of recrystallization. The dominant original textures appear to be colloform banding of the sulphides with much development of framboidal pyrite, perhaps reflecting colloidal deposition. Commonly, however, recrystallization, often due to some degree of metamorphism, has destroyed the colloform banding and produced a granular ore. This may show banding in the zinc-rich section, whereas the chalcopyrite ores are rarely banded. Angular inclusions

Table 14.1 Average grade and tonnage data for VMS deposits of selected regions. (Data from Large *et al.* 1987 and Lydon 1989.)

Region	Dominant deposit type	Number of deposits	Average grade[a] Cu (%)	Zn (%)	Pb (%)	N –	Ag $(g\,t^{-1})$	Au $(g\,t^{-1})$	Tonnage (Mt)
Abitibi Belt, Canada (Archaean)	Cu–Zn	52	1.47	3.43	0.07	47	31.9	0.8	9.2
Norwegian Caledonides (Palaeozoic)	Cu–Zn	38	1.41	1.53	0.05	0	na	na	3.5
Bathurst, New Brunswick, Canada (Palaeozoic)	Zn–Pb–Cu	29	0.56	5.43	2.17	28	62.0	0.5	8.7
Green Tuff Belt, Japan (Tertiary)	Zn–Pb–Cu	25	1.63	3.86	0.92	7	95.1	0.9	5.8
Tasman Geosyncline, Eastern Australia (Palaeozoic)	Cu–Au	42	1.3			5	17.2	2.1	6.4
	Zn–Pb–Cu[b]		0.48	15.15	6.28	4	160.0	3.0	

[a]N = Number of deposits for which data available to calculate Ag and Au grades.
[b]Data in this line are for some Tasmanian deposits only.
na = not available

Table 14.2 Tonnages and grades for some large or high grade VMS deposits

Mined ore + reserves	(Mt)	Cu%	Zn%	Pb%	Sn%	Cd%	Ag g t[-1]	Au g t[-1]
Kidd Creek, Ontario	155.4	2.46	6.0	0.2	r	r	63	–
Horne, Quebec	61.3	2.18	–	–	–	–	–	4.6
Rosebery, Tasmania	19.0	0.8	15.7	4.9	–	–	132	3.0
Hercules, Tasmania	2.3	0.4	17.8	5.7	–	–	179	2.9
Rio Tinto, Spain[a]	500	1.6	2.0	1.0	–	–	r	r
Aznalcollar, Spain	45	0.44	3.33	1.77	–	–	67	1.0
Neves–Corvo Portugal[b]	30.3	7.81	1.33	–	–	–	–	–
	2.8	13.42	1.35	–	2.57	–	–	–
	32.6	0.46	5.72	1.13	–	–	–	–

[a] Rio Tinto, originally a single stratiform sheet, was folded into an anticline whose crest cropped out and vast volumes were gossanized. The base metal values are for the 12 Mt San Antonio section.
[b] The three lines of data represent three ore types and not particular orebodies of which there are four. Ag is present in some sections.
r indicates that this metal is or has been recovered, but the average grade is not available.

of volcanic host rocks are occasionally present and soft sediment structures (slumps, load casts) are sometimes seen. Graded bedding has also been reported from some deposits (Sangster & Scott 1976, Ohmoto & Skinner 1983).

14.2.3 Wall rock alteration

Wall rock alteration is usually confined to the footwall rocks. Chloritization and sericitization are the two commonest forms. The alteration zone is pipe-shaped and contains within it and towards the centre the chalcopyrite-bearing stockwork. The diameter of the alteration pipe increases upward until it is often coincident with that of the massive ore. Metamorphosed deposits commonly show alteration effects in the hanging wall. This may be due to the introduction of sulphur released by the breakdown of pyrite in the orebody. The sulphur, by reacting with pore solutions, could give rise to extensive hydrogen ion production. On the other hand the thermal convection cells extant at the time of deposit formation may have continued for some time afterwards and produced alteration effects in the hanging wall rocks (Solomon *et al.* 1987).

14.2.4 Classifications

These ores show a progression of types. Associated with basic volcanics, usually in the form of ophiolites and presumably formed at oceanic or back arc spreading ridges, we find the Cyprus types. These are essentially cupriferous pyrite bodies. They are exemplified by the deposits of the Troodos Massif in Cyprus and the Ordovician Bay of Islands Complex in Newfoundland. Associated with the early part of the main calc-alkaline stage of island arc formation are the Besshi-type deposits. These occur in successions of mafic volcanics in complex structural settings characterized by thick greywacke sequences. They commonly carry zinc as well as copper and are exemplified by the Palaeozoic Sanbagawa deposits of Japan, and the Ordovician deposits of Folldal in Norway. The more felsic volcanics, developed at a later stage in island arc formation, have a more varied metal association. They are copper–zinc–lead ores often carrying gold and silver. Large amounts of baryte, quartz and gypsum may be associated with them. They are called Kuroko deposits after the Miocene ores of that name in Japan. Similar deposits in the Precambrian are known as Primitive-type.

To some extent corresponding with the above environmental classification a geochemical division into

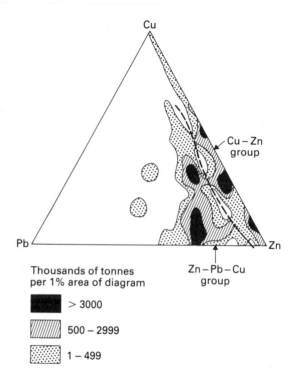

Fig. 14.1 Data from the first four regions of Table 14.1 plotted on a Cu–Pb–Zn ternary diagram weighted in tonnes of ore metal contained in the deposit and then contoured. (Modified from Lydon 1989.)

iron (pyrite deposits used for sulphuric acid production), iron-copper (Cyprus-type), iron–copper–zinc (Besshi-type) and iron–copper–zinc–lead (Kuroko- and Primitive-type) is often employed. Using a different approach from the simple chemical one Lydon (1989) has shown that, if each point on a ternary diagram of the grades of the deposits in the first four regions of Table 14.1 is weighted in terms of tonnes of ore metal within the deposit, then clearly there are just two major groups, Cu–Zn and Zn–Pb–Cu (Fig. 14.1). Indeed, as Lydon stresses, there are few so-called copper deposits without some zinc.

A summary of the nature of the different types of VMS deposits is given in Table 14.3. Morton & Franklin (1987) suggested that the Primitive-type can be divided into two subtypes—the Noranda and Mattabi on the grounds of differences in alteration characteristics, associated volcanic rocks and environments. The Noranda-subtype has a well defined, zoned alteration pipe beneath it, and, if there is a subjacent semiconformable widespread alteration zone, then this is cut by the sharp

Table 14.3 Volcanic-associated massive sulphide deposit types. (Modified from Hutchinson 1980)

Type	Volcanic rocks	Clastic sedimentary rocks	Depositional environment	General conditions	Plate tectonic setting	Known age range
Besshi (= Kieslager) Cu–Zn±Au±Ag	Within plate (intraplate) basalts	Continent-derived greywackes and other turbidites	Deep marine sedimentation with basaltic volcanism	Rifting	Epicontinental or back-arc	Early Proterozoic Palaeozoic
Cyprus Cu(+Zn)±Au	Ophiolitic suites, tholeiitic basalts	Minor or absent	Deep marine with tholeiitic volcanism	Tensional, minor subsidence	Oceanic rifting at accreting margin	Phanerozoic
Kuroko Cu–Zn–Pb±Au±Ag	Bimodal suites, tholeiitic basalts, calc-alkaline lavas and pyroclastics	Shallow to medium depth clastics, few carbonates	Explosive volcanism, shallow marine to continental sedimentation	Rifting and regional subsidence, caldera formation	Back-arc rifting	Early Proterozoic Phanerozoic
Primitive Cu–Zn±Au±Ag	Fully differentiated suites, basaltic to rhyolitic lavas and pyroclastics	Immature greywackes, shales, mudstones	Marine < 1km depth. Mainly developed in greenstone belts	Major subsidence	Much debated: fault-bounded trough, back-arc basin?	Archaean–early Proterozoic

boundaries of the pipe. The volcanics are dominantly mafic and felsic flows and hyaloclastites believed to have formed at water depths greater than 500 m. By contrast the Mattabi-subtype (to which Kidd Creek belongs) has a broad alteration pipe without sharp boundaries that is gradational into the semiconformable alteration below. The footwall succession has a much higher proportion of felsic volcanics which are dominantly fragmental and thought to have been emplaced at depths of less than 500 m. Both subtypes include large and valuable deposits and the recognition of the existence of these two subtypes will be important in any exploration programme.

Stanton (1978) and Vokes (1987) have objected to the division of these deposits into various types. Stanton considered these ores to be part of one continuous spectrum, showing a progressive geochemical evolution which accompanies that of the calc-alkaline rocks in island arcs. It is certainly difficult to assign some deposits to particular types except on a purely geochemical basis. For example, the deposits of West Shasta, California, which are believed to have formed in a Devonian bimodal suite of island arc volcanics (Lindberg 1985, South & Taylor 1985), are, like Kuroko deposits, intimately associated with rhyolite lavas and domes, but they are a Cu–Zn (+Au–Ag) type and lack significant lead. Thus although they have many of the properties of Kuroko-type deposits, a case could be made out for correlating them with the Primitive-type, of which it must be remarked that some deposits, e.g. Kidd Creek, Ontario (0.2%) and Mons Cupri, W.A. (2.5%) do carry recoverable lead. Yes, reader, the situation is confusing!—we may indeed be dealing with a continuous spectrum that we have not yet fully recognized, as Lydon's work suggests (Fig. 14.1), but meanwhile mining geologists will continue to use the terms listed in Table 14.3 because in the present state of our knowledge, they imply useful summary descriptions of the deposits, helpful in formulating exploration models.

It is important to note that precious metals are also produced from some of these deposits; indeed in some Canadian examples of the Primitive-type they are the prime product. Both Besshi and Kuroko types may also produce silver and gold while the Cyprus-type may have by-product gold.

Fig. 14.2 Distribution of dacite lava domes and Kuroko deposits, Kosaka District, Japan. (Modified from Horikoshi & Sato 1970.)

14.2.5 Some important field occurrence features

(a) Association with volcanic domes. This frequent association is stressed in the literature and the Kuroko deposits of the Kosaka district, Japan, are a good example (Fig. 14.2). Many examples are cited from elsewhere, e.g. the Noranda area of Quebec and some authors infer a genetic connexion. Certain Japanese

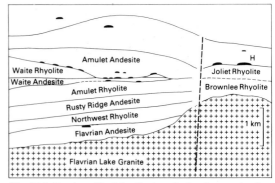

Fig. 14.3 Stratigraphical distribution of volcanic-associated massive sulphide deposits in the Noranda area, Quebec. (After Lydon 1989.) The names on the diagram are those of various volcanic formations each of which may include both flows and pyroclastic deposits. The relative sizes of the orebodies are indicated. The largest, at the Horne Mine (H), is over 60 Mt.

workers, however, consider that the Kuroko deposits all formed in depressions and that the domes are late and have uplifted many of the massive sulphide deposits. In other areas, e.g. the Ambler District, N. Alaska no close association with rhyolite domes has been found (Hitzman *et al.* 1986).

(b) Cluster development. Although the Japanese Kuroko deposits occur over a strike length of 800 km with more than a hundred known occurrences these are clustered into eight or nine districts. Between these districts lithologically similar rocks contain only a few isolated deposits and this tends to be the case, with a few notable exceptions, for massive sulphide occurrences of all ages.

(c) Favourable horizons. The deposits of each cluster often occur within a limited stratigraphical interval. For Primitive- and Kuroko-types, this is usually at the top of the felsic stage of cyclical, bimodal, calc-alkaline volcanism related to high level magma chambers rather than to a particular felsic rock type (Leat *et al.* 1986, Rickard 1987). Sometimes this favourable horizon is hosted by quite relatively thin developments of volcanic rocks, as in the Iberian Pyrite Belt where the volcanic–sedimentary complex underlying the many enormous orebodies of this region is only 50–800 m thick. The Japanese Kuroko mineralization and associated volcanism occurred during a limited period of the Middle Miocene in the Green Tuff volcanic region. By contrast the deposits of the Noranda area, Quebec (Fig. 14.3) lie in a narrow stratigraphical interval within a vast volcanic edifice at least 6 000 m thick. A very good account of the stratigraphical relationships is given in

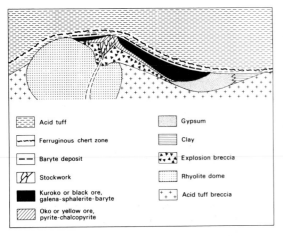

Acid tuff

Ferruginous chert zone

Baryte deposit

Stockwork

Kuroko or black ore,
galena-sphalerite-baryte

Oko or yellow ore,
pyrite-chalcopyrite

Gypsum

Clay

Explosion breccia

Rhyolite dome

Acid tuff breccia

Fig. 14.4 Schematic section through a Kuroko deposit.
(Modified from Sato 1977.)

Kerr & Gibson (1993).

(d) Deposit stratigraphy. As has already been indicated massive sulphide deposits tend to have a well developed zoning or layering. Kuroko deposits have the best and most consistent stratigraphical succession of ore and rock types, and an idealized deposit (Fig. 14.4) contains the following units:

1 hanging wall: upper volcanics and/or sedimentary formation;
2 ferruginous quartz zone: chiefly hematite and quartz;
3 baryte ore zone;
4 Kuroko or black ore zone: sphalerite–galena–baryte;
5 Oko or yellow ore zone: cupriferous pyrite ores; about this level, but often towards the periphery of the deposit, there may be the Sekkoko zone of anhydrite–gypsum–pyrite;
6 Keiko or siliceous ore zone: copper-bearing, siliceous, disseminated and/or stockwork ore;
7 footwall: silicified rhyolite and pyroclastic rocks.

(e) Plate tectonic settings. This subject has been dealt with in section 3.3.2 and will not be repeated here.

14.2.6 Genesis

For many decades VMS deposits were considered to be epigenetic hydrothermal replacement orebodies (Bateman 1950). In the 1950s, however, they were recognized as being syngenetic, submarine-exhalative, sedimentary orebodies, and deposits of this type have been observed in the process of formation from hydrothermal vents (black smokers) at a large number of places along sea floor spreading centres (Rona 1988).

Black smokers were discovered in the late 1970s during ocean floor investigations using a submersible. They are plumes of hot, black, or sometimes white, hydrothermal fluid issuing from chimney-like vents that connect with fractures in the sea floor. The black smoke is so coloured by a high content of fine-grained metallic sulphide particles and the white by calcium and barium sulphates. The chimneys are generally less than 6 m high and are about 2 m across. They stand on mounds of massive ore-grade sulphides (Fig. 14.5) that occur within the grabens and on the flanks of oceanic ridges. The largest mound–chimney deposit so far found is the TAG mound on the Mid-Atlantic Ridge at 26°N which is estimated to contain 4.5 Mt (Rona *et al.* 1986). The mineralogy of the mounds is similar to that of massive sulphide deposits on land with high temperature copper–iron sulphides beneath lower temperature zinc- and iron-rich sulphides, baryte and amorphous silica.

Growth of a sulphide chimney commences with the precipitation of anhydrite from the cold sea water around the hot ascending plume forming a porous wall which continues to grow upward during the life of the plume. Most of the hydrothermal fluid flows up the chimney to discharge as a plume into the surrounding sea water, but a small proportion flows through the porous anhydrite wall. In doing so the fluid moves rapidly from high temperature (>300°C), acidic (pH≈3.5) and reduced (H_2S»SO_4) conditions to conditions close to those of normal sea water at oceanic depths (2°C, alkaline, pH≈7.8, oxidized, SO_4»H_2S) as it meets sea water penetrating the chimney from the outside. Thus, while the anhydrite chimney top grows upwards, the walls of the lower part thicken by the precipitation of sulphides in the interior portions and anhydrite on the outside (Fig. 14.5). This process leads to a concentric zoning with chalcopyrite (±isocubanite ±pyrrhotite) on the inside, an intermediate zone of pyrite, sphalerite, wurzite and anhydrite and an outer zone of anhydrite with minor sulphides, amorphous silica, baryte and other minerals. The zoning is probably caused by the temperature decrease across the wall rather than other factors (Lydon 1989).

After growing upwards at a rate of perhaps 8–30 cm a day chimneys eventually become unstable and collapse forming a mound of chimney debris mixed with anhydrite and sulphides upon which further chimney growth and collapse occur. Once a mound has developed it grows both by accumulation of chimney debris on its upper surface *and* by precipitation of sulphides within the mound. The covering of chimney debris performs the same role as the vertical porous anhydrite chimney

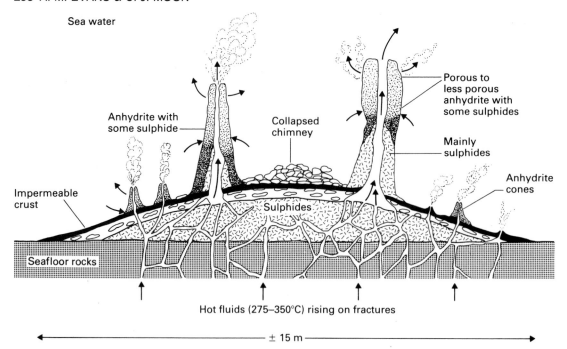

Sea water

Anhydrite with
some sulphide

Collapsed
chimney

Porous to
less porous
anhydrite with
some sulphides

Mainly
sulphides

Anhydrite
cones

Impermeable
crust

Sulphides

Seafloor rocks

Hot fluids (275–350°C) rising on fractures

± 15 m

Fig. 14.5 Formation of chimneys and sulphide mounds on the sea floor. (From Barnes, 1988, *Ores and Minerals*, Open University Press, with permission.)

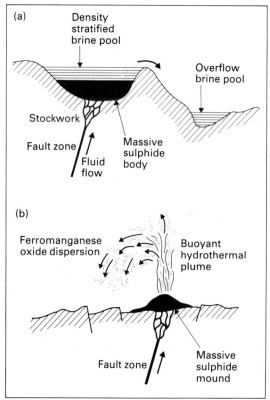

(a)

Density
stratified
brine pool

Overflow
brine pool

Stockwork

Fault zone

Fluid
flow

Massive
sulphide
body

(b)

Ferromanganese
oxide dispersion

Buoyant
hydrothermal
plume

Fault zone

Massive
sulphide
mound

walls producing sulphide and silica precipitation in the outer part of the mound. This decreases the permeability of the mound and forms a crust that constrains fluid escape and leads to considerable circulation of high temperature solutions within the mound. The isotherms within the mound then rise, leading to the replacement of lower temperature mineral assemblages by higher temperature ones, thus producing a similar zoning to that in the chimneys and that of VMS deposits found on land (section 14.2.2).

VMS deposits may have the mound shape of modern massive sulphide deposits or they may be bowl-shaped. The latter type probably developed when hydrothermal solutions, more saline (denser) than the surrounding sea water, vented into a submarine depression (Fig. 14.6). Many Cyprus-type deposits appear to have developed in this way and the available fluid inclusion data is consistent with this model as explained by Rona (1988).

This brief summary of the formation of black smoker deposits, and there is much more important supporting

Fig. 14.6 Sketches showing the development of bowl-shaped deposits (a) when hydrothermal solutions denser than sea water collect in a hollow and mound-shaped deposits (b) when solutions are less dense than sea water and rise buoyantly. (After Rona 1988.)

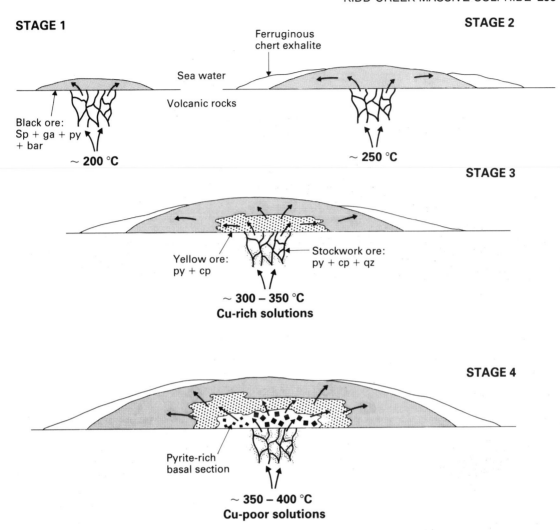

Fig. 14.7 Diagrams to illustrate the first four stages in the formation of VMS deposits as described in the text; bar = baryte, cp = chalcopyrite, ga = galena, py = pyrite, qz = quartz and sp = sphalerite.

evidence in the literature (e.g. see *Can. Mineral., 26*, 429–888 and references therein), shows that our knowledge of their mode of formation confirms the model put forward for the genesis of volcanic-associated massive sulphide deposits by Eldridge *et al.* (1983) and Pisutha-Arnond & Ohmoto (1983) and termed by Huston and Large (1989) the Kuroko model. The principal stages of development of this model are as follows (Fig. 14.7).

1 Precipitation of fine-grained sphalerite, galena, tetrahedrite, baryte with minor chalcopyrite (black ore) by the mixing of relatively cool (≈200°C) hydrothermal solutions with cold sea water.

2 Recrystallization and grain growth of these minerals at the base of the evolving mound by hotter (≈250°C) solutions, together with deposition of more sphalerite, etc.

3 Influx of hotter (≈300°C) copper-rich solutions which replace the earlier deposited minerals with chalcopyrite in the lower part of the deposit (yellow ore). Redeposition of these replaced minerals at a higher level.

4 Still hotter, copper undersaturated solutions then dissolve some chalcopyrite to form pyrite-rich bases in the deposits.

5 Deposition of chert–hematite exhalites above and

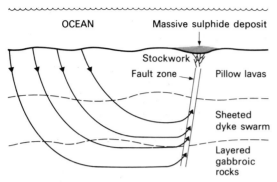

OCEAN   Massive sulphide deposit

Stockwork

Fault zone   Pillow lavas

Sheeted
dyke swarm

Layered
gabbroic
rocks

Fig. 14.8 Diagram showing how sea water circulation through oceanic crust might give rise to the formation of an exhalative VMS deposit.

around the sulphide deposit (not shown in Fig. 14.7). Similar exhalites will have also formed during previous stages. Silica is slow to precipitate, it needs silicate minerals to nucleate on and so, although much may be deposited in the underlying stockwork, the rest is mainly carried through the sulphide body to form exhalites above it.

Note that an evolution with time of the hydrothermal solutions is postulated from cooler to hotter and back to cooler and that the solutions of stages (2) to (4) cool as they move upwards and outwards through the sulphide mound to produce stage (1) type mineralization at the top of the deposit, see Fig. 14.7. The most important evidence on which this model is based, mineral replacement textures and fluid inclusion filling temperature trends, comes from young unaltered deposits and is generally difficult or impossible to find in older (Palaeozoic and Precambrian) deposits which have usually suffered some degree of deformation and metamorphism.

Most workers in this field agree at least on the broad outline of the above mechanisms for the formation of VMS deposits, but they disagree on the origin of the formative solutions with the majority, at the time of writing, favouring the hypothesis that they are composed of sea water which has circulated deep into the crust, where it has been heated, become more saline and dissolved base and precious metals from the volcanic and other rocks it has passed through, and then risen along permeable zones to be exhaled on the sea floor as sketched in Fig. 14.8. These workers point in particular to the evidence from isotopic studies.

14.3 AN EXPLORATION MODEL FOR VMS DEPOSITS

14.3.1 Age

VMS deposits are known from the oldest greenstone belts (say 3 500 Ma) to the modern deposits of oceanic rift zones.

14.3.2 Geological setting

The vast majority of known deposits have been formed near plate boundaries, e.g. modern and Cyprus-type deposits at divergent boundaries and other types at convergent boundaries or continental margins, such as the Iberian Pyrite Belt deposits or the Kuroko deposits of Japan. The plate tectonic environments of the Archaean greenstone belts are less clear but many lines of evidence point to similarities with Phanerozoic ones. These environments can be recognized using the petrochemistry of the associated volcanics which are always submarine. It must be emphasized, however, that VMS deposits are most erratically distributed with respect to submarine vulcanism in general and most of this vulcanism is devoid of such deposits. This suggests that some special concatenation of events is required for their formation. These special circumstances may lead to the subjacent volcanic rocks being strongly enriched in ^{18}O thus producing an isotopically defined regional target (Beaty *et al.* 1988).

14.3.3 Favourable terranes

Intensive exploration may have revealed that certain rock sequences host valuable VMS deposits and, if it is not virgin ground that is to be prospected, then these terranes must receive prime attention. For example all known orebodies of the Iberian Pyrite Belt (southern Portugal and Spain) lie in a particular volcano–sedimentary sequence at the Devonian–Carboniferous boundary which has an exposed strike length of about 250 km in a zone about 40 km wide, (Leca 1990). In the Archaean Abitibi Greenstone Belt of Ontario–Quebec eleven regional volcanic complexes have been identified, but only four of these are known to contain significant VMS deposits (Fig. 14.14). Such terrane may be marked by massive sulphide gossans which have been worked by ancient man as in Iberia, Turkey and other parts of Eurasia, often for gold, or have given rise to nearby alluvial gold deposits as in eastern Ireland.

Many massive sulphide deposits occur within recognized cauldron subsidence areas localized along synvolcanic faults and extracauldron deposits may lie within graben flanked by vents (Kerr & Gibson 1993). Detailed mapping and expertise in the interpretation of volcanic phenomena are obviously essential in the search for deposits in this terrane.

Exploration gelogists working in a favourable terrane should study all the literature on its deposits to refine their exploration model. This will help them to decide on finer points such as whether any newly found deposit is more likely to be flat lying or steeply dipping, little or badly deformed, little or much metamorphosed and to adjust their geochemical, geophysical and drilling programmes accordingly.

14.3.4 **Favourable horizon, clustering, volcanic domes**

VMS deposits tend to occur in clusters within a limited stratigraphical interval and are often, but not always, associated with volcanic domes (section 14.2.5). Sangster (1980) calculated the average area of a cluster to be 850 km^2 with twelve deposits containing in total 94 Mt of ore. The Noranda area (Fig. 14.3) is about 170 km^2 and contains about twenty deposits aggregating about 110 Mt. The four Neves-Corvo deposits lie in an area of 6 km^2 and have 250 Mt of possible ore; more mineralization occurs around Algare just 2.5 km to the north (Leca 1990).

In some districts VMS deposits and associated domes appear to lie along linear fractures which may have controlled their distribution. Synvolcanic basinal areas are important orebody locations in the Iberian Pyrite Belt.

14.3.5 **Morphology, size and mineralogy**

Little deformed VMS deposits take the form of massive sulphide lenses, essentially conformable with the host stratigraphy and underlain by stockworks (Figs 3.8 & 14.7)—(section 14.2.1). They can vary from uneconomic sizes up to 500 Mt of ore and the massive ore consists of upwards of 60% of sulphides (section 14.2.2). In this time of low metal prices large polymetallic orebodies of this type form the high quality deposits much sought after by mining companies (see METAL AND MINERAL PRICES section in 1.2.3). The lateral continuity and shape of VMS deposits usually makes them attractive mining propositions either in open pit or as underground operations using trackless equipment. Mechanization is particularly important in countries with high labour costs, such as Canada, Sweden or Australia; in Canada skilled production miners were earning >$C 100 000 per year in 1991!

It must be remembered that slumping may upset the simple model shape mentioned above and indeed the whole of the massive ore can slide off the stockwork and the two become completely separated. The massive ore is often capped by an exhalative, sulphide-bearing, ferruginous chert which may have a considerably greater areal extent than the orebody and hence be of considerable exploration value.

14.3.6 **Wall rock alteration**

Visible chloritic or sericitic alteration is almost entirely confined to the zone around the stockwork and is only exposed at the surface in cases of near surface, steeply dipping deposits. The alteration halo can be used as a proximity indicator in drill holes that miss the orebody as can the chemical content of the halo (see below).

In several mining districts laterally widespread zones of alteration have been reported as occurring stratigraphically below the favourable horizon. Such zones extend for up to 8 km and vertically over several hundred metres. In a successful attempt to expand the exploration target by mapping a larger part of the causative hydrothermal system, Cathles (1993) carried out a regional whole rock oxygen isotope survey to reveal the extent of the oxygen isotope alteration in the Noranda region.

14.3.7 **Volcanic facies**

The association of massive sulphide with particular parts of the volcanic assemblage has long been recognized. Breccciated rhyolitic volcanics are known as 'mill rock' in some mining camps because of their proximity to processing plants. The breccias are attributed to one of three processes: (1) they formed close to structural magmatic vents which also vented hot springs; (2) they were the result of hydrothermal explosive events and (3) the breccias formed as a result of caldera formation with which the ore forming process was associated (Hodgson 1989). Whichever of these processes is correct it is clear that an understanding of the development of the volcanic pile is important. This knowledge can be obtained by detailed mapping, core logging and comparisation with modern volcanic models. The most useful summary of these models is provided in the book by Cas & Wright (1988) and the reader is referred to this text.

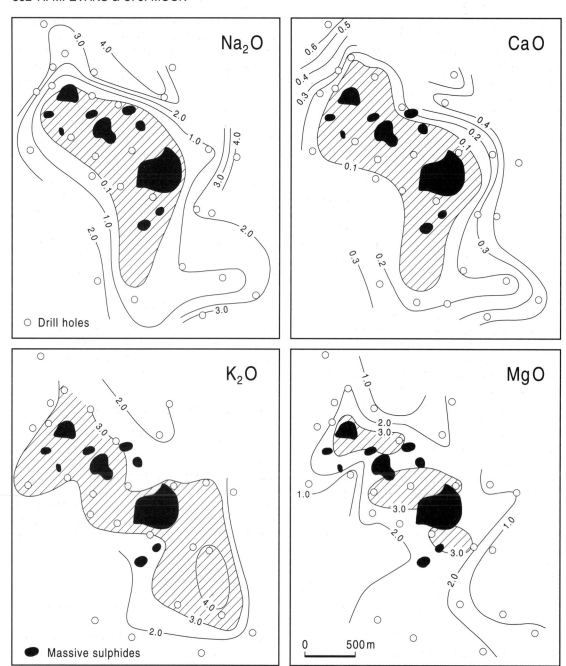

Fig. 14.9 Major element anomalies in the footwall of the Fukasawa Mine, Japan. (From Govett 1983, data from Ishikawa *et al*. 1976.)

14.3.8 **Exploration geochemistry**

Exploration geochemistry has been widely and successfully used in the search for VMS deposits. In Canada the prospective areas have been glaciated and have poor surface response, but rock geochemistry is widely used to delineate drill targets at depth. In the southern hemisphere surface geochemistry has been successful even in areas of intense weathering where e.m. has been ineffective due to conductive overburden.

ROCK GEOCHEMISTRY

The rock geochemical signature of VMS deposits has been investigated by a number of workers but is well summarized by Govett (1989). In a review of a large number of case histories he documents the very consistent response, regardless of age and location. Aureoles are usually extensive, more than 1 km laterally and 500 m vertically with footwall anomalies more extensive than those in the hanging wall, especially in proximal deposits. In Archaean deposits significant hanging wall anomalies appear to be absent. As discussed above Fe and Mg are enriched and Na and Ca depleted in nearly all deposits (Fig. 14.9). Copper tends to be enriched in the footwall and is depleted in the hanging wall; Mn shows the reverse behaviour. Both K and Mn may be relatively depleted near the deposit and enriched further away. The ore elements, such as Zn, are commonly enriched. Copper may be depleted especially in the immediate wall rocks. The behaviour of other trace elements, in particular Au, Cl, Br, As, Sb, may be distinctive but data from case histories are incomplete. Some authors have regressed trace elements against SiO_2 to compensate for fractionation. This is not always effective as some deposits, such as those in the Lac Dufault area, Noranda, have undergone addition of SiO_2 during alteration.

Cherty exhalites have been widely used as indicators of hydrothermal activity but their geochemical signatures are varied. Kalogeropoulos and Scott (1983) have used a major element ratio, $R = (K_2O + MgO) \times 100 / K_2O + Na_2O + CaO + Mg$ to indicate proximity to Kuroko mineralization in Japan. Trace and major elements, in particular zinc and manganese but also volatiles such as arsenic, are enriched sporadically in these horizons but application has had varying degrees of success. A key requirement of any successful application is that the stratigraphical position of the samples must be well understood. It is of little use comparing ratios of samples from different exhalative horizons.

Geochemistry has also been used on a regional scale to attempt to discriminate mineralized sequences of volcanics. However the variation associated with mineralization is usually less than that caused by fractionation and trace element data must be corrected for fractionation. Studies of immobile elements (Ti, Zr, REE), that have not been markedly affected by hydrothermal alteration, may indicate the original chemistry of the volcanics. Campbell et al. (1984) have suggested that mineralized volcanics have flatter REE patterns than barren volcanics which tend to be enriched in light REE and have negative Eu anomalies. Few regional rock geochemical studies have been published but Govett (1989) suggests that many VMS deposits are associated with regional copper depletions. Both Russian and Canadian workers (Allen & Nichol 1984) suggest that analysis of sulphides, separated either by gravity or selective chemical extraction, within volcanics, may give regional anomalies of up to 10 km along strike from VMS deposits.

SURFICIAL GEOCHEMISTRY

The extreme chemical contrast of the sulphides with volcanics gives rise to high contrast transition metal anomalies in soil and stream sediments. A large number of case histories are available.

14.3.9 **Geophysical signatures**

One of the key elements in the exploration model is the geophysical signature of the deposit. The majority of VMS discoveries have resulted from geophysics, notably in the glaciated areas of the Canadian Shield. VMS deposits respond to a number of geophysical techniques, as they contain much sulphide that often shows electrical connection, permitting the use of e.m. techniques, and the bodies are dense, allowing the use of gravity methods.

MAPPING METHODS

Airborne magnetic and e.m. systems are widely used for mapping, both of general geology and of favourable lithologies. The high resolution of modern aeromagnetic systems resulted in the obsolescence of ground magnetics for all but the smallest projects. In addition it is possible to merge aeromagnetic data with remotely sensed information. An excellent example of the use of magnetics in the detailed mapping of Archaean terranes is provided by Isles et al. (1989) for the Yilgarn block in Western Australia. They highlight the use of multi-client surveys as a cost-effective method of mapping major lithologies

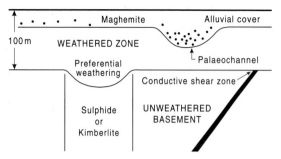

Fig. 14.10 Schematic diagram showing the problems of using magnetics and e. m. in deeply weathered areas, such as the Yilgarn in Western Australia. (After Smith & Pridmore 1989.)

and structural changes. Magnetic units such as banded iron formations are apparent as magnetic highs in contrast to the magnetically less disturbed areas of acid volcanics and sediments. Sometimes secondary maghemite can form in these deeply weathered environments, such as parts of the Yilgarn and confuse the magnetic signature although this can be helpful in detecting palaeochannels as shown in Fig. 14.10. In Canada both the provincial and federal governments have flown large parts of the prospective areas and the data is available at low cost. Much of sub-Saharan Africa has also been flown although availability of data is varied (Reeves 1989).

DIRECT DETECTION OF SULPHIDES

Electrical methods have been very successful, particularly in Canada. Up to the 1970s this involved the use of airborne e.m. systems. Their limited depth penetration (<150 m) and the saturation coverage of prospective areas has resulted in the delineation and drilling of virtually all near surface conductors. More recent geophysical exploration is aimed at finding deposits at depth and has encouraged a switch to down hole methods.

Pemberton (1989) has provided a summary of e.m. responses from a number of Canadian VMS deposits. Figure 14.11 shows the detection of a VMS deposit in the Noranda camp using downhole pulse e.m. from a hole drilled some 100 m from the deposit at 700 m depth.

Airborne geophysics has been ineffective in intensely and deeply weathered areas, such as most of the Yilgarn Shield of Western Australia. The conductive nature of the overburden is responsible for the generation of near surface responses which mask signatures from the VMS.

Options that have been investigated include increasing transmitter power, shortening integration times and increasing the number of time channels used (Smith and Pridmore 1989).

A more general problem is the excellent e.m. response of graphitic sediments. It is extremely difficult, if not impossible, to differentiate graphite from massive sulphides and, as graphitic sediments are often found associated with breaks in the volcanic sequence, this has resulted in many thousands of metres of wasted drilling. However, some VMS deposits, such as the Trout Lake deposit in Manitoba are completely enclosed in graphite and would be missed if graphitic sediments were excluded from consideration.

14.3.10 Integration of exploration techniques

Perhaps the most important part is the incorporation of the factors mentioned above into an integrated exploration programme. A recent example of the successful use of integration is the discovery of the Louvicourt deposit near Val d'Or, Quebec (Fig. 14.12). The programme was aimed at discovering large deep extensions to the near surface Louvem Mine which had produced for about 10 years. Drilling of surface holes to 700 m, costing $C50 000 each, was undertaken on 300 m centres to test prospective horizons at depth (Northern Miner Magazine 1990). The spacing of 300 m was chosen as only large deposits were thought capable of supporting a mine at 600 m depth. Hole 30 intersected a stringer zone containing 0.86% Cu over 118 m but more importantly downhole e.m. surveys showed a significant off-hole conductor and rock geochemistry showed a major element anomaly. This suggested a major zone of hydrothermal alteration associated with a different horizon (the Aur horizon) than that of the Louvem Mine. A further phase of drilling intersected further stringer zone mineralization, leading to massive sulphides grading 11.8% Cu over 36 m in hole 42 and 2.8% Cu and 5.7% Zn over 43 m in hole 44. Resources are estimated at 28 Mt of 4.3% Cu, 2.1 % Zn, 27.8 g t^{-1} Ag and 1.03 g t^{-1} Au. The total cost of the discovery is put at $C 2.5M.

14.4 THE KIDD CREEK MINE

14.4.1 Regional setting

Kidd Creek lies within the Abitibi Belt of the Superior Province of the Canadian Shield (Figs 14.13 & 14).

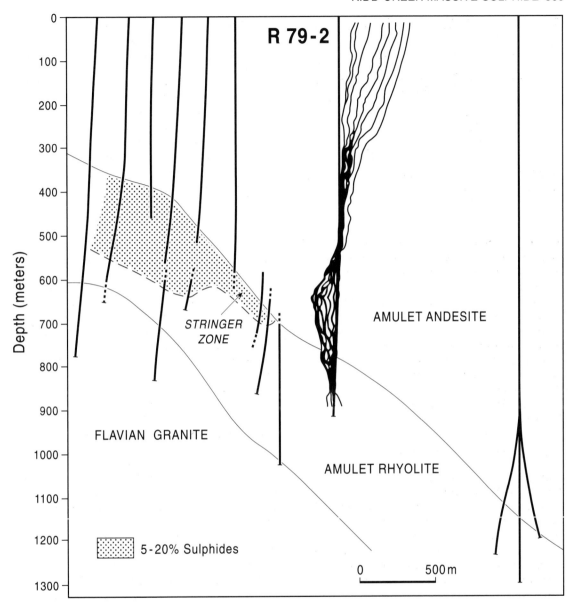

Fig. 14.11 Downhole pulse e.m. survey at the Ribago deposit, Noranda, Quebec. The survey clearly detects the sulphides 130 m from the hole. (After Pemberton 1989.)

With an exposed area of 1.6 million km^2 the Superior Province is the largest Archaean Province in this shield. The rocks are similar to those found at Phanerozoic subduction zones with differences due to higher mean mantle temperature and possibly greater crustal (slab?) melting at mantle depth (Hoffman 1989). The province is composed of subparallel ENE trending belts of four contrasting types: (1) volcanic–plutonic terranes, which

resemble island arcs; (2) metasedimentary belts which resemble accretionary prisms (Percival and Williams 1989); (3) plutonic complexes; and (4) high grade gneiss complexes. In the southern part of the Superior Province linear, ENE trending, greenstone–granite belts alternate, on a 100–200 km scale, with metasedimentary dominated terranes.

Archaean greenstone belts the world over appear to

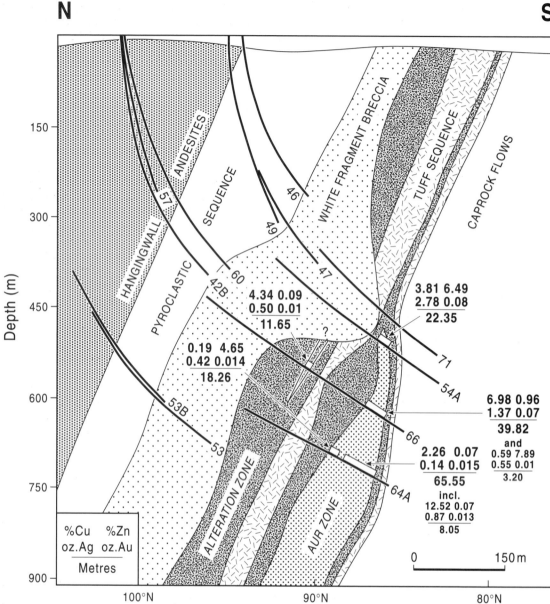

N   **S**

Fig. 14.12 Section of the Louvicourt deposit, Val d'Or . Note the blind nature of the deposit. (After *Northern Miner Magazine* 1990.)

be divisible into a lower dominantly volcanic section and an upper sedimentary section. The lower section may be divided into a lower part, of primarily ultramafic volcanics, and an upper volcanic part in which calc-alkaline or tholeiitic, mafic to felsic rocks predominate. The ultramafic (komatiitic) part also contains some mafic and felsic rocks and is thus bimodal. The tholeiitic

or calc-alkaline part is dominated by basaltic, andesitic and rhyolitic volcanics. With increase in stratigraphical height in greenstone belts there is *inter alia*: (1) a decrease in the amount of komatiite and (2) an increase in the relative amount of andesitic and felsic volcanics. The associated sediments are mainly exhalative chert and iron formation with some thin slates. The upper

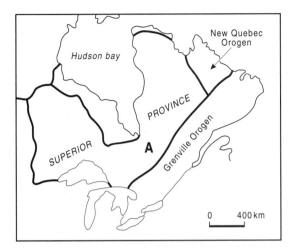

Fig 14.13 Location of the Superior Province of the Canadian Shield. A = position of the Abitibi Belt.

calc-alkaline and tholeiitic rocks and have so far been the best ground for finding VMS deposits. Large areas of the Superior Province formed at 2.7–2.6 Ga.

The Abitibi Belt is 800 km long and 200 km wide making it the world's largest, single, continuous Archaean greenstone belt. It contains a relatively high proportion of supracrustal rocks of low metamorphic grade resulting from an unusually shallow erosion level. This belt has been divided into eleven volcanic complexes (Sangster & Scott 1976) (Fig. 14.14) which are commonly many kilometres thick and contain one or more mafic to felsic cycles. VMS ores occur in just five of these complexes and are exclusively associated with calc-alkaline volcanic rocks, especially the dacitic and rhyolitic members which form 5.5–10% of the total volume of volcanic rocks in this belt and which belong to three different affinities whose importance have been stressed by Barrie *et al.* (1993). Production and reserves of VMS ores at 1990 totalled 424 Mt averaging 4.4 % Zn, 2.1% Cu, 0.1% Pb, 46 g t^{-1} Ag and ≈1.3 g t^{-1} Au (Spooner & Barrie 1993).

14.4.2 Exploration history

The discovery of Kidd Creek is extremely well documented and was a triumph for airborne geophysics and geological understanding. It is worth quoting the eloquent description of the discovery from a series of

sedimentary section is dominantly clastic.

There are regional differences in the volcanic character of greenstone belts that appear to be related at least in part to their age and erosion level. Thus the oldest belts of southern Africa and Australia are typically bimodal and enriched in ultramafics and mafics (which make them prime areas for nickel exploration) whereas the younger Canadian belts have a higher proportion of

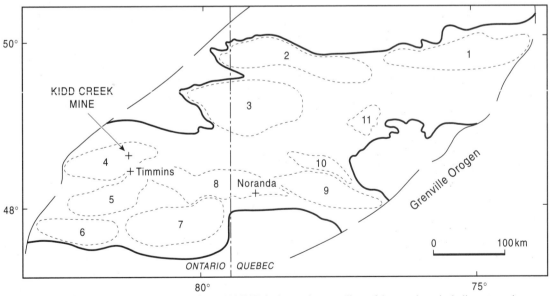

Fig. 14.14 Distribution of volcanic complexes in the Abitibi Belt. Approximate outlines of the complexes including cogenetic intrusions and sediments are shown by the heavy pecked lines. The complexes with significant developments of VMS deposits are 2, 4, 6, 8 and 9. (Modified from Sangster & Scott 1976.)

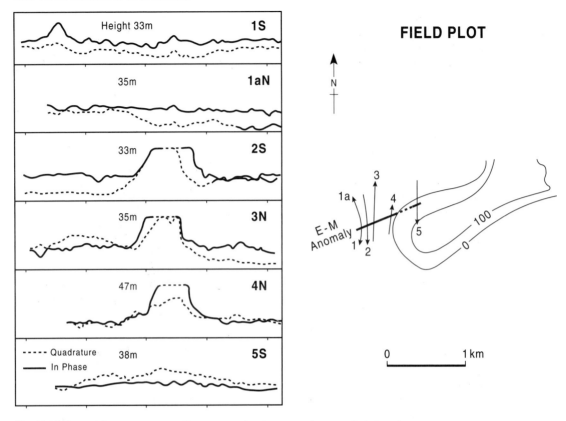

Fig. 14.15 Trace of first airborne anomalies over the Kidd Creek deposit, March 3 1959. (After Donohoo *et al.* 1970.)

papers (The Ecstall Story in *CIM Bull.* May 1974) describing the discovery and establishment of the Kidd Creek or Ecstall Mine, as it was originally known.

'The Ecstall orebody does not owe its discovery to a fantastic stroke of luck but to the foresight, dedication, tenacity and hard work of a Texasgulf exploration team. The orebody was found as a result of years of patient reconnaissance with the latest exploration equipment and using the most advanced technology. It was located only 15 miles north of a gold mining center that had been operating seventeen major mines over a period of fifty years. For most of those years, prospectors had combed every square foot of the area surrounding Timmins looking for gold. In fact, a prospector had his cabin on the rock outcrop that is now occupied by the mine surface crushing plant, just 100 yards from the orebody's southern limit. As most of the gold mines had been found by surface outcrops and simple geological interpretation, the prospector's search for the rich bonanza he felt was nearby was limited to these methods. The layers of muskeg and clay over the Kidd Creek orebody

protected it from these primitive efforts.' (Clarke 1974)

The geological thinking behind the exploration programme stemmed from the successful application of a syngenetic model. At the time of the development of the model (1954) most geologists believed that VMS deposits were replacement deposits and related to granites, although papers from both Germany and Australia had proposed an exhalative origin for these deposits (for discussion see Hodgson 1989, Stanton 1991). Walter Holyk recognized the close association of deposits with the contact between sericite-schists (rhyolites) and inter-volcanic sediment and applied the model successfully in conjunction with airborne e.m. surveys for Texasgulf in the Bathurst area of New Brunswick (Hanula 1982).

Holyk then persuaded his management to apply the same techniques and model to the Canadian Shield. He selected the Timmins area based on the known occurrence of the Kam Kotia deposit at the contact between rhyolite and sediments and the lack of known e.m. coverage. The first stage in the exploration programme was for the exploration geologist (Leo Miller) to make a compilation of the geology based on outcrop and airborne magnetics. The most promising areas were then flown by an e.m. system that had been mounted in a helicopter (Donohoo *et al.* 1970). Although the Kidd Creek anomaly was detected on the first day of the production flying programme (3 March 1959) little further work was done as the exact location was not known due to lack of up-to-date photo mosaic maps and the general area of the anomaly was at the boundary of four privately owned claims, originally granted to veterans of the Boer War. One of the claim owners was approached but would not option the claim and the area was left for the time being. The Kidd Creek anomaly (Fig. 14.15) was further recorded in routine flying in 1959 and 1960; this time accurate location of the anomaly was possible and 65 ha over the main anomaly was optioned from the administrators of a deceased estate. Once this ground had been secured Texasgulf undertook ground e.m. surveys to confirm the location of the conductor and to delineate drill targets. Three conductors were found and drill holes sited on the central anomaly (Fig. 14.16).

The first hole cut 8.37% Zn, 1.24% Cu and 3.9 oz t^{-1} Ag over 177 m in November 1963. It was obvious that a major deposit had been discovered and the rig was moved to the east of the property to confuse scouts from other companies who might get wind of the discovery. In addition the drill crew were kept in the bush until Christmas while lawyers optioned claims adjacent to the discovery site to protect the possible extension of the deposit. When land had been optioned in April 1964 further drilling began and the initial phase lasted until October 1965, totalling 157 holes and 33 500 m and this delineated the deposit to the 335 m (1100 ft) level. The thickness of the deposit and its subcrop rendered it suitable for open pit mining and the initial drilling was aimed at determining the ore reserves for this operation and the upper part of the underground mine. Definition of the continuation of the deposit has been from underground and continues to the present.

The primary phase of exploration drilling was designed to define the ore outlines by drilling on levels 120 m (400 ft) apart. This was followed by secondary drilling to define grade and ore outlines on which the mine layout was based. An example (Fig. 14.17) shows the secondary drilling in the shallow part of the underground mine. The aim was to have an intersection every 55 m (180 ft) vertically; this corresponds to the main levels and every other sub-level. Drill spacing between sections was normally 15m (50ft). All the core was

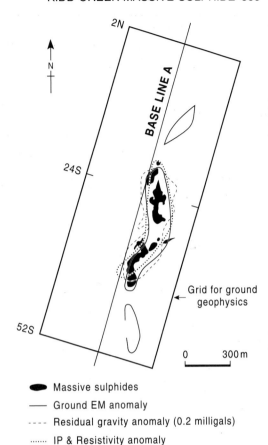

Massive sulphides

—— Ground EM anomaly

‑ ‑ ‑ ‑ Residual gravity anomaly (0.2 milligals)

········ IP & Resistivity anomaly

Fig. 14.16 Composite geophysical map of initial surveys over the Kidd Creek deposit. (After Donohoo *et al.* 1970.)

logged and zones of mineralization assayed (Matulich *et al.* 1974).

Drilling continues on the deposit at depth. In 1990 a total of 23 000 m was drilled. Underground exploration drilling accounted for 7 000 m, delineation drilling 5 500 m and surface exploration drilling 10 500 m (Fenwick *et al.* 1991).

Drilling around the mine proved the existence of a further blind massive sulphide lens, the Southwest Orebody, in 1977. It has subsequently been mined from 785–975 m and is considered to be a distal equivalent of the main north and south orebodies (Brisbin *et al.* 1991).The area around Kidd Creek has been much prospected without the discovery of other deposits and it is still an open question as to whether Kidd Creek is an isolated deposit or whether it belongs to a cluster, whose other members have been eroded away, or remain to be found by deep prospecting.

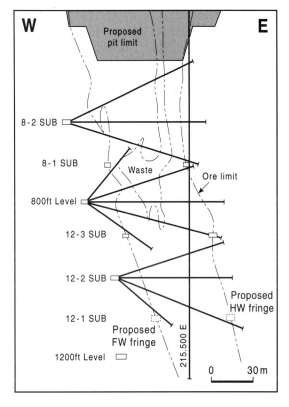

Fig. 14.17 Underground drill patterns used to define reserves. (After Matulich *et al.* 1974.)

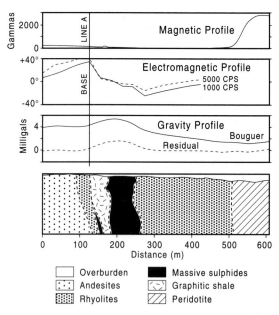

Fig. 14.18 Geophysical section on line 24 S of Fig. 14.16. (After Donohoo *et al.* 1970.)

14.4.3 **Exploration geophysics**

The discovery of the deposit prompted trials of a number of techniques to assess their response over the deposit (Donahoo *et al.* 1970). Besides e.m. with various loop configurations, I.P., gravity and magnetic ground surveys were also made. The ground e.m. surveys confirmed the three anomalies detected in the initial e.m. survey (Figs 14.16 & 14.18). The central anomaly also gave a very high contrast I.P. chargeability and resistivity anomaly as well as a 1.6 milligal residual gravity anomaly. Only magnetics failed to define the deposit and only succeeded in outlining the peridotite.

14.4.4 **Exploration geochemistry**

Although geochemistry was not used during the discovery of Kidd Creek a substantial research programme was undertaken immediately after the discovery to understand if surficial geochemistry would have been of use. Details can be found in Fortescue & Hornbrook

(1969) and there is an excellent summary in more accessible form in Hornbrook (1975).

Kidd Creek is typical of the problems of surficial geochemical exploration in the Abitibi Belt, an area known by soils scientists as the 'Clay Belt'. The area was covered, before mining and logging, by dense forest with swampy peats, known as muskeg. Underlying this is a thick (up to 65 m) sequence of glacial material: consisting of from top to base of clay till, varved clay and lower till. These glacial deposits have a blanketing effect on the geochemical signature and vertical dispersion from the subcrop of the deposit is limited to the lower till (Fig. 14.19) as the varved clays have a lacustrine origin and are not of local derivation. If the lower till is sampled by drilling a distinct dispersion fan can be mapped down ice from the deposit subcrop (Fig. 14.20). The more mobile zinc disperses further than copper.

The blanketing effect of the glacial overburden limits the surface response to zinc anomalies in near surface organic soils. These zinc anomalies are thought to form as a result of the decay of deciduous trembling aspen trees which tap the anomaly at relatively shallow depths (Fig. 14.21).

Recognition of the lack of surface geochemical expression of Kidd Creek and other deposits in the Abitibi

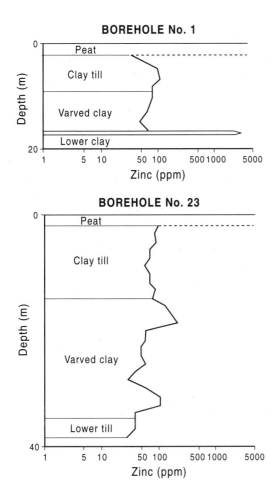

Fig. 14.19 Section through glacial overburden over the deposit and to one side of it; locations are in Fig. 14.20. Note the response is limited to lowest till and in the near surface in borehole 1. (After Hornbroook 1975.)

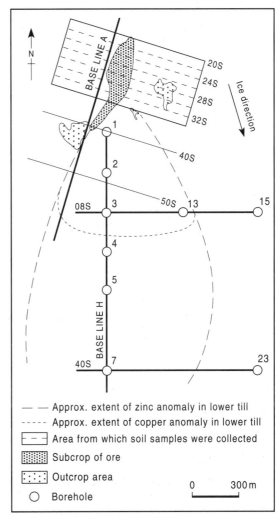

Fig. 14.20 Fan-shaped anomalies in the lower-most till at Kidd Creek. (After Hornbrook 1975.)

Clay Belt led to the widespread use of deep overburden sampling in geochemical exploration. The basal, locally, derived till is sampled by drilling and any dispersion fans can be mapped.

14.4.3 Mine geology

This has been discussed *inter alia* by Matulich *et al.* (1974), Walker *et al.* (1975), Coad (1984 & 1985) and Brisbin *et al.* (1991).

Host rocks. VMS deposits in the Kidd Creek area exemplify the classic case of exhalative deposits developed at, or close to, the top of an Archaean volcanic cycle

where the rhyolitic–dacitic top of the lower cycle is succeeded by a dramatic change to ultramafic and mafic volcanics at the base of the succeeding cycle. Kidd Creek is no exception. This deposit occurs in an overturned volcano–sedimentary sequence near the boundary between what are locally known as the Lower and Upper Supergroups (Figs 14.22 & 23).

The top of the Lower supergroup consists of a number of massive rhyolite sills intruded into a felsic series of pyroclastics containing intercalated rhyolite flows and tuffs. The uppermost massive rhyolites, which have a U–Pb zircon age of 2 717±4 Ma (Nunes & Pyke 1981), have pervasive crackle brecciation interpreted as result-

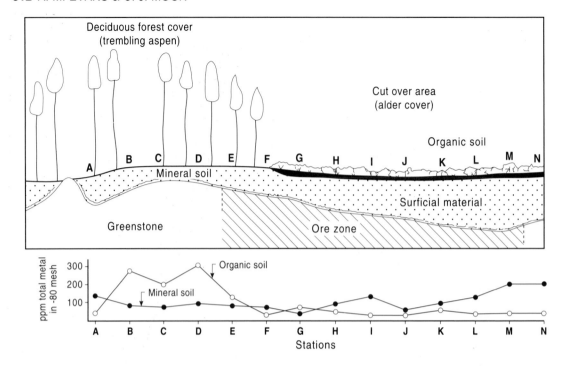

Fig. 14.21 Surface anomaly along line 24S of Fig. 14.20. The surface response in the organic soil is thought to be the result of the tapping of the deep anomaly by trembling aspen trees. (After Hornbrook 1975.)

ing from hydraulic fracturing. A stockwork chalcopyrite zone is present in this crackle-brecciated rhyolite with the massive ore lying immediately above. The orebody is stratigraphically overlain by a minor development of rhyolitic rocks which are succeeded by pillowed metabasalts of the Upper Supergroup. The rocks of the mine area are intruded by three large masses of high Fe tholeiite referred to as andesite–diorite

The rhyolites in the immmediate area around the deposit reach 300 m in thickness, in contrast to 100 m at a distance from the deposit, suggesting control of mineralization by sub-volcanic rifting as do the geometry of the alteration and the shape of the VMS deposit (Fyon 1991).

Mineralization. The orebody has a maximum thickness of 168 m, a strike length of 670 m and is known to extend to a depth of 1525 m. From 1966 to 1989 85 Mt of ore carrying 2.2% Cu, 0.28% Pb, 7.2% Zn and 102 g t^{-1} Ag. were mined. At the beginning of 1990 reserves above the 5600ft (1707m) level were calculated to be 44.6 Mt grading 3.53% Cu, 0.16% Pb, 5.08% Zn and 67 g t^{-1} Ag. For a number of years tin, which may locally reach 3%,

was recovered. There are three major ore types. In the stockwork there is stringer ore in which chalcopyrite stringers ramify crackle-brecciated or fragmental rhyolite. Pyrite or pyrrhotite accompanies the chalcopyrite and this ore averages 2.5% Cu. The massive ore is composed primarily of pyrite, sphalerite, chalcopyrite, galena and pyrrhotite. Silver occurs principally as native grains with accessory acanthite, tetrahedrite–tennantite, stromeyerite, stephanite, pyrargyrite and pearceite. Tin occurs as cassiterite with only a trace of stannite. Digenite, chalcocite and other sulphides occur only in trace amounts.

The massive sulphides are in part associated with a graphitic–carbonaceous stratum containing carbonaceous argillite, slate and pyroclastics. Associated with this horizon are fragmental or breccia ores that are considered to represent debris flows.

The deposit is divided into the North and South Orebodies by the Middle Shear. The orebodies can be divided into a number of zones and, as might be expected, copper-rich areas occur in the base of the massive ore. In these very high silver grades occur (up to 4 457 g t^{-1}) with minor associated gold values

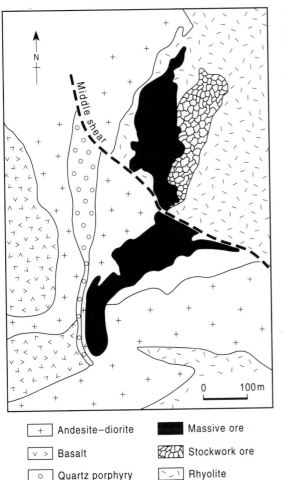

Fig. 14.22 Bedrock geological map, Kidd Creek. (After Beaty *et al.* 1988.)

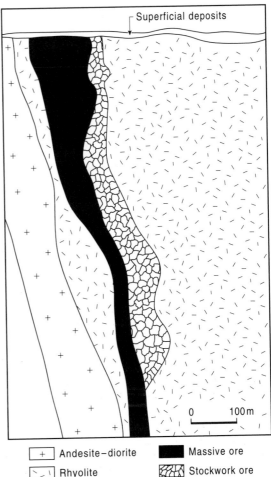

Fig. 14.23 West–east section through the North Orebody. (After Beaty *et al.* 1988.)

(see section 14.2.2).

Wall rock alteration. At both Kidd Creek and the Hemingway Property (a drilled area 1.3 km north of the Kidd Creek orebody containing minor base metal mineralization) early serpentinization affected the ultramafic flows and silicification the rhyolites. Later carbonate alteration was superposed on both by pervasive CO_2 metasomatism (Schandl & Wicks 1993). Strong chloritization and silicification occur in the footwall stockwork zone along the length of the orebody. Morton & Franklin (1987) assign Kidd Creek to the Mattabi-subtype of Primitive-type deposits although there is apparently no recorded evidence of a subjacent,

semiconformable, widespread alteration zone (Slack & Coad 1989).

Genesis. The Kidd Creek Orebody is interpreted as being of exhalative origin and, apart from its large size, is similar to many other deposits of Primitive-type. Beaty *et al.* (1988) have shown that the hydrothermally altered rhyolites are all markedly enriched in ^{18}O compared with most other VMS deposits. Altered rocks associated with the large Iberian Pyrite Belt deposits and Crandon, Wisconsin (62 Mt of ore) show similar ^{18}O enrichment (Munha *et al.* 1986) and for Kidd Creek Beaty *et al.* have suggested that evaporation and enrichment of normal sea water produced a high salinity fluid

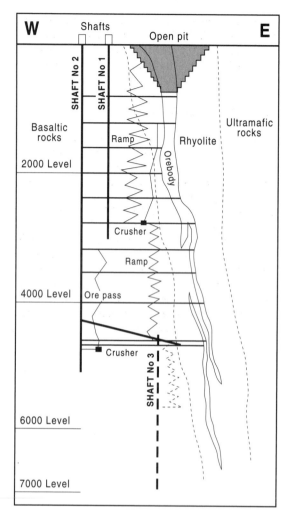

Fig. 14.24 Schematic east–west section through the Kidd Creek Mine showing location of shafts and ramp. (After Brisbin *et al.* 1991.)

enriched in ^{18}O that was capable of transporting large amounts of metals so accounting for the large size of the deposit.

Strauss (1989) has shown that the sulphur isotopic values are similar to those of many other Archaean VMS deposits. Maas *et al.* (1986) using Nd isotopic data suggested that some of the base metals in the ore were leached from the rhyolitic host rocks.

14.4.4 Mining operations

The original operators were Ecstall Mining Ltd, a wholly owned subsidiary of Texasgulf Inc., of the USA.

Following a 1981 takeover offer by Elf Aquitaine (SNEA) Kidd Creek Mines Ltd became a wholly owned subsidiary of the government-owned Canada Development Corp. In 1986 Falconbridge Ltd of Canada acquired Kidd Creek.

When the decision was made in March 1965 to build the open pit–railroad–concentrator complex two of the major problems were soils and transportation. The whole vicinity of the orebody was covered with thick deposits of weak materials: muskeg, till and clay. Only two small outcrops occurred by the pit site and one had to be reserved for the primary crusher. There was no good ground near the pit large enough for the concentrator and no roads or railroads to the pit area at that time. The concentrator was therefore built 27 km east of Timmins near a highway, a railroad, an electricity transmission line and a gas pipeline. It was linked to the mine with a new railroad.

The open pit was designed to have an overall waste-to-ore ratio of 2.58:1 but started with a 4:1 ratio (section 10.2.1). It was 789 by 549 m at the surface, ore was available from the pit on bench 1 by October 1965 and large scale production to supply the concentrator with 8 000 to 9 000 t p.d. was achieved by October 1967.

The pit had a projected production life of about ten years, however it was decided that there were advantages in developing an underground mine fairly quickly. These advantages included: the availability of underground ore for grade control, which would become more difficult as the pit deepened; the chance to develop new systems, equipment and skills, which was possible as a result of revolutionary changes in underground mining techology in the 1960s; and the acquisition of knowledge of local rock behaviour to aid in mine planning. Shaft sinking commenced in March 1970 and what became known as the Upper Mine with a 930 m deep shaft was developed. The 1 550 m deep Lower Mine shaft, 100 m SW of the No. 1 shaft was completed in 1978. The open pit was abandoned in 1979. Blasthole and sublevel caving stopes followed by pillar recovery using blasthole methods have been employed in the Upper Mine. In the Lower Mine a modified blasthole technique with backfill of a weak concrete mix is used so that no pillars are left. Waste is mixed with cement to form backfill at the 790 m level and then trucked to the stopes. The ore crusher station is on the 1 400 m level. A subvertical shaft, the #3 shaft, is being sunk to access the lower part of the deposit to a depth of 2110m. Overall production was 3.7 Mt from the #1 and #2 mines in 1990 and Kidd Creek operations employed 2 420 people (Fenwick *et al.* 1991).

14.4.5 **Rock mechanics**

Considerable use has been made of rock mechanics instrumentation such as extensometers, strain cells and blast vibration monitors to detect and predict ground instability leading to safer operations, the design of more stable underground workings and thus greater productivity.

14.4.6 **The concentrator and smelter**

The highly automated concentrator (section 10.2.2) has undergone a number of developments and improvements and by 1981 its capacity was up to 12 250 t p.d. (about 4.5 Mt p.a.) and it was producing seven different concentrates: copper concentrates with high and low silver contents, two similar zinc concentrates, lead, tin and pyrite concentrates. After about nine years of planning and the testing of various smelting processes, a smelter and refinery complex was built near Timmins and went into production in June 1981. This complex treats two-thirds of the copper concentrates and produces 90 000 t of cathode copper. Most of the rest of the copper concentrate is custom smelted and refined by Noranda Inc. Much of the zinc concentrate is treated in the electrolytic zinc plant which can produce 134 000 t zinc metal p.a. The remaining zinc concentrates and all the lead concentrates are sold to custom smelters.

14.5 **CONCLUSIONS**

VMS deposits can constitute very rewarding targets as they can be large, high grade, polymetallic orebodies like Kidd Creek. They should be sought for using an integrated combination of geological (section 14.2, 14.3 & 14.4.1), geochemical (sections 14.3.8 & 14.4.4), geophysical (sections 7.4, 7.9, 7.10, 14.3.9 & 14.4.3) and drilling (section 14.3.10) methods. In the immediate World War Two years, when geochemical prospecting was in its infancy, the use of airborne e.m. led to many successful discoveries in the Canadian Shield. Experience and research since then has shown that sophisticated geochemical procedures (section 14.4.4) and other geophysical methods can play an important part in the search for more VMS orebodies.

15

Disseminated Precious Metals—
Trinity Mine, Nevada

M. K. G. WHATELEY, T. BELL AND C. J. MOON

Disseminated precious metal deposits, particularly of gold, have been the most popular target for metallic exploration through the 1980s. Their popularity was based on the high price of gold during this period, as discussed in section 1.2.3 METAL AND MINERAL PRICES, the low cost of mining near surface deposits in open pits and improvements in gold recovery techniques, in particular the development of the cheap heap leach method to recover precious metals from low grade oxidized ores. This case history deals with the evaluation of the Trinity silver mine.

15.1 BACKGROUND

15.1.1 Overview of deposit types

A variety of deposit types have been mined in this way but the most important in Nevada (discussed in detail by Bonham 1989) are.
1 Sediment hosted deposits of the Carlin type.
2 Volcanic hosted deposits, such as Round Mountain.
3 Porphyry related systems rich in gold.
4 Hot spring type.
5 Metamorphic veins.

CARLIN TYPE DEPOSITS

These deposits are responsible for the majority of gold production in Nevada, 172 t in 1989. Typically the gold occurs as micron-sized grains which are invisible to the naked eye ('noseeum' gold), within impure limestones or calcareous siltstones. The host rocks appear to be little different from the surrounding rocks and the margins of the deposit are only distinguishable by assay. An excellent summary is given in Bagby and Berger 1985, updated in Berger and Bagby 1991.

Individual deposits occur mainly within major linear trends, up to 34 km in length, the most important of which is the Carlin Trend in NE Nevada. This trend and type of deposit are named after the Carlin Mine, which was the first significant deposit of this type to be recognized, although others had previously been mined. The deposits vary from broadly tabular to highly irregu-

lar within favourable beds and adjacent to structures such as faults that have acted as fluid pathways. The faulting and fluid flow has often resulted in brecciation and several deposits are hosted in breccias.

Silicification is the commonest form of alteration and the occurrence of jasperoid, which replaces carbonate, is common in many deposits. Gold is thought to occur predominantly as native metal although its grain size is very small usually <1 μm and the metal's exact mineralogical form remains unknown in some mines. Gold forms films on sulphide and amorphous carbon and is particularly associated with arsenian pyrite. Pyrite is the most common sulphide and may be accompanied by marcasite and arsenic, antimony and mercury sulphides. The arsenic sulphides realgar and orpiment, which are readily distinguishable by their bright colour, occur in many deposits as does stibnite and cinnabar. The general geochemical association is of As, Sb, Tl, Ba, W and Hg in addition to Au and Ag.

The deposits are spatially associated with granites and many authors favour intrusion driven hydrothermal systems (Fig. 15.1). Berger & Bagby (1991) favour a mixing model in which magmatic fluids have been mixed with meteoric water causing precipitation in favourable host rocks.

VOLCANIC HOSTED SYSTEMS

Much of the bonanza style precious metal mineralization in the western USA is hosted in volcanic rocks, for example, the Comstock Lode which produced 5890 t Ag and 256 t Au at an average grade of about 300 g t⁻¹ Ag and 13 g t⁻¹ Au in the late nineteenth century (Vikre 1989). More recent exploration has resulted in definition of bulk mineable targets such as Round Mountain with reserves of 196 Mt of 1.2 g t⁻¹ Au. Volcanic hosted deposits can be divided in to three general types based on mineralogy and wall rock alteration: acid sulphate (or alunite–kaolinite), adularia sericite and alkalic (Heald *et al.* 1987, Bonham 1989, Henley 1991).

The acid sulphate type generally contains more sulphide and is characterized by the occurrence of the copper sulphide, enargite. Other base metal sulphides

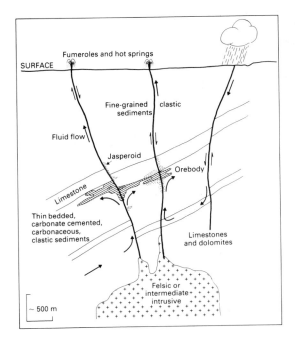

Fig. 15.1 Schema for the formation of Carlin style deposits.(After Sawkins 1984.)

are frequently present as is pyrite. Advanced argillic alteration is common as is alunite ($K_2Al_6(OH)_{12}(SO_4)_4$) and kaolinite. The adularia sercite type generally has less sulphide and sercitic to intermediate argillic alteration. Alunite may be present but it is only of supergene origin. The structural setting of both types may be similar (Fig. 15.2) but the differences between the two types probably reflects their distance from the heat source (Heald *et al.* 1987). There are obvious parallels between the formation of these deposit and modern geothermal systems. Evidence from modern day geothermal systems, such as the deposition of gold on pipes used to tap the systems for geothermal power, suggests that the gold may have been deposited rapidly. Several major gold deposits occur in alkalic volcanics that may be spatially associated with alkalic porphyries. The deposits are characterized by the occurrence of the gold as tellurides with quartz–carbonate–fluorite–roscoelite (vanadium mica)–adularia alteration and regional propylitic alteration (Bonham 1989).

GOLD RICH PORPHYRIES

Gold is of major economic importance in several porphyry deposits that produce copper as their main product, such as Ok Tedi (section 10.7.3). More recent

research (Sillitoe 1991) suggests that there may be related copper deficient systems that contain economic gold contents. The deposits are associated with granitic intrusives that vary in composition but with quartz monzonite and diorite as the most important hosts. Alteration shows a characteristic concentric zoning from inner potassic through phyllic to propylitic zones. Gold generally correlates with copper grade and is associated with K alteration (Sillitoe 1991). In some cases high sulphidation gold systems are mined at higher levels than the porphyry although a direct connection is hard to prove. For further discussion of porphyry copper deposits see Evans (1993).

HOT SPRING DEPOSITS

These appear to be a variant of the volcanic type and are characterized by the occurrence of a recognizable palaeosurface that is usually associated with a siliceous sinter zone interbedded with hydrothermal breccia (Bonham 1989). The sinter passes downward into a zone of silicification and stockwork with a zone of acid leaching. The precious metals are generally restricted to within 300 m of the palaeosurface and probably reflect rapid deposition. There is an overall spatial association with major centres of andesitic to rhyolitic volcanism.

METAMORPHIC VEINS

Vein hosted deposits have formed the basis for several bulk mineable deposits in the western USA particularly in the Mother Lode system of NE California where the Jamestown Mine was based on reserves of 12.7 Mt of 2.47 g t⁻¹ Au. The veins are associated with major structures and hosted in a variety of lithologies, metavolcanics, slates and mafic intrusives. The veins tend to be limited in size with a chlorite–carbonate–sericite alteration assemblage. Associated elements are As, Sb, Te, W and minor Pb, Zn and Cu. Isotopic evidence indicates deposit formation from deeply circulating meteoric fluids although magmatic fluids derived from felsic plutonism may also be present.

15.1.2 **Geochemistry**

One of the major features in the discovery of disseminated gold deposits has been the further developments in gold geochemistry. The recognition of 'noseeum' gold deposits depends on the ability to detect fine-grained gold. As this is usually in the micron size class it does not often form grains large enough to be panned. Much gold

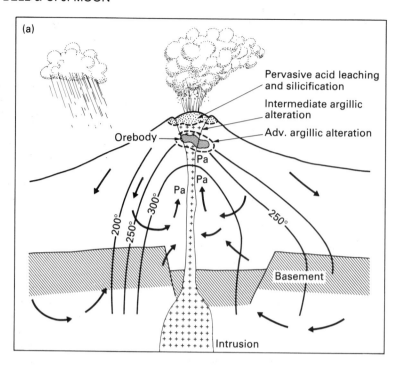

(a)

Pervasive acid leaching
and silicification

Intermediate argillic
alteration

Adv. argillic alteration

Orebody

Pa

Pa

Pa

Pa

200°
250°
300°

250°

Basement

Intrusion

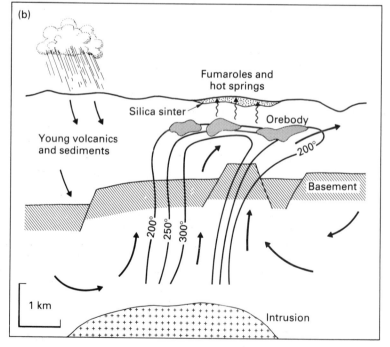

(b)

Fumaroles and
hot springs

Silica sinter

Orebody

Young volcanics
and sediments

200°

Basement

200°
250°
300°

1 km

Intrusion

Fig. 15.2 Schemata for the formation
of two types of epithermal precious
metal deposit in volcanic terranes,
based on Heald *et al.* (1987) with
modifications from Henley (1991). (a)
Acid sulphate type. Pa = propylitic
alteration. Note that the mineralization
occurs within the heat source.
(b) Adularia–sericite type. The
upwelling plume of hydrothermal fluid
is outlined by the 200°C isotherm. The
mushroom-shaped top reflects fluid
flow in the plane of major fracture
systems, a much narrower thermal
anomaly would be present perpendicular to such structures. The heat source
responsible for the buoyancy of the
plume is shown as an intrusion several
kilometres below the mineralized zone.
In both schemata the arrows indicate
circulation of meteoric water. (a) is
drawn to the same scale as (b).

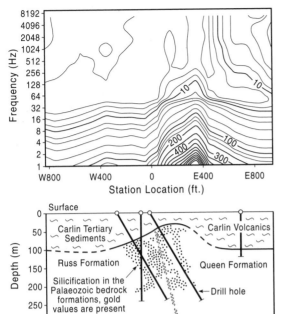

Fig. 15.3 Section showing the application of a CSAMT survey in detcting areas of silification under Tertiary cover. Top: contoured data; bottom: drill section. (After Paterson & Hallof 1991.)

remains encased in fine-grained silica and therefore sampling and analytical techniques must be devised that accurately determine all the gold present. This analytical problem has largely been overcome, firstly by the improvement of graphite furnace AAS and more latterly by the application of Neutron Activation Analysis and ICP–MS (section 8.2.3). Graphite furnace AAS methods have relied on the separation of gold from the rock or soil matrix using an organic solvent. This allows analysis with detection limits of <10 ppb.

As gold is present as discrete grains within the sample either encased in the matrix or as free gold, sampling methods have to be devised that will allow the analytical aliquot to represent accurately the original sample. Investigations by Clifton *et al.* (1969), well summarized by Nichol *et al.* (1989), have shown that it is necessary to have 20 gold grains in each sample to achieve adequate precision. In real terms this means taking stream sediment samples of around 10 kg and carefully subsampling. The geologist who takes a 1 kg sample can not expect to make an accurate interpretation of analytical data. An alternative strategy, which has been used where rocks have been oxidized, is to take a large sample and leach it with cyanide; this technique is known as Bulk

Leach Extractable Gold (BLEG). This technique has found wide acceptance in Australia and The Pacific Rim countries.

Most but not all gold deposits have concentrations of pathfinder elements associated with the gold. In Nevada the most widely used elements are As, Sb, Hg, Tl and W for sedimentary hosted deposits with the addition of base metals for most deposits. Although pathfinder elements can be extremely informative as to the nature of the gold anomaly they should not be used without gold as pathfinder elements may be displaced from gold. In addition some gold deposits, particularly adularia–sericite-type deposits have no pathfinder signature and the exploration geologist must rely largely on gold geochemistry. An example is the Ovaçik deposit in Turkey (section 4.2.1) that was discovered by the follow-up of BLEG geochemistry.

In Nevada large stream sediment samples and extensive soil sampling have proved effective where overburden is residual. However in some areas the overburden has been transported and surface sampling is ineffective. Chip sampling of apparently mineralized rocks, particularly jasperoids, has been successful although jasperoidal anomalies are often slightly displaced from the associated deposits.

There are a large number of case histories on gold exploration and the reader is advised to read Zeegers & Leduc (1991) and case histories on Nevada in Lovering and McCarthy (1978).

15.1.3 Geophysics

Disseminated gold deposits are difficult geophysical targets. Electrical and e.m. methods have been used to map structure and identify high grade veins, for example, at Hishikari, Japan (Johnson & Fujita 1985) but are not usually effective in detecting mineralization except indirectly. On occasion unoxidized sulphides are present allowing the use of I.P. One of the more successful uses of geophysics in Nevada has been the detection of silicified bedrock under Tertiary cover (Fig. 15.3). The section shows the results of a Controlled Source Audio Magneto Telluric Survey that clearly detect a zone of silicification under 50 m of cover. Further case studies can be found in Paterson & Hallof (1991).

15.1.4 Mining and metallurgy

The recognition that disseminated precious metal deposits could be successfully bulk mined and processed, sometimes with spectacular returns, has been responsible for the allocation of the large exploration budgets spent in Nevada in the 1980s. The operations are invariably open pits mining oxidized deposits, usually with low stripping ratios. One of the keys to a successful operation is effective grade control (section 15.5) as different grade ores are treated differently in most operations. The higher grade material is usually milled and extracted with cyanide in a conventional manner whereas the lower grade ore is crudely crushed and placed on leach pads. These leach pads are built on plastic liners, sprinkled with cyanide, using a system similar to a garden sprinkler, and the cyanide allowed to percolate through the crushed ore. The cyanide is then pumped to a central treatment plant where the gold is extracted by passing the cyanide through activated carbon. Marginal ore and waste are stockpiled separately to allow for the later treatment of marginal material if the gold price increases.

One of the interesting developments in Nevada is the recognition that some of the low grade material has previously unsuspected high grade material with different metallurgical characteristics beneath it. For example at the Post deposit on the Carlin trend high grade material in which gold is difficult to extract, or refractory, underlies the main surface deposit. The grade of this material is so much higher that it is worthwhile investing in underground mining and pressure leach extraction to recover the gold.

15.2 TRINITY MINE, NEVADA

The Trinity Silver Mine was an open pit, heap leach, silver mining operation extracting rhyolite hosted, disseminated, hydrothermal, silver oxide mineralization. The mine was brought into production on 3 September 1987 and mining ceased on 29 August 1988. The heap leach operation continued for just over one more year. Our choice of a silver mine rather than a gold producer has been governed entirely by the availability of data. Most of our discussion is of course applicable to any disseminated precious metal deposit. US measures are used throughout this chapter as the mining industry in the USA has yet to metricate. Thus oz t^{-1} are troy ounces per short ton of 2 000 avoirdupois pounds (lb).

This case study focuses upon the statistical assessment of blast hole assay data, for the purpose of determining the distribution and quality of mineralization. It is essential for reliable ore reserve calculations to derive the best grade estimate for an orebody. The actual mined reserves and estimated reserves calculated by US Borax and Chemical Corporation (US Borax) from exploration

borehole data are also described. The Surpac Mining System Software was used for data management and statistical evaluation. This work was undertaken as an MSc dissertation at Leicester University (Bell 1989) under the auspices of US Borax.

The Trinity Mine is in Pershing County, NW Nevada (Fig. 15.4), on the NW flanks of the Trinity Range, 25 km NNW of Lovelock. The area is covered by the SW part of the USGS Natchez Spring 7.5' topographic map and Pershing County Geological Map. Elevations vary from 1200 m to 2100 m (3900 to 7000 ft). The mineralization at Trinity principally lies between 1615 m and 1675 m (5300 and 5500 ft).

15.3 EXPLORATION

15.3.1 Previous work

Mineralization was first discovered in the Trinity Range by G. Lovelock in 1859 and limited, unrecorded production from Pb–Ag and Ag–Au–Cu–Zn veins occurred between 1864–1942 (Ashleman 1983). Evidence of minor prospecting during the 1950s was found within Willow Canyon (Fig. 15.5) (Ashleman 1988). Geophysical exploration and trenching during the 1960s was carried out by Phelps Dodge within Triassic sediments to the north of Willow Canyon. Later a geochemical exploration programme aimed at locating gold mineralization was undertaken by Knox–Kaufman Inc. on behalf of US Borax, but they were unable to locate previously reported gold anomalies. However in 1982 a significant silver show was found within altered rhyolites. As a result, agreements were reached between five existing claim holders, Southern Pacific Land Company (SPLC), holders of the surface rights and US Borax, after which the Seka claims were staked. The Trinity joint venture between Sante Fe Pacific Mines Inc., the land holder and a subsidiary of SPLC, and Pacific Coast Mines Inc., a subsidiary of US Borax, was operated by US Borax and mining was contracted to Lost Dutchman Construction.

15.3.2 US Borax exploration

Various stages of exploration were completed by US Borax during the period 1982 to 1986. In 1982 and 1983 mapping, geochemical and geophysical techniques were used to located payable sulphide mineralization. In 1983 drilling commenced within an area known to host sulphide mineralization. Exploration continued from 1985 to 1986 in an attempt to find additional reserves. Although assessment of the sulphide ore continued with

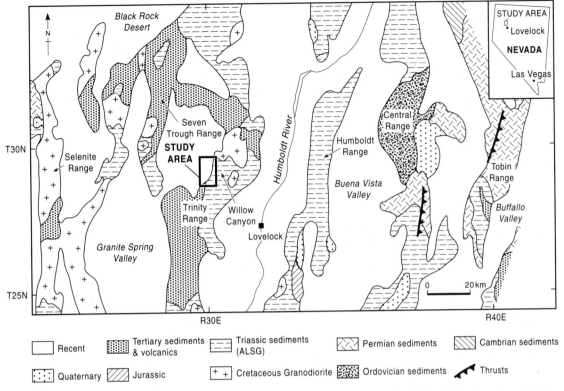

Fig. 15.4. Sketch map of Pershing County, Nevada showing the outline geology and location of the Trinity silver deposit. (After Ashleman 1988.) ASLG = Auld Lang Syne Group.

metallurgical testing and feasibility studies, the possibility of a high grade oxidized zone to the south-west of the main target area became the focus of attention. Efforts were thereafter concentrated upon the evaluation of this south-western extension (Fig. 15.5).

15.3.3 Mapping

Initial mapping of the claim area at 1:500 was followed by detailed mapping of the sulphide zone at a scale of 1:100. Mapping of structure and lithology helped define the surface extent of mineralization and alteration, and the controls of the distribution of mineralization. When the potential of the silver mineralization within the oxidized zone was established in mid 1986, detailed mapping of the south-western extension commenced. Mapping showed the host rock, of both oxide and sulphide mineralization, to be fractured rhyolite porphyry and defined an area of low grade surface mineralization intruded by barren rhyolitic dykes (Fig. 15.5). These barren dykes divide the SW extension into NE and

SW lobes. The rhyolite dykes were seen to be intruded along E and NNE faults and to disrupt the stratigraphical sequence within the SW lobe.

15.3.4 Geochemistry and geophysics

A geochemical soil survey was completed over the oxide and sulphide zones and samples were analysed for Pb, Zn and Ag. The mobility of Ag and Zn in the soil horizons placed constraints upon the significance of anomalies reflected by these elements. Lead is more stable and was therefore used as a pathfinder element for Ag mineralization. Significant Pb anomalies were located over the sulphide zone. Anomalies of greater than 100 ppm defined potential mineralized areas and higher levels (>1000 ppm) coincided with the surface intersection of Ag mineralization. Surface rock samples were taken as part of a reconnaissance survey, and contour maps of the results helped define the extent of surface mineralization.

Between 1985 and 1986 rock geochemistry was ap-

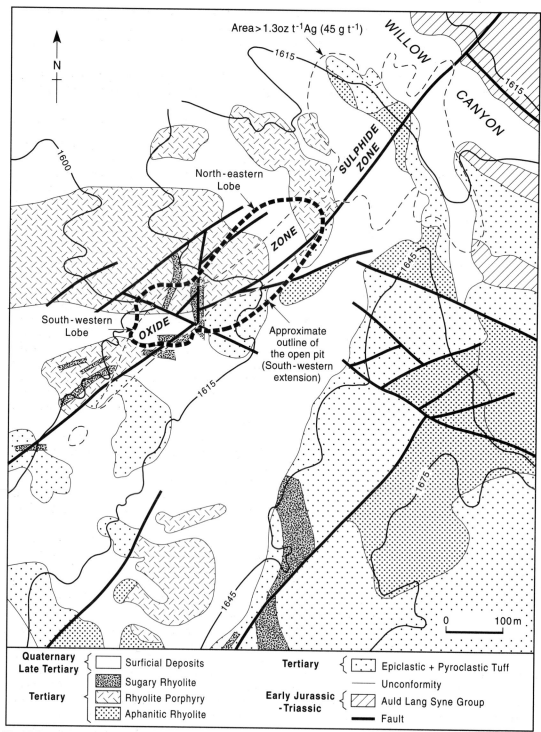

Fig. 15.5 Sketch map of the geology in the vicinity of the Trinity silver deposit, Nevada. (After Ashleman 1988.)

plied to the oxide zone in the SW extension. Rock chips from drilling, trenching and surface exposure were analysed for Pb, Zn, Ag and Au. High grade samples containing 30–50 ppm Ag were reported and localized samples containing 40–90 ppm Ag were recorded as the zone of oxidation was traced southward along strike. Anomalies from the rock and soil geochemical surveys were used in the planning and layout of the drilling programme within the sulphide zone.

Geophysical surveys, including gradient array IP, magnetic and gamma ray spectrometry, were carried out concurrently over the sulphide mineralization. The gradient I.P. survey required three arrangements of transmitting electrodes with a pole spacing of 12 000 ft (3 650 m). Readings were taken every 500 ft (150 m) along lines spaced 1000 ft (300 m) apart (Ashleman 1984). A time domain chargeability anomaly coincided with the area of sulphide mineralization. A NNW trending belt of high resistivity was located adjacent to the eastern margin of an area of known subsurface mineralization. The magnetic survey proved to be of little value due to the poor contrast in magnetic values. Gamma ray spectrometry using K, Th and U channels was tested, but only subtle differences were seen between altered and unaltered rocks and the suvey was discontinued (Ashleman 1984).

15.3.5 Drilling

Successive stages of drilling on a 100 ft (33 m) grid orientated 11° west of true north (Fig. 15.6) totalled 29 880 m (98 031 ft) drilled (Table 15.1). Most of the drilling was within areas of principal sulphide mineralization, focusing on the surface exposure of silver mineralization, lead and I.P. anomalies and areas of favourable geology. Drilling defined a body of silver mineralization dipping to the west.

The drilling grid was extended to explore for additional reserves and high grade silver oxide mineralization was intercepted to the SW (Fig. 15.5) In 1985 drilling was concentrated within the SW extension. Ninety-two of the holes were drilled to define the oxide zone. Although mineralization was located south of the SW extension along strike and at depth, it was noted to be discontinous.

Diamond drill holes were sampled every 1–3 ft (0.3–0.9 m) for metallurgical testing, while percussion holes were sampled as 5 ft (1.5 m) composites (Ashleman 1986). Samples were sent to the laboratory and analysed for Ag, Au, Pb, Zn and As. Analyses on the samples from the exploration holes were carried out using atomic absorption spectrometry (AAS), which gave the total silver content. Samples recording Ag in excess of 1000 ppm were reanalysed using fire assay.

15.4 THE GEOLOGY OF PERSHING COUNTY AND THE TRINITY DISTRICT

15.4.1 Regional setting—Pershing County

The Trinity Mining District of Pershing County is part of the Pacific Rim Metallogenic Belt. The mountain ranges of Pershing County reflect the surface expression of major NE–SW structural lineaments sequentially repeated (basin and range) across Nevada. Much of the present geological setting is the response to regional compression subsequently modified by extension. The regional geology is outlined in Fig. 15.4.

Cambrian to Permian shallow marine carbonates and clastics are exposed to the NE and E of Pershing County. The Harmony Formation (U. Cambrian) is composed of quartzite, sandstone, conglomerate and limestone which pass laterally into clastic and detrital sediments of the Valmy Formation (Ordovician). Carboniferous to Permian sediments (limestone, sandstone and chert) and volcanics (andestic flows and pyroclastics) form most of the Tobin Range to the east. During the Mesozoic fault bounded, deep water, sedimentary basins dominated the area, with associated widespread volcanic and silicic intrusive activity.

Triassic cover is widespread in central Pershing County. The lower units to the east consist of interbedded andesites, rhyolites and sediments, locally intruded by leucogranites and rhyolite porphyries. To the SE the Middle Trias is exposed as calcareous detrital and clastic sediments, limestone and dolomite. The upper unit which forms the major sedimentary cover within the Trinity, South Humbolt and Severn Troughs Range, defines cyclic units of sandstone and mudstone with limestone and dolomite. Adjacent to Cretaceous granodiorite intrusions, the Triassic to Jurassic sediments are locally metamorphosed to slates, phyllites, hornfels and quartzite.

The Auld Lang Syne Group (ALSG) forms the majority of the Triassic sedimentary sequence and reflects the shallow marine conditions of a westerly prograding delta (Johnson 1977). Tertiary basaltic and andestic volcanics pass into a thick, laterally variable, silicic, volcanoclastic sequence consisting of a complex pile of rhyolite domes, plugs, flows and tuffs. The Tertiary rhyolitic volcanics are overlain and interdigitate with lacustrine and shallow lake sediments, later intruded by

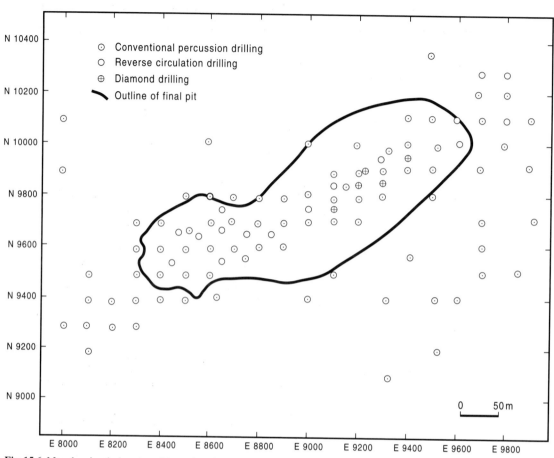

Fig. 15.6 Map showing the location of the exploration boreholes in relation to the final outline of the Trinity mine open pit from which the oxide mineralization was extracted.

Table 15.1 Summary of drilling exploration programmes

Drilling methods	Number of holes	Total meterage	
Percussion	199	22 857	(74 990 ft)
Reverse Circulation	39	6 257	(20 530 ft)
Cored	10	765	(2 511 ft)
Total	**258**	**29 879**	**(98 031 ft)**

silicic dykes, sills and stocks. The Tertiary sequence is covered by Quaternary to Recent gravel and alluvial deposits. The Miocene to Pliocene deposits form a thick stratified, well bedded sequence of tuffs, shales, sandstone, clay and gravel, locally interbedded with basaltic and andestic flows.

Various structural phases are apparent within Pershing County (Johnson 1977). Pre-Cenozoic large-scale compressional folding and thrusting during the Sonoma Orogeny (mid Jurassic) resulted in the south and south-

Table 15.2 Outline of silver deposits of Nevada, including a comparison with the Trinity deposit. (Adapted from Smith 1988.)

	Disseminated	Vein	Trinity
Host rock	Tertiary felsic vocanics & clastic sediments, Devonian limestone	Tertiary andesites and rhyolites	Teritary porphyry rhyolite, minor occurrence in tuff & argillite
Type of mineralization	Discrete grains and vein stockworks	Banding in quartz veins	Disseminations, microfractures, veinlets & breccia infill (vein stockwork)
Morphology	Tabular, podiform bodies, defined by assay cut-off of <300 g t^{-1}	Veins defined by fault contacts with assay cut-off between 200–300 g t^{-1}. Change in cut-off has little effect on tonnage	Lenticular & stacked tabular, sharp assay boundaries of < 30 g t^{-1}
Mean grade	270 g t^{-1}	500 – 800 g t^{-1}	160 – 250 g t^{-1}
Geological control	Structural, rock permeability & proximity to conduit	Structural & fracture density, intersection of fault & fractures	Structural, rock permeability, fault intersections, percentage fracture density
Zoning	Increased mineral content with depth, high Pb–Zn–Mn, low Au, Ag:Au>100	Sharp vertical zoning, sulphides increase with depth, Ag:Au>50	Sulphide-oxide zoning, high Ag:Au
Mineralogy	Acanthite, tetrahedrite, native Ag, pyrite & Pb–Zn base metals	Acanthite, low pyrite & sulphide content	Sulphides: friebergite–pyragyrite, chalcopyrite, pyrite, stannite. Oxides
Alteration	Silica–sericite (cabonate, clay, chlorite, potassic feldspar)	Extensive & variable, hematite–magnetite, propylitic, adularia & illite	Silicification, adularia–quartz–sericite, minor propylitic, illite & kaolinite
Process of formation	Hypogene	Hypogene, hydrofracturing & brecciation	Epithermal & hydro-tectonic fracturing
Hydrothermal solutions	165–200°C, 5 wt% NaCl, low salinity ground water mixing with oilfield brines	200—300°C, meteoric solutions, precipitation induced by mixing, boiling & wall rock altertation	Not available

east movement of basinal sediments and volcanics over shelf sediments. These folded, imbricate, thrust slices were later offset by dextral strike–slip faulting that has NW arcuate and discontinuous trends (Stewart 1980). Such thrusting is apparent to the E and SE within the Central and Tobin Ranges (Fig. 15.4). Crustal extension during the Cenozoic resulted in northerly trending basin and range faulting. Associated with the basin and range faulting are effusive phases of bimodal volcanism. Stewart (1980) records both pre- and post-basin and range NE and EW fault sets, and implies possible strike slip components to the normal faults.

Tertiary silver mineralization is reviewed by Smith (1988), who classifies the silver bearing deposits into three categories; disseminated type, vein type and carbonate hosted. The former two types are outlined and compared with the Trinity deposit in Table 15.2. A further classification into four subtypes (Bonham 1988), includes volcanic hosted epithermal, silver-base metal limestone replacement deposits, sedimentary hosted stockworks and volcanic hosted stockworks. Pershing County is dominated by Pb–Zn–Ag–Au veins hosted within and adjacent to Cretaceous granodiorites, with minor occurrences in Triassic sediments. Mineralization tends to be concentrated in shear zones and along hornfels dykes forming pinch and swell structures along oblique joint sets (Bonham 1988).

15.4.2 The Geology of the Trinity Mine

STRATIGRAPHY

The local stratigraphy of the Trinity Mine and surrounding area is outlined in Fig.15.5 and Table 15.3. The ALSG is described by Johnson (1988) as a fine-grained clastic shelf and basin facies with interbedded turbidites. The local silicic and calcareous units of the ALSG (Ashleman 1988) have been subjected to both low grade regional metamorphism and contact metamorphism. The latter is associated with the intrusion of Cretaceous granodiorite dykes and stocks located to the NE of the mine (Fig. 15.4). Outcrops of the ALSG to the south are dominated by phyllites and slates which grade into coarse-grained facies (siltstones, sandstones and quartzites) northwards. To the east the unit is represented by carbonaceous siltstones and impure limestones (Ashleman 1988). The transition between the Tertiary rhyolite volcanics and the ALSG is marked by an argillite breccia. The breccia is fine-grained and matrix supported, it contains semi-angular to angular argillite clasts and is closely associated with faulting.

Tertiary volcanism resulted in a complex sequence of rhyolitic flows superseded by epiclastic and pyroclastic deposits. The Tertiary rhyolites dominate the geology of the Trinity area, with extensive exposure to the south and flanking the Trinity Range to the north. Latitic and 'sugary' rhyolitic dykes disrupt the sequence, with effusive phases and breccias spatially associated with rhyolite domes. The volcanics are capped by Pliocene to Pleistocene fluvial sediments and locally by basaltic flows. Quaternary alluvium infills valleys and flanks the Trinity Range forming channel fill and alluvial fan deposits.

STRUCTURAL GEOLOGY

Polyphase deformation involving compressional tectonics and granodiorite intrusions resulted in two phases of metamorphism of the Mesozoic sediments: 1) low grade regional and 2) contact metamorphism. Pre-Tertiary folding and faulting of the ALSG resulted in isoclinal folds (Nevadian Orogeny). Tertiary deformation produced large scale, north trending open folds and low to high angle faulting.

Mid Miocene tectonic activity marked the onset of extensional tectonics, resulting in NNE and NE trending, high angle, basin and range faulting. This was preceded by N, NNW and WNW fault sets and localized thrust zones (Ashleman 1988). Faulting appears to postdate the Tertiary rhyolites and predate alteration and mineralization; however age relationships may not be so easily defined. Fault displacement and reactivation in conjunction with localized NE thrusts and oblique ENE shear zones (Johnson 1977, Stewart 1980) confuses the structural history of the area. The structural history of the Trinity area appears to be one of initial NE trending block faults and associated thrusts and shear zones, later offset by NW trending normal faults.

ALTERATION

The ALSG was not generally receptive to Tertiary hydrothermal alteration, except locally along faults and breccia zones. Tertiary volcanics were receptive to hydrothermal fluids which resulted in varying degrees of alteration and hydrofracturing. Rhyolitic tuffs and porphyritic flows have been extensively altered, defining an alteration halo that extends 2.5 km beyond the main mineralized zone (Fig. 15.7). Sericitization, silicification and quartz–adularia–sericite (QAS) alteration tends to be most intense along faults, within permeable lithologies and breccia zones. Minor propylitic

Table 15.3 Stratigraphy of the Trinity area. (From Bell 1989.)

Age	Unit name	Lithologies	Intrusions
Quaternary	Surficial deposits	Gravel, alluvium & colluvium	
Pleistocene	Volcanics, fluvial deposits, Upper Tuff	Basalt, fanglomerates & channel sands, phreatic-clastic welded & unwelded, tuff	Latitic & rhyolitic dykes
Tertiary (early Pliocene)	Rhyolite porphyry	Rhyolite flows, agglomerates & breccias	Rhyolitic, exogenous domes, porphyritic to aphanitic
	Lower tuff	Epiclastic & pyroclastic airfall, reworked tuff	
Cretaceous			Granodiorite, medium-grained, hypidiomorphic (90 Ma)
	Argillite breccia	Sub-angular to angular argillite clasts (1–10 cm) in fine-grained matrix. Breccia is gradational, tectonic in origin & fault associated	
Triassic—early Jurassic	Auld Lang Syne Group	Shelf & basin facies. Shale, siltstone, slate. phyllites & argillites. Siliceous & carbonaceous sandstones & limestones. Quartzites	

alteration and kaolinite–illite clays are also reported (Ashleman 1988). QAS alteration is most extensive within the rhyolitic porphyry and tuffs. Silicification is restricted to discrete W dipping lenticular zones within the upper zone of the deposit, with minor extensions along dykes and veins. Silicification is concentrated within the NE lobe extending from and parallel to NE and ENE faults. Patchy Fe staining and limonite is spatially associated with high grade Ag mineralization.

MINERALIZATION

Silver mineralization in Pershing County is reported to be hosted within breccias peripheral to rhyolite domes with mineralization concentrated in microfractures (Johnson 1977). However, the Trinity silver orebody is described as a hydrothermal, volcanic hosted, silver-base metal (Cu–Pb–Zn–As) deposit (Ashleman 1988). Bonham (1988) describes it as occurring in stockwork breccias formed in and adjacent to a rhyolite dome of Miocene age. Both sulphide and oxide ore is present, but high grade zones amenable to cyanide leaching are restricted to the oxide zone. The principal sulphide

phases identified in pan concentrates from rotary drill cuttings consist of pyrite–marcasite–arsenopyrite–sphalerite–galena with lesser amounts associated with chalcopyrite–pyrrhotite–stannite. They were deposited with silica along stockwork fractures. The silver-bearing sulphide phases consist of freibergite (silver-rich tetrahedrite) and pyrargyrite. Acanthite and native silver are also present in trace amounts. Silver in this zone is fine-grained (0.01 to 0.5 mm), locked within sulphide phases as inclusions and intergrowths, imposing significant metallurgical problems (section 15.6).

Only the SW oxide zone, on which this chapter concentrates, has been exploited. The Ag mineralogy is essentially bromine rich cerargyrite ($Ag(Cl,Br)$) with minor amounts of acanthite, often closely associated with limonite and Fe staining. There are some chloride and sulphide minerals present too. Payable mineralization is mainly restricted to porphyritic rhyolite flow units (75% of total) as disseminations, fracture infills, veinlets and replacements of potassic feldspars. Weak mineralization occurs within rhyolitic tuffs, breccia and argillites (25% of total) associated with QAS alteration. High grade zones (>10 oz t^{-1} (>343 g t^{-1})) exhibit a

Fig. 15.7 Aerial photograph of Trinity Mine. The light areas in the centre represent waste and low grade ore stockpiles. The medium tone areas in the foreground show the extent of the alteration in the area. The rectangular area is the heap on which the ore is placed for cyanide leaching. The ponds in which the pregnant solutions collect are seen next to the mine buildings. (Photo supplied by US Borax.)

spatial relationship with areas of strong jointing, sericitic alteration and limonite (Ashleman 1988).

Base metals (Pb–Zn–Cu–As) exhibit a spatial association with the distribution of silver mineralization, i.e. a gradual increase to the south. Lead shows the strongest association with silver in the oxide zone while zinc values tend to be low.

A sharp redox boundary within the oxide zone marks the transition from upper oxide ore into lower sulphide ore. The oxide orebody is subdivided into a continuous NE lobe and fragmented SW lobe, the former containing the highest grades. The NE lobe trends NE to ENE defined by sharp but irregular boundaries, and dips steeply (55–70°) to the W and NW. The SW lobe is disrupted by post mineralization faulting and dykes. The

oxide orebody defines a linear feature 75 × 610 m (250 × 2000 ft) extending to depths of 90–135 m (300–450 ft) forming lenticular to stacked tabular bodies.

Mineralization appears to be confined to the NW side of a NE trending fault zone with the southern boundary defined by a steep, possibly listric, normal fault which brought tuff into contact with the rhyolitic porphyry. NW trending offshoots cross the footwall contact into the rhyolitic tuffs and parallel NW faults. The hanging wall, or northern contact is not marked by any significant lithological change other than a brecciated form of the rhyolitic porphyry at a high stratigraphical level. This boundary has been inferred to reflect a change in host rock permeability or a high angle reverse fault consistent with basin and range faulting.

North-easterly trending basin and range faulting off-set by a cogenetic or later, minor, NW fault system probably acted as feeders for the invasion of early Pliocene hydrothermal fluids. Hydrofracturing and brittle failure resulted in the formation of a conjugate fracture zone. Hydrothermal fluids exploited the fracture zone and, in conjunction with alteration, made the host rock amenable to mineralization. Mineralization is concentrated at fault intersections (NW offshoots) and in a zone parallel but displaced from the fault planes.

The sharp footwall contact is either the function of mineralization being dissipated by a highly porous tuffaceous rhyolite or ponded by an impermeable fault gouge. Alternatively, both the sharp N and S boundaries may reflect the effects of fault reactivation, i.e. dextral strike slip displacing mineralization along strike and producing further uplift and erosion. This however is complicated by NW–SE offshoots that show no displacement.

15.5 DEVELOPMENT AND MINING

The exploration drilling programme defined the ore zone on which the block model for the reserve estimation and the open pit mine plan were based. Drilling continued during mining to confirm ore reserves. Pit modification was required during mining to improve ore control (section 15.8.4), slope stability, ore shoot excursion and operational access. The final pit dimensions were 1400 × 500 × 246 ft (427 × 152 × 75 m) (Fig. 15.6), with 15 ft (4.5 m) benches attaining slopes of 60–72°. Pit slopes were 52° to the west and 43° to the east, and safety benches were cut every 45 ft (14 m). Production haulage ramps were 55 ft (16.5 m) wide with 4 ft (1.2 m) berms and had a maximum gradient of 10%.

Standard mining methods of drilling and blasting, front end loading and haulage by truck were carried out by the contractor. A total of 16 500 t of ore and waste were removed daily in a 10 h shift, five days a week. Drilling for blasting, using a 5 5/8 in (143 mm) down-the-hole hammer, was carried out at 15 ft (4.5 m) centres, and at 12 ft (3.7 m) centres in areas of blocky silicification.

Grade control was based upon the blast hole samples and pit geology . The orebody was mined principally as a vein type deposit (Perry 1989). The eastern boundary (hanging wall contact) was taken to be a steeply dipping range front fault and the western boundary was an assay boundary. Blast holes were sampled, by sample pan cut or manual cone cuts, each representing a 15 ft (4.5 m) composite. In contrast to the AAS analytical method

used on the exploration borehole samples, the blast hole samples were analysed by a cyanide (CN) leach method, giving a more representative extractable silver content. Statistical and financial analyses had shown that the silver grades could be subdivided into four groups: (1) ore at >1.3 oz t^{-1} (45 g t^{-1}); (2) low grade ore at 0.9–1.3 oz t^{-1} (31–45 g t^{-1}); (3) lean ore at 0.5–0.9 oz t^{-1} (17–31 g t^{-1}); and (4) waste at <0.5 oz t^{-1} (<17 g t^{-1}). All conversions of ounces per short ton to grammes per tonne have been made using a factor of 34.285 (Berkman 1989). Each bench was divided into blocks, each block centred on a blast hole. The assayed grade for each blast hole was assigned to the block. The grades were colour coded and each block was marked on the bench in the pit by the surveyors with the appropriate colour coded flags (Fig. 15.8). This ensured that waste was correctly identified and sent to the waste pile, while the different grade materials were sent either to the cyanide leach heap or to one of the two low grade stockpiles.

Sulphide ore was not amenable to heap leaching and where encountered was selectively mined and hauled to low and high grade sulphide ore stockpiles (Fig. 15.8) (Perry 1989). Sulphide ore was differentiated from oxide ore on the basis of colour (yellow oxide ore and grey sulphide ore) and a percentage comparison of total AAS assays (or fire assay) and cyanide leachable assays i.e.

$$\frac{\text{AAS–CN} \times 100}{\text{AAS}}$$

a percentage difference of 6–17% defined the oxide ore and the sulphide ore was marked by a 24–40% difference.

15.6 MINERAL PROCESSING

Metallurgical testing of core samples determined that cyanide leaching was the most cost-effective method of treating the oxide ore. An excellent overview of heap leach technology is given by Dorey *et al.* (1988). Sulphide ore resulted in high cyanide consumption and required fine grinding to give a 78–84% recovery. The oxide ore was amenable to direct leaching resulting in a 94–97% recovery. Flotation tests on the sulphide ore liberated 90–95% of the silver and 90% of the Pb and Zn providing the pH values of the collectors, which suppressed Fe and As, were high. Oxide ore recoveries by flotation were low (50–60%). Laboratory tests indicated that it would be possible to heap leach the oxide ore, but the sulphide ore had to be stockpiled with a view to possible flotation extraction at a later date. A flow sheet of the process is shown in Fig. 15.9.

Fig. 15.8 Photograph showing the colour coded flags used to delineate ore from waste in the pit. The dark flags denote ore blocks and the pale flags denote waste blocks.

Mined ore was hauled and dumped on to an ore surge pile and fed in to a primary jaw crusher (Fig. 15.9). The minus 3/4" (minus 19 mm) product passed over a two-deck vibrating screen on to an in-line agglomerator. Secondary crushing of the oversized ore was completed by cone crushers in parallel circuit. The crushed ore was automatically sampled and weighed.

Ten lb (4.5 kg) of cement was added to one tonne of ore prior to agglomeration. The agglomerated ore was screened to remove the minus 3/8" (minus 10 mm) fines, and was then spread on the leach pad with a slough stacker. The 700×1100 ft (210×335 m) leach pad was lined with 60 mm heavy duty polyurethane (HDPE) and divided into six cells. Each cell had a 183 t capacity when stacked to 41 ft (12.5 m). Primary and secondary leaching utilized drip feed emitters providing a flow rate of 0.005 to 0.008 gallons per minute per square foot (200–350 ml min^{-1} m^{-2}). Sprinklers were used during the final stage of leaching and rinsing.

Pregnant solutions passed into the pregnant pond and were then processed by a Merrill–Crowe Zn precipitator plant with a 1000 gallons min^{-1} (3 785 litres min^{-1}) capacity. After initial filtering the precipitate was dried, analysed, fluxed and smelted (Fig. 15.9), to produce silver doré bullion (99% silver). It was estimated that 75% of the silver in the oxide ore was recovered by cyanide leaching.

15.7 ENVIRONMENTAL CONSIDERATIONS

Any mining project which will result in the alteration of the surface of a property, either through the addition of mine buildings or because of mining, is required to submit a Planning Application (in the UK) or a Plan of

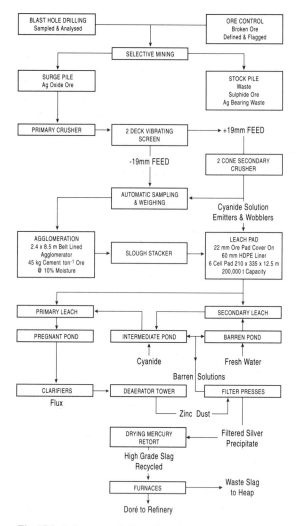

Fig. 15.9 A diagrammatic flow sheet of the mining and processing of the ore at the Trinity Mine, Nevada.

Operation (in the USA) to the authority responsible for administering the area. This application should describe the proposed mining operation and the infrastructure requirements and include a schedule of the proposed activities (Thatcher *et al.* 1988). Often the operator will include a plan of the procedures he proposes to implement to protect the environment during and after mining (reclamation).

In the case of a precious metal heap leach operation, additional permits and approvals are required before a heap leach operation can be commissioned. Thatcher *et al.* (1988) discuss the various regulatory aspects and permitting requirements for precious metal heap leach operations. These requirements include air quality, and surface and groundwater quality permits as well as cyanide neutralization requirements. In the latter case, Nevada specifies detailed heap rinsing procedures and analytical methods for cyanide detection as operating conditions in their water quality or discharge permits (Thatcher *et al.* 1988). In Nevada operators are expected to obtain cyanide neutralization levels of 0.2 mg per litre free CN^- as a target concentration, but most operators are unable to meet these standards. Similarly, there should be only 1 ppm cyanide in the residual weak acid digest (WAD) or <10 ppm CN in the soil. More realistically, they expect the company to attempt to meet these criteria and then show that the cyanide remaining behind will not affect or put at risk the groundwater.

The Trinity operators had several alternatives to consider with regard to cyanide degradation and heap detoxification. These are described in some detail by Smith (1988), and the various methods are outlined here. The methods fall into two categories, namely natural processes or chemical treatment.

15.7.1 **Natural degradation and detoxification**

1 *Passive abandonment.* High UV levels and strong winds for 9 months of the year would make this a possible alternative. The natural degradation of the cyanide would be fast in the near surface layers, but would slow rapidly with depth.
2 *Rinsing with barren solutions.* Rinsing to clean the entire heap has been tested empirically in the laboratory. A column in the laboratory was rinsed with fresh water and the cyanide levels were reduced from 700 ppm to 50 ppm. Unfortunately this took between 12 and 21 months to achieve and extremely large quatities of water (of the order of 4 500 litres tonne^{-1}) were required.

15.7.2 **Chemical treatment**

Five types of chemical treatment are currently available and were considered by US Borax.
1 *Acid leach.* The leachate thus produced would be collected in the ponds, neutralized with lime and the metals which precipitated would be removed mechanically.
2 *Sulphur dioxide and air oxidation method.* (Patented by INCO). These are added in conjunction with lime and a copper catalyst, but special equipment is required and this method is expensive.
3 *Alkaline chlorination.* Either chlorine gas or calcium hypochlorite are used in the oxidation process but these methods have very high reagent costs and careful pH control with the second method is required to prevent the formation of toxic cyanogen chloride.
4 *Hydrogen peroxide oxidation.* This method is clean and non-toxic, but the H_2O_2 is expensive. This method is used by OK Tedi in PNG where H_2O_2 is added to the recirculation pond and the cyanide-free water is then recirculated through the heap using the original sprinkler system.
5 *Ferrous sulphate.* Chemical destruction of the cyanide is used whereby 1 mole of ferrous sulphate ties up 6 moles of cyanide to produce Prussian Blue. This reaction takes place in the pond so that the recirculated water is free from cyanide. Both the last two methods are quick acting, especially with the initial flush. Thereafter there is a longer term diffusion mechanism which completes the flushing procedure.

15.7.3 **Summary**

There are advantages and disadvantages with each process as outlined above. The State of Nevada discourages the use of hypochlorite and peroxide neutralization methods (Thatcher *et al.* 1988), but the INCO SO_2–air and the peroxide processes are thought to be more technically efficient (Smith 1988). However the final conclusion may well be that each of the major detoxification processes in use has its application in a particular set of circumstances which each operator must establish for each property.

15.8 **GRADE ESTIMATION**

It is essential for reliable ore reserve calculations to derive the best Ag grade estimate for an orebody. The 5220 bench of the Trinity silver deposit was chosen to illustrate the problems and procedures of grade estimation. The 5220 level exhaustive data set contains 1 390 blast hole samples. No spatial bias or clustering effects are evident due to the regular sample pattern (15 ft (4.5 m) centres).

For ore reserve calculations a global estimate of mean silver is required. The global estimate must however, be controlled by local estimates because of the spatial variation of silver values. Measures of variability are important in evaluating the accuracy of a grade estimate, especially within a spatial context and have strong implications for mine planning.

In many geological environments the problem of grade estimation is intrinsically linked to the type of statistical distribution. Many problems can be solved from an assessment of the univariate statistics, histograms and probability plots (Davis 1986) but this assessment can not be divorced from the spatial context of the data (Isaaks & Srivastrava 1989). The following sections explore the problems of grade estimation and derive simple solutions for modelling skewed distributions.

15.8.1 Univariate statistics

For the purpose of grade estimation the distribution of assay values within the population is a prime requirement, especially regarding frequencies above a specified lower limit, e.g. cut-off grades. Most statistical parameters are applied to the Gaussian distribution, yet within a geological environment a normal distribution is not always immediately evident. It is common to find a large number of small values and a few large values, the lognormal distribution is then a good alternative (Isaaks & Srivastrava 1989). The histogram (Fig. 15.10a) describes the character of the distribution. The clustering of points at low values and the tail extending toward the high values indicates that the assays on the 5220 bench do not conform to a normal distribution. Most of the cumulative frequencies for the lower values plot in a relatively straight line, but departures occur toward the higher values (Fig. 15.10b). This is further evident from the curved nature of the normal probability plot (Fig. 15.10c).

The departure from a straight line at lower values on the log normal probability plot (Fig 15.10d) indicates that the data also do not fit a log-normal distribution. Although it is possible to model a 'best fit' distribution to the overall population, errors will occur during grade estimation. This is evident from the summary statistics (Table 15.4). It is clear from the frequency distributions that a global estimate based upon a normal distribution will be biased toward high values (overestimated). Similarly the log transformed data will be biased by low values (underestimated) (Fig. 15.10d). Due to the mathematical complications that arise from a log transformation, such a transformation is best avoided. It is a preferred alternative to try to establish Gaussian distributions within the exhaustive data set in order to improve grade estimation.

The important features of the distribution are captured by the univariate statistics (Table 15.4) and illustrate immediately the problem of grade estimation for

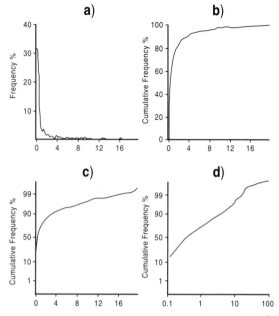

Fig. 15.10 (a) Histogram of (cyanide leach) silver values (oz t^{-1}) obtained from blast hole samples on the 5220 bench, Trinity Mine, Nevada, (b) cumulative frequency plot, (c) normal probability plot, (d) log-normal probability plot.

Table 15.4 Univariate statistics of the silver values derived from cyanide leach analysis of the blast hole samples on the 5220 level, Trinity Mine, Nevada

Silver values are in oz t^{-1} (g t^{-1}).

	Total Population	Excluding >20 oz t^{-1} Ag	1.3 to 20 oz t^{-1} Ag
Number of assays	1390	1385	309
Mean	1.43 (49)	1.30 (45)	4.65 (159)
Median	0.33 (11)	0.33 (11)	2.30 (79)
Trimean	0.48(16)	0.48(16)	3.45(118)
Sichel-T	1.28	1.22	4.52
Variance	13.59	6.93	16.36
Standard Deviation	3.69	2.63	4.04
Standard Error	0.10	0.07	0.36
Skewness	10.62	3.89	1.84
Coefficient of Variance	2.58	2.02	0.87
Lower Quartile	0.16 (5)	0.16 (5)	1.90(65)
Upper Quartile	1.11 (38)	1.09 (37)	5.84(200)
Interquartile Range	0.95 (33)	0.93 (32)	3.93 (135)

the 5220 bench. The summary statistics provide measures of location, spread and shape. The mean Ag content of the bench is recorded as 1.43 oz t^{-1} (49 g t^{-1}), yet other estimates of central tendency, the median (0.33) and the trimmed mean (0.48), invoke caution in using the mean as an accurate global estimate. The high variance and standard deviation describe the strong variability of the data values. The implication is that insufficient confidence can be placed upon the mean as an accurate global estimate. A strong positive skew with a long tail of high values to the right is evident from the high positive skewness (+10.62). The degree of asymmetry is also supported by the coefficient of variance (2.58). Knudsen (1988) implies that if the coefficient of variance is greater than 1.2 a log-normal distribution could be modelled. The Sichel-T estimator can be used to estimate the mean of a log normal distribution, and is calculated to be 1.28.

The frequency histogram (Fig. 15.10a) illustrates how the data are proportioned. It is important to note that even though the data range is 0 to 85 oz t^{-1} Ag, only 5% is > 7.2, 2% > 12.5 and 0.5% > 20 oz t^{-1} Ag. The upper quartile indicates that 75% of the data lie between a restricted range of 0 to 1.1 oz t^{-1} Ag, and the mean which is higher (1.43) does not reflect the majority of the data. It is obvious therefore that the magnitude of the positive skew will have a disproportionate effect upon the mean.

The initial statistical evaluation illustrates a potential error in the global estimate resulting in possible overestimation and bias toward erratic high values. More detail is required given that the mean is only a robust estimate if applied to a Gaussian distribution and a log transformation is not adopted. It is important at this stage to investigate the distribution further. Having ascertained that the data do not conform to a normal or log-normal distribution, the next step is to evaluate potential modality and departures from a continuous distribution.

The probability plot and histogram are very useful in checking for multiple populations. Although breaks in the graphs do not always imply multiple populations, they represent changes in the character of the distribution over different class intervals. The initial step is to explode the lower end of the histogram by making the class interval smaller (Fig. 15.11) and searching for potential breaks in the population. Breaks in the distribution are chosen on the basis of significant changes in the frequency between class limits or repeated patterns reflecting polymodality. If subpopulations are found, explanations must be sought and may depend upon sample support, geological control or population mixing. The positively skewed distribution of the data may be a function of the overprinting of multiple populations. Defining subgroups which may approximate normal distributions could improve the global estimate.

Such an evaluation reveals a number of breaks within the population of silver grade on the 5220 level. The following class breaks have been interpreted: 0.4, 0.8, 1.3, 5, 10 and 20 oz t^{-1} Ag (14, 27, 45, 171, 343 and 686 g t^{-1}). Those values greater than 20 oz t^{-1} are considered outliers (5 out of 1390 samples) and are not evaluated. To assess the statistical significance of the subgroups univariate statistics are calculated for each group (Table 15.5). In all instances normal distributions are approximated with significant reductions in the variance, coefficient of variance and skewness. Estimates of central tendency lie within statistically acceptable limits indicating improved confidence in the mean of each group. The 1.3 to 5 oz t^{-1} group is slightly skewed and a further split possible. However, improved detail may not result in enhanced accuracy or confidence in the mean.

The exhaustive data set can be split into subgroups according to grade classes and the grade estimate for each group is improved. The statistical and financial evaluation undertaken by US Borax resulted in a similar grouping (section 15.5). US Borax applied a 1.3 oz t^{-1} cut-off to the data such that everything above 1.3 oz t^{-1} was mined as ore. It is important to evaluate the effect of cut-off on the global estimate and the implications in relation to the subgroups.

15.8.2 Outlier and cut-off grade evaluation

Those values in excess of 20 oz t^{-1} (686 g t^{-1}) Ag have been classed as anomalously high values, and if rejected have a significant effect upon the univariate statistics and grade estimate (Table 15.4) despite only representing 0.5% of the data. The mean is reduced by 10% to 1.3 oz t^{-1} (45 g t^{-1}) Ag while the variance and skewness are reduced by up to half. However the distribution is still strongly skewed and the problems of estimation still remain. If the 1.3 oz t^{-1} cut-off is then also applied the distribution becomes flatter, the global estimate more precise, but accuracy is reduced because of increased variability (Table 15.4). The 1.3 oz t^{-1} cut-off also reflects a major break in the population and improved estimates may be a function of a separate population approximating to a normal distribution. With a reduced skewness the variability of values become more symmetrical about the mean.

A further investigation of outlier removal (Fig. 15.12) shows that as the tail effect of the distribution is reduced

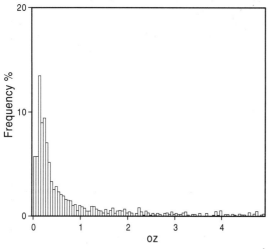

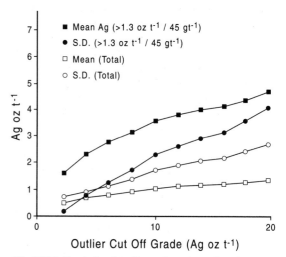

Fig. 15.11 Histogram of (cyanide leach) silver values (oz t^{-1}) obtained from blast hole samples on the 5220 bench, Trinity Mine, Nevada, with class intervals reduced to 0.05 oz t^{-1}. The higher values (>5 oz t^{-1}) are omitted for clarity.

Fig. 15.12 Graph showing changes in mean grade and standard deviation when different outlying (high) grades are removed from the data base of the silver values from the 5220 bench, Trinity Mine, Nevada.

the mean grade and standard deviation decreases in a linear fashion. The decline is more gradual for the total population than when the 1.3 g t^{-1} cut-off is applied. It is obvious that as values are included or excluded the magnitude of the mean varies. This must be viewed in relation to the number of samples, the effect of which is evident with values above or below 20 oz t^{-1} Ag. Although the magnitude of the grade estimate can be increased by including outliers the rate of improvement

in the ore reserve estimation is reduced. A grade estimate must also evaluate how the values fluctuate around the mean, therefore the relationship of the standard deviation to the mean is important.

Within a mining environment the confidence of the grade estimate is derived from the standard deviation. The value of the standard deviation however, is subject to the group of values from which the estimate was obtained and the spatial context. The standard deviation

Table 15.5 Univariate statistics of the subpopulations of the silver values derived from cyanide leach analysis of the blast hole samples on the 5220 level, Trinity Mine, Nevada. Silver values are in oz t^{-1} (g t^{-1}).

	0.0 to 0.4 oz t^{-1} Ag	0.4 to 0.8 oz t^{-1} Ag	0.8 to 1.3 oz t^{-1} Ag	1.3 to 5 oz t^{-1} Ag	5 to 10 oz t^{-1} Ag	10 to 20 oz t^{-1} Ag
Number of assays	762	212	1032	215	731	
Mean	0.18 (6)	0.57 (20)	1.03 (35)	2.56 (88)	7.46 (256)	14.37 (493)
Median	0.17 (6)	0.55 (19)	1.01 (35)	2.25 (77)	7.34 (252)	13.60 (466)
Trimean	0.17 (6)	0.56 (19)	1.02 (35)	2.35 (81)	7.44 (255)	13.93 (478)
Sichel-T	0.20	0.57	1.03	2.56	7.46	14.37
Variance	0.01	0.01	0.02	1.04	2.48	10.79
Standard Deviation	0.10	0.11	0.15	1.02	1.57	3.28
Standard Error	0.00	0.01	0.02	0.07	0.21	0.58
Skewness	0.19	0.28	0.19	0.74	0.08	0.25
Coefficient of Variance	0.55	0.19	0.15	0.40	0.21	0.23
Lower Quartile	0.10 (3)	0.47 (16)	0.89 (31)	1.71 (59)	5.98 (205)	11.18 (383)
Upper Quartile	0.26 (9)	0.67 (23)	1.16 (40)	3.20 (110)	9.09 (312)	16.70 (573)
Interquartile Range	0.16 (6)	0.20 (7)	0.27 (9)	1.49 (51)	3.11 (107)	5.52 (189)

expresses confidence in symmetrical intervals, therefore when the standard deviation exceeds the mean the magnitude of underestimates is significantly different from the magnitude of overestimates. This is illustrated in Fig. 15.12 for the total population such that the global estimate would exceed the local estimate in the majority of cases. The degree of variability and error increase in magnitude as the tail effect of the distribution is more prominant, therefore the above effect becomes more pronounced.

The relationship of the standard deviation to the mean changes when the same evaluation is applied to ore grade material. The graph (Fig. 15.12) indicates that the estimate for the ore grade material is more reliable and subject to less fluctuation, although as the distribution expands, variability increases. The grade cut-off curves reinforce the relationship of the mean to the standard deviation for the total population (Fig. 15.13a) and the effect of removing the 20 oz t^{-1} outliers (Fig. 15.13b). As extreme values of the distribution are altered the degree of variability declines. A point is achieved where the number of local underestimates and overestimates as compared with the global estimate balance each other. As low or high values are included or removed the mean changes accordingly and when an approximate normal distribution is modelled the variability reaches a constant (Fig. 15.13b). At this point there is no improvement in the confidence of the grade estimate and, in terms of reserve estimation, it is the spatial distribution of silver that becomes important.

15.8.3 Spatial distribution

Although it is possible to improve grade estimates by evaluating the statistical distribution, the spatial features of the data set are important. A spatial evaluation will focus upon the location of high values, zoning, trends and continuity. A contour map of the 5220 level (Fig. 15.14) shows an E–W trend to the data which fragments and changes to a NW–SE trend to the west. Zoning occurs and discrete pods of ore grade material are evident. Low grade values less than 0.8 oz t^{-1} (27 g t^{-1}) cover a greater area. The closeness of the contours especially at the 1.3 oz t^{-1} Ag (45 g t^{-1}) contour indicates a steep gradient from high to low values. The abrupt change from waste to ore grade material marked by a steep gradient will be reflected by a high local variance.

It is obvious that the subgroups previously defined are not randomly distributed but conform to a zonal arrangement. The indicator maps (Fig. 15.15) support the spatial integrity of the subgroups, illustrating the con-

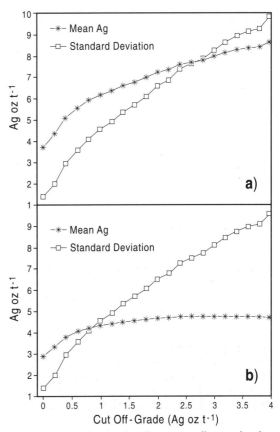

Fig. 15.13 Graphs showing the average silver grade when varying cut-offs are applied to the data base of the silver values for the 5220 bench, Trinity Mine, Nevada. (a) total population, (b) with values >20 oz t^{-1} removed.

centric zoning and isolated groupings of the data. The statistics of each subgroup can then be applied to specific areas of the 5220 level and local estimates can be applied to the contoured zones. A further assessment of local estimates can be achieved by using moving windows (Isaaks & Srivastava 1989, Hatton 1994). Statistics were calculated for 100 ft (30 m) square windows overlapping by 50 ft (15 m) (Fig. 15.16). The contour plots of the mean and standard deviation for each window (Fig. 15.17) show that the average silver values and variability change locally across the area. Zones of erratic ore grades can be located and flagged for the purpose of mine planning and grade control.

In general the change in variability reflects the change in the mean although at the edge of the high grade zone to the east, the variability increases at a greater rate. High variability is a function of the mixing of two

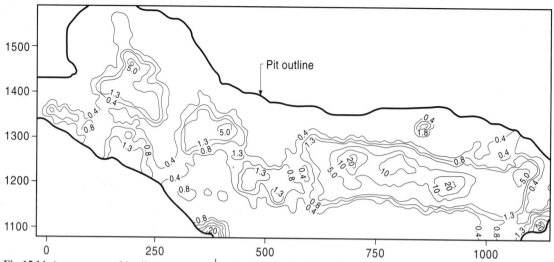

Fig. 15.14 A contour map of the silver grades (oz t^{-1}) on the 5220 bench, Trinity Mine. Contours are coded according to grade classes.

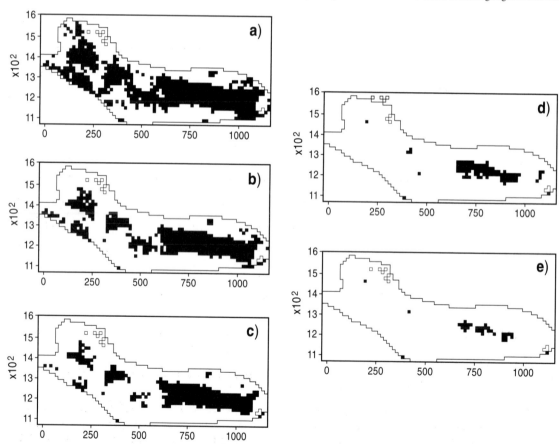

Fig. 15.15 Indicator maps of the Ag grades (oz t^{-1}) of the 5220 bench demonstrating the spatial integrity of the data. The grade boundaries are at (a) 0.4, (b) 0.8, (c) 1.3, (d) 5.0 and (e) 10.0 oz t^{-1} (14, 27, 45, 171 and 343 g t^{-1}).

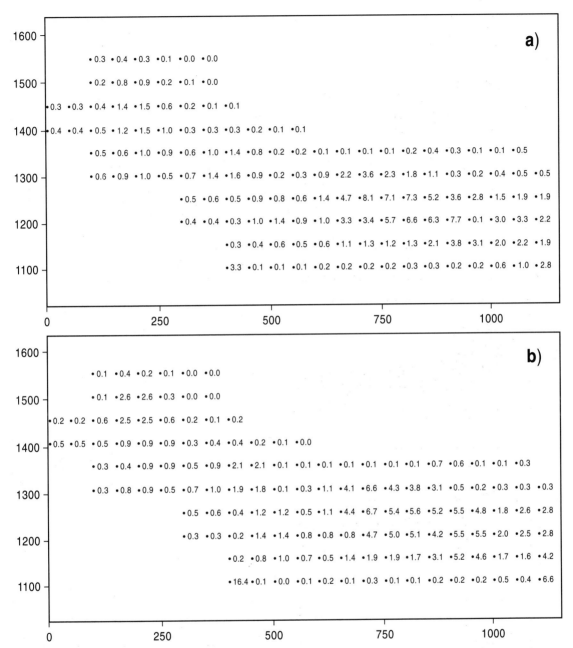

Fig. 15.16 Window statistics performed on the silver grades of the 5220 bench, based on a transformed grid oriented east–west with 30 m square, moving window overlapping by 15 m. (a) Mean Ag oz t^{-1}, (b) standard deviation.

populations or the transition from one population or zone to another. This is seen in the profile taken along N1300 (Fig. 15.18a). A profile passing through the high grade zone (N1200, Fig. 15.18b) shows a reduced change in variability compared with a pronounced increase in the magnitude of the mean. The strong relationship between the local mean and variability (Fig. 15.18) is referred to as a proportional effect which implies that the variability is predictable although the data do not reflect a normal distribution. The contour plots (Fig. 15.17) show that in

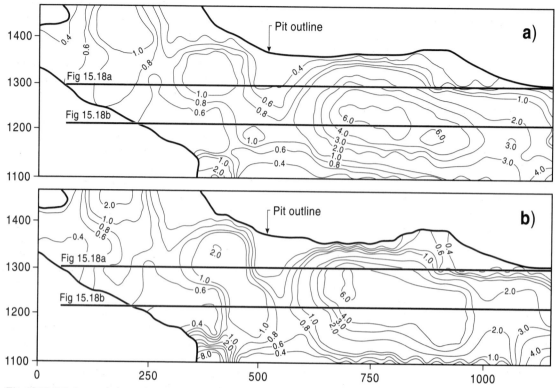

Fig. 15.17 Window statistics performed on the silver grades of the 5220 bench, based on a transformed grid oriented east–west with 30 m square, moving window overlapping by 15 m. (a) Contours of mean Ag oz t^{-1}, (b) contours of standard deviation. (From Bell & Whateley 1994.)

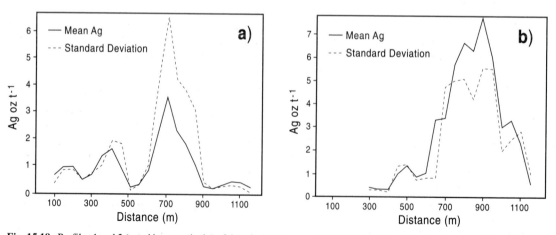

Fig. 15.18 Profiles 1 and 2 (a and b respectively) of the window mean and standard deviation showing the proportional effect and highlighting areas of expected high error. (From Bell & Whateley 1994.)

certain locations the change in local variance mirrors or is greater than the change in the mean (the proportional effect). However elsewhere the variability remains roughly constant whilst the local mean fluctuates.

15.8.4 Semi-variograms

Semi-variograms are used to quantify the spatial continuity of the data (Chapter 9). The range (a) of the semi-variogram defines a radius around which the local estimate has the least variability. The omnidirectional semi-variogram of the 5220 bench (Fig. 15.19a) shows a range of 120 ft (35 m). When samples >20 oz t^{-1} (686 g t^{-1}) were excluded (Fig. 15.19b) the sill was reduced by a third from 13 to 8 oz Ag^2. This again demonstrates the effect that the few high values have on the variability.

Directional semi-variograms show different ranges in different directions (Fig. 15.19a–d). Continuity was greater in the E–W direction (300 ft) (Fig. 15.19c) compared with all other directions. The minimum continuity was 90 ft in a N–S direction (Fig. 15.19a). An E–W elipse measuring 300 × 90 ft (Fig. 15.19f) would be the optimal moving window shape and size to give the lowest variability for a local estimate.

The NW–SE directional semi-variogram (Fig. 15.19d) shows a 'hole effect' (Journel & Huijbregts 1978, Clark 1979) which results from alternate comparison of high and low zones in the NW half of the deposit.

15.8.5 Conclusions

Attempting a global estimate of the grade for the entire bench is irrelevant in a mining situation because geologists and mining engineers are only interested in ore grade material. The skewed population means that ore grade material will be underestimated. It is essential to understand what constitutes the population, in order that sensible subdivisions into subpopulations can be made. It is incorrect to evaluate an exhaustive data set if the data are derived from different episodes of mineralization, when their physical and chemical characteristics differ. If the global estimate is used to make local estimates of grade, then the grade will be overestimated, because no account is made of the spatial relationship of the samples or of the high variability.

To improve the grade estimation it is necessary to try to establish normal distributions within the whole skewed population, using geology, structure, mineralization,

alteration, etc., rather than transforming the whole population. The variability within each subpopulation is thus reduced, which increases the confidence in the subpopulation mean.

The high values cause extreme variability which influences the mean. By excluding them the mean and variance are reduced and the estimation becomes more reliable.

By applying these techniques to the data above cut-off grade (omitting outliers), a population which approximates a normal distribution is achieved. This improves the confidence in the estimation of the mean. By subdividing the data above cut-off into subpopulations, variability is further reduced.

Indicator maps show that subgroups, as defined by statistical parameters, are spatially arranged. High local variance is a function of boundaries between these populations. Application of the moving window statistics should therefore be viewed in terms of the spatially distributed groups, trying to avoid crossing subpopulation boundaries.

From the univariate statistics and spatial distribution of the data the conclusion is that a global estimate must be derived from local estimates rather than the whole population.

In hindsight, the best ore reserve estimation could be derived using the average of the estimates from discrete zones or the average of the estimates derived from the areas between contours of the grades. Bell & Whateley (1994) undertook global estimation of the grade by applying a number of estimation techniques to the original exploration borehole data of the 5220 level. They compared the results with the blast hole data set and concluded that linear interpolation provided grade estimates with the lowest variance.

15.9 ORE RESERVE ESTIMATION

During the life of the project, from the exploration stages to the feasibility and mine design studies, different ore reserve calculations were carried out, each with a different confidence level. The confidence in the reserve estimate grew as the amount of borehole data increased.

15.9.1 Evaluation of initial exploration data

Immediately after exploration was initiated in early 1982, it was realized that a significant tonnage of disseminated silver existed. During the 1982–83 explo-

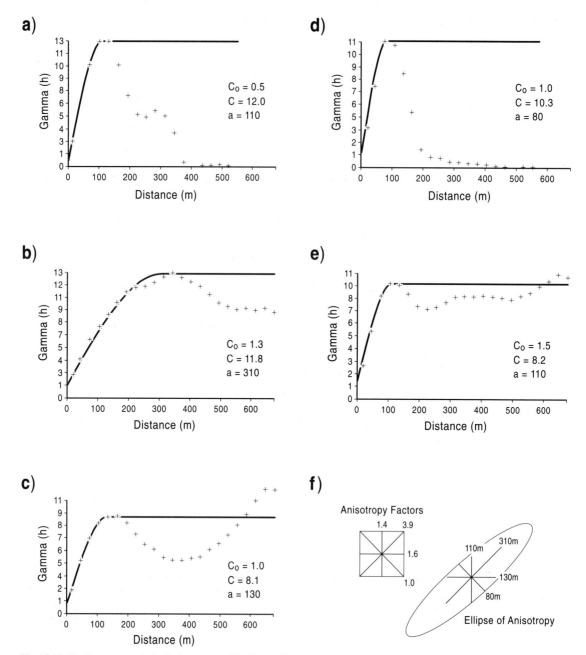

Fig. 15.19 Semi-variograms (spherical scheme models) for the silver assays of the 5220 bench. (a) N–S direction, (b) NE–SW direction, (c) E–W direction, (d) SE–NW direction, (e) omni-directional, (f) ellipse of anisotropy and anisotropy factors. (From Bell & Whateley 1994.)

Table 15.6 Summary of the ore reserve calculation for the silver oxide zone, Trinity Mine, Nevada

Method	Tonnage factor $ft^3 t^{-1}$ (SG)	Composite length ft (m)	Ag cut-off grade $oz\,t^{-1}(g\,t^{-1})$	Tonnage $\times 10^6$	Average Ag grade $oz\,t^{-1}(g\,t^{-1})$	Total Ag oz (g) $\times 10^6$
Polygons–USB	13.3 (2.41)	20 (6.1)	1.5 (50)	0.967	6.95 (238)	6.22 (213)
N–S Cross Sections, USB	13.3 (2.41)	10 (3.0)	1.5 (50)	1.304	6.16 (211)	8.03 (275)
E–W Cross Sections, USB	13.3 (2.41)	10 (3.0)	1.5 (50)	1.293	5.90 (200)	7.63 (262)
N–S Cross Sections, USB	13.3 (2.41)	10 (3.0)	1.5 (50)	0.932	7.69 (264)	7.17 (246)
Polygons, Santa Fe	13.3 (2.41)	20 (6.1)	3.0 (100)	0.669	9.10 (310)	6.09 (209)
N–S Cross Sections, Santa Fe & USB	13.3 (2.41)	10 (3.0)	2.0 (69)	0.870	8.00 (274)	6.96 (240)

Table 15.7 Reserve estimates on the silver oxide zone, using the $1/D^2$ block model, Trinity Mine, Nevada

Tonnage $\times 10^6$	Tonnage factor $ft^3 t^{-1}$ (SG)	Composite length ft (m)	Ag cut-off grade $oz\,t^{-1}(g\,t^{-1})$	Average Ag grade $oz\,t^{-1}(g\,t^{-1})$
2.65	13.7 (2.34)	15 (4.5)	0.5 (17)	3.05 (105)
2.05	13.7 (2.34)	15 (4.5)	1.0 (34)	3.76 (129)
1.65	13.7 (2.34)	15 (4.5)	1.5 (50)	4.46 (153)
1.26	13.7 (2.34)	15 (4.5)	2.0 (69)	5.19 (178)
0.88	13.7 (2.34)	15 (4.5)	3.0 (100)	6.33 (217)
0.63	13.7 (2.34)	15 (4.5)	4.0 (137)	7.52 (258)

Table 15.8 Updated reserve estimate on the silver oxide zone, Trinity Mine, Nevada

Method	Tonnage factor $ft^3 t^{-1}$ (SG)	Composite length ft (m)	Ag cut-off grade $oz\,t^{-1}(g\,t^{-1})$	Tonnage $\times 10^6$	Average Ag grade $oz\,t^{-1}(g\,t^{-1})$	Total Ag oz (g) $\times 10^6$
N–S Cross Sections	13.7 (2.34)	15 (4.5)	1.5 (50)	0.899	6.94 (238)	6.24 (214)
Bench Plans *	13.7 (2.34)	15 (4.5)	1.6 (55)	0.859	7.33 (251)	6.29 (216)
Blocks 15 × 15 × 15 ft (4.5 × 4.5 × 4.5 m)	13.7 (2.54)	15 (4.5)	1.6 (55)	0.963	6.55 (225)	6.31 (217)

* These estimates exclude the first three benches which had already been mined out.

ration programme the oxide mineralization had not been fully evaluated, and the initial reserve estimates only took the sulphide mineralization into account.

15.9.2 Evaluation of additional exploration data

Additional exploration drilling took place in 1986 with the intention of identifying additional sulphide mineralization, and to collect samples for metallurgical and engineering studies. In mid 1986 it became apparent that a small area of high grade silver oxide mineralization could possibly support a heap leach operation. The reserve potential of the oxide zone was calculated by several members of the joint venture, using different methods and cut-off grades. Table 15.6 summarizes the results.

The tonnage calculations vary from 0.932 to 1.304 Mt where a 1.5 oz Ag t^{-1} (50 g t^{-1}) cut-off was used. Where the cut off grade was raised to 2 and 3 oz t^{-1} (69 and 103 g t^{-1}), as one would expect, the tonnage was reduced but the average grade increased. These differences in tonnage and average grade were accounted for when thin, lower grade intersections used in earlier calculations, were omitted, and modified polygons were used. No reference was made to the geology of the deposit in any of the above calculations.

This demonstrates the difficulty in calculating reserve estimates in the exploration phases of a programme (Bell & Whateley 1994). Different methods, by different people using differing cut-off grades, produces a variety of estimates. It is not possible to quantify the error of estimation using these two-dimensional manual methods of reserve estimation. It is not even practical to compare the early reserve estimates with the final tonnage and grade extracted, because in the final estimate a cut-off grade of 1.3 oz t^{-1} (45 g t^{-1}) was used.

15.9.3 Reserve estimate for mine planning

In 1987 a computerized three-dimensional block modelling technique was used to estimate reserves and the data thus generated were used for mine planning. A conventional two-dimensional inverse distance squared (1/D^2, section 9.5.1) technique was used to interpolate grades from borehole information to the block centres, using a 150 ft (46 m) search radius (Baele 1987). The blocks were given a 25 × 25 ft (7.6 × 7.6 m) area and were 15 ft (4.6 m) deep. Borehole assays were collected at 5 ft (1.5 m) intervals, but were composited by a weighted average technique over 15 ft (4.6 m) to conform to the

block model. Geology and alteration were incorporated into the block model by assigning a rock and alteration code to each block. Tonnage and grade was estimated at various cut-off grades (Table 15.7).

Despite the geological control, this method of reserve estimation appears to have overestimated the reserve potential (in comparison with the early ore reserve estimates). This may be partially a result of an edge effect where the majority of any given block at the edge of the deposit may lie outside the mineralized area, but because of the shape of the boundary they have been included in the reserve estimate. An additional factor is the use of 13.7 cu ft t^{-1} (2.34 t m^{-3}) as the ore and waste specific gravity, to calculate *in situ* tonnage compared with only 13.3 cu ft t^{-1} (2.41 t m^{-3}) used in the exploration evaluation.

15.9.4 Updated tonnage and grade estimate

The grade and tonnage of the ore mined from the first three benches was significantly different from that predicted by the 1/D^2 block model. An updated reserve estimate was hand-calculated using cross-sections to evaluate the discrepancy. The geology and structure from the mined out area were plotted on the cross sections and used to project the orebody geometry downwards (Reim 1988). Reim felt that the unsatisfactory grade control given by the 1/D^2 method could be accounted for by the inadequate structural control in the 1/D^2 model, variability in grade and paucity of sampling data within the SW portion of the oxide ore body. Additional drilling increased the sample density in poorly represented areas and this helped define the orebody geometry. The reserve estimate calcluated using the updated cross-sections is shown in Table 15.8.

For mine planning purposes a mineable reserve was calculated on a bench by bench basis. Mineralized zones were transferred from the cross-sections on to 15 ft (4.5 m) bench plans. Polygons were constructed around each borehole and a reserve calculated (Table 15.8).

15.9.5 Grade control and ore reserve estimation using blastholes

Mine grade control was exercised by sampling and assaying the blast hole cuttings. Mined blocks were 15 × 15 × 15 ft (4.6 × 4.6 × 4.6 m) and only blocks >1.3 oz t^{-1} (45 g t^{-1}) were mined as ore. A final reserve was calculated using these blocks and a slightly higher reserve with a lower grade was estimated (Table 15.8).

15.9.6 **Factors effecting ore reserve calculations**

A number of factors influenced the validity of ore reserve estimates, namely geological control, sampling method, cut-off grade, sample mean, analytical technique, specific gravity and dilution factors.

1 *Geological control.* It is most important to establish the geological and structural control of mineralization as early in a project as possible. In the example used here, the $1/D^2$ block model had good geological control, but the structural control was lacking. This resulted in overestimation of reserves.

2 *Sampling methods.* It is important to control the sampling method to ensure consistent and reliable sampling, e.g. down-hole variation in silver values derived from either reverse circulation or percussion drilling may show disparity.

3 *Cut-off grade.* This varied between estimation methods from 3.0 oz t^{-1} (103 g t^{-1}) in the exploration assessment to 1.3 oz t^{-1} (45 g t^{-1}) during mining (Tables 15.6, 15.7 & 15.8). Comparisons of reserve estimates using different cut-off grades can be achieved if grade–tonnage graphs are constructed. These are difficult to construct during the exploration phase as data are sparse. Normally economic criteria are used to establish the cut-off grade (Lane 1988). US Borax assumed a commodity price of \$6.50 per oz and expected silver recoveries between 65 and 79% (amongst other criteria) to establish the 1.3 oz t^{-1} (45 g t^{-1}) cut-off grade used during mining.

4 *Sample mean.* The mean grade is dependent upon the method used to estimate it (section 15.8.2). The arithmetic mean probably overestimated the mean in this study but this could have been improved by establishing subpopulations with normal distibutions and averaging the estimates from these discrete zones.

5 *Analytical technique.* In order to compare samples used to estimate reserves using exploration boreholes and blast holes, a simple linear regression of cyanide extractable silver on total recoverable silver was performed. A correlation coefficient of 0.94 indicated a strong positive relationship. Above 10 oz t^{-1} (343 g t^{-1}) Ag, the conversion of total silver to cyanide leachable silver was less accurate. Changing analytical procedure during an exploration or feasibility exercise, without undertaking an overlapping series of analyses which can be used for correlative purposes, introduces an unnecessary risk of error.

6 *Specific gravity.* Initially, densities of 13.0 and 13.3 ft^3 t^{-1} (2.34 and 2.41 g cm^{-3}) were used to calculate tonnages. Density was determined on a number of different rock types and an average of 13.7 ft^3 t^{-1} (2.34 g t^{-1}) was calculated, which was used for the final tonnage calculations.

7 *Mining dilution.* Dilution of ore by waste during mining is inevitable but careful grade control will minimize the risk of dilution. Dilution usually occurs in ore deposits where grade is highly variable, as at the Trinity Mine. The assumption that a blast hole in the centre of a $15 \times 15 \times 15$ ft block ($4.5 \times 4.5 \times 4.5$ m) is representative of that block is not necessarily true. The arbitrary boundary between a block showing a grade less than cut-off and a high grade block is drawn half way between the two holes. The low grade ore may extend farther into the high grade block than one expects, thus adding to the mining dilution problem. US Borax initially expected a 20% dilution, but this was later modified to 7% after further assessment during mining.

15.10 **SUMMARY AND CONCLUSIONS**

The Trinity oxide orebody defined a small but economic silver deposit, amenable to cyanide heap leaching. The orebody was fairly continuous to the NE but became fragmented and fault controlled to the SW. The ore zone was defined by stacked tabular to lenticular bodies dipping steeply to the W and NW and offset by N–S and NW–SE cross-faulting. Mineralization tended to be controlled by faulting, alteration and fracture density. The sharp ore grade boundaries along the footwall and hanging wall suggest geological control. These boundaries may reflect change in host rock permeability or sharp fault contacts.

The Trinity deposit illustrates the estimation problems encountered in deposits which have highly skewed data. The problem is compounded when trying to estimate grade and resource potential from a relatively low number of exploratory boreholes. Recognizing the importance that a few very high values may have upon the estimation of the global mean is important. If the variability of the data can be reduced the confidence in the mean goes up.

16 Narrow Vein Deposits—Wheal Jane, Cornwall

C. J. MOON, M. K. G. WHATELEY AND A.M. EVANS

16.1 INTRODUCTION

An orebody which is longer in two directions and is restricted in the other is considered to be tabular. Veins are typically discordant, regularly shaped, tabular orebodies, and are sometimes called lodes, especially in SW England. They are often inclined and, in mining terms, they have a footwall and hanging wall. Veins are usually developed in fracture systems and therefore show regularities in their orientation. However, veins frequently pinch and swell as they are followed both along strike and down dip. These pinch-and-swell structures make it very difficult to estimate the shape and resources from exploration drilling data. They also create difficulties during mining because often only the swells are workable. The infilling of veins may consist of one mineral but more usually it consists of an intergrowth of ore and gangue minerals (Evans 1993). The orebody boundaries may be the vein wall or they can be assay boundaries within the veins. These boundaries must be established before detailed reserve assessment can be undertaken. Vein deposits show great variations in their properties, e.g. thickness, which can vary from a few millimetres to more than a hundred metres. Veins no longer hold their pre-eminent position in the mining world but are still important for their production of gold, tin, uranium and industrial minerals such as baryte and fluorite.

Veins can be found in practically all rock types and situations, although they are often grouped around plutonic intrusions as in Cornwall and Devon, England. This area has produced tin and base metals since prehistoric times, initially from placers but predominantly from bedrock, and was the world's main producer of tin in the early to mid 1800s. Production came from a large number of relatively small mines and a few larger mines using a large, low productivity, labour force. The mines were organized on a cost book system in which the adventurers (shareholders) shared the costs and profits, with little held in reserve for capital development or periods of low commodity prices. Few mines survived a price crash in the 1890s, caused by increased production from alluvial deposits in Malaysia, and only two (Geevor and South Crofty) became much larger, more mechanized mines that operated through until the 1990s. Increases in tin prices in the 1960s caused renewed interest in tin production and prompted a revival in exploration. This phase of exploration resulted in the opening of two new mines, Pendarves and Wheal Jane. The exploration and production history of Wheal Jane is typical of vein mines in mining districts with long mining histories and is described in detail in this case history. It should be noted that Wheal is the Cornish language word for mine and Wheal Jane Mine would be a tautology.

16.2 ECONOMIC BACKGROUND

The supply and demand of tin and therefore its price has had a very different history from that of other metals. This has resulted from the concentration of tin production in a few less developed countries (Malaysia, Indonesia, Bolivia, Thailand and Nigeria) and severe fluctuations in its price caused by changes in demand in tin's main uses, in cans to preserve food and in solder. In an attempt to minimize these price fluctuations and the knock-on effects on export earnings some important producing and consuming countries formed an organization, or cartel, to intervene in the international tin markets. As part of this agreement they set production quotas for each country to limit supply, and appointed a buffer stock manager with authority to buy tin if the price reached a pre-determined floor price and to sell if it reached a ceiling. This worked successfully for several agreements over the period 1930–1985 (Robertson 1982). The sixth post-war agreement, however, could not survive a decrease in world consumption from 200 000 t tin in concentrates in 1979 to 140 000 t in 1982 and the emergence of a major new producing country, Brazil, which refused to be limited by quotas or to join the cartel. This was entirely understandable as Brazilian production is from very low cost eluvial and alluvial deposits in the Amazonian province of Rondônia and the country was desperate for foreign exchange to repay its debts. As a result world supply exceeded demand, forcing prices down and the buffer stock manager was forced to buy large quantities of tin at the floor price, eventually running out of cash in October 1985 and

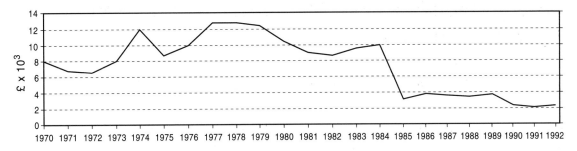

Fig. 16.1 Average tin price t[-1] in 1984 values.

becoming bankrupt. This caused the demise of the international tin agreement and the reversion to the free market accompanied by a drop in the tin price (Fig. 16.1) to £3600 t[-1] in May 1986. By 1990 Brazil was the dominant producing country with 40 000 t production out of a total market economy production of 158 000 t (Shaw 1991).

16.3 GEOLOGY

16.3.1 Geological setting and regional mineralization

The Cornubian Orefield of SW England lies within folded and thrusted upper Palaeozoic, lower greenschist facies, sedimentary rocks with subordinate mafic volcanics which form much of the north-western portion of the European Variscan (Hercynian) Orogen (Alderton 1993). These rocks are intruded by a chain of six major granitoid plutons and several minor ones which unite at depth to form a large batholith (Fig. 16.2). The sediments are argillaceous and arenaceous, continent-derived rocks deposited in a deltaic–marine environment (Willis-Richards & Jackson 1989). Their original thickness was about 5 km but the succession was much thickened by thrusting during the Hercynian Orogeny. These Devonian–Carboniferous rocks probably lie on a late Precambrian basement and the present total crustal thickness is 27–30 km.

Most of the lavas have continental intraplate chemical characteristics and many workers favour a back-arc ensialic setting for this region and suggest that it lay to the north of a southward dipping subduction zone (Holder & Leveridge 1986). Shackleton *et al.* (1982) and other workers have interpreted the complex structural patterns of the region in terms of thin-skinned tectonics in which the Palaeozoic rocks were detached from their basement and pushed northward on low angle thrusts to form a stack of nappes. This implies that the whole crust, before

erosion and crustal thinning by normal faulting, cannot have been more than 33–34 km thick. This is of critical importance when the petrogenesis of the granites is considered because the majority of workers favour a crustal melting hypothesis. The thin-skinned model does not allow such a simple idea of magma generation, with purely vertical movement in response to tectonic thickening and radiogenic heating, to be entertained. Shackleton *et al.* (1982) therefore, proposed that the granites were generated further south and injected northwards as a sheet-like body along a substrate whose present depth varies from 10–15 km and which appears to rise northwards at about 45 m km[-1]. The batholith is mainly composed of a tourmaline-bearing biotite or two-mica, peraluminous monzogranite. There is a minor development of alkali–feldspar granite. The granites can be classified as S-type and high heat-producing and they crystallized about 295–270 Ma ago.

Chen *et al.* (1993) obtained concordant U–Pb ages from monazites which indicate that the intrusion of the six principal plutons was diachronous, ranging from 293.1 to 274.5 Ma. ^{40}Ar–^{39}Ar spectra suggest that the mineralization followed magmatic emplacement of the younger plutons. Emplacement of the major plutons was essentially passive and there was much large scale stoping.

The base metal deposits occur mainly as steeply dipping veins in the roof of the batholith and its surrounding contact metamorphic aureole. The majority of the Sn, Cu, W and As production came from veins along the axis of the batholith with the western end enriched in Sn, Cu and Zn relative to the eastern end (Willis-Richards & Jackson 1989). Most of the Sn, W, Cu, As and Zn ores occur in fractures running parallel to the batholithic axis, apart from the western end where there is a strong development of radially oriented veins. The main episode of vein formation has been shown to coincide with late magmatic events during which microgranite dykes (locally known as elvans) were injected. These are generally pre-vein in time of formation.

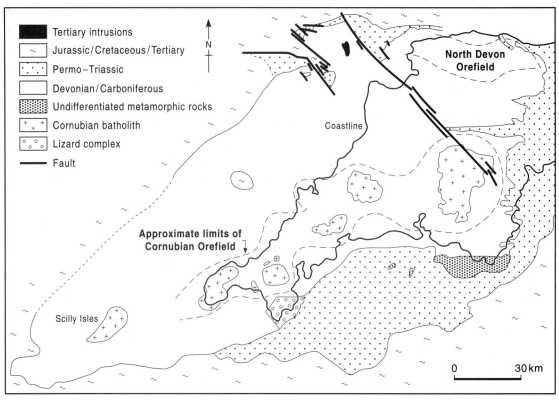

Fig. 16.2 Regional geology of south-west England and the location of the orefields. (After Bromley 1989.)

There is a general tendency for Sn, W, Cu and As mineralization to be nearer to, or in, the granite bosses and for Pb, Zn, Ag and Au to be farther away (Fig. 16.3). In the nineteenth century these and other observations led to the idea of a simple zonation which was held by many to reflect continual mineral deposition from monoascendant fluids passing upwards from the granite into cooler crustal zones. In some instances mining revealed such simple zoning, e.g. at Dolcoath Mine, where the upper zone of copper mineralization graded downward through a Cu–Sn zone to dominantly Sn in what was the richest Cu–Sn vein worked anywhere in the world. Elsewhere a gap with uneconomic mineralization occurs between the Cu and Sn zones. This simple picture, although valid in some places, cannot be extrapolated to all parts of the orefield and there are many areas where zoning of this simple form is not present. In some instances the component parts of a supposed zonal sequence have been shown to be of different ages and a younger mineral suite has been superimposed on an older one. These and other observations do not support the simple, steady state, monoascendant hypothesis outlined above.

Besides showing only a tenuous or even non-existent spatial relationship to the granite bosses, the Pb–Zn mineralization is generally in veins running at high angles (cross-courses) to the Sn–W–Cu veins. Fluid inclusion work shows that these two vein groups were formed from different hydrothermal solutions. The Sn–Cu formed from dilute to moderately saline (10–30 eq. wt% NaCl) aqueous fluids in the minimum range 200–500°C, whereas the Pb–Zn cross-course mineralization was from often high salinity (generally 20–27 eq. wt% NaCl), low temperature (<200°C) solutions comparable with those from which limestone-hosted Pb–Zn–F–Ba type deposits were formed (Alderton 1978).

In addition to the above complications there are a number of different vein types, the most important on the regional scale being tourmalinite breccia veins. Some persistent veins of this type pass gradationally upwards into chloritic breccia veins and into extensional, fissure filled veins with central filling of sulphide minerals (Bromley 1989). The gangue association is chlorite–fluorite–quartz and early cassiterite is followed by a later

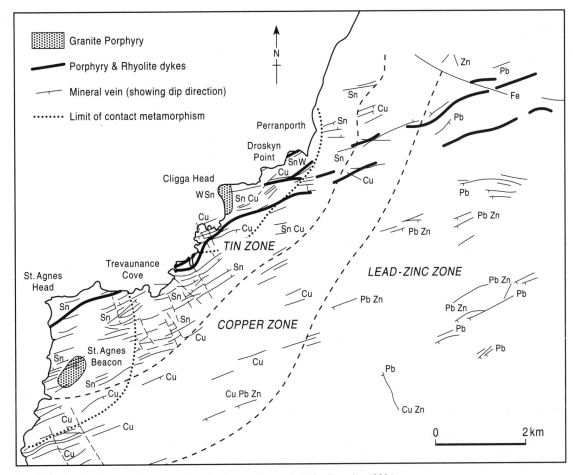

Fig. 16.3 Metal zoning around two small granites in west Cornwall. (After Bromley 1989.)

complex sulphide association which includes stannite, arsenopyrite, chalcopyrite, sphalerite, pyrite and marcasite. This subsequent mineralization often accompanied strong shearing which destroyed the primary depositional textures. These features are well developed at Wheal Jane.

16.4 EXPLORATION PROGRAMMES IN CORNWALL AND DEVON

16.4.1 Exploration models

There is no exploration for tin in SW England at the time of writing (1993) as the result of the low tin price. However, the more favourable prices during the 1960s–1980s encouraged considerable exploration. The basic premise was that a relatively large-scale mechanized

mine would be profitable. Analysis of the size and grade of major underground vein and open pit stockwork deposits gives some idea of the deposit to be sought. According to Menzie *et al.* (1988) the two different types of deposit form distinct areas on a grade–tonnage plot (Fig. 16.4) although Bromley & Holl (1986) suggested that these may be more of a continuum in SW England. It can be seen that a deposit of >2 Mt of ore at better than 1% Sn (approximately 20 000 t of contained tin metal) would have been of interest as an underground proposition or >40 Mt at >0.2 %Sn as an open pit operation. An underground operation with 2 Mt would support a reasonably sized mining operation for 10 years. If the distribution of historically mined tin in SW England is plotted (Fig. 16.5) it can be seen that very few deposits produced at this level historically and a new underground mine would either involve the mining of a number of veins from one set of underground workings or a major structure which for some reason was impossible to mine in the past. An open pit mine would involve bulk mining of disseminated mineralization, probably within a granite cupola.

A basic assumption made by all companies was that most veins that cropped out had been discovered although they might not have been evaluated at depth and might not have been amenable to mining or recovery using 19th century technology. Most companies concentrated within the emanative centres defined by Dines (1956). These are discrete areas forming a small proportion of the granite outcrop, which he recognized were responsible for the bulk of production and probably represent favourable areas for the deposition of tin mineralization. Besides outcrops of tin veins, companies sought either veins which did not crop out or which changed from copper to tin at depth as at Dolcoath. This was the scene of a major exploration success in the 1840s when Captain Charles Thomas persuaded shareholders to test his theory that tin would be present at depth beneath copper mineralization when copper grades decreased sharply with depth. After 100 m of poor values the richest single tin-bearing structure in Cornwall was discovered. On a more regional scale sections through the edges of plutons (Fig. 16.6) demonstrate that tin zones with distinct tops and bottoms can be identified. Thicknesses of the tin zone vary considerably between plutons but in general are around 400 m and the dip of the zone is in the same direction as the more steeply dipping granite–slate contact. The values used for the thickness need to be treated with caution and cannot be extrapolated from one district to another, reflecting differences in tin deposition (Walsham 1967). In addition, tin is not necessarily present below all major copper bearing structures as demonstrated at Devon Great Consols, one of the major copper producers in the area (Cominco, open file report) and at Geevor copper is at a maximum below the tin zone.

Although exploration models can identify areas with potential for tin-bearing structures it is much more difficult to predict grade and overall payability. Tin grades vary greatly in a typical narrow vein system and much of the structure, perhaps on average 70%, has been subeconomic even in times of good tin prices. Detailed examination by Garnett (1966) of the variation in grades at Geevor Mine provided a number of methods for analysing this variation and predicting grades. He used contours of grade and grade–thickness product to deduce controls on overall payability by comparison with contours of thickness, dip and distance of the vein from a fixed reference plane (Connolly diagrams). In fissure fill veins he discovered the main control on payability to be

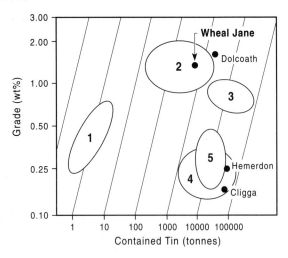

Fig. 16.4 Grade–tonnage plots for tin deposits. (From Bromley & Holl 1986). Areas: 1 Mexican rhyolite deposits, 2 Hydrothermal vein deposits, 3 Carbonate replacement deposits, 4 Greisens and stockworks, 5 Skarns.

changes in dip and intersections with other veins.

The difficulty in evaluating grade is reflected in the relative importance of underground exploration as an exploration technique. An analysis of exploration programmes partly subsidized by the UK government over the period 1971–1981 shows that approximately 35% of projects were interested in generating new targets, 60% used diamond drilling and 40% used underground exploration. Target selection has been guided by literature study of old mines, prospecting and geochemistry. The literature on old mining in the area is voluminous although fortunately there are good summaries by industrial archeologists (e.g. Jenkin 1961) and by the British Geological Survey (Dines 1956). Prospecting and surface mapping are of limited use as exposure is poor and most of the area is intensively cultivated. One technique which is of use is the mapping of vein material in field boundary walls, which are generally built of rock of local origin.

16.4.2 Exploration geochemistry

Exploration geochemistry has been used extensively to define the location of prospective veins and stockworks although interpretation is difficult as a result of extensive contamination from previous mineral workings. Stream sediment sampling is ineffective in defining targets because of this contamination and the redistribution of placer cassiterite along Miocene and Pliocene shorelines as shown by the transported anomaly in Fig. 16.7. A

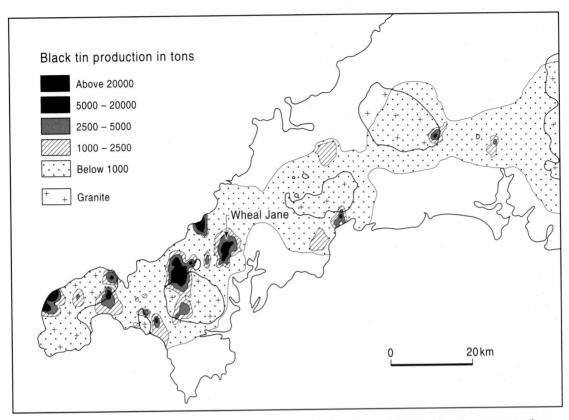

Fig. 16.5 Tin production in Cornwall and west Devon. The data are mainly from Burt *et al.* (1987) and are tin concentrates (long tons km⁻²) after smoothing.

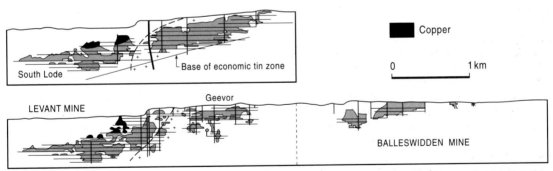

Fig. 16.6 Tin zones at Geevor and Levant mines. (After Mount 1985). Note the dip of the mineralized zones relative to that of the granite–sediment contact. The sharp change in topography in both section is the sea cliff— most production from Levant was from under the sea.

detailed investigation of this anomaly by Dunlop & Meyer (1978) demonstrates that anomalies derived from placers are anomalous in tin but not in copper or arsenic whereas virtually all tin vein outcrops have anomalies in tin, copper and arsenic and therefore it is possible to distinguish between the two types of anomalies. Hosking (1971) provided a comprehensive review of the different types of contamination, the five principal ones being mine dumps, adit discharges, mill wastes, smelter wastes, and agricultural contamination. Of these, adit wastes present the most severe problem in stream sediment sampling. The other types of contamination are likely to be more localized and probably more subtle with the exception of arsenic contamination from smelting. One example of subtle contamination is the use by farmers of cheap and plentiful mine waste for building hedges between fields.

Soil geochemistry is in general more reliable than stream sediment chemistry and has been widely used. An example is a large block of ground investigated by Consolidated Gold Fields (CGF) in an attempt to extend reserves at Wheal Jane. Orientation studies by Hosking (1971) and by CGF had shown that the subcrop of the main Wheal Jane B Lode is well defined by tin and copper soil geochemistry (Fig. 16.8) with samples taken at depths of 60 cm in relatively thin soils with little layering. A regional survey was mounted in an area of about 20 km^2 around the mine where access could be obtained from landowners. Samples were collected using a 300 by 16 m grid, orientated at 90° to the general strike of the veins. The summary results for tin are shown in Fig. 16.8 and clearly detect the Wheal Jane structure as well as three other major anomalies on the edge of the village of Chacewater, at Cusgarne and along the Carnon River. A number of smaller anomalies are also shown which define the subcrop of parallel vein sets to the north and south. However, the large amount of contaminated land makes assessment of these anomalies extremely difficult. The Carnon River anomaly is easily explained by large-scale placer workings but the Chacewater and Cusgarne anomalies required follow-up by deep sampling. This revealed that the Cusgarne anomaly increased in intensity with depth (Fig. 16.9), whereas that at Chacewater decreased with depth. The Chacewater anomaly seems to be the result of contamination of unknown origin whereas the Cusgarne anomaly is from a natural source. The strike of the Cusgarne anomaly correlates with the strike extension of local veins and of an induced polarization anomaly but to the authors' knowledge has not been drilled due to difficulties in obtaining exploration rights.

16.4.3 Geophysics

Geophysical exploration has not been very successful in locating narrow vein deposits in this orefield. On a regional basis gravity surveys have proved useful in modelling the sub-surface distribution of the granite batholith (Willis-Richards & Jackson 1989) and thus of detecting buried cusps of granite which may be associated with mineralization. The exception to the generally poor geophysical response of vein deposits are pyrite rich ones as at Wheal Jane. CGF used induced polarization in their regional survey around the mine and were able to detect the main structure from the high chargeability of the sulphides and the high resistivity of the hanging wall quartz porphyry dyke.

16.4.4 Drilling

Diamond drilling has been extensively used in Cornwall and Devon to investigate beneath previously mined structures and to test newly discovered veins. Although there is no doubt about diamond drilling's ability to delineate structure there is considerable debate on its effectiveness in determining the grade and distribution of tin within these structures. If only 30% of the structure is economic then only one borehole in three might be expected to return payable values. However, the consensus (Walsham 1967) is that the situation is somewhat better than that.

Walsham provided a detailed account of the use of diamond drilling by a joint venture led by Union Corporation. He defined three broad types of drill programmes; extension drilling, scout drilling and long-structure drilling as shown in Fig. 16.10 a–c. Extension drilling was aimed at proving further reserves in an area previously mined and can be accomplished by a short programme with regular intersections to the intended mining depth. Scout drilling was concerned with delineating favourable structures, first testing the most promising of them. Long structure drilling was designed to test a belt of ground previously worked. A number of structures were usually tested with the most promising areas located in the centre of the belt where veins are intersected by cross courses.

As drilling was aimed at reaching the main lode system and in testing veins in its hanging wall, boreholes were normally drilled at an inclination of >45°. The major problem was that the holes tended to droop and deviate from the planned course as shown in Fig. 16.10d. This wandering can be reduced by casing the holes but all holes required surveying. In general core recovery was good. The vein material was logged in detail and sampled over intervals of 15–30 cm as shown in Fig. 16.11. Cores were split into two halves and one half assayed.

16.4.5 Environmental impact

Cornwall and Devon contain some of the most scenic areas in England and attract a large number of holiday makers each year. This holiday industry is responsible for a large proportion of local income and the possibility of new mining which would detract from the local landscape with tailings and industrial buildings is in many people's opinions an unwelcome one. This especially applies to

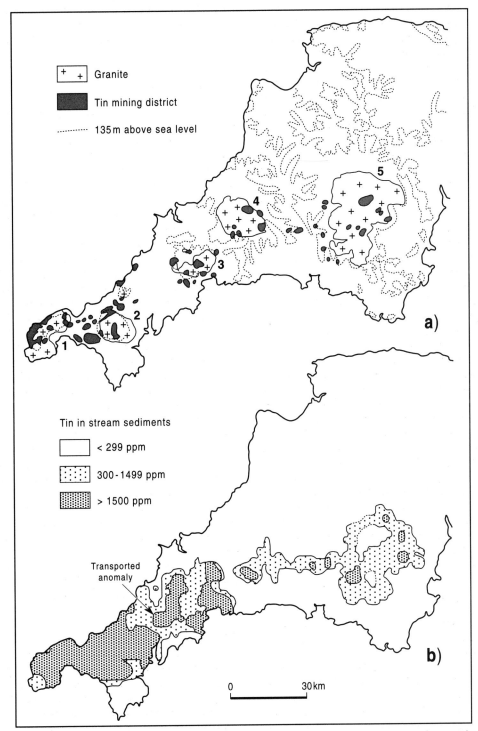

Fig. 16.7 Regional stream sediment anomalies for tin in comparison with main tin mining areas (1–5). Note the general correspondence with the exception of a transported anomaly related to regional placer development of Miocene-Pliocene age. (After Dunlop & Meyer 1978.)

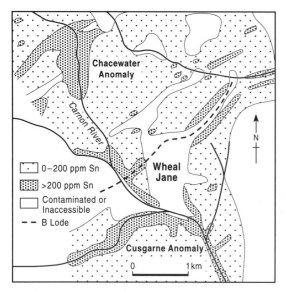

Fig. 16.8 Tin soil geochemistry around Wheal Jane. Data from Consolidated Gold Fields open file reports at the British Geological Survey. The survey was conducted after the re-opening of Wheal Jane in the search for similar deposits. Grid size 300 by 16 m.

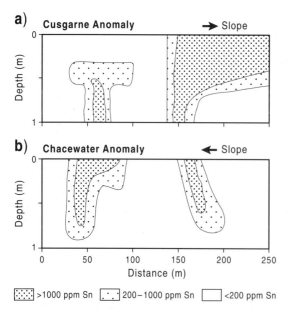

Fig. 16.9 Follow-up soil sampling over two of the major anomalies in Fig. 16.8. Note the general decrease of tin with depth at Chacewater (b) compared with discrete sources at Cusgarne (a). Consolidated Gold Fields open file data.

exploration in the coastal areas, for example Geevor Mine is at the edge of a very scenic cliff line. Although this was not a major problem in the 1960s, the pressure to prevent mining has become much more severe since then. This has to some extent been counterbalanced by the stated policy of Cornwall County Council to reduce a high unemployment rate and sustain traditional industries. Much of the mineralized ground in Devon is within a national park and it is unlikely that any mining would be permitted. However development in Cornwall within old mining areas is likely to be approved.

The process of obtaining consent for mining involves a detailed planning application to the relevant district authority, which is examined at a county level with an option for national discussion. The proposal will be advertised and opportunity provided for submissions by the general public. If the proposal is controversial there is likely to be an open public inquiry, conducted by an inspector from the Department of the Environment. Examples of these were the inquiries into the development of an open pit tungsten operation at Hemerdon, 10 km NE of Plymouth in West Devon and the expansion of the Lee Moor kaolin open pit in the same area. Both would infringe on the Dartmoor National Park, although both pits are outside the park and both were conditionally approved. Fuller details of the Lee Moor inquiry are given in Blunden (1985).

16.5 WHEAL JANE EXPLORATION

An example of a successful exploration programme is the discovery of the Wheal Jane deposit by Consolidated Gold Fields in the late 1960s which is described in detail by Willson (1969).

16.5.1 Desk study

The programme began with a literature study in 1964, based on the assumption that all tin deposits that cropped out had been discovered and that previous, mainly nineteenth century, explorers had not missed the opportunity of working any deposits that could be worked profitably at that time. This resulted in the selection of a variety of target types.

1 *A low grade stockwork* capable of providing a few thousand tonnes of ore each day providing economies of scale not previously possible.

2 *A well developed vein* or group of veins which could not be worked profitably by previous mineral processing techniques.

3 *A strong group of veins* worked for copper down to the

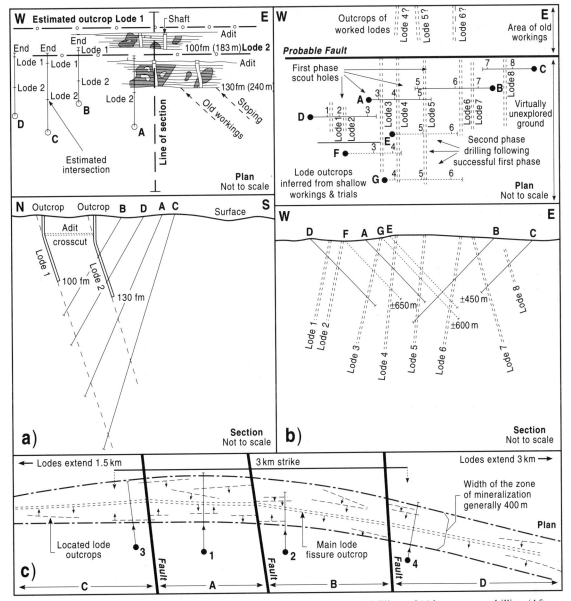

Fig. 16.10 Typical drill hole patterns in Cornwall : (a) extension drilling, (b) scout drilling and (c) long structure drilling. (After Walsham 1967.)

top of the tin zone but not developed for tin.

4 *Veins under the sea* which had not been worked.

5 *Hidden veins* or groups of veins which did not crop out but which could be located by geochemical or geophysical techniques.

6 *Mines* reputedly abandoned due to excessive water inflow.

16.5.2 Legal problems

Following the selection of an area based on one or more of the target types, initial attempts were made to obtain options to take mining leases at an agreed royalty (generally 1/34 of the value of mine production). This proved to be a major problem as mineral rights in England are

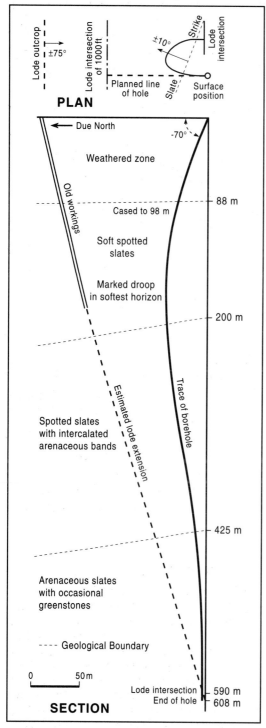

Fig. 16.10d Extreme droop in a drill hole in weakly metamorphosed sediments. (After Walsham 1967.)

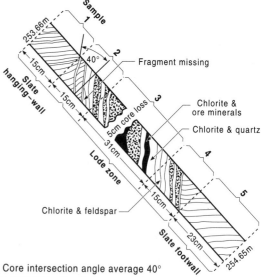

Core intersection angle average 40°

Sample Number	Depth from(m)	to(m)	Recovery (m)	(%)	True width (cm)	%Sn	cm%
1	253.66	253.81	0.15	100	12	0.03	0.36
2	253.81	253.96	0.15	100	12	0.56	6.72
3	253.96	254.27	0.25	81	20	4.75	95.00
4	254.27	254.42	0.15	100	12	1.27	15.24
5	254.42	254.65	0.23	100	18	NIL	–

$$\text{Weighted average lode value} = \frac{116.96\,\text{cm}\%}{44\,\text{cm (true thickness)}} = 2.66\%\,\text{Sn}$$

at a weighted average core recovery of 91%

Fig. 16.11 Typical intersection of a tin-bearing vein showing sampling intervals and the uncertainty introduced by incomplete core recovery. (After Walsham 1967.)

privately held (section 4.2.3), have often been separated from the surface rights and subdivided among various heirs. In addition, it was often difficult to find mineral rights owners due to mass emigration at the end of the nineteenth century and even when located the owners were sometimes unwilling to permit exploration. This unwillingness was not surprising as some landowners, under the then prevailing tax regime, would have seen over 95% of their royalties disappearing in tax. Although CGF began without an experienced land agent they soon found it of great advantage to employ one; indeed legal fees represented a large part of the exploration budget where the land was held by several owners. As a result of these problems the company only considered prospects where the mineral and surface rights were owned by estates with large blocks of land.

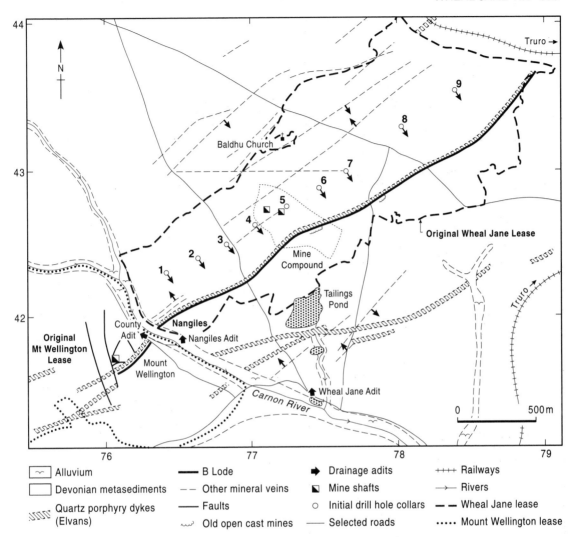

Fig. 16.12 Sketch plan of the Wheal Jane area

16.5.3 **Target selection**

Following the acquisition of mineral options a thorough search was made of available literature, plans and previous geological investigations of the occurrences. These reports and plans were useful, although not always complete and reporting was often overly optimistic as most reports were used to raise capital for further work. Soil geochemistry was used to define a number of targets, as was trenching, but geophysics was of little effect. Gold Fields investigated 40 prospects on paper, sought the mineral owners of ten and used geochemistry on five; of these, three prospects were taken to the drilling stage.

16.5.4 **Target testing**

Of the three prospects drilled only that at Wheal Jane was brought into production. Unlike most previously worked mines it is situated at a considerable distance (5 km) to the west of the granite outcrop and the main mineralized structure is intimately associated with a quartz porphyry dyke which intrudes pelitic metasediments. The Wheal Jane prospect (Fig. 16.12) had been previously worked at various times as several different mines. The first recorded efforts were in the early 1700s but most working took place in the period 1854–1895, when around 4300 t of tin concentrate and 60 000 t of pyrite together with some arsenic, copper and silver concentrates were recovered. The main vein continues to the south-west of the Carnon River (Fig. 16.12) where attempts were made to work it in the Mount Wellington area over the period 1930–1941. The main vein at Mount Wellington was described in the available literature (Dines 1956) as a complex tourmaline breccia, with a quartz porphyry hanging wall, up to 6 m in width and averaging 1% SnO_2, although only 0.35% SnO_2 was recoverable in the mill during the 1930–41 working. During this working two drill holes were also sunk on the north side of the Carnon River in the old Wheal Jane area to test veins at depth. One drill hole of 275 m proved 3 m of 1–1.5% SnO_2 at the base of a quartz porphyry dyke at 60 m depth and the other inclined hole cut 3.3 m of 1.5–2% SnO_2 at 52 m. These intersections were considered interesting enough to re-timber an old shaft and drive a crosscut from it, but it was not completed. These reports were sufficiently encouraging for CGF to investigate the mineral rights ownership.

The mineral rights to the Wheal Jane area are owned by the estate of Lord Falmouth and it was a relatively easy matter for the CGF land agents to negotiate an agreement. The Mount Wellington rights were held separately and were optioned to another company, Cornwall Tin Mining and Smelting. The Wheal Jane and Mount Wellington areas, although contiguous, were investigated separately but at roughly the same time.

As the target was well defined, no geochemistry or geophysics was used by CGF before the Wheal Jane rediscovery and the first stage was diamond drilling on the basis of previous underground mapping. Holes were spaced at 260 m along strike and were designed to cut the wide flat lode at 130 m down dip from the adit but most drill holes were continued to 300 m to test other steeply dipping veins. After six holes a number of intersections had produced grades of >0.7% Sn over 1.6 m immediately in the footwall of the best developed quartz porphyry dyke. This vein was termed B Lode as it was the second vein from surface. Other intersections of interest were made in the hanging wall of the dyke (A Lode) and at depth (C and D Lodes). Structure contours suggested that B Lode was a continuous structure and 12 further drill holes were drilled to test its strike and dip extensions. The structure became more complicated to the east but was continuous at depth. At this stage CGF decided that underground exploration was required in view of general Cornish experience of grade variability.

16.5.5 Underground evaluation

This was initially done at Wheal Jane by clearing an old

shaft, Nangiles Shaft, and the adit prior to developing along the structure at adit level (45 m below surface). Old pyritic stopes with good tin values were examined by development and diamond drilling for 240 m. Further drifting for 800 m proved continuity of structure although much of the development was in uneconomic ground as anticipated from the drilling. At the same time a pilot metallurgical study was initiated to investigate if good tin recovery was possible from the complex sulphidic mineralization. A key sample was made up of sample rejects from the relevant drill cores. It was found that the poor tin recovery experienced by previous operating companies was due to the very fine grain size of the cassiterite and intermixing with sulphides of much of the tin. This fine-grained tin was not amenable to recovery by gravity methods but good recoveries could be achieved by adding a froth flotation circuit, a novel approach in the processing of these ores which was pioneered by CGF.

Early in 1968 further underground development was undertaken at Clemow's Shaft to test B Lode at depth (>120 m). In general the vein was found to be encouragingly mineralized although somewhat more structurally complex than indicated by drilling. Of particular interest were the high grades found at the intersection of B Lode with a steeply dipping vein structure, called M Lode after Moor Shaft in which it was intersected. Relative tin values are highest in the area (Fig. 16.13) around Clemow's and Tippet's Shafts and were interpreted from the data available as being controlled by the thinning of the quartz porphyry and associated thickening of B Lode (Rayment *et al.* 1971). The overall mineralization was interpreted as five types as summarized in Fig. 16.14.

1 *Mineralization associated with the footwall and hanging wall contacts* of quartz–porphyry dykes, such as B Lode.

2 *Tin mineralization in steeper fractures* such as M Lode.

3 *Massive sulphides* in restricted areas of B Lode.

4 *Narrow chalcopyrite-sphalerite veins* (caunter lodes) dipping 50–70° south and of different strike to the dykes.

5 *Late pyrite and galena fillings* in N–S trending faults, termed cross courses.

Within B Lode, which was interpreted as a shear zone, five zones of tin mineralization were identified.

1 *A footwall zone* of sericitized sheared slate.

2 *Siliceous tourmalinized slate* with consistent concentration of pyrite and arsenopyrite.

3 *Pyrite rich vein material* with tourmaline fragments.

4 *Massive pyrite* and sphalerite with relatively low tin values.

5 *Massive quartz* with discrete wolframite, cassiterite

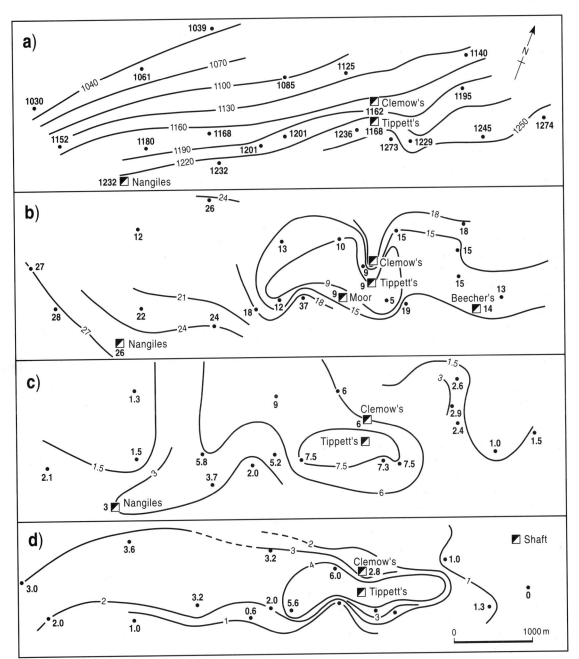

Fig. 16.13 Plots of CGF drill results. (a) Structure contours of the base of B quartz porphyry dyke (arbitrary datum), (b) isopachs of B quartz porphyry, (c) isopachs of B lode, (d) relative tin values (% x a factor). (After Rayment *et al.* 1971.)

and other sulphides in the footwall.

Zone 2 was the dominant source of tin and was recognized as the most mineralogically complex with very fine-grained (100–200 μm) intergrowths of cassiterite and

tourmaline. Cassiterite in zone 5 was much coarser from around 700 μm to 5 mm and was amenable to gravity recovery.

Drilling and underground development delineated re-

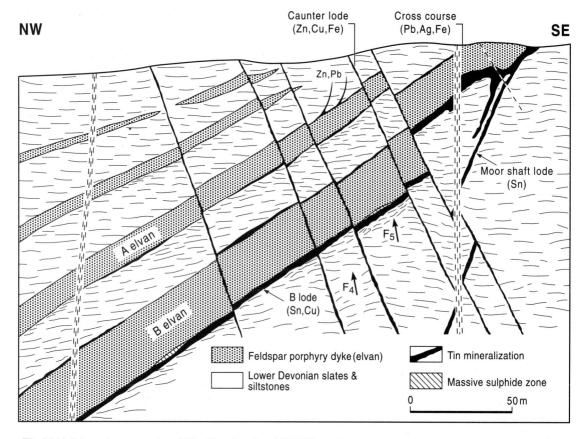

Fig. 16.14 Schematic cross-section of Wheal Jane based on CGF drilling. (After Rayment *et al.* 1971.)

sources of 5 Mt at 1.25% Sn, which were sufficient to support a 200 000 t a^{-1} for 25 years. Exploration costs to April 1968 were £325 000. The total capital cost to bring the mine into production was £6M of which £2M was funded by a UK government grant. Clemow's Shaft was used as an access while a second shaft was sunk to 350 m below surface. This was achieved in two years despite several cases of intersecting water under high pressure in holes drilled ahead of shaft development.

Negotiations with the County and District Councils for planning permission had begun in January 1966 with an application for permission to drill in the lease area. Informal consultations over the possibility of obtaining planning permission for mining were started in January 1969 and formal planning permission was applied for in July of that year. Few problems were encountered in obtaining permission to mine or even in obtaining permission for a tailings dam which is situated in a small valley to the south-east of the main shafts.

On the western side of the Carnon Valley, the Mount

Wellington property, Cornwall Tin Mining and Smelting defined resources along strike from the Wheal Jane structures as the result of 69 holes, totalling around 16 000 m, 46 of which intersected the mineralization beneath the main quartz porphyry dyke. Shaft sinking began in August 1969 and planning permission for mining was granted in January 1972. Resources were put at 5 Mt of 1.37% Sn. Mining on this property began in 1976, after a 20 month construction phase.

16.6 PRODUCTION HISTORY

Production started at Wheal Jane in October 1971. Almost immediately the mine ran into problems making enough ore available from stopes and with poor tin recovery in the mill. By 1976 the mine was running at a loss and a further capital expenditure of £4M was required.

Mount Wellington was also beset by a number of problems, notably poor recovery and heavy water flows

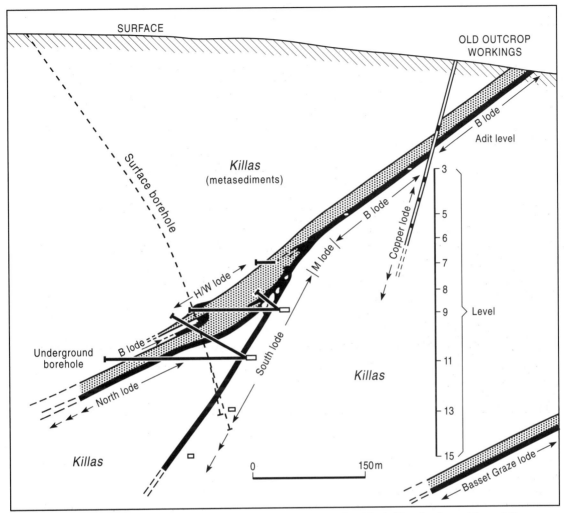

Fig. 16.15 Section of Wheal Jane in the early 1980s. The main working was on South Lode. Note that the spatial association between mineralization and the dykes does not persist with depth. (After *Mining Magazine* 1982.)

and by early 1978 had accumulated a loss of £2.75M. It was decided to close the mine in April 1978. As the two mines were linked hydrologically, if not by development, this water flowed into the Wheal Jane workings. Within a few weeks of the flooding of Mount Wellington, the cost of pumping at Wheal Jane became prohibitive and CGF decided to close the mine. However, the pumps were kept going and negotiations were begun with companies interested in re-opening the mines. During June–August of 1979 RTZ together with J.S. Redpath evaluated the orebodies and decided to open a unified mine incorporating both the Wheal Jane and Mount Wellington properties (Davis & Battersby 1985). Their studies had shown

that an increase in production to 360 000 t a⁻¹ could reduce the fixed costs per tonne by reducing manpower needed per tonne and spreading the pumping costs over a larger tonnage. These pumping costs alone accounted for 17% of costs. Even though the mine had been kept pumped a considerable programme of rehabilitation was required in all the underground development as much of the underground metal work had been destroyed by the extremely acid mine waters. The mining method was changed to sublevel stoping using an increasing amount of trackless equipment to reduce labour costs. The mill was also refurbished to improve the poor recovery which had dogged its operation.

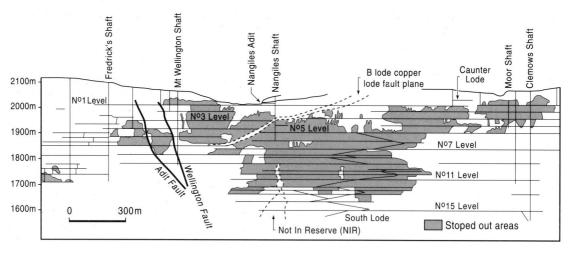

Fig. 16.16 Longitudinal section of Wheal Jane 1991. (Courtesy Carnon Consolidated.)

This unified mine produced at the anticipated production levels until the crash in the price of tin in 1985, when production became uneconomic and the mine was sold by RTZ to a management buy out, Carnon Consolidated. Although the mine was kept going until June 1991 it traded at a loss and was only sustained by government grants and savage cuts in the labour force and fixed costs. Finally the mine was allowed to flood in June 1991.

16.6.1 **Geology of Wheal Jane**

Although the original mining plan was to mine B Lode it soon became clear that other structures, which are not closely related to the quartz porphyry, carried economic tin grades and that B Lode splits into a complex series of veins with depth. B Lode rapidly attenuates with depth and Moor Lode, which splits from the footwall of B Lode, itself splits into two: North Lode which follows the footwall of the quartz–porphyry, at around 22° dip, and South Lode which has a steeper dip (Fig. 16.15). In general North Lode was uneconomic. South Lode is a complex, brecciated shear zone dipping to the NNW at 50–70° which underwent several phases of reactivation and mineralization (Hopper 1990). It is generally 3 m wide, though it reaches 10 m at lode intersections, and the payable zone plunged steeply to the west where it diverged with depth into what may be two discrete payshoots (Fig. 16.16). The subeconomic area between the split was known as the 'not in reserve' pillar (NIR), and its distribution with depth was of considerable importance as regards the strike extent, and hence ore reserves, of the payable lode down dip (Hopper 1990).

16.6.2 **Mineralization**

At depth all the lodes are mineralized shear zones which consist of altered and brecciated host rock clasts and a variety of hydrothermal minerals. Shearing stress parallel to the lode contacts produced tension and eventually fracturing between adjacent lodes, with the result that this fractured area was also mineralized (Hackett 1988). At depth mineralization involved five episodes of hypothermal to mesothermal deposition (Holl & Bromley 1988). A simplified model of the mineralizing episodes is as follows.

1 *Greisenization* with a wolframite, cassiterite, lollingite, native bismuth and quartz assemblage (greisen phase).
2 *Tourmalinization* of the shear zone with deposition of quartz and cassiterite in a very fine-grained assemblage as a result of rapid deposition and shearing (tourmaline phase).
3 *Extensive chloritization* of the shear zone and/or wall rocks with further introduction and remobilization of quartz, cassiterite and sulphides (arsenopyrite, stannite, chalcopyrite, sphalerite, pyrite and quartz), (chlorite phase).
4 *Sulphide mineralization* as veining of earlier ore and as massive banded pyrite, sphalerite and chalcopyrite (replacement veins).
5 *Vughy quartz* and barren sulphide veins cross-cutting the original fabric.

Due to the polyphase nature of deposition and remobilization under differing PT conditions, the monoascendent model of Cornish zoning does not apply and the telescoping and overprinting of the phases have given rise to a comparatively even distribution of tin. With depth it seems that the phases could be separating into more discrete feeders and a drop in sulphide content and rise in tin grade was beginning to emerge as the underlying granite was approached.

16.7 UNDERGROUND EXPLORATION

16.7.1 Grade

The remobilization of the tin mineralization resulted in a relatively uniform distribution of tin throughout the mine. This is in contrast to other Cornish vein deposits which characteristically have localized high grade zones in structures of overall low profitability. Wheal Jane produced an average of 160 000 t of ore p.a. at an average grade of 0.85% Sn, 2.5% Zn and 0.35% Cu. This gives a tin equivalent grade of 1.34%, wherein 1% Zn is represented as 0.13% Sn equivalent, and 1% Cu is equated with 0.35% Sn. In addition minor amounts of fluorite and silver were produced. The ore reserves (unclassified) for July 1989 were 1.59 Mt at 1.34% Sn equivalent.

Sampling was the basis for grade control and two types of samples were taken at Wheal Jane, namely channel and grab samples. Channel samples were taken at regular intervals, usually between 4–5 m apart, along every drive on lode (Table 16.1). These samples were used for ore reserve estimation and grade prediction. Grab samples were collected in stopes and development headings and formed the basis for grade control and recording production statistics. These samples were used in comparisons with the mill feed grades. Diamond drilling from surface and from special underground drilling chambers was also used for exploration sampling. All samples thus collected were analysed for tin, zinc, copper and sulphur.

The Geology Department at Wheal Jane found that in order to achieve parity with the mill it was necessary to truncate (or cut) high values to predetermined levels. The values used were: Sn 3.00%, Zn 12.00% and Cu 0.70%. These fixed grades were used to try to reduce the effect of anomalously high values on grade and reserve estimations. They were determined by constructing cumulative frequency graphs of the raw channel sample data. The fixed upper limit was determined by drawing the ninety-five percentile. The geological staff considered that the top 5% of analyses were anomalously high.

It has been known for some years (Kuscevic et al. 1972) that the grade distributions in Cornish mines show a log-normal distribution. All the calculations of the ninety-five percentiles were done on the log-normal data. Mean grades for a series of levels were also determined from the cumulative frequency graphs and some examples are shown on Table 16.1 (Hackett 1988).

16.7.2 Reserve estimation

Clearly the estimation of ore reserves is inextricably linked to geological interpretations, which should ideally be as quantifiable as possible. Any doubt in the geological models casts uncertainty upon the reserve estimates, and that in turn upon the mine's future (Hopper 1990). Normally, confidence in the extension of the lode with depth was low. The Geology Department at Wheal Jane calculated the ore reserves every quarter using the US Bureau of Mines classification (USBM 1976). The measured reserves included ore for which tonnage was computed from dimensions revealed in workings and boreholes, and for which the grade was computed from the results of detailed channel sampling. Indicated reserves were computed partly from specific measurements and sample data, and partly from borehole core samples. The reserves were projected for a distance of 30 m above and below the level concerned. Inferred reserves were based largely on borehole core samples and a broad knowledge of the geological character of the deposit, and upon the assumed continuity for which there was geological evidence. The exercise was repeated quarterly to provide a rolling three year reserve estimate of approximately 280 000 t a^{-1}. The first year's reserves were considered as measured while years two and three were considered as indicated.

16.8 MINING

The orebodies were mined using the No.2 and Clemows Shafts (Fig. 16.17). The Clemows Shaft, which extends to the 17 Level, was the main ore hoisting shaft, while the

Table 16.1 Some examples of mean grades of the major elements at Wheal Jane

Level	Sn %	Zn %	Cu %
6 Sublevel	0.56	3.92	0.27
7 Sublevel	0.54	3.07	0.27
8 Sublevel	0.48	2.06	0.23
9 Sublevel	0.64	1.84	0.24
11 level	0.49	1.52	0.22
A Sublevel	0.55	2.11	0.26
B Sublevel	0.49	1.93	0.20
C Sublevel	0.44	2.16	0.20
D Sublevel	0.39	2.99	0.28
Averages	0.51	2.40	0.24

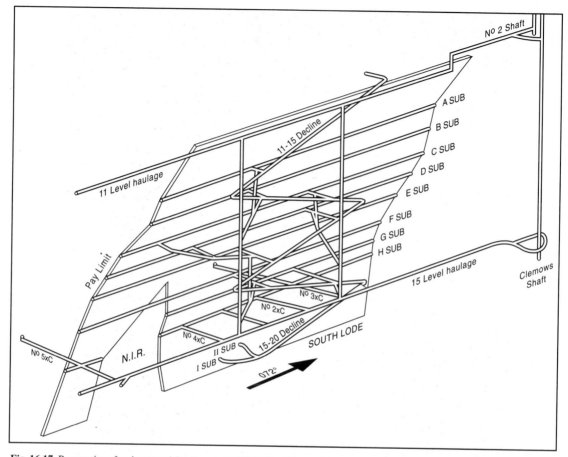

Fig. 16.17 Perspective of underground development 1990. (After Hopper 1990.)

No.2 shaft was used for men and materials and only reaches the 11 Level. The 11 and 15 Levels were the main footwall haulage drives along which ore was trammed in 5 t Granby wagons. The levels are connected by the 11–15 Decline from which the lode was reached by crosscuts. Upon intersecting the lode, drives were pushed out from the crosscut east and west to the pay-limit. Mining then proceeded back to the crosscut on a retreat basis, ore being mucked to ore-passes in the crosscut which in turn fed the ore-chutes on the 15 Level haulage. Mining was conducted using either long-hole sublevel (sub) open stoping or semi-shrinkage stoping (Figs 16.18 & 16.19). Main haulage and decline development was done using a Minimatic twin-boom Tamrock.

A major fixed cost was due to pumping, with 1215 Ml of water being pumped to the surface each month (Hackett 1988). Furthermore the acidity of the water, which had a pH as low as 2 in the 15 West Level, required constant maintenance of piping and pumps costing over £1M p.a. (Gidley *pers. comm.* in Hopper 1990). The aquiferous nature of the orebody caused numerous mining problems, requiring dewatering of the lode prior to driving into it (pressures of 1.38 Mpa have been encountered while drilling) and made Wheal Jane one of the wettest mines per tonne of ore, in the world (Operations 1985). After closure in 1991, the mine was not kept on a care and maintenance programme, due to the costs of dewatering, and within six months of closure the workings were totally flooded.

An approximate breakdown of operating costs is shown in Table 16.2.

In a mine using sublevel stoping methods, on-lode development is required in advance of mining. It is important that it proceeds far enough in advance of production to provide an abundance of new working faces, otherwise productivity falls. The development

Fig. 16.18 Underground view of a typical stope with scoop tram. (Courtesy P.R. Deakin.)

strategy is obviously entirely dependent upon the distribution and quality of the ore reserves. Essentially the establishment of sufficient reserves entailed the driving of a crosscut normal to the lode drive and the establishment of a drill chamber at its end (Fig. 16.20). A programme of underground diamond drilling was then conducted to prove the existence of mineralization above and below the existing drive.

An alternative option, that was frequently used to establish the presence of reserves immediately before mining, was to advance the drive, parallel to the lode, into uncharted territory. Holes would still normally be drilled from the drive to assess the geology above and below the drive. The option would also be available to drive crosscuts to the lode to develop it (Table 16.3, Fig. 16.20). The main disadvantage of this method is the cost if the lode proves to be uneconomic. On the other hand, if the lode

is well mineralized, considerable time will have been saved and ore is already developed and ready for mining.

16.9 MINERAL PROCESSING

Before the slump in tin prices in the mid to late 1980s, there were three operating underground tin mines in Cornwall, each with its own mill, namely South Crofty, Geevor and Wheal Jane (Turner 1984). The mills at South Crofty and Geevor treated ore derived mainly from veins with only small quantities (1–2% by weight) of sulphides. The cassiterite was relatively coarse-grained (0.8–1.5 mm) and it was possible to separate the ore and gangue using heavy medium separation.

In contrast the mill at Wheal Jane treated ore which usually contained 15–20% sulphides. Pyrite formed the bulk of the sulphides, but there were also small amounts

a)

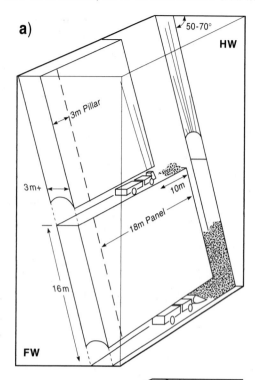

b)

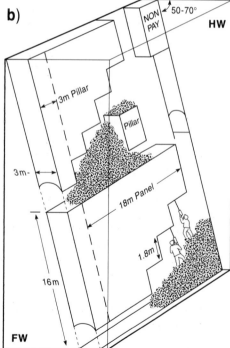

Fig. 16.19 Mining: comparison of mechanized and manual techniques. (a) long hole sub-level open stoping, (b) semi-shrinkage stoping . (After Hopper 1990.)

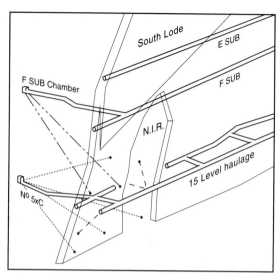

Fig. 16.20 Exploration drilling in advance of development. (After Hopper 1990.)

Table 16.2 Operating costs. (After Operations 1985.)

By Activity:	
Mining:	
Hoisting & Crushing	8%
Development	12%
Stoping	7%
Tramming	4%
Pumping	10%
Mine Services	8%
Mining sub total	49%
Treatment	19%
Services and supervision	14%
Total operating costs	82%
By Element :	
Labour	39%
Materials	21%
Power	13%
Contractors	3%
Exploration	3%
Fixed costs amounted to roughly	65%
Variable costs amounted to roughly	35%

Table 16.3 Costs for establishing a typical underground drilling chamber. (Gatley *pers. comm.* in Hopper 1990)

Drivage for tracked mining	£450 m^{-1}
Drivage for trackless mining	£350 m^{-1}
Drilling	£ 27 m^{-1}

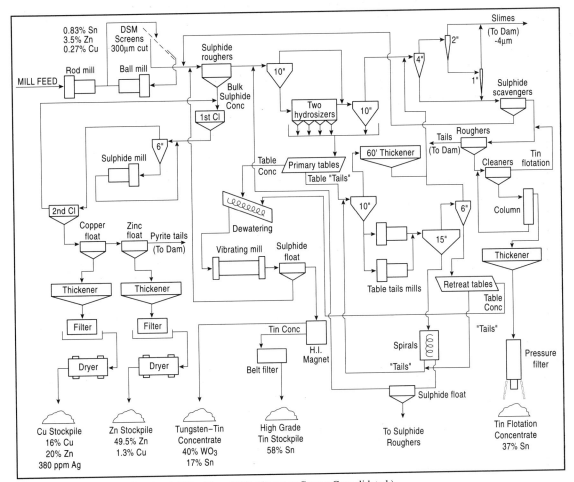

Fig. 16.21 Mineral processing circuit, Wheal Jane 1990. (Courtesy Carnon Consolidated.)

of sphalerite, chalcopyrite, arsenopyrite and traces of other sulphides. The ore was washed in the crushing plant and classified into several size fractions (Fig. 16.21). A gravity section separated the coarse (>40 mm) and fine fractions. The fine fraction went to the flotation section, where sulphides and cassiterite were separated. The waste was deposited in a tailings pond. The other mines deposited their tailings in streams or directly into the sea.

Recovery of cassiterite reached a maximum of 72% toward the end of the mine's life and is related to the its mineralogy (Holl & Bromley 1988). The greisen associated primary cassiterite was relatively coarse and amenable to gravity recovery. Secondary cassiterite derived from replacement of stannite by chalcopyrite was however not amenable to recovery. The tourmalinite-associated cassiterite was typically fine-grained (<100 μm) but recoverable by flotation. The chlorite assemblage con-tained relatively coarse cassiterite and sulphides. Recovery problems occurred as a result of the replacement of cassiterite by sulphides along fractures. In addition minor amounts of the tin–silver sulphide, canfieldite, were present and contributed to the tin lost in processing. Recovery of separate copper and zinc concentrates was extremely difficult due to intimate intergrowths of the two minerals. However the overall recovery of copper and zinc was 85% and the combined concentrate contained substantial silver of unknown mineralogical form.

16.10 THE ENVIRONMENTAL IMPACT OF MINE CLOSURE

Although March 1991 marked the end of mining, it was the not the end of the Wheal Jane story, particularly as far publicity was concerned. At the end of mining, as mentioned above, the mine was allowed to flood by stopping pumping.

By November 1991 the water level had almost reached the surface and there were indications of leakage into the Carnon River revealed during monitoring by the regulatory authority in the UK, the National Rivers Authority (NRA). Although this leakage was not very significant as the Carnon had been polluted by acid mine drainage since 1750, there was concern about a major new influx and the pollution of ground water. The Carnon River discharges, 3 km downstream from Wheal Jane, into an estuarine area that is a popular sailing area and shell fishery. The main potential problems were the acidity of mine water in Wheal Jane (pH 2–3) and the high concentrations of metals, Cd, Zn, Cu, Pb, As and Fe. These metals are released from the oxidation of *in situ* sulphides as a result of the low pH generated by weathering of pyrite. A decision was therefore made to control the final rise of the water level to pre-mining level by pumping water from the shafts and to treat the water by liming in order to raise the pH and precipitate metals such as Zn and Cd. This treatment was financed by the NRA in collaboration with Carnon Consolidated.

The pumping continued until 4 January 1992 when it was stopped due to stormy conditions. After nine days the build up of water within the mine caused a concrete plug in the Nangiles adit to burst and there was a major outflow of untreated acid water into the Carnon river. This released 320 Ml of water into the estuary, resulting in a major pollution plume which discoloured the estuary and deposited metals. Cadmium levels in the river reached 600 mg l⁻¹ relative to a UK water quality standard of 1 mg l⁻¹. In addition there were significant concentrations of copper and zinc (Fig. 16.22). The immediate public outcry caused a resumption of pumping and water treatment at Wheal Jane main shaft. Pumping has continued to date while a sustainable solution to the problem is sought. It is likely that this will involve the neutralization of acid waters by passage through limestone tanks and extraction of metals using vegetation. The crisis at Wheal Jane also highlighted the other sources of acid drainage in the Carnon Valley and these will be treated at the same time. Under UK law, as the mine was abandoned, the mining company was not liable and the entire budgeted cost of £8M has been borne by the UK government.

16.11 SUMMARY

In this chapter we have tried not only to illustrate some of the difficulties encountered in the exploration for, and exploitation of, vein deposits, but also the complications

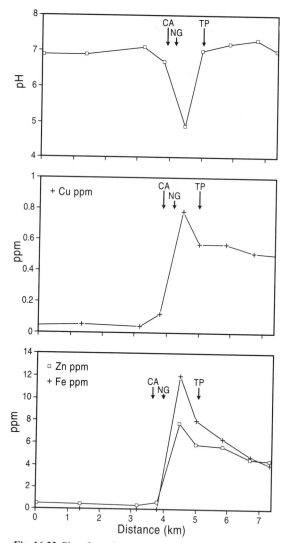

Fig. 16.22 Plot of metal content and pH of the River Carnon, December 1993. The main sources of metals are: CA= County Adit, NG= Nangiles Adit and TP= Tailings pond. The central part of the plot is located in Fig. 16.12.

that may arise when exploring in an old mining field in a country possessing somewhat archaic mining legislation and strong environmentalist lobbies. In such an orefield, as here in south-west England, much time must be devoted in the initial stages to literature researches and, once targets have been selected, competent lawyers must be engaged to determine whether the owners of the mineral and surface rights of the target areas can be located.

In an orefield like that of south-west England soil geochemistry can help in target location but geophysics may be of little value. In the case of Wheal Jane the mineralization had been discovered back in the 1700s but much of the deposit remained unexploited because of the poor recoverability during mineral processing. Thus the literature survey suggested the possibility of much untouched ore being present and drilling revealed that this indeed was the case. Mineral processing research by CGF, which greatly improved the recoverability, and later on the application of large tonnage mining procedures by the RTZ Corporation, made this a profitable operation in the early 1980s up to the time of the crash in the price of tin in 1985; despite the excessive water problems that resulted from working in an old mining district riddled with interconnecting tunnels driven during previous underground workings.

References

Abrams M.J., Conel J.E. & Lang H.R. (1984) Joint NASA/GEOSAT Test Case Report, *Am. Assoc. Petroleum Geol.*, 3 vols.

Adams S. S. (1985) Using geological information to develop exploration strategies for epithermal deposits. In Berger B.R. & Bethke P.M. (eds), *Geology and Geochemistry of Epithermal Systems*, 273–298. Economic Geology Pub. Co.

Alderton D.H.M. (1978) Fluid inclusion data for lead–zinc ores from south-west England. *Trans. Instn Min. Metall. (Sect. B Appl. Earth Sci.)*, **87**, B132–135.

Alderton D.H.M. (1993) Mineralization associated with the Cornubian Granite Batholith. In Pattrick R.A.D. & Polya D.A. (eds), *Mineralization in the British Isles*, 270–354. Chapman & Hall, London.

Ali J.W. & Hill I.A. (1991) Reflection seismics for shallow geological investigations: a case study from central England. *J. Geol. Soc. London,* **148**, 219–222.

Allen M.E.T. & Nichol I. (1984) Heavy-mineral concentrates from rocks in exploration for massive sulphide deposits. *J. Geochem. Explor.*, **21**, 149–165.

Allum J.A.E. (1966) *Photogeology and Regional Mapping,* Pergamon Press, Oxford.

Andrew R.L. (1978) *Gossan Evaluation.* Short Course, Rhodes University, Grahamstown.

Anglo American Corp. (1991) *Anglo American Corporation of South Africa Ltd Annual Report for 1991.*

Annels A.E. (1991) *Mineral Deposit Evaluation : A Practical Approach.* Chapman and Hall, London.

Anon. (1977) Consolidated Rambler finds zinc bearing mercury. *World Mining*, **March**, 79.

Anon. (1985a) Politics and mining—when risk becomes reality. *World Mining Equipment*, **August**, 24–26.

Anon. (1985b) Where are new developments worth the risk? *World Mining Equipment*, **July**, 44–55.

Anthony M. (1988) Technical change—the impending revolution. In *Australasian Institute of Mining and Metallurgy Sydney Branch, Minerals and Exploration at the Crossroads*, 19–25, Sydney.

Antrobus E.S.A. (ed.) (1986) *Witwatersrand Gold—100 Years.* Geol. Soc. S. Afr., Johannesburg.

Armitage M.G. & Potts M.F.A. (1994) Some comments on the classification of resources and reserves. In Whateley M.K.G. & Harvey P.K. (eds), *Mineral Resource Evaluation II: Methods and Case Histories*, 11–16. Spec. Publ. **79**, Geol. Soc. London.

Ashleman J.C. (1983) *Trinity Silver Project 1982–1983, Progress Summary Report*, Unpublished Report EX84–10, US Borax.

Ashleman J.C. (1986) *Trinity Venture Project, 1986 Summary Report*, Unpublished Report, EX 87–3, US Borax.

Ashleman J.C. (1988) *The Trinity Silver Deposit, Pershing County, Nevada*, Unpublished Field Guide.

Australian Drillers Guide (1985) Published by Australian Drilling Training Committee; H.F. Eggington (ed.), P.O. Box 11545, Macquarie Centre, New South Wales 21113, Australia.

Austr. Inst. Geoscientists (1987) *Meaningful Sampling in Gold Exploration.* Bull. **7**, Duncan and Associates, Leederville.

Austr. Inst. Geoscientists (1988) Sample Preparation and Analyses for Gold and Platinum Group Elements. Bull. **8**, Duncan and Assoc., Leederville.

Austr. Inst. Geoscientists (1990) Strategies for Grade Control. Bull. **10**, Duncan and Assoc., Leederville.

Azaroff L.V. & Buerger M.J. (1958) *The Powder Method in X-ray Crystallography.* McGraw-Hill, New York.

Baele S.M. (1987) *Silver Ore Reserve Estimation, Trinity Heap Leach Project, Pershing County, Nevada.* Unpublished Report, MD 87–3, US Borax.

Bagby W.C. & Berger B.R. (1985) Geologic characteristics of sediment-hosted, disseminated precious-metal deposits in the western United States. In Berger B.R. and Bethke P.M. (eds), *Geology and Geochemistry of Epithermal Systems*, 169–202. Reviews in Economic Geology **2**, Econ Geol. Pub. Co.

Bailey D.C. & Hodson D.I. (1991) The Neves-Corvo Project—the successes and challenges. *Trans. Instn Min. Metall., Sect. A*, **100**, A1–A10.

Barnes J.W. (1981 & 1993) *Basic Geological Mapping.* Open University Press, Milton Keynes.

Barrett W.L. (1987) Auger Drilling. *Trans. Instn Min. Metall., Sect. B*, **96**, B165–169.

Barrett W.L. (1992) A case history of pre-extraction site investigation and quarry design, Cliffe Hill Quarry, Leicestershire. In Annels A.E. (ed.) *Case Histories and Methods in Mineral Resource Evaluation*, 69–76. Spec. Publ. **63**, Geol. Soc. London.

Barrie C.T., Ludden J.N. & Green T.H. (1993) Geochemistry of volcanic rocks associated with Cu–Zn and Ni–Cu deposits in the Abitibi Subprovince. *Economic Geology*, **88**, 1341–1358.

Bateman A.M. (1950) *Economic Mineral Deposits*. Wiley, New York.

Beaty D.W., Taylor H.P. Jr & Coad P.R. (1988) An oxygen isotope study of the Kidd Creek, Ontario, volcanogenic massive sulfide deposit: evidence for a high ^{18}O ore fluid. *Econ. Geol.*, **83**, 1–17.

Beckinsale R.D. (1979) Granite magmatism in the tin belt of South-east Asia. In Atherton M.P. & Tarney J. (eds), *Origin of Granite Batholiths*, 34–44. Shiva, Orpington.

Bell A.R. & Hopkins D.A. (1988) From farmland to tarmac. *Extractive Industry Geology 1985*, 19–26.

Bell T. (1989) *Ore Reconciliation and Statistical Evaluation of the Trinity Silver Mine, Pershing County, Nevada*. Unpublished MSc Dissertation, University of Leicester.

Bell T. & Whateley M.K.G. (1994) Evaluation of grade estimation techniques. In Whateley M.K.G. & Harvey P.K. (eds), *Mineral Resource Evaluation II, Methods and Case Histories*, 67-86. Spec. Publ. **79**, Geol. Soc. London.

Bennett M.J., Beer K.E., Jones R.C., Turton K., Rollin K.E., Tombs J.M.C. & Patrick D.J. (1981) *Mineral Investigations Near Bodmin, Cornwall. Part 3—The Mulberry and Wheal Prosper Area*. British Geological Survey Mineral Reconnaissance Report **48**, Keyworth.

Berger B.R. & Bagby W.C. (1991) The geology and origin of Carlin-type gold deposits. In Foster R.P. (ed) *Gold Metallogeny and Exploration*, 210-48. Blackie, Glasgow.

Berkman D.A. (1989) *Field Geologist's Manual*, 3rd Edition. Australasian Institute of Mining and Metallurgy, Parkville.

Berry L.G., Mason B. & Dietrich R.V. (1983) *Mineralogy*. Freeman, San Francisco.

Beus A.A. & Grigorian S.V. (1977) *Geochemical Methods for Mineral Deposits*. Applied Publishing, Wilmette, lllinois.

Bieniawski Z.T. (1976) Rock mass classification in rock engineering. In Bieniawski Z.T. (ed.), *Exploration for Rock Engineering*, **1**, 97–106. A.A Balkema,

Cape Town.

Billingsley P. & Locke A. (1941) Structures of ore districts in the continental framework. *Trans. Am. Inst. Min. (Metall.) Eng.*, **144**, 9–21.

Blanchard, R. (1968) *Interpretation of Leached Outcrops*. Nevada Bureau of Mines Bull. **66**, Reno.

Blanchard R. & Boswell P.F. (1934) Additional limonite types of galena and sphalerite derivation. *Econ. Geol.*, **29**, 671–690.

Blunden J. (1985) *Mineral Resources and their Mangement*. Longman, Harlow.

Bonham H.F. (1988) Silver deposits of Nevada. In *Silver: Exploration, Mining and Treatment*, 25–31. Instn Min. Metall., London.

Bonham H.F. Jr (1989) Bulk mineable gold deposits of the Western United States. In Keays R.R., Ramsay W.R.H. & Groves D.I. (eds) *The Geology of Gold Deposits—The Perspective in 1988*, 193–207. Econ. Geol. Monograph **6**, El Paso.

Bonham-Carter G.F. (1994) *Geographic Information Systems for Geoscientists: Modelling with GIS*. Pergamon, New York.

Boyle R.W. (1974) *Elemental Associations in Mineral Deposits and Indicator Elements of Interest in Geochemical Prospecting*. Paper **74–45**, Geol. Surv. Canada.

Bradshaw P.M.D. (ed.) (1975) Conceptual models in exploration geochemistry—the Canadian Shield and Canadian Cordillera. *J. Geochem. Explor.*, **4**, 1–213.

Brady B.H.G. & Brown E.T. (1985) *Rock Mechanics for Underground Mining*. George Allen and Unwin, London.

Brassington R. (1988) *Field Hydrogeology*. Geological Society of London Professional Handbook Series, Wiley, Chichester.

Brinkmann R. (1976) *Geology of Turkey*. Elsevier, Amsterdam.

Brisbin D., Kelly V. & Cook R. (1991) Kidd Creek Mine. In Fyon J.A. & Green A.H. (eds), *Geology and Ore Deposits of the Timmins District, Ontario*, 66–71. Geological Survey Of Canada Open File Report **2161**

British Standard 812 (1975) *Method for Sampling and Testing of Mineral Aggregates, Sands and Fillers*. British Standards Institution, London.

Bromley A. V. (1989) *Field Guide: the Cornubian Orefield. Sixth International Symposium on Water-Rock Interaction*. Camborne School of Mines, Camborne.

Bromley A. V. & Holl C. J. (1986) Tin mineralisation in southwest England. In Wills B.A. and Barley R.W.

(eds), *Mineral Processing at a Crossroads,* 195–262. Martinus Nijhoff, Dordrecht.

Brooks R.R. (1983) *Biological Methods of Prospecting for Minerals.* Wiley, New York.

Brown E.T. (ed.) (1981) Rock characterization, testing and monitoring, *International Society for Rock Mechanics,* Pergamon Press, Oxford.

Burke K., Kidd W.S.F. & Kusky T.M. (1986) Archaean foreland basin tectonics in the Witwatersrand, South Africa. *Tectonics,* **5,** 439–456.

Burke K. & Dewey J.F. (1973) Plume-generated triple junctions: key indicators in applying plate tectonics to old rocks. *J. Geol., 81,* 406-433.

Burt R., Waite P. & Burnley R. (1987) *Cornish Mines.* University of Exeter, Exeter.

Butt C.R.M. & Smith R.E. (1980) Conceptual models in exploration geochemistry—Australia. *J. Geochem. Explor.,* **12,** 89–365.

Butt C.R.M. & Zeegers H. (eds) (1992) *Regolith Exploration Geochemistry in Tropical and Subtropical Terrains.* Handbook of Exploration Geochemistry Vol. **4.,** Elsevier, Amsterdam.

Cady J.W. (1980) Calculation of gravity and magnetic anomalies of finite length right polygonal prisms. *Geophysics,* **45,** 1507–1512.

Cameron R.I. & Middlemis H. (1994) Computer modelling of dewatering a major open pit mine: case study from Nevada, USA. In Whateley M.K.G. & Harvey P.K. (eds), *Mineral Resource Evaluation II: Methods and Case Histories,* 205–215. Spec. Publ. **79,** Geol. Soc. London.

Camisari-Calzolari F.A.G.M., Ainslie L.C. & Van Der Merwe P.J. (1985) *Uranium in South Africa 1985.* South African Atomic Energy Corporation of South Africa, Pelindaba.

Camm T.W. (1991) *Simplified Cost Models for Prefeasibility Mineral Evaluations. Bureau of Mines Information Circular* **9298**, United States Dept. of Interior.

Campbell A.S. (ed.) (1971) Geology and History of Turkey. *Petroleum Exploration Society of Lybia,* 13th Annual Conference.

Campbell, G. & Crotty, J.H. (1990) 3-D seismic mapping for mine planning purposes at the South Deep prospect. In Ross-Watt, R.A.J. & Robinson, P.D.K. (eds), *Technical Challenges in Deep Level Mining,* 569–597. S. Afr. Inst. Min. Metallurgy, Johannesburg.

Campbell I.H., Lesher C.M., Coad P., Franklin J.M., Gorton M.P. & Thurston P.C. (1984) Rare-earth mobility in alteration pipes below Cu–Zn–sulphide

deposits. *Chemical Geology,* **45,** 181–202.

Cann J.R. (1970) New model for the structure of the ocean crust. *Nature,* **266,** 928–930.

Carnon (1989) *Carnon Consolidated Internal Newspaper,* No. **6.**

Cartwright A.P. (1967) *Gold Paved the Way.* Purnell, Johannesburg.

Carver R.N., Chenoweth L.M., Mazzucchelli R.H., Oates C.J. & Robbins T.W. (1987) Lag—a geochemical; sampling medium for arid regions. *J. Geochem. Explor.,* **28,** 183–199.

Cas R.A.F. & Wright J. (1987) *Volcanic Successions: Modern and Ancient.* Allen & Unwin, London.

Cathles L.M. (1993) Oxygen isotope alteration in the Noranda Mining District, Abitibi Greenstone Belt, Quebec. *Economic Geology,* **88,** 1483–1511.

Chaussier J.-B. & Morer J. (1987) *Mineral Prospecting Manual.* Elsevier, Amsterdam.

Chayes F. (1956) *Petrographic Modal Analysis, An Elementary Statistical Appraisal.* Wiley, New York.

Chen W. (1988) Mesozoic and Cenozoic sandstone-hosted copper deposits in South China. *Mineral. Deposita,* **23,** 262–267.

Chen Y., Clark A.H., Farrar E., Wasteneys H.A.H.P., Hodgson M.J. & Bromley A.V. (1993) Diachronous and independent histories of plutonism and mineralization in the Cornubian Batholith, southwest England. *J. Geol. Soc. London,* **150,** 1183–1192.

Chilingarian G.V. & Vorabutr P. (1981) *Drilling and Drilling Fluids.* Elsevier, Amsterdam.

Clark I. (1979) *Practical Geostatistics.* Elsevier Applied Science, Amsterdam.

Clark R.J. McH., Homeniuk L.A. & Bonnar R. (1982) Uranium geology in the Athabaska and a comparison with other Canadian Proterozoic Basins. *CIM Bull.,* **75** (April), 91–98.

Clarke P.R. (1974) The Ecstall Story: Introduction. *Can. Inst. Min. Metall. Bull.,* **77,** 51–55.

Clifford J.A., Meldrum A.H., Parker, R.H.T. & Earls G. (1990) 1980–1990: A decade of gold exploration in Northern Ireland and Scotland. *Trans. Instn Min. Metall., Sect. B,* **99,** B133–138.

Clifford J.A., Earls G., Meldrum A.H. & Moore N. (1992) Gold in the Sperrin mountains, Northern Ireland: an exploration case history. In Bowden A.A., Earls G., O'Connor P.G. & Pyne J.F. (eds) *The Irish Minerals Industry 1980–1990,* 77–87. Irish. Assoc. Econ. Geology, Dublin.

Clifton H.E., Hunter R.E., Swanson F.J. & Phillips R.L. (1969) *Sample Size and Meaningful Gold Analysis.* US Geol. Surv. Prof. Paper **625–C.**

Closs L.G. & Nichol I. (1989) Design and planning of geochemical programs. In Garland G.D. (ed.) *Proceedings of Exploration '87*, 569–583. Spec. Vol. **3**, Ontario Geological Survey, Toronto.

Coad P.R. (1984) Kidd Creek Mine. In Gibson H.L. *et al.* (eds) Surface geology and volcanogenic base metal massive sulphide deposits and gold deposits of Noranda and Timmins. Geological Association of Canada Field Trip Guide Book **14**.

Coad P.R. (1985) Rhyolite geology at Kidd Creek—a progress report. *Can. Inst. Min. Metall. Bull.*, **78**, 70–83.

Cohen A.D., Spackman W. & Raymond R. (1987) Interpreting the characteristics of coal seams from chemical, physical and petrographic studies of peat deposits. In Scott A.C. (ed.), *Coal and Coal-bearing Strata: Recent Advances*, 107–125. Spec. Publ. **32**, Geol. Soc. London.

Coker W.B., Hornbrook E.H.W. & Cameron E.M. (1979) Lake sediment geochemistry applied to mineral exploration. In Hood P.J. (ed.), *Geophysics and Geochemistry in the Search for Metallic Ores*, 435–447. Econ. Geol. Rept **31**, Geol. Surv. Canada.

Collins L. & Fox R.A. (eds) (1985) *Aggregates; Sand, Gravel and Crushed Rock Aggregates for Construction*. Geol. Soc. London.

Colman T.B. (1990) *Exploration for Metalliferous and Related Minerals: a Guide*. British Geological Survey, Keyworth.

Colwell R.N. (ed.) (1983) *Manual of Remote Sensing*, 2nd Edition, 2 Vols, Am. Soc. Photogrammetry, Falls Church, Virginia.

Cook D.R. (1986) Analysis of significant mineral discoveries in the last 40 years and future trends. *Min. Engrg*, **Feb**, 87–94.

Cook D.R. (1987) A crisis for economic geologists and the future of the Society. *Econ. Geol.*, **82**, 792–804.

Cook P.J. (1984) Spatial and temporal controls on the formation of phosphate deposits. In Nriagu J.O. & Moore P.B. (eds), *Phosphate Minerals*, 242–274. Springer-Verlag, Berlin.

Coope J.A. (1991) *Monitoring Laboratory Performance with Standard Reference Samples*. Abstract, 15th Geochemcial Exploration Symposium, Reno.

Cooper P.C. & Stenberg M. (1988) Deep directional core drilling with Navi-Drill. *Eng. and Min. J.*, **189**, (7), 48–49 and 59.

Corner B. & Wilshire W.A. (1989) Structure of the Witwatersrand Basin derived from the interpretation of aeromagnetic and gravity data. In Garland G.D. (ed.), *Proceedings of Exploration '87*, 532–545.

Spec. Vol. **3**, Ontario Geological Survey, Toronto.

Cox D.P., & Singer D. (1986). *Mineral Deposit Models*. US Geol. Surv. Bull. **1693**, Reston.

Cox F.C., Bridge D.McC. & Hull J.H. (1977) *Procedure for the Assessment of Limestone Resources*. Mineral Assessment Report **30**, Institute of Geological Sciences, London.

Craig J.R. & Vaughan D.J. (1981) *Ore Microscopy and Ore Petrography*. Wiley, New York.

Craig Smith, R. (1992) PREVAL: Prefeasibility Software Program for Evaluating Mineral Properties. Information Circular **9307**, United States Dept. of Interior.

Crosson C.C. (1984) Evolutionary development of Palabora. *Trans. Instn Min. Metall., Sect. A*, **93**, A58–A69.

Crowson P.C.F. (1988) A perspective on worldwide exploration for minerals. In Tilton J.E., Eggert R.G. & Landsberg H.H. (eds), *World Mineral Exploration*, 21–103. Resources for the Future, Washington.

Cummings J.D. & Wickland A. P. (1985) *Diamond Drill Handbook*. J.K. Smit and Sons Ltd., Toronto.

Curtin G.C., King H.D. & Moiser E.L. (1974) Movement of elements into the atmosphere from coniferous trees in the subalpine forests of Colorado and Idaho. *J. Geochem. Explor.*, **3**, 245–263.

Danchin P.D. (1989) Obituary of J. Dalrymple. *Geobulletin*, 2nd Quarter, 21–22.

David M. (1988) *Handbook of Applied Advanced Geostatistical Ore Reserve Estimation*. Developments in Geomathematics, **6**, Elsevier, Amsterdam.

Davidson C.F. (1965) The mode of origin of banket orebodies. *Trans. Instn Min. Metall.*, **74**, 319–338.

Davies M.H. (1987) Prospects for copper. *Bull. Institn Min. Metall.*, **September**, 3–6.

Davis C.G. (1986) Narrow vein mining, the geologist's role. *Can. Inst. Min. Metall.*, Spec.Vol. **7**, 186–195.

Davis C.J. and Battersby G.W. (1985) Reopening of Wheal Jane mine. *Trans. Instn Min. Metall., Sect. A*, **94**, A135–A147.

Davis J.C. (1986) *Statistics and Data Analysis in Geology*. Wiley, New York.

De Beer J.H. & Eglington B.M. (1991) Archaean sedimentation on the Kaapvaal Craton in relation to tectonism in granite–greenstone terrains: geophysical and geochronological constraints. *J. Afr. Earth Sci.*, **13**, 27–44.

Deutsch M., Wiesnet D.R. & Rango A. (eds) (1981) *Satellite Hydrogeology*. American Water Resources Association, Minneapolis, Minnesota.

Dewey H. (1935). *British Regional Geology: South-*

West England. Geol. Surv. & Museum, HMSO, London.

Dickinson R.T., Jones C.M. & Wagstaff J.D. (1986) MWDatanet—acquisition and processing of down-hole and surface data during drilling. *Trans. Instn Min. Metall., Sect. B,* **95**, B140–148.

Dines H.G. (1956) *The Metalliferous Mining Region of South-west England.* HMSO, London.

Dononhoo H.V., Podolsky G. & Clayton R.H. (1970) Early geophysical exploration at Kidd Creek Mine. *Mining Congress Journal,* **May**, 44–53.

Dorey R., Van Zyl D. & Kiel J. (1988) Overview of heap leaching technology. In Van Zyl D., Hutchison I. & Kiel J. (eds), *Introduction to Evaluation, Design and Operation of Precious Metal Heap Leaching Projects,* 3–22. Society of Mining Engineers, Denver.

Drury S.A. (1986) Remote sensing of geological structure in European agricultural terrains. *Geol. Mag.,* **123**, 113–121.

Drury S.A. (1987) *Image Interpretation in Geology.* Allen and Unwin, London.

Drury S.A. (1993) *Image Interpretation in Geology,* 2nd edition. Allen & Unwin, London.

Dunlop A.C. & Meyer W.T. (1978) Detrital tin patterns in stream sediments and soils in mid-Cornwall. *J. Geochem. Explor.,* **10**, 259–276.

Dunn C.E. (1989) Developments in biogeochemical exploration. In Garland G.D. (ed.), *Proceedings of Exploration '87,* 417–438. Spec. Vol. **3**, Ontario Geological Survey, Toronto.

Dunn P. R., Battey G. C., Miezitis Y. & McKay A. D. (1990a). The distribution and occurrence of uranium. In Glasson K.R. & Rattigan J.H. (eds), *Geological Aspects of the Discovery of Some Important Mineral Deposits in Australia,* 455–462. Australasian Institution of Mining and Metallurgy, Melbourne.

Dunn P. R., Battey G. C., Miezitis Y. & McKay A. D. (1990b). The uranium deposits of the Northern Territory. In Glasson K.R. & Rattigan J. H. (eds), *Geological Aspects of the Discovery of Some Important Mineral Deposits in Australia,* 463–476. Australasian Institute of Mining and Metallurgy, Melbourne.

Du Toit A.L. (1954) *The Geology of South Africa.* Oliver and Boyd, Edinburgh.

Eckstrand O. R. (ed.) (1984) *Canadian Mineral Deposit Types: a Geological Synopsis.* Geol. Surv. Canada Econ. Rept **36**, Ottawa.

Edwards R.P. & Atkinson K. (1986) *Ore Deposit Geology.* Chapman & Hall, London.

Eggert R. G. (1988) Base and precious metal exploration by major corporations. In Tilton J.E., Eggert R.G. & Landsberg H.H. (eds), *World Mineral Exploration,* 105–144. Resources for the Future, Washington D.C.

Eimon P.I. (1988) Exploration for and evaluation of epithermal deposits. In Gold Update 87–88, Epithermal Gold, 68–80. Course Manual, Department of Geology, University of Southampton, Southampton.

Eldridge C.S., Barton P.B. Jr & Ohmoto H. (1983) Mineral textures and their bearing on formation of the Kuroko orebodies. In Ohmoto H. & Skinner B.J. (eds), *Kuroko and Related Volcanogenic Massive Sulfide Deposits,* Econ. Geol. Monogr., **5**, 241–281.

Ellis D. (1988) *Well Logging for Earth Scientists.* Elsevier, Amsterdam.

Elmore R.D. (1984) The Copper Harbor Conglomerate: a late Precambrian fining-upward alluvial fan sequence in northern Michigan. *Bull. Geol. Soc. Am.,* **95**, 610–617.

Emerson D.W. (1980) *The Geophysics of the Elura Orebody.* Australian Society of Exploration Geophysicists, Sydney.

Engineering Group Working Party (1977) *Quart. J. Eng. Geol.,* **10**, Geol. Soc., London.

Englebrecht C.J. (1986) The West Wits Line. In Antrobus E.S.A. (ed.), *Witwatersrand Gold—100 Years,* 199–225. Geol. Soc. S. Afr., Johannesburg.

Englebrecht C.J., Baumbach G.W.S., Matthysen J.L. & Fletcher P. (1986) The West Wits Line. In Annhaeusser C.R. & Maske S. (eds), *Mineral Deposits of Southern Africa,* 599–648. Geol. Soc. S. Afr., Johannesburg.

Erseçen N. (1989). *Known Ore and Mineral Resources of Turkey.* M.T.A., Ankara.

Evans A.M. (1968). Charnwood Forest. In Sylvester-Bradley P.C. & Ford T.D. (eds), *The Geology of the East Midlands,* 1–12. Leicester University Press, Leicester.

Evans A.M. (1993) *Ore Geology and Industrial Minerals—An Introduction.* Blackwell Scientific Publications, Oxford.

Farmer I. (1983) *Engineering Behaviour of Rocks,* 2nd Edition. Chapman & Hall, London.

Fauth H., Hindel R., Siewers U. & Zinner J. (1985) *Geochemischer Atlas Bundesrepublik Deutschland.* Bundesanstalt für Geowissenschaften und Rohstoffe, Hannover.

Feather C.E. & Koen G.M. (1975) The mineralogy of the Witwatersrand reefs. *Miner. Sci. Eng.,* **7**, 189–224.

Feiss P.G. (1989) Metallogenic provinces. In Carr D.D. & Herz N. (eds), *Concise Encyclopedia of Mineral*

Resources, 2996–3000. Pergamon Press, Oxford.

Fenwick K.G., Newsome J.W. & Pitts A.E. (1991) *Report of Activities 1990*. Ontario Geological Survey Misc. Paper **152**, Toronto.

Fletcher W.K. (1981) *Analytical Methods in Exploration Geochemistry*. Handbook of Exploration Geochemistry Vol. **1**, Elsevier, Amsterdam.

Fletcher W.K. (1987) Analysis of soil samples. In Fletcher W.K., Hoffman S.J., Mehrtens M.B., Sinclair A.J. & Thomson I. (eds), *Exploration Geochemistry: Design and Interpretation of Soil Surveys*, 73–96. Reviews in Economic Geology Vol. **3**, Society of Economic Geologists, El Paso.

Fletcher W.K., Hoffman S.J., Mehrtens M.B., Sinclair A.J. & Thomson I. (1987) *Exploration Geochemistry: Design and Interpretation of Soil Surveys*. Reviews in Economic Geology Vol. **3**, Society of Economic Geologists, El Paso.

Fluor Mining and Metals Inc. (1978) *The Use of Geostatistics in Exploration and Development*. Unpublished report, Fluor Metals Inc.

Forstner U. & Wittman G.T.W. (1979) *Metal Pollution in the Aquatic Environment*. Springer-Verlag, Berlin.

Fortescue J.A.C. & Hornbrook D.H.W. (1969) Two quick projects: one at a massive sulphide orebody near Timmins, Ontario and the other at a copper deposit in Gaspé Park, Quebec. In *Progress Report on Biogeochemical Research at the Geological Survey of Canada*, 39–63. Geol. Surv. Canada Paper **67–23**, Ottawa.

Fraser D. C. & Fogaresi S.L. (1993) The basics of image processing. In van Blaricom R. (ed.), *Practical Geophysics II*, 537–559. Northwest Mining Association, Spokane.

Friend T. (1990) The Vegetation Unit—caring for the environment. *Mining Survey*, **1990 No.1**, 18–22.

Frost A., McIntyre R.C., Papenfus E.B. & Weiss, O. (1946) The discovery and prospecting of a potential goldfield near Odendaalsrust in the Orange Free State, Union of South Africa. *Trans. Geol. Soc. S. Afr.*, **49**, 1–24.

Frutos J. (1982) Andean metallogeny related to the tectonic and petrologic evolution of the Cordillera. Some remarkable points. In Amstutz G.C., El Goresy A., Frenzel G., Khuth C., Moh G., Wauschkuhn A. & Zimmerman R.A. (eds), *Ore Genesis: the State of the Art*, 493–507. Springer-Verlag, Berlin.

Fry N. (1993) *The Field Description of Metamorphic Rocks*. Wiley, Chichester.

Fyon A. (1991) Volcanic–associated massive base metal sulphide mineralization. In Fyon J.A. & Green (eds), Geology and Ore Deposits of the Timmins District, Ontario (Field Trip 6), 51. Geol. Surv. Canada Open File Rept **2161**.

Gableman J.W. & Conel J.E. (1985) Section 7, Uranium Commodity Report. In Abrams M.J., Conel J.E. & Lang H.R. (eds), *The Joint NASA/Geosat Test Case Report* 3 vols. American Association of Petroleum Geologists, Tulsa.

Gardiner P. R. (1989) The recent Irish experience in promoting mineral exploration. In Chadwick J.R. (eds), *Mineral Exploration Programmes '89*, Paper **28**. International Mining, Madrid.

Garland G.D. (ed.) (1989) *Proceedings of Exploration '87*. Spec. Vol. **3**, Ontario Geological Survey, Toronto.

Garnett R.H.T. (1966) Distribution of cassiterite in vein tin deposits. *Trans. Instn Min. Metall., Sect. B*, **75**, B245–273.

Garrett R. G. (1989) The role of computers in exploration geochemistry. In Garland, G.D. (ed.) *Proceedings of Exploration '87*, 586–608. Spec. Vol. **3**, Ontario Geological Survey, Toronto.

Gatzweiler R., Scheling B. & Tan B. (1981) Exploration of the Key Lake uranium deposits, Saskatchewan, Canada. In IAEA (eds), *Uranium Exploration Case Histories*, 195–220. International Atomic Energy Agency, Vienna.

Gentry D.W. (1988) Minerals Project Evaluation—An Overview. *Trans. Instn Min. Metall., Sect. A*, **97**, A25–A35.

Gilluly J., Waters A.C. & Woodford A.O. (1959) *Principles of Geology*. Freeman, San Francisco.

Glasson K.R. & Rattigan J.H. (eds) (1990), *Geological Aspects of the Discovery of Some Important Mineral Deposits in Australia*, 455–462. Australas. Inst. Min. Metall., Melbourne.

Gocht W.R., Zantop H. & Eggert R.G. (1988) *International Mineral Economics*. Springer-Verlag, Berlin.

Goetz A.F.H., Rock B.N. & Rowan L.C. (1983) Remote sensing for exploration: an overview. *Econ. Geol.*, **78**, 573–590.

Goetz A.F.H. & Rowan L.C. (1981). Geologic remote sensing. *Science*, **211**, 781–791.

Gökçen N. (1982) The Ostracod Biostratigraphy of the Denizli–Mugla Neogene Sequence. *Bull. Inst. Earth Sci.*, Hacettpep University Beytepe, Ankara, **9**, 111–131.

Golder Associates (1983) *Feasibility Report on the Soma–Isiklar Lignite Deposit, Manisa Province, Western Turkey*. Unpl. Rept for Turkiye Komur

Isletmeleri Kurumu, 8 Vols.

Goldspear (1987) *Gold Panning. Practical and Useful Instructions.* Goldspear (UK) Ltd, Beaconsfield.

Goldstein J.I., Newbury D.E., Echlin P., Joy D.C., Fiori C.E. & Lifshin E. (1981) *Scanning Electron Microscopy and X-ray Microanalysis.* Plenum, New York.

Goode A.J.J. & Taylor R.T. (1988) Geology of the country around Penzance. *Memoir of the British Geological Survey,* Sheets 351 and 358 (England and Wales).

Goode J.R., Davie M.J., Smith L.D. & Lattanzi C.R. (1991) Back to basics: the feasibility study. *Can. Inst. Min. Metall. Bull.,* **84**, 953, 53–61.

Goodman R.E. (1976) *Methods in Geological Engineering in Discontinuous Rocks.* West Publishing Co., St Paul.

Goodman R.E. (1989) *Introduction to Rock Mechanics,* 2nd edition. Wiley, New York.

Gorman P.A.(1994) A review and evaluation of the costs of exploration, acquisition and development of copper related projects in Chile. In Whateley M.K.G. & Harvey P.K. (eds), *Mineral Resource Evaluation II: Methods and Case Histories,* 123–128. Spec. Publ. **79**, Geol. Soc. London.

Govett G.J.S. (1983) *Rock Geochemistry in Mineral Exploration.* Handbook of Exploration Geochemistry, Vol. **3**, Elsevier, Amsterdam.

Govett G.J.S. (1989) Bedrock geochemistry in mineral exploration. In Garland G.D. (ed.) *Proceedings of Exploration '87,* 273–299. Spec. Vol. **3**, Ontario Geological Survey, Toronto.

Grant F.S. & West G.F. (1965) *Interpretation Theory in Applied Geophysics.* McGraw-Hill Inc., New York.

Grant J.N., Halls C., Sheppard S.M.F. & Avila W. (1980) Evolution of the porphyry tin deposits of Bolivia. *Min. Geol.* Special Issue, **8**, 151–73.

Graton L.C. (1930) Hydrothermal origin of the Rand gold deposits, Part I, Testimony of the conglomerates. *Econ. Geol.,* **25**, Supp. to No. 3. 185pp.

Greathead C. & Graadt van Roggen J.F. (1986) The Orange Free State Goldfield. In Antrobus E.S.A. (ed.), *Witwatersrand Gold—100 Years,* 225–280. Geol. Soc. S. Afr., Johannesburg.

Green C.J.B. (1989) Metal markets and the LME. *Mining Annual Review,* C5–9, Mining Journal, London.

Gribble P. (1993) Fault interpretation from coal exploration borehole data using Surpac software. *Abstract volume,* Mineral Resource Evaluation 1993 Conference, Leicester University.

Gribble P. (1994) Fault interpretation from coal explo-

ration borehole data using Surpac software. In Whateley M.K.G. & Harvey P.K. (eds), *Mineral Resource Evaluation,* 29–35. Spec. Publ. **79**, Geol. Soc., London.

Guild P.W. (1978) Metallogenesis in the western United States. *J. Geol. Soc. London,* **135**, 355–376.

Gulson B.L. (1986) *Lead Isotopes in Mineral Exploration.* Elsevier, Amsterdam.

Gunn A.G. (1989) Drainage and overburden geochemistry in exploration for platinum-group element mineralisation in the Unst Ophiolite, Shetland, U.K. *J. Geochem. Explor.,* **31**, 209–236.

Gy M. (1992) *Sampling of Heterogeneous and Dynamic Material Systems.* Elsevier, New York.

Hackett C. (1988) *Geostatistical Study of South Lode, Wheal Jane, Cornwall.* Unpublished BSc (Mining Geology) Dissertation, University of Leicester.

Hale, M & Plant J.A. (1994) *Drainage Geochemistry.* Handbook of Exploration Geochemistry, Vol. **6.**, Elsevier, Amsterdam.

Hallbauer D.K. (1975) The plant origin of Witwatersrand carbon. *Miner. Sci. Eng.,* **7**, 111–131.

Hallberg J.A. (1984) A geochemical aid to igneous rock identification in deeply weathered terrain. *J. Geochem. Explor.,* **20**, 1–8.

Hallof P.G. (1993). Resistivity and induced polarization: the spectral induced polarization method. In van Blaricom R. (ed.), *Practical Geophysics II,* 139–176. Northwest Mining Association, Spokane.

Handley G.A. & Carrey R. (1990) Big Bell gold deposit. In Hughes, F. (ed.), Geology of the Mineral deposits of Australia and Papua New Guinea, 217–220. Australas. Inst. Min. Metall. Monograph **14**, Melbourne.

Hannington M.D. & Scott S.D. (1988) Gold and silver potential of polymetallic sulphide deposits on the sea floor. *Mar. Min.,* **7**, 271–282.

Hanula M.R. (ed.) (1982) *The Discoverers.* Pitt Publishing, Toronto.

Harben P.W. & Bates R.L. (1984) *Geology of the Nonmetallics.* Metal Bulletin, New York.

Hartman H.L. (1987) *Introductory Mining Engineering.* Wiley,

Hatton W. (1994) INTMOV. A program for the interactive analysis of spatial data. In Whateley M.K.G. & Harvey P.K. (eds), *Mineral Resource Evaluation II: Methods and Case Histories,* 37–43. Spec. Publ. **79**, Geol. Soc., London.

Hawkes H. E. (1976) The downstream dilution of stream sediment anomalies. *J. Geochemical Exploration,* **6**, 345–358.

Hawkes H. E. (1982) *Exploration Geochemistry Bibliography*. Spec. Vol. **11**, Association of Exploration Geochemists, Elsevier, Amsterdam.

Hawkes H. E. (1985) *Exploration Geochemistry Bibliography*. Spec. Vol. **11**, Supplement 1, Association of Exploration Geochemists, Elsevier, Amsterdam.

Hawkes H. E. (1988) *Exploration Geochemistry Bibliography*. Spec. Vol. **11**, Supplement 2, Association of Exploration Geochemists, Elsevier, Amsterdam.

Hawkins M.A. (1991) Large fine fraction stream samples—a practical method to maximize catchment size, minimize nugget effect and allow detection of coarse or fine gold. *Abstract*, 15th International Geochemical Symposium, Reno.

Heald P., Foley N.K. & Hayba D.O. (1987) Comparative anatomy of volcanic-hosted epithermal deposits: acid sulphate and adularia–sercite types. *Econ. Geol.*, **82**, 1–26.

Henley R.W. (1991) Epithermal gold deposits in volcanic terranes. In Foster R.P. (ed.) *Gold Metallogeny and Exploration*, 133–164. Blackie, Glasgow.

Henley S. (1981) *Nonparametric Geostatistics*. Applied Science Publishers, London.

Heyl A.V. (1972) The 38th Parallel Lineament and its relationship to ore deposits. *Econ. Geol.*, **67**, 879–894.

Hill I.A., (1990) *Shallow Reflection Seismic Methods in the Extractive and Engineering Industries*. Unpublished Workshop Notes, University of Leicester.

Hinde, C. (ed.) (1993) *Mining and The Environment—The Berlin Guidelines*. Mining Journal Books, London.

Hitzman M.W., Proffett J.M. Jr, Schmidt J.M. & Smith T.E. (1986) Geology and mineralization of the Ambler District, Northwestern Alaska. *Econ. Geol.*, **81**, 1592–1618.

Hodgson C. J. (1989) Use (and abuses) of ore deposit models in mineral exploration. In Garland G.D. (ed.), *Proceeedings of Exploration '87*, 31–45. Spec. Vol. **3**, Ontario Geological Survey, Toronto.

Hoek E. & Bray J. (1977) *Rock Slope Engineering*. Instn Min. Metall., London.

Hoek E. & Brown E.T. (1980) *Underground Excavation in Rock*. Instn Min. Metall., London.

Hoeve J. (1984) Host rock alteration and its application as an ore guide at the Midwest Lake Uranium deposit, northern Saskatchewan. *Can. Inst. Min. Metall. Bull.*, **77**, (August), 63–72.

Hoffmann P.F. (1989) Precambrian geology and tectonic history of North America. In Bally A.W. & Palmer A.R. (eds), *The Geology of North America: An Overview*, 447–512. Geol. Soc. Am., Boulder.

Hoffman S.J., (1986) *Writing Geochemical Reports—Guidelines for Surficial Surveys*. Spec. Vol. **12**, Association of Exploration Geologists, Toronto.

Hoffman S.J. (1987) Soil sampling. In Fletcher W.K, Hoffman S.J., Mehrtens M.B., Sinclair A.J. & Thomson I. (eds), *Exploration Geochemistry: Design and Interpretation of Soil Surveys*, 39–77. Reviews in Economic Geology Vol. **3**, Society of Economic Geologists, El Paso.

Holder M.T. & Leveridge B.E. (1986) A model for the tectonic evolution of south Cornwall. *J. Geol. Soc. Lond.*, **143**, 125–134.

Holl C.J. & Bromley A.V. (1988) Application of mineralogy to the beneficiation of a tin–sulphide ore. *Miner. Eng.* **1**, 19–30.

Holmes S. W. (1977) Some find mines and some do not—why? *Northern Miner,* **52**(51), C1.

Hopper D. (1990) *15 level west, Wheal Jane. A case history of the development and geological interpretationn of a narrow vein tin deposit in Cornwall.* Unpublished BSc (Applied Geology) Dissertation, University of Leicester.

Horikoshi E. & Sato T. (1970) Volcanic activity and ore deposition in the Kosaka Mine. In Tatsumi T. (ed.), *Volcanism and Ore Genesis*, 181–95. University of Tokio Press, Tokio.

Hornbrook D.H.W. (1975) Kidd Creek Cu–Zn–Ag deposit Ontario (case history). *J. Geochem. Explor.*, **4**, 165–168.

Hornbrook E.H. (1989) Lake sediment geochemistry: Canadian applications in the eighties. In Garland G.D. (ed.), *Proceedings of Exploration '87*, 405–416. Spec. Vol. **3**, Ontario Geological Survey, Toronto.

Hosking K. F. G. (1971) Problems associated with the application of geochemical methods of exploration in Cornwall, England. In Boyle, R.W. (ed.), *Geochemical Exploration*, 176–189. Spec. Vol. **11**, Can. Inst. Min. Metall., Toronto.

Howarth R.J. (1984) Statistical applications in geochemical prospecting: a survey of recent developments. *J. Geochem. Explor.*, **21**, 41-61.

Howarth R.J. (ed.) (1982) *Statistics and Data Analysis in Exploration Geochemistry,* Vol **2**. Handbook of Exploration Geochemistry. Elsevier, Amsterdam.

Huston D.L. & Large R.R. (1989) A chemical model for the concentration of gold in volcanogenic massive sulphide deposits. *Ore Geology Reviews*, **4**, 171–200.

Hutchinson R.W. (1980) Massive base metal sulphide

deposits as guides to tectonic evolution. In Strangeway D.W. (ed.), *The Continental Crust and Its Mineral Deposits*, 659–684. Geol. Assoc. Canada, Spec. Pap. **20**.

Hutchinson R.W. & Viljoen R.P. (1988) Re-evaluation of gold sources in Witwatersrand ores. *S. Afr. J. Geol.*, **91**, 157–173.

Hutchison C.S. (1974) *Laboratory Handbook of Petrographic Techniques*. Wiley, New York.

Hutchison C.S. (1983) *Economic Deposits and Their Tectonic Setting*. Macmillan, London.

IG (1989) *Code of Practice for Geological Visits to Quarries, Mines and Caves*. Institute of Geologists, London.

Ineson P.R. (1989) *Introduction to Practical Ore Microscopy*. Longman Scientific & Technical, Harlow.

Institution of Mining and Metallurgy (1987) Finance for the Mining Industry. *Trans. Instn Min. Metall., Sect. A*, **96**, A7–A36.

Isaaks E.H. & Srivastava R.M. (1989) *An Introduction to Applied Geostatistics*. Oxford University Press, Oxford.

Ishikawa Y., Sawaguchi T., Iwaya S. & Horiuchi M. (1976) Delineation of prospecting targets for Kuroko deposits based on modes of volcanism of the underlying dacite and alteration halos. *Min. Geol.*, **26**, 105–117. (In Japanese).

Isles D.J., Harman P.G. & Cunneen J.P. (1989) The contribution of high resolution aeromagnetics to Archaean gold exploration in the Kalgooorlie region, Western Australia. In Keays R.R., Ramsay W.R.H. & Groves D.I. (eds) *The Geology of Gold Deposits - The Perspective in 1988*, 389–397. Econ. Geol. Monograph **6**, El Paso.

Ivosevic S.W. (1984) *Gold and Silver Handbook*. Ivosevic, Denver.

Jacks J. (1989) Gold. *Mining Annual Review*, C11–21. Mining Journal, London.

Jackson D.G. & Andrew R.L. (1990) Kintyre uranium deposit. In Hughes, F. (ed.), *Geology of the Mineral Deposits of Australia and Papua New Guinea*, 653–658. Australas. Inst. Min. Metall. Monograph **14**, Melbourne.

Jackson N. J., Willis-Richards J., Manning D.A.C. & Sams M.S. (1989) Evolution of the Cornubian ore field, southwest England: II mineral deposit and ore forming processes. *Econ. Geol.*, **84**, 1101–1133.

Jacobsen J.B.E. & McCarthy T.S. (1976) The copper-bearing breccia pipes of the Messina District, South Africa. *Mineral. Deposita*, **11**, 33–45.

Jagger F. (1977) Bore Core Evaluation for Coal Mine Design. In Whitmore R.L. (ed.), *Coal Borehole Evaluation*, 72–81. Symposium, Australas. Inst. Min. Metal., Parkville.

Jakubiak Z. & Smakawski T. (1994) Classification of mineral reserves in the former Comecon countries. In Whateley M.K.G. & Harvey P.K. (eds), *Mineral Resource Evaluation II: Methods and Case Histories*, 17–28. Spec. Publ. **79**, Geol. Soc.. London.

Janisch P.R. (1986) Gold in South Africa. *J. S. Afr. Inst. Min. Metall.*, **86**, 273–316.

Jankovic S. & Petrascheck W.E. (1987) Tectonics and metallogeny of the Alpine–Himalayan Belt in the Mediterranean area and western Asia. *Episodes*, **10**, 169–175.

Jeffery R.G., Sampsom D.B., Seymour K.M. & Walker I.W. (1991) A review of current exploration and evaluation practices in the search for gold in the Archaean of Western Australia. In *World Gold '91*, 313–322. Australas. Inst. Min. Metall., Cairns.

Jenkin A.K.H. (1961–1970) *Mines and Miners of Cornwall,* 16 vols. Truro Bookshop.

Jenner J.W. (1986) Drilling and the Year 2000. *Trans. Instn Min. Metall., Sect. B*, **95**, B77–144.

Johnson I.M. & Fujita M. (1985) The Hishikari gold deposit: an airborne EM discovery. *CIM Bull.*, **78**, 61–66.

Johnson M.G. (1977) *Geology and mineral deposits of Pershing County, Nevada.* Bull. **80**, Nevada Bureau of Mines and Geology.

Jones M.P. (1987) *Applied Mineralogy*. Graham & Trotman, London.

Jordaan M. & Austin M. (1986) *Gold-placer Sedimentology.* Excursion guide 1A Geocongress '86. Geol. Soc. S. Afr., Johannesburg.

Journel A.G. & Huijbregts C.-H.J. (1978) *Mining Geostatistics.* Academic Press, London.

Kalogeropolous S.I. & Scott S.D. (1983) Mineralogy and geochemistry of tuffaceous exhalites (Tetsuekiei) of the Fukazwa Mine, Hokuroku District, Japan. In Ohmoto H. & Skinner B.J. (eds), *Kuroko and Related Volcanogenic Massive Sulfide Deposits*, Econ. Geol. Monogr., **5**, 412–432.

Kaplan R.S. (1989) Recycling of metals: technology. In Carr D.D. & Herz N. (eds), *Concise Encyclopedia of Mineral Resources*, 296–304. Pergamon Press, Oxford.

Kauranne, K., Salimen, R. & Eriksson, K. (1992)*Regolith Exploration Geochemistry in Arctic and Temperate Terrains.* Handbook of Exploration Geochemistry, Vol. **5**., Elsevier, Amsterdam.

Keary P. & Brooks M. (1991) *An Introduction to Geophysical Exploration*, Second Edition. Blackwell Scientific Publications Ltd., Oxford.

Kernet C. (1991) *Mining Equities: Evaluation and Trading.* Woodhead Publishing Ltd, Cambridge.

Kerr D.J. & Gibson H.L. (1993) A comparison of the Horne volcanogenic massive sulfide deposit and intracauldron deposits of the Mine Sequence, Noranda, Quebec. *Econ. Geol.*, **88**, 1419–1442.

Kerrich R. (1986) Fluid transport in lineaments. *Philos. Trans. R. Soc. London,* **A317**, 219–251.

Kettaneh Y.A. & Badham J.P.N. (1978) Mineralization and paragenesis at Mount Wellington Mine, Cornwall. *Econ. Geol.*, **73**, 486–495.

Khramann R (1936) The geophysical magnetometric investigation on the West Rand between Randfontein and Potchesfstroom. *Trans. Geol. Soc. S. Afr.*, **39**, 1–44.

Kim Y.C., Martino F. & Chopra I. (1981) Application of geostatistics in a coal deposit. *Min. Eng.*, **33** (10), 1476–1481.

Klemd R. & Hallbauer D.K. (1987) Hydrothermally altered peraluminous Archaean granites as a provenance model for Witwatersrand sediments. *Mineral. Deposita*, **22**, 227–235.

Knudsen H.P. (1988) *A Short Course on Geostatistical Ore Reserve Estimation.* Unpublished Rept, Montana Tech., Butte.

Knudsen H.P. & Kim Y.C. (1978) A comparative study of the geostatistical ore reserve estimation method over the conventional methods. *Min. Eng.*, **30** (1), 54–58.

Koch G.S. & Link R.F. (1971) *Statistical Analysis of Geological Data*, 2 Vols. Wiley, New York.

Koch G.S. & Link R.F. (1980) *Statistical Analysis of Geological Data*, 2 vols. Dover Publications, New York.

Kreiter V.M. (1968) *Geological Prospecting and Exploration*, Mir, Moscow.

Krige D.G. (1978) *Log Normal—De Wijsian Geostatistics for Ore Evaluation.* South African Institute of Mining and Metallurgy Monograph **1**, Johannesburg.

Krynine D. & Judd W.R. (1957) *Principles of Engineering Geology and Geotechnics.* McGraw-Hill, New York.

Kunasz I.A. (1982) Foote Mineral Company—Kings Mountain Operation. In Cerny P. (ed.), *Short Course in Granitic Pegmatites in Science and Industry*, 505–512. Mineralogical Society of Canada, Winnipeg.

Kuscevic B., Thomas T.L. & Penberthy J. (1972) Sampling distributions in South Crofty Tin Mine, Cornwall. *Trans. Instn Min. Metall., Sect. A*, **81**, A180–187.

Lafitte P. (ed.) (1970) *Metallogenic Map of Europe*, 9 Sheets. UNESCO, Paris.

Lambert I.B., Knutson J., Donnelly T.H. & Etminan H. (1987) Stuart Shelf–Adelaide Geosyncline Copper Province, South Australia. *Econ. Geol.*, **82**, 108-23.

Lane K.F. (1988) *The Economic Definition of Ore— Cut-off Grades in Theory and Practice.* Mining Journal Books, London.

Large R.R., Herrmann W. & Corbett K.D. (1987) Base metal exploration of the Mount Read Volcanics, western Tasmania: Pt I. Geology and exploration, Elliott Bay. *Econ. Geol.*, **82**, 267–90.

Larson L.T. (1989) Geology and gold mineralization in west Turkey. *Min. Eng.*, **41**, 1099–1102.

Leat P.T., Jackson S.E., Thorpe R.S. & Stillman C.J. (1986) Geochemistry of bimodal basalt–subalkaline/peralkaline rhyolite provinces within the southern British Caledonides. *J. Geol. Soc. London*, **143**, 259–73.

LeBrun S. (1987) *The Development of a Menu Driven Computer Program, "The Geostats Program", to Carry Out Geostatistical Analysis of Geological Data*, 2 Vols. Unpubl. MSc Dissertation, University of Leicester.

Leca X. (1990) Discovery of a Concealed massive sulphide deposit at Neves-Corvo, southern Portugal—a case history. *Trans. Instn Min. Metall. Sect B*, **99**, B139–52.

Lednor M. (1986) The West Rand Goldfield. In Antrobus E.S.A. (ed.), *Witwatersrand Gold—100 Years*, 49–110. Geol. Soc. S. Afr., Johannesburg.

Levinson A.A. (1980) *Introduction to Exploration Geochemistry.* Applied Publishing, Wilmette, Illinois.

Levinson A.A., Bradshaw P.M.D. & Thomson I. (eds) (1987) *Practical Problems in Exploration Geochemistry.* Applied Publishing, Wilmette, Illinois.

Lillesand T.M. & Kieffer R.W. (1979) *Remote Sensing and Image Interpretation.* Wiley, New York.

Lindberg P.A. (1985) A volcanogenic interpretation for massive sulfide origin, West Shasta District, California. *Econ. Geol.*, **80**, 2240–2254.

Lindley I.D. (1987) The discovery and exploration of the Wild Dog gold–silver deposit, east New Britain, Papua New Guinea. In *Pacific Rim Congress '87*, 283–286, Australasian Institution of Mining and Metallurgy, Melbourne.

Lo C.P. (1986) *Applied Remote Sensing.* Longman Scientific & Technical, New York.

Lock, N.P. (1987) Kimberlite exploration in the Kalahari region of Southern Botswana with emphasis on the Jwaneng Kimberlite Province. *In Prospecting in Areas of Desert Terrain*, 183–190. Instn Min. Metall., London.

Lofty G.J., Hillier J.A., Burton E.M., Mitchell D.C., Cooke S.A. & Linley K.A. (1989) *World Mineral Statistics*. British Geological Survey, Keyworth.

Lovell J.S. & Reid A.R. (1989) Carbon dioxide/oxygen in the exploration for sulphide mineralization. In Garland, G.D. (ed.) *Proceedings of Exploration '87*, 457–472. Spec. Vol. **3**, Ontario Geological Survey, Toronto.

Lovering T.G. & McCarthy J.H. Jr (eds) (1977) Conceptual models in exploration geochemistry—the Basin and Range Province of the western United States and Northern Mexico. *J. Geochem. Explor.*, **9**, 89–365.

Lydon J.W. (1989) Volcanogenic massive sulphide deposits parts 1 & 2. In Roberts R.G. & Shehan P.A. (eds), *Ore Deposit Models*, 145–181. Geol. Assoc. Can., Memorial University, Newfoundland.

Maas R., McColloch M.T., Campbell I.H. & Coad P.R. (1986) Sm–Nd and Rb–Sr dating of an Archaean massive sulfide deposit: Kidd Creek, Ontario. *Geology*, **14**, 585–588.

Machamer J.F., Tolbert G.E. & L'Esperance R.L. (1991) Discovery of Serra dos Carajas. In Hutchinson R.W. & Grauch R.I. (eds), *Historical Perspectives of Genetic Concepts and Case Histories of Famous Discoveries*, 275–285. Monograph **8**, Econ. Geol. Pub. Co., El Paso.

Mackenzie B. & Woodall R. (1988) Economic productivity of base metal exploration in Australia and Canada. In Tilton J.E., Eggert R.G. & Landsberg H.H. (eds), *World Mineral Exploration*, 21–104. Resources for the Future, Washington.

Magri E.J. (1987) Economic optimization of boreholes and deflections in deep gold exploration. *J. S. Afr. Inst. Min. Metall.*, **87**, 307–321.

Maguire D.J., Goodchild M.F. & Rhind D.W. (1991) *Geographical Information Systems, Principles and Applications*. 2 vols. Longman, London.

Mason A.A.C. (1953) The Vulcan tin mine. In Edwards A. B. (ed.), *Geology of Australian Ore Deposits*, 718–721. Australas. Inst. Min. Metall., Melbourne.

Mather P.H. (1987) *Computer Processing of Remotely Sensed Images*. John Wiley & Sons, Chichester.

Matulich A., Amos A.C., Walker R.R. *et al.* (1974) The Ecstall Story; the geology department. *Can. Inst. Min. Metall. Bull.*, **77**, 228–235.

Maynard J.B., Ritger, S.D. & Sutton, S.J. (1991) Chemistry of sands from the modern Indus River and the Archean Witwatersrand Basin: implications for the composition of the Archean atmosphere. *Geology*, **19** 265–268.

Mazzucchelli R.H. (1989) Exploration geochemistry in areas of deeply weathered terrain: weathered bedrock geochemistry. In Garland G.D. (ed.), *Proceedings of Exploration '87*, 300–11. Spec. Vol. **3**, Ontario Geological Survey, Toronto.

McCabe P.J. (1984) Depositional environments of coal and coal–bearing strata. In Rahmani R.A. & Flores R.M. (eds), *Sedimentology of Coal and Coal-bearing Strata*, 13–42. Spec. Publ. **7**, International Association of Sedimentologists.

McCabe P.J. (1987) Facies studies of coal and coal-bearing strata. In Scott A.C. (ed.), *Coal and Coal-bearing Strata: Recent Advances*, 51–66. Spec. Publ. **32**, Geol. Soc. London.

McCabe P.J. (1991) Geology of coal: environments of deposition. In Gluskoter H.J., Rice D.D. & Taylor R.B. (eds), *Economic Geology, US.* Vol. P–2 of the Geology of North America, 469–482. Geol. Soc. Amer., Boulder, Colorado.

McCarl H.N. (1989) Prices of industrial minerals: history. In Carr D.D. & Herz N. (eds), *Concise Encyclopedia of Mineral Resources*, 285–288. Pergamon Press, Oxford.

McConachy T.F. & Scott S.D. (1987) Real-time mapping of hydrothermal plumes over southern Explorer Ridge, NE Pacific Ocean. *Mar. Min.*, **6**, 181–204.

McConchie D. (1984) A depositional environment for the Hamersley Group: palaeogeography and geochemistry. In Muhling J.R., Groves D.I. & Blake T.S. (eds), *Archaean and Proterozoic Basins of the Pilbara, Western Australia. Evolution and Mineralization Potential*, 144–177. **Publ. No. 9,** Geol. Dept & Univ. Extension, Univ. of Western Australia, Nedlands.

McLean A.C. & Gribble C.D. (1985) *Geology for Civil Engineers*, 2nd Edition. George Allen & Unwin, London.

McMullan S. R., Matthews R. B. & Robertshaw P. (1989). Exploration geophysics for Athabasca uranium deposits. In Garland G. (Ed.), *Proceeedings of Exploration '87*, 547–568. Ontario Geological Survey, Toronto.

McVey H. (1989) Industrial Minerals—can we live without them? *Industrial Minerals, 259,* 74–5.

Mead R.D., Kesler S.E., Foland K.A. & Jones L.M. (1988) Relationship of Sonoran tungsten mineralization to the metallogenic evolution of Mexico. *Econ.*

Geol., **83**, 1943–1965.

Mellor E.T. (1917) *The Geology of the Witwatersrand: Map and Explanation.* Spec. Pub. **3**, Geol. Surv. S. Afr., Johannesburg.

Menzie W.L., Reed B.L. & Singer D.A. (1988) Models of grades and tonnages of some lode tin deposits. In Hutchison C.S. (ed.) *Geology of Tin Deposits in Asia and the Pacific*, 73–88. Springer-Verlag, Berlin.

Metals Economics Group (1992) Current exploration spending. *Metals Economics Group Strategic Report* **5**, 1–9. Metals Economics Group, Halifax, Nova Scotia.

Meyer F.M., Tainton S. & Saager R. (1990) The mineralogy and geochemistry of small-pebble conglomerates from the Promise Formation in the West Rand and Klerksdorp areas. *S. Afr. J. Geol.*, **93**, 103–117.

Miller C.F. & Bradfish L.J. (1980) An inner cordilleran belt of muscovite-bearing plutons. *Geology*, **8**, 412–416.

Milsom J. (1989) *Field Geophysics.* Open University Press, Milton Keynes.

Min. Mag. (1982) Metal mining in the UK. *Mining Magazine*, **147**, 398–407.

Minter W.E.L., Hill W.C.N., Kidger R.J., Kingsley C.S. & Snowden P.A. (1986) The Welkom Goldfield. In Annhaeusser C.R. & Maske S. (eds), *Mineral Deposits of Southern Africa*, 497–539. Geol. Soc. S. Afr., Johannesburg.

Mitcham T.W. (1974) Origin of breccia pipes. *Econ. Geol.*, **69**, 412–413.

Mitchell A.H.G. & Garson M.S. (1981) *Mineral Deposits and Global Tectonic Settings.* Academic Press, London.

Mitchell A.H.G. & Reading H.G. (1986) Sedimentation and tectonics. In Reading H.G. (ed.), *Sedimentary Environments and Facies*, 471–551. Blackwell Scientific Publications, Oxford.

Moeskops P.G. (1977) Yilgarn nickel gossan geochemistry—a review with new data. *J. Geochem. Explor.*, **8**, 247–258.

Moon C.J. (1973) *A Carborne Radiometric Survey of the Prince Albert–Beaufort West area, Cape Province.* Open File Report Geol. Surv. S. Africa, **G227**, Pretoria.

Moon C.J. & Whateley M.K.G. (1989) A retrospective analysis of commercial exploration strategies for uranium in the Karoo Basin, South Africa. In Chadwick J.R. (ed.), *Mineral Exploration Programmes '89*, Mining International, Madrid.

Moore P.D. (1987) Ecological and hydrological aspects of peat formation. In Scott A. (ed.), *Coal and Coal-bearing Stata*, 17–24. Spec. Publ. **32**, Geol. Soc. London.

Morgan J.D. (1989) Stockpiling. In Carr D.D. & Herz N. (eds), *Concise Encyclopedia of Mineral Resources*, 288–292. Pergamon Press, Oxford.

Morris R.O. (1986) Drilling—Potential Developments. *Trans. Instn Min. Metall., Sect. B,* **95**, B77–78.

Morrissey C.J. (1986) New trends in geological concepts. *Trans. Instn Min. Metall., Sect. B,* **95**, B54–57.

Morton R.L. & Franklin J.M. (1987) Two-fold classification of Archaean volcanic–associated massive sulfide deposits. *Econ. Geol.*, **82**, 1057–1063.

Moseley F. (1981) *Methods in Field Geology.* Freeman, Oxford.

Mount M. (1985) Geevor Mine: a review. In Halls C. (ed.), *High Heat Production (HHP) Granites, Hydrothermal Circulation and Ore Genesis*, 221–238. Instn Min. Metall., London.

Mular A.L. (1982) *Mining and Mineral Processing Equipment Costs and Preliminary Cost Estimations.* Spec. Vol. **25**, Can. Inst. Min. Metall., Montreal.

Mullins M.P. (ed.) (1986) *Gold Placer Sedimentology.* Excursion Guidebook 1A. Geol. Soc. S. Afr., Johannesburg.

Munha J., Barriga F.J.A.S. & Kerrich R. (1986) High $^{18}O/^{16}O$ ore-forming fluids in volcanic-hosted base metal massive sulphide deposits: geologic, $^{18}O/^{16}O$, and D/H evidence from the Iberian Pyrite Belt; Crandon, Wisconsin; and Blue Hill, Maine. *Econ. Geol.*, **81**, 530–552.

Murphy G.J. (1982) Some Aspects of Sampling in Terms of Mineral Exploration and Mine Geology. In *Sampling and Analysis for the Minerals Industry*, Symposium, Instn Min. Metall., London.

NAMHO (1985) *NAMHO Guidelines.* National Association of Mining History Organizations, Matlock.

Nash C.R., Boshier P.R., Coupard M.M., Theron A.C. & Wilson T.G. (1980) Photogeology and satellite image interpretation in mineral exploration. *Miner. Sci. Eng.*, **12**, 216–244.

Nebert K. (1978) Das Braunkohlenfuhrende Neogengebiet von Soma, West Anatolien. *Bull. Mineral Research and Exploration, Institute of Turkey*, **90**, 20–72.

Niblak W. (1986) *An Introduction to Digital Image Processing.* Prentice Hall, New York.

Nichol I., Closs L.G. & Lavin O.P. (1989) Sample representativity with reference to gold exploration. In Garland G.D. (ed.), *Proceedings of Exploration '87*, 609–624. Spec. Vol. **3**, Ontario Geological Survey, Toronto.

Noetstaller R. (1988) *Industrial Minerals: a Technical Review*. The World Bank, Washington.

Northern Miner (1990) Buried treasures. *Northern Miner Magazine*, **March 1990**, 32–37.

Nunes P.D. & Pyke D. (1981) Time–stratigraphic correlation of the Kidd Creek Orebody with volcanic rocks south of Timmins, Ontario, as inferred from zircon U–Pb ages. *Econ. Geol.*, **76**, 944–951.

Obert L. & Duval W.I. (1967) *Rock Mechanics and the Design of Structures in Rock*. Wiley, New York.

O'Driscoll E.S.T. (1986) Observations of the lineament–ore relations. *Phil. Trans. R. Soc. London,* **A317**, 195–218.

O'Hara T.A. (1980) Quick Guide to the Evaluation of Orebodies. *Can. Inst. Min. Metall. Bull.*, Feb., 87–99.

Ohmoto H. & Skinner B.J. (1983) The Kuroko and Related Volcanogenic Massive Sulfide Deposits. Economic Geology, Monograph **5**.

Olade M.A. (1980) Plate tectonics and metallogeny of intracontinental rifts and aulacogens in Africa—a review. In Ridge J.D. (ed.), *Proc. 5th Quadriennial IAGOD Symp.,* **Vol.I,** 81–89. E. Schweizer-bart'sche Verlagbuchhandlung, Stuttgart.

O'Neil T.J. (1982) Mine Evaluation in a Changing Investment Climate. *Min. Eng.*, **34**, Nov., 1563–1566.

Operations (1985) *Operations at Wheal Jane*. Unpublished company report.

Ottley D.J. (1966) Gy's Sampling Slide Rule. *World Mining*, **19**, 9, 40.

Papenfus J.A. (1964) The Black Reef Series within the Witwatersrand Basin with special reference to the occurrence at Government Gold Mining Areas. In Haughton S.H. (ed.), *The Geology of Some Ore Deposits in Southern Africa,* 191–218. Geol. Soc. S. Afr., Johannesburg.

Parkinson W.D. (1983) *Introduction to Geomagnetism*. Scottish Academic Press, Edinburgh.

Paterson N.R. & Halof P.G. (1991) Geophysical exploration for gold. In Foster R.P. (ed.), *Gold Metallogeny and Exploration,* 360–398. Blackie, Glasgow.

Pearce J.A., Lippard S.J. & Roberts S. (1984) Characteristics and tectonic significance of suprasubduction zone ophiolites. In Kokelaar B.P. & Howells M.F. (eds), *Marginal Basin Geology,* 77–94. Blackwell Scientific Publications, Oxford.

Pemberton R.H. (1989) Geophysical response of some Canadian massive sulphide deposits. In Garland G.D. (ed.), *Proceedings of Exploration '87,* 517–531. Spec. Vol. **3**, Ontario Geological Survey, Toronto.

Percival J.A. & Williams H.R. (1989) Late Archaean Quetico metasedimentary belt, Superior Province, Canada. *Can. J. Earth Sci.*, **26**, 677–693.

Perry G. (1989) *Trinity Mine Operations Report*. Unpublished Report for US Borax.

Peters E.R. (1983) The use of Multispectral Satellite Imagery in the Exploration for Petroleum and Minerals. *Phil. Trans. Roy. Soc. Lond.*, **A309**, 243–255.

Peters W.C. (1987) *Exploration and Mining Geology* (2nd ed.) Wiley, New York.

Peterson U. (1970) Metallogenic provinces in South America. *Geol. Rundsch.*, **59**, 834–897.

Phillips G.N., Myers R.E & Palmer J.A. (1987) Problems with the placer model for Witwatersrand gold. *Geology*, **15**, 1027–1030.

Pintz W. (1984) *Ok Tedi—Evaluation of a Third World Mining Project*. Mining Journal Books, Edenbridge, Kent.

Piper D.P. and Rogers P.J. (1980) *Procedure for the Assessment of the Conglomerate Resources of the Sherwood Sandstone Group*. Mineral Assessment Rept **56**, Institute of Geological Sciences, Keyworth.

Pirajno F. (1992) *Hydrothermal Mineral Deposits: Principles and Fundamental Concepts for the Exploration Geologist*. Springer-Verlag, Berlin.

Pirow H. (1920) Distribution of the pebbles in the Rand banket and other features of the rock. *Trans. Geol. Soc. S. Afr.,* **23**.

Pisutha-Arnond V. & Ohmoto H. (1983) Thermal history, chemical and isotopic compositions of the ore-forming fluids responsible for the Kuroko massive sulfide deposits in the Hokuroku District of Japan. In Ohmoto H. & Skinner B.J. (eds), *Kuroko and Related Volcanogenic Massive Sulfide Deposits*, Economic Geology Monograph **5**, 523–558.

Pitcher W.S. (1983) Granite: typology, geological environments and melting relationships. In Atherton M.P. & Gribble C.D. (eds), *Migmatites, Melting and Metamorphism,* 277–285. Shiva, Nantwich.

Plant J.A., Hale M. & Ridgway J. (1988) Developments in regional geochemistry for mineral exploration. *Trans. Instn Min. Metall., Sect. B,* **97**, 116–140.

Popoff C. (1966) *Computing Reserves of Mineral Deposits; Principles and Conventional Methods*. US Bureau of Mines Information Circular **8283**.

Potts D. (1985) Guide to the Financing of Mining Projects. *Trans. Instn Min. Met., Sect. A,* **94**, A127–A133.

Prain K.A.R. (1989) Application of MWD in the Oil

Field Drilling Industry. *Miner. Ind. Int.*, **987**, March, 10–12.

Pretorius C.C., Jamison A.A. & Irons C. (1989) Seismic exploration in the Witwatersrand Basin, Republic of South Africa. In Garland G.D. (ed.), *Proceedings of Exploration '87*, 241–253. Spec. Vol. **3**, Ontario Geological Survey, Toronto.

Pretorius D.A. (1991) The sources of Witwatersrand gold and uranium: a continued difference of opinion. In Hutchinson R.W & Grauch R.I. (eds), *Historical Perspectives of Genetic Concepts and Case Histories of Famous Discoveries*, 139–163. Monograph **8**, Econ. Geol.

Price M. (1985) *Introducing Ground Water*. Chapman & Hall, London.

Rahmani R.A. & Flores R.M. (1984) *Sedimentology of Coal and Coal-bearing Sequences*. Spec. Publ. **7**, International Association of Sedimentologists, Blackwell Scientific Publications, Oxford.

Ramboz C. & Charef A. (1988) Temperature, pressure, burial history and paleohydrology of the Les Malines, Pb–Zn deposit: reconstruction from aqueous inclusions in barite. *Econ. Geol*, **83**, 784–800.

Rayment B. D., Davis G.R. & Willson J (1971) Controls to the mineralisation at Wheal Jane, Cornwall. *Trans. Instn Min. Metall., Sect. B*, **80**, B224–237.

Reed S.J.B. (1993) *Electron Microprobe Analysis*. Cambridge University Press, Cambridge.

Reedman J.H. (1979) *Techniques in Mineral Exploration*. Applied Science Publishers, London.

Reeves C.V. (1989) Geophysical mapping of Precambrian granite–greenstone terranes as an aid to exploration. In Garland G.D. (ed.), *Proceedings of Exploration '87*, 254–266. Spec. Vol. **3**, Ontario Geological Survey, Toronto.

Reeves, P.L. & Beck, L.S. (1982). Famous mining camps: a history of uranium exploration in the Athabasca Basin. In Hanula, M.R. (ed.), *The Discoverers*, 153–161. Pitt Publishing, Toronto.

Regan M. D. (1971) *Management of Exploration in the Metals Mining Industry*. MS thesis, Mass. Inst. Tech., Cambridge.

Regan R.D. (1983) Current Status of the IGRF and Its Relation to Magnetic Surveys. *Geophysics*, **48**, 997–998.

Reim K. (1988). *Trinity Ore Reserve, Task Force Report*, Unpublished Report for US Borax.

Reimann C. (1989) Reliability of geochemical analysis: recent experiences. *Trans. Instn Min. Metall., Sect. B*, **98**, B123–129.

Reimer T.O (1984) Alternative model for the derivation of gold in the Witwatersrand Group. *J. Geol. Soc. Lond.*, **141**, 263–272.

Reimer T.O. & Mossman D.J. (1990) Sulfidization of Witwatersrand black sands: from enigma to myth. *Geology*, **18**, 426–429.

Reinecke L. (1927) The location of payable ore-bodies in the gold-bearing reefs of the Witwatersrand. *Trans. Geol. Soc. S. Afr.*, **30**, 89–119.

Rickard D. (1987) Proterozoic volcanogenic mineralization styles. In Pharaoh T.C., Beckinsale R.D. & Rickard D. (eds), *Geochemistry and Mineralization of Proterozoic Volcanic Suites*, 23–35. Spec. Publ. **33**, Geol. Soc. London.

Riddihough P. (1971) Diurnal corrections to magnetic surveys—an assessment of errors. *Geophysical Prospecting*, **19**, 551–67.

Riddler G. P. (1989). The impact of corporate and host country strategy on mineral exploration programmes. In Chadwick J.R. (ed.), *Paper 15, Mineral Exploration Programmes '89*. International Mining, Madrid.

Riddler G.P., 1994. What is a mineral resource? In Whateley M.K.G. & Harvey P.K. (eds), *Mineral Resource Evaluation II: Methods and Case Histories*, 1–10. Spec. Publ. **79**, Geol. Soc. London.

Ritchie W., Wood M., Wright R. & Tait D. (1977) *Surveying and Mapping for Field Scientists*. Longman, New York.

Robb L.J., Davis D.W., Kamo N.L. *et al.* (1990) U–Pb ages on single detrital zircon grains from the Witwatersrand Basin, South Africa: constraints on the age of sedimentation and the evolution of granites adjacent to the basin. *J. Geol.*, **98**, 311–328.

Robb L.J. & Meyer F.M. (1990) The nature of the Witwatersrand hinterland: conjectures on the source area problem. *Econ. Geol.*, **85**, 511–536.

Roberts R.G. & Sheahan P.A. (1988) *Ore Deposit Models*. Geol. Assoc. Canada Reprint Series 3, St Johns, Newfoundland.

Robertson W. (1982) *Tin: its Production and Marketing*. Croom Helm, Beckenham, Kent.

Robinson S.C. (1952) Autoradiographs as a means of studying distribution of radioactive minerals in thin section. *Am. Miner.*, **37**, 544–547.

Rogers, P.J., Chaterjee, A.K. & Aucott, J.W. (1990) Metallogenic domains and their reflection in regional lake sediment surveys from the Meguma Zone, southern Nova Scotia. *J. Geochem. Explor.*, **39**, 153–174.

Rona P.A. (1988) Hydrothermal mineralization at oceanic ridges. *Can. Mineral.*, **26**, 431–465.

Rona P.A., Klinkhammer G., Nelson T.A., Tefrey J.H. & Elderfield H. (1986) Black smokers, massive sulphides and vent biota at the Mid-Atlantic Ridge. *Nature*, **321**, 33–37.

Rose A.W., Hawkes H.E. & Webb J.S. (1979) *Geochemistry in Mineral Exploration.* Academic Press, London.

Ross G.M., Parrish R.R., Villeneuve M.E. & Bowring S.A. (1991) Geophysics and geochronology of the crystalline basement of the Alberta Basin, western Canada. *Can.. J. Earth Sci.*, **28**, 512–522.

Rothschild L. (1978) Risk. *The Listener*, 30 November, 715.

Roux A.T. (1969) The application of geophysics to gold exploration in South Africa. In *Mining and Groundwater Geophysics*, 425–438. Econ. Geol. Rept **26**, Geol. Surv. Canada.

Saager R. (1981) Geochemical studies of the origin of detrital pyrites in the conglomerates of the Witwatersrand Goldfields, South Africa. In Armstrong F. C. (ed.), *Genesis of Uranium– and Gold–Bearing Precambrian Quartz–pebble Conglomerates* L1–L17. Prof. Paper **1161**, US Geol. Surv.

Sabins F.F. (1987) *Remote Sensing. Principles and Interpretation*, 2nd Edition. Freeman, New York.

Sangster D.F. (1976) Carbonate–hosted lead–zinc deposits. In Wolf K.H. (ed.), *Handbook of Stratabound and Stratiform Deposits, Volume 6*, 447–456. Elsevier, Amsterdam.

Sangster D.F. (1977) Some grade and tonnage relationships among Canadian volcanogenic massive sulphide deposits. *Geol. Surv. Can. Rept Pap.* **77–1A**, 5–12.

Sangster D.F. (1980) Quantitative characteristics of volcanogenic massive sulphide deposits: 1. metal content and size distribution of massive sulphide deposits in volcanic centres. *Can. Inst. Min. Metall.*, **73**, 74–81.

Sangster D.F. & Scott S.D. (1976) Precambrian, stratabound, massive Cu–Zn–Pb sulphide ores in North America. In Wolf K.H. (ed.), *Handbook of Strata-Bound and Stratiform Deposits, Volume* **6**, 129–222. Elsevier, Amsterdam.

Sato T. (1977) Kuroko deposits: their geology, geochemistry and origin. In *Volcanic Processes in Ore Genesis*, Spec. Publ. **7**, Geol. Soc., London.

Sawkins F.J (1984 & 1990) *Metal Deposits in Relation to Plate Tectonics.* Springer-Verlag, Berlin.

Schandl E.S. & Wicks F.J. (1993) Carbonate and associated alteration of ultramafic and rhyolitic rocks at the Hemingway Property, Kidd Creek Volcanic Complex, Timmins, Ontario. *Econ. Geol.*, **88**, 1615–1635.

Schowengerdt R.A. (1983) *Techniques for Image Processing and Classification in Remote Sensing*, Academic Press, New York.

Schuiling R.D. (1967) Tin belts on continents around the Atlantic Ocean. *Econ. Geol.*, **62**, 540–550.

Scott A.C. (ed.) (1987) *Coal and Coal-bearing Strata: Recent Advances.* Spec. Pub. **32**, Geol. Soc. London.

Scott F. (1981). Midwest Lake uranium discovery, Saskatchewan, Canada. In IAEA (eds), *Uranium Exploration Case Histories,* 221–239. International Atomic Energy Agency, Vienna.

Scott-Russell H., Wanblad G.P. & De Villiers J.S. (1990) The impact of modified oilfield technology on deep level mineral exploration and ultimate mining systems. In Ross-Watt R.A.J. & Robinson P.D.K. (eds), *Technical Challenges in Deep Level Mining*, 429–439. S. Afr. Inst. Min. Metall., Johannesburg.

Select Committee (1982) Memorandum by the Inst. Geol. Sci., Strategic Minerals. In *Strategic Minerals Select Committee,* HL Rep; 1981–2 (217) xii.

Selley R.C. (1989) Deltaic reservoir prediction from rotational dipmeter patterns. In Whateley M.K.G. & Pickering K.T. (eds), *Deltas: Sites and Traps for Fossil Fuels*, 89–95. Spec. Publ. **41**, Geol. Soc. London.

Sengor A.M.C., Gorur N. & Saroglu F. (1985) Strike-slip faulting and related basin formation in zones of tectonic escape: Turkey as a case study. In Biddle K.T. & Christie-Blick N. (eds), *Strike-slip Deformation Basin Formation and Sedimentation*, 227–64. Spec. Publ. **37**, Soc. Econ. Paleo. Mineral.

Settle M., Abrams M.J., Conel J.E., Goetz A.F.H. & Lang H.R. (1984) Sensor assessment report. In Abrams M.J. (ed.), *Joint NASA/Geosat Test Case Report*, 2–1 to 2–24. Am. Assoc. Petrol. Geol.

Severin P.W.A., Knuckey M.J. & Balint F. (1989) The Winston Lake, Ontario, massive sulphide discovery—a successful result of an integrated exploration program. In Garland G.D. (ed.), *Proceedings of Exploration '87*, 60–69. Spec. Vol. **3**, Ontario Geological Survey, Toronto.

Shackleton R. M., Ries A.C. & Coward M.P. (1982) An interpretation of the Variscan structures in SW England. *J. Geol. Soc. London*, **139**, 533–541.

Shaw A. (1991) Tin. In *Metals and Minerals Annual Review*, 52–53. Min. J., London.

Sichel H.S. (1966) The estimation of means and associated confidence limits for small samples from log normal populations. In *Symposium on Mathematical Statistics and Computer Applications in Ore Valuation*, 106–122. S. Afr. Inst. Min. Metall., Johannesburg.

Siegal B.S. & Gillespie A.R. (1980) *Remote Sensing in Geology*, Wiley, New York.

Siegel S. (1956) *Nonparametric Statistics for the Behavioural Sciences*. McGraw-Hill, New York.

Sillitoe R.H. (1972a) Formation of certain massive sulphide deposits at sites of sea-floor spreading. *Trans. Instn Min. Metall. Sect. B*, **81**, B141–B148.

Sillitoe R.H. (1972b) A plate tectonic model for the origin of porphyry copper deposits. *Econ. Geol.*, **67**, 184–197.

Sillitoe R.H. (1976) Andean mineralization: a model for the metallogeny of convergent plate margins. In Strong D.F. (ed.), *Metallogeny and Plate Tectonics*, 59–100. Geol. Assoc. Can. Spec. Pap. **14**.

Sillitoe R.H. (1980) Strata-bound ore deposits related to Infracambrian rifting along northern Gondwanaland. In Ridge J. D. (ed.), *Proc. 5th IAGOD Symp. Vol. 1*, 163–172. E. Sch-weizerbart'sche Verlagsbuchhandlung, Stuttgart.

Sillitoe R.H. (1991) Intrusion–related gold deposits. In Foster R.P. (ed), *Gold Metallogeny and Exploration* 165–209. Blackie, Glasgow.

Sinclair A.J. (1976) *Probability Graphs*. Spec. Publ. **4**, Assoc. Explor. Geochemists, Toronto.

Sinclair A.J. (1991) A fundamental approach to threshold estimation in exploration geochemistry: probability plots revisited. *J. Geochem. Explor.*, **41**, 1–22.

Slack J.F. & Coad P.R. (1989) Multiple hydrothermal and metamorphic events in the Kidd Creek volcanogenic massive sulphide deposit, Timmins, Ontario: evidence from tourmalines and chlorites. *Can. J. Earth Sci.* **26**, 694–715.

Slade M.E. (1989) Prices of metals: history. In Carr D.D. & Herz N. (eds), *Concise Encyclopedia of Mineral Resources*, 343–356. Pergamon Press, Oxford.

Smith A. (1988) Cyanide degradation and detoxification in a heap leach. In Van Zyl D. , Hutchison I. & Kiel J. (eds), *Introduction to Evaluation, Design and Operation of Precious Metal Heap Leaching Projects*, 293–305. Society of Mining Engineers, Denver.

Smith, B.H. (1977) Some aspects of the use of geochemistry in the search for nickel sulphides in lateritic terrain in Western Australia. *J. Geochem. Explor.*, **8**, 259–281.

Smith D.M. (1988) Geology of silver deposits along the western cordilleras. *In Silver Exploration, Mining and Treatment*, 11–24. Instn Min. Metall., London.

Smith J. (1991) How companies value properties. *Can. Inst. Min. Metall. Bull.*, **84**, 953, 50–52.

Smith R.J. & Pridmore D.F. (1989) Electromagnetic exploration for sulphides in Australia. In Garland G.D. (ed.), *Proceedings of Exploration '87*, 504–516. Spec. Vol. **3**, Ontario Geological Survey, Toronto.

Snow G. G. & McKenzie B. W. (1981). The environment of exploration: economic, organizational and social constraints. In Skinner B.J. (ed.), *Economic Geology: Seventy–Fifth Anniversary Volume*, 871–896. Society of Economic Geologists.

Solomon M. (1976) 'Volcanic' massive sulphide deposits and their host rocks—a review and an explanation. In Wolf K.H. (ed.), *Handbook of Strata-Bound and Stratiform Deposits, Volume 6*, 21–54. Elsevier, Amsterdam.

Solomon M., Walshe J.L. & Eastoe C.J. (1987) Experiments on convection and their relevance to the genesis of massive sulphide deposits. *Australian J. Earth Sci.*, **34**, 311–323.

Solovov, A.P., (1987) *Geochemical Prospecting for Mineral Deposits*. Mir, Moscow.

Souch B.E., Podolsky T. & Geological Staff (1969) The sulphide ores of Sudbury: their particular relationship to a distinctive inclusion-bearing facies of the nickel irruptive. In Wilson H.D.B. (ed.), *Magmatic Ore Deposits Symposium*, 252–261, Econ. Geol. Monogr. **4**.

South B.C. & Taylor B.E. (1985) Stable isotope geochemistry and metal zonation at the Iron Mountain Mine, West Shasta District, California. *Econ. Geol.*, **80**, 2177–2195.

South Deep (1990) *Prospectus for Flotation of South Deep Exploration Company Ltd.*, South Deep Exploration Company Ltd, Johannesburg.

South Deep (1991) *Annual Report for 1990*, South Deep Exploration Co. Ltd., Johannesburg.

South Deep (1992) *Annual Report for 1991*. South Deep Mining and Exploration Co., Johannesburg.

Speight J.G. (1983) *The Chemistry and Technology of Coal*. Marcel Dekker Inc., New York.

Spock L.E. (1953) *Guide to the Study of Rocks*. Harper, New York.

Spooner E.T.C. & Barrie C.T. (1993) A special issue devoted to Abitibi ore deposits in a modern context. *Econ. Geol.*, **88**, 1307–1322.

Sprigg G. (1987) Advances in Image Analysers. *Microscopy and Analysis*, **September**, 11–13.

Springett M. (1983a) Sampling and Ore Reserve Estimation for the Ortiz Gold Deposit, New Mexico, USA. In *AIME Precious Metal Symposium*, Sparks, Nevada, USA.

Springett M. (1983b) Sampling Practices and Problems. Preprint **83–395**, *Society of Mining Engineers of AIME*, New York.

Stach E. (1982) *Stach's Textbook of Coal Petrology*. Borntraeger, Berlin.

Staff Bureau of Mines (1987) *Bureau of Mines Cost Estimating System Handbook* (in two parts). Bureau of Mines Information Circulars **9142** & **9143**, US Dept. of Interior.

Stanistreet I.G. & McCarthy T.S. (1991) Changing tectono-sedimentary scenarios relevant to the development of the late Archean Witwatersrand Basin. *J. Afr. Earth Sci.*, **13**, 65–81.

Stanton R.L. (1978) Mineralization in island arcs with particular reference to the south-west Pacific region. *Proc. Australas. Inst. Min. Metall.*, **268**, 9–19.

Stanton R.L. (1991) Understanding volcanic massive sulfides—past, present and future. In Hutchinson R.W & Grauch R.I. (eds), *Historical Perspectives of Genetic Concepts and Case Histories of Famous Discoveries*, 82–95. Monograph **8**, Econ. Geol., El Paso.

Stanworth C.W. & Badham J.P.N. (1984) Lower Proterozoic red beds, evaporites and secondary sedimentary uranium deposits from the East Arm, Great Slave Lake, Canada. *J. Geol. Soc. London*, **141**, 235–242.

Steiger R. & Bowden A. (1982) Tungsten mineralization in southeast Leinster, Ireland. In Brown A.G. (ed.), *Mineral Exploration in Ireland, Progress and Developments 1971–1981*, 108–114. Irish Assoc. Econ. Geol., Dublin.

Steiger R. & Bowden A. (1986) Tungsten mineralization in SE Leinster, Ireland. In Andrew C.J., Crowe R.W.A., Finlay S., Pennell W.M. & Pyne J.F. (eds), *Geology and Genesis of Mineral Deposits in Ireland*, 211–216. Irish Assoc. Econ Geol., Dublin.

Stephens J.D. (1972) Microprobe applications in mineral exploration and development programmes. *Miner. Sci. Eng.*, **3**, 26–37.

Stewart B.D. (1981) Exploration of the uranium reefs of Cooke Section, Randfontein Estates Gold Mining Company (Witwatersrand) South Africa. In International Atomic Energy Agency (eds), *Uranium Exploration Case Histories*, 141–170. International Atomic Energy Agency, Vienna.

Stewart J.H. (1980) Geology of Nevada. Spec. Publ. **4**, Nevada Bureau of Mines and Geology.

Strauss H. (1989) Carbon and sulfur isotope data for carboaceous metasediments from the Kidd Creek massive sulfide deposit and vicinity, Timmins, Ontario. *Econ. Geol.*, **84**, 959–962.

Strauss G.H. & Beck J.S. (1990) Gold mineralizations in the SW Iberian Pyrite Belt. *Mineral. Deposita*, **25**, 237–245.

Sumner J.S. (1976) *Principles of Induced Polarisation in Geophysical Exploration*. Elsevier, Amsterdam.

Sutherland D.G., & Dale M.L. (1984) Methods of establishing the minimum size for sampling alluvial diamond deposits. *Trans. Instn Min. Metall., Sect. B*, **93**, B55–58.

Talwani M., Sutton G.H. & Landisman M. (1959) Rapid gravity computation for two-dimensional bodies with application to the Mendocino Submarine Fracture Zone. *J. Geophys. Res.*, **64**, 49–59.

Tankard A.J., Jackson M.P.A., Eriksson K.A., Hobday D.K., Hunter D.R. & Minter W.E.L. (1982) *Crustal Evolution of Southern Africa*. Springer-Verlag, Berlin.

Taylor H.K. (1989) Ore reserves—a general overview. *Miner. Ind. Int.*, **990**, 5–12.

Telford W.M., Geldart L.P., Sheriff R.E. & Keys D.A. (1976) *Applied Geophysics*. Cambridge University Press, Cambridge.

Thatcher J., Struhsacker D.W. & Kiel J., (1988) Regulatory aspects and permitting requirements for precious metal heap leach operations. In Van Zyl D., Hutchison I. & Kiel J. (eds), *Introduction to Evaluation, Design and Operation of Precious Metal Heap Leaching Projects*, **40–58**. Society of Mining Engineers.

The Management of Drilling Projects (1981) A Short Course given and published by Northwest Mining Association, 414 Peyton Building, Spokane, Washington 99201, USA.

Theron J. C. (1973) Sedimentological evidence for the extension of the African continent during the late Permian–early Triassic times. In Campbell K.S.W. (ed.), *Gondwana Geology*, 61–71. A.N.U. Press, Canberra.

Thiann R. Yu. (1983) Rock Mechanics to Keep a Mine Productive. *Can. Min. J.*, April, 61–66.

Thomas L. (1992) *Handbook of Practical Coal Geology*. Wiley, Chichester.

Thompson M. (1982) Control procedures in exploration geochemistry. In Howarth R.J. (ed.) *Statistics and Data Analysis in Exploration Geochemistry*, 39–58. Handbook of Exploration Geochemistry, Vol. **2**., Elsevier, Amsterdam.

Thompson M. & Walsh N. (1989) *Handbook of Inductively Coupled Plasma Spectrometry*. Blackie, London.

Thomson I. (1987) Getting It Right. In Fletcher W.K, Hoffman S.J., Mehrtens M.B., Sinclair A.J. & Thomson I. (eds), *Exploration Geochemistry: Design and Interpretation of Soil Surveys*, 1–17. Reviews in Economic Geology Vol. **3**, Society of Economic Geologists.

Thorpe R.S. & Brown G.C. (1985) *The Field Description of Igneous Rocks*. Open University Press, Milton Keynes.

Thorpe R. S. & Brown G.C. (1993) *The Field Description of Igneous Rocks*. Wiley, Chichester.

Tilton J. E. Eggert R.G. & Landsberg H.H. (eds) (1988) *World Mineral Exploration*. Resources for the Future, Washington.

Tona F., Alonso D. & Svab M. (1985) Geology and mineralization in the Carswell Structure—a general approach. In Laine R, Alonso D. & Svab M. (eds), *The Carswell Structure Uranium Deposits*, 1–18. Geol. Assoc. Canada. Spec. Paper **29**, St Johns, Newfoundland.

Travis G.A., Keays R.R. & Davison R.M. (1976) Palladium and iridium in the evaluation of nickel gossans in Western Australia. *Econ. Geol.*, **71**, 1229–1242.

Tregoning G.W & Barton V.A. (1990) Design of a wide orebody mining system for a deep-level mine. In Ross-Watt R.A.J. & Robinson P.D.K. (eds), *Technical Challenges in Deep Level Mining*, 601–623. S. Afr. Inst. Min. Metall., Johannesburg.

Trueman D.L., Pedersen J.C., de St Jorre L. & Smith D.G.W. (1988) The Thorr Lake rare-metal deposits, Northwest Territories. In Taylor R.P. & Strong D.F. (eds), *Recent Advances in the Geology of Granite-related Mineral Deposits*, 280–290. Spec. Vol. **39**. Can. Inst. Min. Metall., Montreal.

Tucker M.E. (1988) *Techniques in Sedimentology*. Blackwell Scientific Publications, Oxford.

Tucker M.E. (1993) *The Field Description of Sedimentary Rocks*. Wiley, Chichester.

Tucker R. F. & Viljoen R. P. (1986) The geology of the West Rand Goldfield. In Annhaeusser C.R. & Maske S. (eds), *Mineral Deposits of Southern Africa*, 649–698. Geol. Soc. S. Afr., Johannesburg.

Turner B.R. (1985) Uranium mineralization in the Karoo Basin, South Africa. *Econ. Geol.*, **80**, 256–269.

Turner J.F. (1984) Developments in mineral processing in the Cornish tin industry. *Camborne School of Mines J.*, **84**, 50–55.

Tweedie E.B. (1986) The Evander Goldfield. In Annhaeusser C.R. & Maske S. (eds), *Mineral Deposits of Southern Africa*, 705–730. Geol. Soc. S. Afr., Johannesburg.

USBM (1976). *Coal Resource Classification System of the U.S. Bureau of Mines and U.S. Geological Survey*. US Geol. Surv. Bull., **1450–B**.

Van Blaricom R. (1993) *Practical Geophysics II*. Northwest Mining Association, Spokane.

Vearncombe J.R., Cheshire P.E., de Beer J.H., Killick A.M., Mallinson W.S., McCourt S. & Stettler E.H. (1988) Structures related to the Antimony Line, Murchison Schist Belt, Kaapraal Craton, South Africa, *Tectonophysics*, **154**, 285–308.

Vikre P.G. (1989) Fluid mineral relations in the Comstock Lode. *Econ. Geol.*, **84**, 1574–1613.

Viljoen R.P. (1990) Deep level mining—a geological perspective. In Ross-Watt, R.A.J. & Robinson, P.D.K. (eds), *Technical Challenges in Deep Level Mining*, 411–427. S. Afr. Inst. Min. Metall., Johannesburg.

Viljoen R.P., Viljoen M.J., Grootenboer J. & Longshaw T.G. (1975). ERTS–1 imagery—an appraisal of applications on geology and mineral exploration. *Min. Sci. Eng.*, **7**, 132–168.

Vokes F.M. (1987) Caledonian stratabound sulphide ores and the factors affecting them. *Geol. Surv. Finland, Spec. Pap.*, **1**, 15–26.

von Stackelberg U. & SONNE 49 Cruise, Shipboard Scientific Party (1988) Active hydrothermalism in the Lau back-arc basin (SW-Pacific); first results from the SONNE 49 cruise (1987). *Marine Mining*, **7**, 431–442.

Walker R.R., Matulich A., Amos A. C., Watkins J.J. & Mannard G.W. (1975) The geology of the Kidd Creek Mine. *Econ. Geol.*, **70**, 80–89.

Walsham B.T. (1967) Exploration by diamond drilling for tin in west Cornwall. *Trans. Instn Min. Metall., Sect. A*, **76**, A49–56.

Wanless R.M. (1982) *Finance for Mine Management*. Chapman and Hall, London.

Ward C.R. (ed.) (1984) *Coal Geology and Coal Technology*. Blackwell Scientific Publications, Oxford.

Weaver T.A., Freeman S.H., Broxton D.E. & Bolivar S.L. (1983) *The Geochemical Atlas of Alaska*. Los Alamos National Laboratory, Los Alamos.

Webb J.S., Howarth R.J., Thompson M. *et al.* (1978) *Wolfson Geochemical Atlas of England and Wales*. Clarendon Press, Oxford.

Werdmuller V.W. (1986) The Central Rand. In Antrobus E.S.A. (ed.), *Witwatersrand Gold—100 Years*, 7–48. Geol. Soc. S. Afr., Johannesburg.

West G. (1991) *The Field Description of Engineering*

Soils and Rocks. Open University Press, Milton Keynes.

West G.F., Macnae J.C. & Lamontagne Y. (1984) A time-domain EM system measuring the step response of the ground. *Geophysics*, **49**, 1010–1026.

Western Mining (1993) *News Release to Shareholders*, Western Mining Corp., Adelaide.

Whateley M.K.G. (1991) Geostatistical determination of contour accuracy in evaluating coal seam parameters: an example from the Leicestershire Coalfield, England. Bull. Soc. Géol. de France, **162** (2), 209–218.

Whateley M.K.G. (1992) The evaluation of coal borehole data for reserve estimation and mine design. In Annels A.E. (ed.), *Mineral Resource Evaluation*, 95–106. Spec. Publ. **63**, Geol. Soc. London.

Whateley M.K.G. & Harvey P.K. (eds) (1994) *Mineral Resource Evaluation II: Methods and Case Histories.* Spec. Publ. **79**, Geol. Soc. London.

Whateley M.K.G. & Jordan G. (1989). Fan delta-lacustrine sedimentation and coal development in the Tertiary Ombilin Basin, W. Sumatra. In Whateley M.K.G. & Pickering K.T. (eds), *Deltas: Sites and Traps for Fossil Fuels*, 317–332. Spec. Publ. **41**, Geol. Soc. London.

Whateley M.K.G. & Spears D.A. (eds) (1995) *European Coal Geology.* Spec. Publ. **82**, Geol. Soc. London.

Wheat T.A. (1987) Advanced ceramics in Canada. *Can. Inst. Min. Metall. Bull.*, **80**, (April), 43–48.

Whitchurch K.D., Gillies Saunders, A.D. & Just G.D. (1987) A geostatistical approach to coal reserve classification. *Pacific Rim Congress*, **87**, 475–482.

White C.J. (1989) *A Summary of the Mineral Sands Industry of Western Australia Including an Exploration and Feasibility Project.* Unpublished BSc Dissertation, Leicester University.

Whitely R.J. (ed.) (1981) *Geophysical Case Study of the Woodlawn Orebody, Australia.* Pergamon Press, Oxford.

Williams H.R. & Williams R.A. (1977) Kimberlites and plate tectonics in West Africa. *Nature,* **270**, 507–508.

Willis R.P.H. (1992) The integration of new technology in South African gold mines as a survival strategy. In Jones M.J. (ed.), *Mines, Materials and Industry*, 509–524. Instn Min. Metall., London.

Willis-Richards J. & Jackson N. J. (1989) Evolution of the Cornubian ore field, southwest England: I batholith modeling and ore distribution. *Econ. Geol.*, **84**, 1084–1100.

Wills B.A. (1985 & 1988) *Mineral Processing Technology.* Pergamon Press, Oxford.

Willson J.D. (1969) Consolidated Gold Fields Ltd., prospecting in Cornwall. In Jones, M.J. (ed.), *Exploration and Mining Geology,* 627–645. Proceedings Commonwealth Mining and Metallurical Congress, Vol. **1**, London.

Wilson J.T. (1949) Some major structures of the Canadian Shield. *Can. Inst. Min. Metall. Bull.,* **52**, 231–242.

Windley B.F. (1984) *The Evolving Continents.* Wiley, London.

Woodall R. (1984). Success in mineral exploration. *Geoscience Canada,* **11**, 41–46, 83–90, 127–133.

Woodall R. (1992) Challenge of minerals exploration in the 1990s. *Min. Eng.*, **July**, 679–684.

Zarraq G. (1987) *Comparison of Quality and Reserve Estimation Methods on Soma Lignite Deposit, Turkey.* Unpublished MSc Dissertation, University of Leicester.

Zeegers H. & Leduc C. (1991) Geochemical exploration for gold in temperate, arid, semi-arid and rain forest terrains, 309–335. In Foster R.P. (ed.), *Gold Metallogeny and Exploration.* Blackie, Glasgow.

Zussman J. (1977) *Physical Methods in Determinative Mineralogy.* Academic Press, London.

Index

Note: page numbers in *italic* and **bold** refer to figures and tables respectively. The letter 'B' after a page number refers to material in boxes.